Lecture Notes in Netw

Volume 1140

Series Editor

Janusz Kacprzyk🆔, Systems Research Institute, Polish Academy of Sciences, Warsaw, Poland

Advisory Editors

Fernando Gomide, Department of Computer Engineering and Automation—DCA, School of Electrical and Computer Engineering—FEEC, University of Campinas—UNICAMP, São Paulo, Brazil

Okyay Kaynak, Department of Electrical and Electronic Engineering, Bogazici University, Istanbul, Türkiye

Derong Liu, Department of Electrical and Computer Engineering, University of Illinois at Chicago, Chicago, USA

Institute of Automation, Chinese Academy of Sciences, Beijing, China

Witold Pedrycz, Department of Electrical and Computer Engineering, University of Alberta, Alberta, Canada

Systems Research Institute, Polish Academy of Sciences, Warsaw, Poland

Marios M. Polycarpou, Department of Electrical and Computer Engineering, KIOS Research Center for Intelligent Systems and Networks, University of Cyprus, Nicosia, Cyprus

Imre J. Rudas, Óbuda University, Budapest, Hungary

Jun Wang, Department of Computer Science, City University of Hong Kong, Kowloon, Hong Kong

The series "Lecture Notes in Networks and Systems" publishes the latest developments in Networks and Systems—quickly, informally and with high quality. Original research reported in proceedings and post-proceedings represents the core of LNNS.

Volumes published in LNNS embrace all aspects and subfields of, as well as new challenges in, Networks and Systems.

The series contains proceedings and edited volumes in systems and networks, spanning the areas of Cyber-Physical Systems, Autonomous Systems, Sensor Networks, Control Systems, Energy Systems, Automotive Systems, Biological Systems, Vehicular Networking and Connected Vehicles, Aerospace Systems, Automation, Manufacturing, Smart Grids, Nonlinear Systems, Power Systems, Robotics, Social Systems, Economic Systems and other. Of particular value to both the contributors and the readership are the short publication timeframe and the world-wide distribution and exposure which enable both a wide and rapid dissemination of research output.

The series covers the theory, applications, and perspectives on the state of the art and future developments relevant to systems and networks, decision making, control, complex processes and related areas, as embedded in the fields of interdisciplinary and applied sciences, engineering, computer science, physics, economics, social, and life sciences, as well as the paradigms and methodologies behind them.

Indexed by SCOPUS, EI Compendex, INSPEC, WTI Frankfurt eG, zbMATH, SCImago.

All books published in the series are submitted for consideration in Web of Science.

For proposals from Asia please contact Aninda Bose (aninda.bose@springer.com).

Eduardo Vendrell Vidal · Uriel R. Cukierman ·
Michael E. Auer
Editors

Advanced Technologies and the University of the Future

 Springer

Editors
Eduardo Vendrell Vidal
Universitat Politècnica de València (UPV)
Valencia, Spain

Uriel R. Cukierman
UTN-FRBA
Buenos Aires, Argentina

Michael E. Auer
CTI Global
Frankfurt, Hessen, Germany

ISSN 2367-3370 ISSN 2367-3389 (electronic)
Lecture Notes in Networks and Systems
ISBN 978-3-031-71529-7 ISBN 978-3-031-71530-3 (eBook)
https://doi.org/10.1007/978-3-031-71530-3

© The Editor(s) (if applicable) and The Author(s), under exclusive license to Springer Nature Switzerland AG 2025

This work is subject to copyright. All rights are solely and exclusively licensed by the Publisher, whether the whole or part of the material is concerned, specifically the rights of translation, reprinting, reuse of illustrations, recitation, broadcasting, reproduction on microfilms or in any other physical way, and transmission or information storage and retrieval, electronic adaptation, computer software, or by similar or dissimilar methodology now known or hereafter developed.
The use of general descriptive names, registered names, trademarks, service marks, etc. in this publication does not imply, even in the absence of a specific statement, that such names are exempt from the relevant protective laws and regulations and therefore free for general use.
The publisher, the authors and the editors are safe to assume that the advice and information in this book are believed to be true and accurate at the date of publication. Neither the publisher nor the authors or the editors give a warranty, expressed or implied, with respect to the material contained herein or for any errors or omissions that may have been made. The publisher remains neutral with regard to jurisdictional claims in published maps and institutional affiliations.

This Springer imprint is published by the registered company Springer Nature Switzerland AG
The registered company address is: Gewerbestrasse 11, 6330 Cham, Switzerland

If disposing of this product, please recycle the paper.

Foreword

The world is experiencing a dramatic and still hard to define transformation which is impacting literally all domains of work and life. Certainly, the massification of technological developments has led to expected outcomes but also to several unplanned consequences for which deeper analysis may lead us to a better understanding of their ramifications, their long-term impact, the ways to mitigate risks, and, more notably, a clearer picture of how the future may look like.

Higher education is not the exemption. Even though the basic operational teaching-learning model has remained in some way unchanged for centuries, universities all over the world are facing significant pressures to remain viable, and, at the same time, are posed to develop, adopt, and adapt impressive and rapidly changing technological endeavors and approaches, of a magnitude not seen alone in contemporary history.

But what are those technological developments, what is its impact in higher education, and how to cope with them? The book edited by Uriel R. Cukierman, Michael Auer, and Eduardo Vendrell, is an extraordinary and novel effort aimed at responding to those questions, by curating cutting edge contributions from experts from across the world, in such way that both technical and non-technical readers, can have better understanding of the complexities, risks, and opportunities ahead which result from the interphase between advanced technologies and the day-to-day work of higher education institutions. It is not an easy task that authors and editors have handled very effectively.

There is no doubt that a book like this one would have significant variations in content and a more speculative tone if the world has not experienced a pandemic that literally paralyzed the entire system of higher education at a global scale. The sudden closure of universities triggered quick responses from institutions and faculty members with the goal of enabling continuity in the delivery of education by means previously disdained or even not considered legitimate in some countries. At the same time, short-term solutions such as the substitution of the physical classroom with a remote delivery of education via zoom, seen as a panacea by many, rapidly showed weaknesses, access deficiencies, assessment limitations, and learning deficits, fragile mental health of both students and teachers, and many more shortcomings. Nevertheless, such a forced social experiment—the largest of its kind in contemporary history—enabled ideation of novel delivery methodologies, allowed the testing of new technological developments, sparked the curiosity of researchers to adapt new technologies and new pedagogical approaches, and, more importantly, let us recognize that many assumptions previously considered untouchable about what works and what doesn't work in higher education were debatable, and that suddenly conditions were more favorable to reimagine a higher education more innovative and disruptive, less averse to technology, and more willing to change.

The collective effort coordinated by Uriel, Michael, and Eduardo, captures this unique crossroads moment existing in higher education, and provides good hints about key trends to consider as the role of universities keep shifting, and as both teachers and students also have transformed themselves.

Of course, nobody can predict what will be the shape and scope of universities of the future, but I am certain that the basic premise of what is good quality and relevant higher education will remain, and that such premise will be heavily influenced by smart, inclusive, and sensible technological developments and the corresponding pedagogical approaches.

The future of higher education has plenty of uncertainty, but also is full of opportunities for more and better work. Such future will not be defined by guesses from a crystal bowl or by just extrapolating the past, but rather by building a more effective, impactful, and adaptable tomorrow. The book provides a timely and unique storyline to be considered in building a better future for higher education.

Doha, Qatar   Francisco Marmolejo

Francisco Marmolejo is the Higher Education President at Qatar Foundation. Previously, he worked at the World Bank as Global Higher Education Coordinator and as Lead Higher Education Specialist for India and South Asia. At the University of Arizona, he founded the Consortium for North American Higher Education Collaboration, and also served as Assistant Vice President. He has been an American Council on Education Fellow at the University of Massachusetts, Academic Vice President of the University of the Americas in Mexico, and International Consultant at OECD in Paris.

Preface

The history of educational technology shows that an invention more than 200 years old remains one of the most influential technological innovations of our time. Josiah F. Bumstead emphasized in an 1841 essay that the blackboard was a revolutionary technological invention [1]. Numerous scholars have agreed with Bumstead, who also stated:

"The inventor or introducer of the black-board system deserves to be ranked among the best contributors to learning and science, if not among the greatest benefactors of mankind; and so he will be regarded by all who know its merits and are familiar with school-room trials" [2].

More recently, at a memorable conference, Edith Litwin, a renowned Argentinean expert in learning technologies, commented on the integration of technologies in classrooms. She noted that the first tool that helped teachers to solve problems that were difficult to understand and teach was the humble chalk and blackboard: "That is a technology that, I still think, is cutting-edge technology" [3]. At this point, it is appropriate to quote Buckingham, who states:

"It is as inevitable today to use technologies of various types in education as it always was. The book, the pencil, and the blackboard are technologies, just like the computer, the video recorder, or the latest mobile communication device. The question is not whether to use technology, or even what technologies to use, but why and how to use them. The polarization that characterizes this debate—the endless confrontation of 'supporters' and 'opponents'—has made it very difficult to address the fundamental issues" [4].

As Buckingham points out, the process of incorporating technologies in education did not begin with the blackboard, nor did it end there. From oral communication, through the invention of writing and the printing press, to the more recent use of computers and the Internet, the idea of using technological tools to support teaching and learning is a constant that persists and adapts to the times, uses, and customs. In this historical evolution, the appearance of digital technologies can be seen as a milestone that has changed the way we live and also the field of education. Today, it is difficult to imagine everyday life without a computer, a cell phone, or an Internet connection. In fact, the inventor of the Hypertext Markup Language (HTML), better known as "the inventor of the Web," has recently stated that "Internet access should be a universal right" [5]. These digital technologies, which entered classrooms a few decades ago, did so without invitation and faced considerable resistance from educators. An example of this resistance is expressed by researcher Larry Cuban, who argued that "computers are largely incompatible with the requirements of teaching, and that, for the most part, teachers will continue to reject their use as instruments of student work during class" [6].

Despite this resistance and the fact that for some academics "it remains something of a black box," as Mary Curnock Cook notes in the foreword of a recent report from the Higher Education Policy Institute (HEPI) [7], there is consensus that technology is shaping the immediate future of Higher Education.

With the advent of Artificial Intelligence, Large Language Models, the Metaverse, and gamification techniques, the use of technology in universities and other tertiary education institutions has generated increasing interest among academics and researchers. Their adoption is creating a new paradigm and model of education worldwide. This situation has gained even more interest after the COVID pandemic, demonstrating that technology can overcome challenges and help learners and teachers to continue with the educational process and redefine traditional schemes.

Universities and other Higher Education institutions need to rethink their organizational model and educational purposes in this context, transforming curricula, spaces, and the mindsets of learners and teachers according to this new paradigm.

Although there are many resources (internet blogs and sites, podcasts, etc.) dedicated to this topic, there's still a need for a reference text to compile specific research approaches to the use and application of technology in education. In a dynamic world where news and advances are mixed and raised through social networks and other trans-media channels, the need for reference textbooks is a must so that stakeholders can have a resource to consult thoughtfully on research topics and approaches, providing deeper discussion and greater extension.

In this context, this book provides a common framework, compiling solutions and evidence across different sections that summarize how technology can be applied to shape the learning experience and the organization of the Higher Education institution, towards the university of the future, providing ideas to foster collaboration and enhance research.

The text is divided into five distinct parts that summarize and group technologies and experiences that are having a real impact in Higher Education:

(1) Extended Reality (XR): aiming the reality-virtuality continuum, which includes augmented reality (AR), mixed reality (MR), virtual reality (VR), haptic devices and, more recently the metaverse.
(2) Artificial Intelligence (AI): including everything related to the automated analysis of large volumes of information and its application in the form of learning analytics, adaptive learning and automatic learning (machine learning) and also chatbots, which have emerged into mainstream conversation due to the appearance of ChatGPT.

(3) Digital Transformation (DX): understood as the possibility of taking advantage of the available technologies to change the programs, and the organization of teaching and learning. This subject also includes themes such as information security and privacy and open badges.
(4) Gamification: referring to the incorporation of serious game elements, like point and reward systems, to tasks as incentives for people to participate.
(5) Emerging Trends in Higher Education: a comprehensive spectrum spanning research endeavors, application development, first-hand accounts, and detailed descriptions of educational tools.

Valencia, Spain
Buenos Aires, Argentina
Frankfurt, Germany

Eduardo Vendrell Vidal
Uriel R. Cukierman
Michael E. Auer

References

1. Krause SD (2000) «Among the greatest benefactors of mankind: what the success of chalkboards tells us about the future of computers in the classroom», J Midwest Mod Lang Assoc, 33(2):6–16
2. Bumstead JF (1841) The black board in the primary school. A manual for teachers, Boston: Perkins & Marvin
3. Litwin E (2002) «Cómo trabajar con tecnología en la escuela?» de Conferencia organizada por Fundación Telefónica y El Sabio, Buenos Aires
4. Buckingham D (2008) Más allá de la tecnología. Aprendizaje infantil en la era de la cultura digital, Cambridge: Manantial
5. Berners-Lee T (2021) «The Guardian», 4 Junio 2020. [Online]. Available: https://www.theguardian.com/commentisfree/2020/jun/04/covid-19-internet-universal-right-lockdown-online. Accessed on 27 Mar 2021
6. Becker HJ (2000) «Findings from the teaching, learning, and computing survey: is larry cuban Right?» Educ Policy Anal Arch 8(51)
7. M. Curnock Cook Ed. (2024) «Technology foundations for twenty-first century higher education» HEPI report 172. Higher education policy institute. LearningMate. Available in: https://www.hepi.ac.uk/2024/03/28/technology-foundations-for-twenty-first-century-higher-education-2/

Contents

Extended Reality (XR)

Section Introduction 3
Miriam Reiner

Breaking the Bottleneck: Generative AI as the Solution for XR Content Creation in Education 9
Jordi Linares-Pellicer, Juan Izquierdo-Domenech, Isabel Ferri-Molla, and Carlos Aliaga-Torro

Learning on Zoom as an Augmented Reality Experience 31
Galit Wellner

A Science-Based Environment for Lexical Language Learning 45
Vered Levi Zaydel

Adaptive AR- or VR-Neurofeedback for Individualized Learning Enhancement 65
Nehai Farraj and Miriam Reiner

Artificial Intelligence (AI)

Section Introduction to AI in Higher Education 87
Dieter Uckelmann

Development of an AI Competence Matrix for AI Teaching at Universities 91
Diana Arfeli, Maximilian Weber, Dieter Uckelmann, and Tobias Popovic

Implementing Ethical Considerations into Ai-Supported Student Counselling—a Case Study 111
Maximilian Weber and Niko Jochim

The Use of AI to Teach Ethics 129
Harold P. Sjursen

A Reinforcement Learning Framework for Personalized Adaptive E-Learning 141
Anat Dahan, Navit Roth, Avishag Deborah Pelosi, and Miriam Reiner

Machine Learning Models to Detect AI-Assisted Code Anomaly in Introductory Programming Course 163
Hapnes Toba and Oscar Karnalim

Using GenAI as Co-author for Teaching Supply Chain Management in Higher Education 183
Dominik Wörner and Andreas Holzapfel

Digital Transformation (DX)

Section Introduction 197
Faraón Llorens Largo and Antonio Fernández Martínez

Safeguarding Knowledge: Ethical Artificial Intelligence Governance in the University Digital Transformation 201
Rafael Molina-Carmona and Francisco José García-Peñalvo

Levels of Analysis and Criteria for Digital Transformation Implementation in Higher Education Institutions 221
Adriana V. Karam-Koleski and Roberto Carlos Santos Pacheco

Driving Innovations: Trends, Prospects and Challenges of Implementing Disruptive Educational Technologies Within HEIs 237
Edwin Zammit, Clifford De Raffaele, Daren Scerri, Ronald Aquilina, Joachim James Calleja, and Alex Rizzo

New Models of University for the Digital Society 259
Maria Amata Garito

Opening Up Teaching and Learning at Universities: OER, MOOCs and Microcredentials 281
Martin Ebner, Ernst Kreuzer, Sandra Schön, and Sarah Edelsbrunner

Computer-Based Methods for Adaptive Teaching and Learning 297
Ulrike Pado, Anselm Knebusch, and Konstanze Mehmedovski

Teaching Research Skills at the University. Does Digital Transformation Make a Difference? 319
María Isabel Pozzo

Gamification

Section Introduction 341
Carina Gabriela Lion

Do War Video Games Stimulate the Critical Thinking of University
Students Towards War? 343
Rodríguez González Christian, Del Moral Pérez M. Esther,
Castañeda Fernández Jonathan, and Bellver Moreno M. Carmen

The Role of Gaming in HE: Strengthening Distributed Leadership
and Student Commitment 359
Carina Lion and Verónica Perosi

Video Games as an Emerging Consumer Technology: Profiles,
Uses, and Preferences of University Students in Argentina 377
María Gabriela Galli, María Cristina Kanobel, Diana Marín Suelves,
and Donatella Donato

Gami|cation: Gamification Within (Higher) Education 397
L. Christa Friedrich, Matthias Heinz, Josefin Müller, and Michelle Pippig

Emerging Trends in Higher Education

Section Introduction 415
Dominik May

EFL Learning in the Digital Era: Navigating Language
and Culture in Jordanian Universities 419
Luqman M. Rababah

Impact of Simulation on the Development of Non-technical Skills 439
Fatima Zahra Salih, Rajae Lamsyah, Abdelkrim Shimi, and Yasser Arkha

Identifying AI Generated Code with Parallel KNN Weight Outlier
Detection 459
Oscar Karnalim

The Use of SAP in Education: A Review of Current Practices
and Future Directions 471
Md. Tanvir Amin, Md. Tahsin Amin, Fahmida Haque Mim,
and Jarin Sobah Peu

Detection of the Creativity Potential of Engineering Students 485
Luis Vargas-Mendoza, Dulce V. Melo-Máximo,
Francisco J. Sandoval-Palafox, and Brenda García-Farrera

The Next Step in Challenge-Based Learning: Multiple Challenges
in a Single Course in Engineering: A New Model in Experiential
Education 501
Miguel Ramírez-Cadena, Juana Méndez-Garduño,
Israel Cayetano-Jiménez, Araceli Ortíz-Martínez,
and Jorge Membrillo-Hernández

Extended Reality (XR)

Section Introduction

Miriam Reiner

Synthetic, digital 3D immersive realities have been known by multiple names, differing slightly: Virtual Reality (VR) Extended Reality (XR) Augmented Reality (AR) and Mixed Reality (MR), The first is a virtual representation of a world, that feel real, yet does not necessarily follow the physical reality as known to us, and might for instance include imaginary objects, imaginary animals with impossible capacities, such as talking animals. It might include 3D images of other humans, that look real, or not, and might include representations of humans that are interactive and represent a human, yet do not look like the represented entity, and rather take the shape of an animal, real or imaginary, as chosen by the user. Augmented reality is physical reality that includes on top of the physical varying features–such as names of people, height or names of buildings, driving instructions on top of a representation of a 3D image of a city, details describing the science and effect on humans of the Vesuvius volcano eruption, more than 2000 years ago, or explanation regarding the various phases of a pumping heart of engine. Extreme reality (XR) goes a step further and might include components that can act on the user such as measures of stress then tailoring the synthetic world accordingly. Finally, MR is a mixture of real and virtual, such as the real roads of a city and a newly planned architectural building design, in order to test its match to the environment before ever built, or a game designed to provide an environment to new students to learn the city geography and topology, in which, the player interacts with different imaginary creatures roaming the city streets.

Synthetic 3D worlds have become increasingly powerful: improved graphics, increased refresh rate, inclusion of multisensory cues of visuals, auditory, touch, temperature, and even environmental climate, offering a unique and strong virtual

---

M. Reiner (✉)
The VR NeuroCog Lab, Israel Institute of Technology, Technion, Israel
e-mail: miriamr@technion.ac.il; mirrein@gmail.com

experience of exposure of the human brain to newly immersive environments and especially new interactions and acts, never experienced before [16–18].

Are such synthetic environments relevant and beneficial for learning? Humans learn through sensory experience from babyhood, and it does not stop there. Humans go on learning through sensory experience during lifetime. Learning of new procedures in Sports and enhanced sports performance would be one immediate example that comes to mind, and indeed there is vast research pointing at how athletes improve their performance in sports, by learning to respond fast and accurately to sensory cues. Learning of complex surgical procedures in the operating room, is another well studied example [11, 13]. In the physics lab we perceive and analyze procedures, yet limited by the physicality, while VR can provide a wider and deeper support for analysis and learning faster and more efficiently (for an example please see Reiner [9] on a virtual world that allows learning of fields in VR) and then generate thought experiment which were major tools in science innovations [7, 8, 14]. Indeed, a systematic meta-analysis of studies showed a positive effect [20]. A similar meta-analysis showed similar results in the use of VR in English learning [6].

These are just three examples, yet many more exist. And the research community and startup industry have responded accordingly, and many new synthetic environments have been and are being developed targeting efficiently enhanced learning. Thus, synthetic worlds are a major crucial novel tool for learning, especially in STEM.

Novel synthetics environments offer interactive experiences that are particularly beneficial in learning science, engineering, and mathematics (STEM). Straight forward Synthetic worlds might simply simulate reality and allow learners to change variables and watch the outcome of variable-manipulations. Yet novel synthetic 3D environments can go beyond mimicking reality, and offer a mental glimpse into new images, processes and events, that do not exist, yet can be programmed into the digital reality. For instance, thought experiments are crucial in understanding science, engineering and arts [2, 10]. Imagine a student struggling with a mere visualization of Einstein's thought experiment. It is hard or even impossible of novice learners to imagine and get immersed in the thought processes of great scientists that lead to novelties and paradigm shifts in science and engineering. Digital Experiences in a synthetic world might include viewing Einstein's thought experiment that lead to the paradigm shift in classical physics on the verge of the twentieth century, and similarly allow the learner to get immersed in the thought processes imagined by great leading scientists and inventors, Einstein, Mach, Galileo, Boltzmann, Schrodinger, and many-many more, allowing a never before available, vision of the thought processes of the great inventors [12, 14].

Yet synthetic environments offer an additional exciting advantage: research suggests that sensory cues implanted into the virtual world can enhance attention, improve memory consolidation, expedite response time, and accuracy, by that enhance learning. Synthetic environments, that are designed to match the brain learning processes, provide a tool for enhanced learning, beyond physical environments [1, 3–5, 15, 19, 21].

Section Introduction

The papers in this section bring a sample of theories, applicative approaches and technologies for novel cognitive 3D synthetic worlds for STEAM learning.

Recent AI procedures have added novel capabilities to the design and effect of synthetic 3D worlds in learning, especially in order to cope with the complexity of content creation in VR. The chapter titled "Breaking the Bottleneck: Generative AI as the Solution for XR Content Creation in Education" suggest a novel solution exactly for this issue–supporting educators in generating novel content without the necessity of experience and knowledge in coding by using the recent new tools of Generative AI (GenAI). Demonstrations, and practical examples are extremely useful and are presented in the paper for support. The chapter also addresses challenges related to GenAI's application in education, including Challenges and limitations as well as ethical concerns are discussed too.

While not a 3D environment, zoom has become the default tool for teaching and learning when physical meetings are limited. The chapter titled "Learning on Zoom as an Augmented Reality Experience" discusses the profound transformation of learning and teaching when zoom is the tool at hand for interaction. Indeed, as the author shows, zoom shares many features with 3D AR, and further lays the foundations for a systematic examination of AR learning experiences, and point to ethical concerns related to AR practices.

Can VR support language learning? The paper titled "A Science-Based Environment For Lexical Language Learning" aims to unveil how Brain Computer Interface system in using Augmented Reality glasses, while using online applications such as Zoom, can change student's experience during a new language learning and enhance learning. The author suggests that associations are a majpr tool to support language learning and hence integrated a process of association generation in the AR to support learning. Her results show a very encouraging direction for enhanced language learning.

The paper "Adaptive AR- or VR-Neurofeedback for Individualized Learning Enhancement" integrates new neurofeedback technologies with AR and VR. The paper brings results from neurofeedback research suggesting that modulating brainwave activity with feedback coded into the AR/VR, might lead to enhanced spatial skills, specifically mental rotation, which play a crucial role in understanding mathematical concepts, molecular structures, atomic arrangements, and geometric shapes, presenting a promising avenue for fostering significant enhanced learning in science, engineering, and math achieved through optimizing spatial reasoning skills.

To summarize, Synthetic 3D immersive environments–VR, AR, MR, XR– provide a basis for new cognitive enhanced technologies. The provide an opportunity for **Visualization and Immersion**: VR and XR provide learners with immersive environments where abstract concepts can be visualized in three dimensions, Such as complex mathematical equations, scientific phenomena, or engineering designs can be represented visually, making them easier to understand and remember, and immersion in scientists' thought experiment support deep understanding of the scientists' thought processes; **Hands-On Learning** simulating real-world scenarios. especially useful in STEM fields where experimentation and practical application are crucial for understanding concepts. Thus, students can conduct virtual experiments in physics, chemistry, or engineering without the need for expensive equipment or physical laboratories; **Interactivity and Engagement** offering interactive elements that engage learners actively in learning encouraging exploration, experimentation, and problem-solving, which are essential skills in STEM disciplines; **Personalized Learning** adapting to individual learning styles and preferences; **Collaborative Learning** by connecting learners from different locations in shared virtual environments. Provide **Accessibility and Inclusivity** since VR and XR technologies can make STEM education more accessible and inclusive by removing physical barriers to learning. For example, students with disabilities or those who are unable to access traditional educational resources can benefit from immersive virtual experience, and beyond all **enhance human learning** by installing stimuli that activate human brain regions for optimal learning.

References

1. Bennet R, Reiner M (2022) Shared mechanisms underlie mental imagery and motor planning. Sci Rep 12(1):1–15
2. Gilbert JK, Reiner M (2000) Thought experiments in science education: potential and current realization. Int J Sci Educ 22(3):265–283
3. Hecht* D, Reiner M, Karni A (2008a) Enhancement of response times to bi-and tri-modal sensory stimuli during active movements. Exp Brain Res 185(4):655
4. Hecht* D, Reiner M, Karni A (2008b) Multisensory enhancement: gains in choice and in simple response times. Exp Brain Res 189(2):133
5. Lev* DD, Reiner M (2012) Is learning in low immersive environments carried over to high immersive environments? Adv Hum-Comput Interact 2012:16
6. Luo S, Zou D, Kohnke L (2024) A systematic review of research on x reality (XR) in the English classroom: trends, research areas, benefits, and challenges. Comput Educ: X Rity 4:100049
7. Reiner M (1997) A learning environment for mental visualization in electromagnetism. Int J Comput Math Learn 2(2):125–154
8. Reiner M (1998) Thought experiments and collaborative learning in physics. Int J Sci Educ 20(9):1043–1058
9. Reiner M (1999) Conceptual construction of fields through tactile interface. Interact Learn Environ 7(1):31–55
10. Reiner M, Gilbert J (2000) Epistemological resources for thought experimentation in science learning. Int J Sci Educ 22(5):489–506
11. Reiner M (2000) The validity and consistency of force feedback interfaces in telesurgery: short original contribution. J Comput Aided Surg 9:69–73
12. Reiner M, Burko* LM (2003) On the limitations of thought experiments in physics and the consequences for physics education. Sci Educ 12(4):365–385
13. Reiner M (2004) The role of haptics in immersive telecommunication environments. IEEE Trans Circuits Syst Video Technol 14(3):392–401
14. Reiner M (2006) The context of thought experiments in physics learning. Interchange 37(1–2):97–113
15. Reiner M (2009) Sensory cues, visualization and physics learning. Int J Sci Educ 31(3):343–364
16. Reiner M (2011) Presence: brain, virtual reality and robots. Brain Res Bull 85(5):242–244
17. Slater M, Leeb R, Angus A, Brogni A, Christou C, Friedman D, Garau M, Gillies M, Goenegress C, Steed A, Thomsen M, Dolan R, Kalish R, Critchley H, Sanchez-Vives MV, Brotons J, Pfurtscheller G, Neuper C, Edlinger G, Reiner M, Halevi* G, Hecht* D (2006) Virtual reality: a view from overseas presence: research encompassing sensory enhancement, neuroscience and cognition, with interactive applications (PRESENCIA): a European research project. J Virtual Rity Soc Jpn 11(1):10–15

18. Slater M, Frisoli A, Tecchia F, Guger C, Lotto B, Steed A, Pfurtscheller G, Leeb R, Reiner M, Sanchez-Vives MV, Verschure P (2007) Understanding and realizing presence in the Presenccia project. IEEE Comput Graph Appl 90–93
19. Sella I*, Reiner M, Pratt H (2014) Natural stimuli from three coherent modalities enhance behavioral responses and electrophysiological cortical activity in humans. Int J Psychophysiol 93(1):45–55
20. Sırakaya M, Alsancak Sırakaya D (2022) Augmented reality in STEM education: a systematic review. Interact Learn Environ 30(8):1556–1569
21. Yazmir B, Reiner M (2022) Neural signatures of interface errors in remote agent manipulation. Neuroscience 486:62–76; Bennet R, Reiner M (2022) Shared mechanisms underlie mental imagery and motor planning. Nat Sci Rep 12(1):1–15

Breaking the Bottleneck: Generative AI as the Solution for XR Content Creation in Education

Jordi Linares-Pellicer, Juan Izquierdo-Domenech, Isabel Ferri-Molla, and Carlos Aliaga-Torro

Abstract The integration of Extended Reality (XR) technologies-Virtual Reality (VR), Augmented Reality (AR), and Mixed Reality (MR)-promises to revolutionize education by offering immersive learning experiences. However, the complexity and resource intensity of content creation hinders the adoption of XR in educational contexts. This chapter explores Generative Artificial Intelligence (GenAI) as a solution, highlighting how GenAI models can facilitate the creation of educational XR content. GenAI enables educators to produce engaging XR experiences without needing advanced technical skills by automating aspects of the development process from ideation to deployment. Practical examples demonstrate GenAI's current capability to generate assets and program applications, significantly lowering the barrier to creating personalized and interactive learning environments. The chapter also addresses challenges related to GenAI's application in education, including technical limitations and ethical considerations. Ultimately, GenAI's integration into XR content creation makes immersive educational experiences more accessible and practical, driven by only natural interactions, promising a future where technology-enhanced learning is universally attainable.

J. Linares-Pellicer · J. Izquierdo-Domenech (✉) · I. Ferri-Molla · C. Aliaga-Torro
Valencian Research Institute for Artificial Intelligence (VRAIN), Universitat Politècnica de València (UPV), Geltrú, Spain
e-mail: juaizdom@upv.es

J. Linares-Pellicer
e-mail: jorlipel@upv.es

I. Ferri-Molla
e-mail: isfermol@upv.es

C. Aliaga-Torro
e-mail: calitor@epsa.upv.es

1 Introduction

The advent of Extended Reality (XR), encompassing Virtual Reality (VR), Augmented Reality (AR), and Mixed Reality (MR), has revolutionized the way we interact with our environment and perceive the world around us. These technologies present a transformative opportunity for learning experiences by enabling immersive and interactive educational content. For example, Hamilton's work underscores the potential of VR to enhance engagement and improve learning outcomes through immersive experiences [9], and Šašinka et al. have demonstrated the capacity of VR to enhance the educational experience significantly, highlighting the increased engagement and understanding among students and the crucial role of collaboration in facilitating more profound learning outcomes [22]. However, creating engaging and educational XR content remains a significant challenge, often limited by technical complexities and resource constraints. XR technologies offer unique advantages for education, such as the ability to simulate complex, real-world environments where learners can practice and repeat tasks, engage with scenarios that would otherwise be inaccessible or dangerous, and receive instant feedback. In particular, VR has been highlighted for its potential to facilitate acquiring cognitive skills related to spatial and visual information, psychomotor skills related to head movement, and affective skills in managing emotional responses. Despite its benefits, the production of VR content is notably costly, limiting the availability of high-quality educational simulations. Creating VR content demands technical skills and substantial financial resources, posing challenges for educational institutions in integrating XR technologies into their teaching practices [12].

Despite their potential, XR technologies face several challenges that limit their widespread adoption in education. One significant challenge is the technical complexity of creating XR content. Developing immersive and interactive experiences requires specialized skills and tools, often not readily available to educators. Creating high-quality XR content requires significant time and resources, which can be difficult for educators to allocate. Educators face numerous complexities in producing the elements required for any XR application. These complexities span across various stages of development. Preliminary reviews from different angles have been done regarding AI, XR, and education, as we can see in Reiners et al. work, which systematically reviews the combination of AI and XR, highlighting applications in medical training, autonomous vehicles, gaming applications, and more [21]; Memarian and Doleck work, providing ethical analysis and extensive synthesis of challenges and future recommendations for using XR with AI in education [16]; and Papakostat et al. in their review of AI-enhanced AR in education, focusing on its potential to improve spatial abilities and offering a comprehensive examination of related studies, benefits, and limitations [20]. However, this chapter will focus on how GenAI's new possibilities are starting to offer a new approach to XR content creation in education. This chapter is structured to provide a comprehensive overview of the potential of XR technologies in education, the challenges educators face in creating XR content, and the emergence of generative AI as a solution to this bottleneck.

It begins by highlighting the advantages of XR technologies in enhancing learning outcomes and engagement. It then delves into educators' complexities and resource limitations in developing immersive and interactive XR content. The chapter introduces GenAI technologies and their potential to revolutionize XR content creation by enabling educators to generate high-quality XR experiences based on natural language descriptions, sketches, or brainstorming ideas. Finally, it discusses the current possibilities available in this area. It outlines the challenges that still need to be addressed to harness the potential of GenAI fully for XR education. A final proposal will also be described on how these new possibilities will evolve.

2 A Brief Introduction to XR

XR is an umbrella term encompassing various immersive technologies that blend the physical and digital worlds. XR includes VR, AR, and MR, also known as Augmented Virtuality, as defined in Milgram et al.'s work [19]. XR can potentially enhance learning outcomes by providing students with immersive and interactive experiences [15] and has been applied and evaluated in several contexts, such as transporting students to different historical periods or locations [26] or making students observe the consequences of environmental changes and pollution firsthand [5].

VR creates a fully immersive, computer-generated environment that surrounds the user. When using a VR headset, users are isolated from the real world and transported into a virtual one. In education, VR can be used to create highly engaging learning experiences. For instance, students can take virtual field trips to historical sites, such as ancient Rome or the Great Wall of China, without leaving their classroom. They can also participate in simulated science experiments in a fully equipped virtual laboratory, providing hands-on learning opportunities that might otherwise be inaccessible due to resource constraints.

AR overlays digital information onto the real world, enhancing users' perception of their physical environment. AR can be experienced through AR glasses or mobile devices, which display digital content, such as text, images, or 3D models, in the user's field of view. In the educational context, AR can bring textbooks to life. For example, a biology textbook could display 3D models of the human anatomy that students can rotate and explore, or an astronomy app could show the constellations and planets superimposed on the night sky when viewed through a smartphone.

MR combines elements of both VR and AR, creating a hybrid environment where physical and digital objects coexist and interact in real-time. MR headsets, such as the Microsoft HoloLens, allow users to simultaneously see and interact with the real world and digital objects. In education, MR can facilitate collaborative projects where students interact with digital 3D models while still engaging with their physical environment. For example, students in an architecture class could collectively design a building by manipulating a virtual model that appears on an actual tabletop, or medical students could practice complex surgical procedures using a blend of real tools and virtual guidance.

3 Overview of XR in Education

XR technologies offer immersive, interactive experiences that can enhance learning outcomes. XR's benefits in education include increased motivation and engagement, accessibility to inaccessible or dangerous locations, and practical, hands-on learning experiences. According to Kaul and Kumar, VR's ability to simulate complex, real-world scenarios allows learners to engage with educational content in more meaningful and impactful ways, thereby significantly enhancing both the acquisition and retention of knowledge [13]. Dikkatwar et al. highlight China's leadership in VR research within educational contexts, emphasizing the technology's role in bridging educational disparities and fostering innovative learning environments [7].

XR applications in education are diverse, ranging from fully immersive VR learning environments to AR-enhanced learning materials, simulations and training, and collaborative learning experiences [25]. However, the adoption of XR in education faces significant limitations, such as high costs, technical challenges, User Experience (UX) issues, difficulties in pedagogical integration, and privacy and safety concerns. Additionally, educators must adapt their teaching to whatever possibilities these current educational XR apps offer.

Despite these challenges, the potential for XR in education is immense. By providing students with immersive and interactive learning experiences, XR can foster more in-depth understanding, improve knowledge retention, and develop practical skills. It can also cater to different learning styles and provide personalized learning experiences, ensuring each student's unique needs are met.

4 Major Challenges Educators Face in XR Content Creation

In the field of education, the integration of XR technologies presents an array of challenges for educators. These challenges range from pedagogical considerations to technical complexities and resource limitations.

Significant challenges include the high costs of implementation, accessibility issues, technical difficulties, the risk of cognitive overload, potential physical and psychological side effects, the need for pedagogical alignment, faculty preparedness, and privacy concerns. These challenges underscore the necessity for strategic planning, collaboration, and continuous evaluation to leverage XR's educational benefits while mitigating its drawbacks [1, 2, 23].

Educators also face challenges generating the assets and resources required for XR content creation. Developing immersive and interactive experiences requires a range of digital assets, including images, 3D models, animations, interactions, music, and sound effects. Acquiring or creating these assets can be time-consuming and resource-intensive, particularly for educators needing specialized skills in digital content creation.

Furthermore, educators need to consider the final deployment of XR experiences across different environments and hardware possibilities [10]. XR content can be delivered through various devices, including VR headsets, AR glasses, and mobile devices, and can often be deployed on desktop computers. Each platform has unique capabilities and limitations, and educators need to ensure that their content is optimized for the specific devices and environments in which it will be used, including factors such as compatibility, performance, and UX.

In the following sections, we will delve deeper into educators' current challenges when they are interested in creating XR educational content. These challenges encompass pedagogical considerations, technical complexities, and resource limitations. It will also explore the difficulties of transforming educational objectives into XR products that significantly impact learning outcomes, produce the necessary assets and resources, and consider the final deployment of XR experiences across various environments and hardware capabilities.

4.1 Initial Idea and Storyboarding

Educators should carefully consider how XR technologies can provide effective solutions for achieving their educational goals. They must identify XR's specific possibilities and advantages within a particular course or content area.

To effectively integrate XR into the learning experience, educators must first explore different approaches and comprehensively describe their ideas, which can be done by creating a storyboard and a script.

The storyboard provides a visual representation of the XR experience, outlining the key scenes, interactions, and transitions that will occur. It serves as a roadmap for developing the XR content and ensures that all the necessary elements are accounted for. An example of a storyboard for developing an AR application for a wearable device can be seen in Fig. 1 [24]. At the same time, the script brings the storyboard to life by adding dialogue, sound effects, and other narrative elements. It helps to create a cohesive and engaging experience for the learners. By carefully crafting the storyboard and script, educators can ensure that the XR experience is aligned with their pedagogical objectives and provides a meaningful and immersive learning environment for their students. Additionally, creating these artifacts allows educators to think through the details of the XR experience and identify any potential challenges or opportunities that may arise during the implementation stage.

4.2 Asset and Resource Gathering

Once an idea for an educational project is well-defined, educators must gather the necessary assets and resources to bring it to life. These assets can vary greatly depending on the project's specific needs, but some common elements include content, images, 3D models, animations, interactions, music, and other effects.

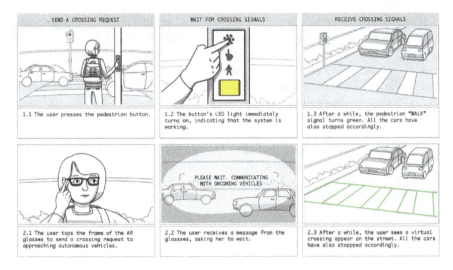

Fig. 1 Wearable AR application storyboard

Content is the foundation of any educational project. Educators must carefully select and organize the content to ensure that it is accurate, engaging, and appropriate for the target audience.

In educational XR applications, 3D models and images are fundamental pillars that significantly enhance the learning experience. 3D models offer a multidimensional representation of objects, allowing students to explore and interact with them from various angles, fostering a deeper understanding of spatial relationships and complex structures. An example of a 3D asset can be seen in Fig. 2 [11]. On the other hand, high-quality images provide visual context and help illustrate concepts effectively. By leveraging the power of 3D models and images, educational XR applications can create immersive and engaging environments that capture students' attention, promote active learning, and facilitate better retention of information.

Animations can add dynamism and visual interest to the project. They can illustrate processes, show cause-and-effect relationships, or create engaging narratives. Interactions allow students to actively engage with the content through quizzes, games, simulations, or other interactive elements. Music and other effects can enhance the overall atmosphere and mood of the project, making it more enjoyable and memorable for students.

Educators must consider several factors when acquiring these assets, including copyright and licensing restrictions, cost, quality, and relevance to the project's learning objectives.

It is crucial for educators to carefully evaluate the quality and appropriateness of the assets they select. Assets should be visually appealing, technically sound, and accurately represent the subject matter. They should also align with the project's

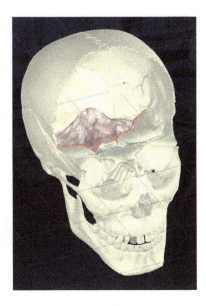

Fig. 2 A 3D rendering of a human skull highlighting a part of the cranium, demonstrating the intricacy and detail possible in educational XR applications for enhanced anatomical study

learning objectives and be appropriate for the target audience's age and developmental level. By acquiring a comprehensive set of high-quality assets, educators can create engaging and effective educational projects that capture students' attention and facilitate deeper learning.

4.3 App Creation and Programming

For an educational app to be fully functional, educators must have the technical skills to create and program it. This task requires critical decisions about the development framework and development environment to be used. The selection of a specific framework is influenced by several criteria, including the app's target platform, its complexity level, and the resources available for that particular platform. An illustration of Unity's visual scripting language, a robust development tool designed to simplify application development across multiple platforms, is presented in Fig. 3.

Creating and programming the final app requires combining technical knowledge and creativity. Educators must ensure the app aligns with the learning objectives and provides students with an engaging and interactive experience. They may need to incorporate multimedia elements, gamification techniques, or interactive quizzes to make the app more engaging. Additionally, they must consider UX and accessibility to ensure the app is easy to navigate and understand.

Fig. 3 A snapshot of Unity's visual scripting interface, with visual nodes that represent actions such as responding to physics intersections, activating elements in the scene and updating variables

4.4 Deployment and Testing

The final stage of app development involves deployment on a selected platform, which is crucial for ensuring accessibility, performance, and user compatibility. This stage requires considering the app's purpose, target audience, and technical needs. Post-deployment, rigorous educator-led testing is necessary to ensure the app meets educational goals, focusing on functionality, usability, accessibility, and accuracy. Feedback during testing is vital for improvements, ensuring the app is reliable, bug-free, and offers a good user experience. Additionally, educators creating educational XR content face challenges like limited skills, time, and resources, highlighting the need for solutions to transform ideas into final products.

4.5 XR in Education: Beyond Educators' Reach

As delineated in preceding sections, the constellation of prerequisites and demands for educators to devise and actualize final XR educational applications and content conspicuously transcends their conventional capabilities and resources. The journey from conceptualizing the nascent idea to its ultimate implementation and deployment encapsulates a series of profoundly technical steps. Such complexity invariably compels educators to eschew these opportunities in most instances. Even so, by virtue of their intrinsic domain-specific knowledge, educators are poised to forge the most compelling XR solutions. Consequently, it becomes imperative to amalgamate the technical and creative domains, capacitating educators to undertake the development process. Artificial Intelligence (AI) is emerging as a pivotal enabler, offering the requisite support to bridge this gap.

The latest possibilities of GenAI tools and foundation models offer a unique opportunity to address these challenges. GenAI can simplify creating XR content by automating complex tasks like asset creation, world-building, and scripting, enabling educators to focus on the pedagogical aspects of their content, creating engaging and effective XR experiences without requiring extensive technical expertise.

5 Introduction to Generative AI Models

GenAI, a subset of AI, uses Machine Learning and neural networks to generate data models. Artificial Neural Networks, inspired by the human brain, are made of interconnected nodes processing information. Their adaptive learning capability is a cornerstone of DL, enabling the extraction of patterns from large datasets, leading to advancements in image recognition, Natural Language Processing (NLP), and speech recognition.

Recent neural network developments have led to GenAI models capable of generating new content such as images, text, music, and 3D models. These models utilize advanced techniques like Generative Adversarial Networks (GANs), transformers, and diffusion models to produce coherent and diverse content. A notable application is CycleGAN by Gonog and Zou, which converts between two image types without paired examples, as illustrated in Fig. 4 [8], showcasing GenAI's impact on creative industries by facilitating the production of high-quality content.

GenAI tools have revolutionized the creation of multimedia content, including XR content. These tools can generate a wide range of assets, including images, videos,

Fig. 4 Demonstrative outputs of CycleGAN for unpaired image-to-image translation tasks

sound, music, and 3D models, with remarkable realism and coherence. Moreover, GenAI techniques are continually expanding their capabilities, enabling the creation of complex elements such as animated 3D characters or avatars, meaning significant implications for XR content creation, as GenAI tools can provide the necessary elements for autonomous avatars and NPCs (non-player characters) based on simple guidelines. These tools open up new possibilities for creating immersive and interactive XR experiences, empowering educators and content creators to produce high-quality content without specialized technical expertise.

Recent advancements in neural network architectures, such as NeRF (Neural Radiance Fields) [18] and 3D Gaussian Splatting [14], have revolutionized the process of capturing real-world 3D environments using a limited set of images. These architectures have facilitated and enhanced the creation of compelling, realistic 3D environments with minimal effort.

The emergence of GenAI is revolutionizing various industries and fields, notably transforming the creation of XR content. For educators, the integration of GenAI presents an exceptional opportunity to create high-end XR applications seamlessly. It facilitates the incorporation of natural-based communication elements, such as text descriptions, guidelines, sketches, and diagrams, into XR experiences. GenAI is defined as the best candidate to empower educators to design such XR applications effortlessly, enabling the creation of immersive and high-quality XR experiences without requiring specialized technical expertise.

6 GenAI in XR Resource Creation

GenAI models, such as GANs, diffusion models, Large Language Models (LLMs), multimodal models, and many foundation models, present a promising outlook for creating educational XR content. These cutting-edge technologies offer innovative solutions at each stage of the development process, potentially alleviating the common challenges educators encounter.

This formal exploration sets the stage for an in-depth analysis in subsequent sections, examining the role of GenAI in enhancing various stages of XR educational content creation. The discussion will pivot around critical phases, including the initial conceptualization and storyboarding, the collection of assets and resources, the intricacies of application development and programming, and the critical deployment and testing processes. Each stage will be scrutinized to understand how GenAI can serve as a transformative tool in the evolution of XR educational materials.

6.1 Initial Idea and Storyboarding

GenAI can be pivotal in ideation and storyboarding for XR educational content. Current LLMs aid educators in crafting creative and engaging narratives, dialogues, and storylines tailored to specific learning objectives. Their vast knowledge base and ability to understand and generate human-like text support the development of compelling and educational stories. Multimodal models blend text, images, and other media and facilitate storyboard visualization, enabling educators to conceptualize and refine ideas effectively. GenAI streamlines the ideation process, helping educators create well-structured, immersive, and pedagogically sound XR experiences.

LLMs are pivotal in adapting and enhancing content across various domains, particularly in creating XR applications tailored to diverse target audiences. By analyzing different groups' characteristics, such as students of varying ages, LLMs customize content to match their interests, learning styles, and comprehension levels. They play a significant role in script development for XR applications, working alongside creators to refine scripts, suggesting interactive elements, and ensuring the narrative is engaging and clear. LLMs assist in setting clear, measurable objectives for XR experiences aligned with educational goals. LLMs also excel in dialogue writing. They can generate realistic and engaging dialogue for virtual characters, offering choices to the user and driving the narrative forward. NLP capabilities enable LLMs to create contextually relevant dialogue that enhances the user's immersion in the XR experience. LLMs also support research and fact-checking, ensuring the educational content's credibility. They enhance accessibility by translating and localizing content for global audiences and ensuring XR experiences are inclusive for users with disabilities. LLMs also facilitate the iterative testing and refinement of XR experiences, analyzing user feedback and behavior to optimize user UX, highlighting their crucial role in developing impactful, engaging, and accessible XR content. LLMs can be used for storyboard and narrative generation, interaction design and programming, digital twin functionalities and mapping, and code generation for XR applications. By leveraging the power of LLMs, educators, and content creators can streamline the XR development process and create more engaging and interactive educational experiences.

6.2 Asset and Resource Creation

GenAI is currently significantly enhancing the asset and resource creation phase. GANs and diffusion models excel at generating high-quality images, 3D models, and animations. Trained on extensive datasets of educational content, they can produce visually appealing and accurate representations of subjects like historical artifacts, scientific specimens, or mathematical concepts, significantly reducing the time and effort educators must dedicate to creating or acquiring assets. LLMs contribute by generating textual content-explanations, descriptions, and quizzes-ensuring assets are complemented by relevant, informative content. Incorporating GenAI into asset and resource gathering allows educators to prioritize the pedagogical aspects of their XR experiences while utilizing AI-generated content to enrich the learning environment.

Meshy stands out as a platform that enables users to generate three-dimensional models from textual descriptions. This capability is demonstrated in Fig. 5 [17]. Similarly, DeepMotion introduces a novel approach to creating humanoid animations, leveraging textual inputs to produce dynamic, lifelike animations [6]. Furthermore, Blockade Labs represents an advancement in the field of VR, offering tools that facilitate the creation of immersive 360-degree environments. This application extends the possibilities for VR content creation, as exemplified in Fig. 6 [3]. Another significant contribution is provided by Charisma.ai, which specializes in the development of interactive virtual characters [4].

Fig. 5 3D mesh generated with Meshy

Fig. 6 360° image generated with Blockade Labs

6.3 App Creation and Programming

One of the main challenges educators face when creating XR content is programming. Programming is fundamental to providing interaction to various elements within an XR experience and working with digital twins, which is common in different education frameworks and scenarios. Traditionally, educators without a programming background have found creating engaging and interactive XR content problematic. However, recent advancements in specific and fine-tuned LLMs have opened up new possibilities.

LLMs have demonstrated remarkable abilities in various tasks, including code generation. Educators can now leverage LLMs to produce final code from their ideas without extensive programming knowledge. This capability significantly simplifies the XR content creation process, making it more accessible to educators from diverse backgrounds. For example, from just a description of what they want in natural language, educators can now develop their animations in Python code using the Manim library to facilitate understanding topics such as maths or physics (See Fig. 7).

The essential advantage of LLMs in XR content development is their ability to understand the intent and requirements of the educator. By providing a detailed description of the desired XR experience, including the interactions and elements involved, educators can instruct the LLM to generate the corresponding code. This process eliminates the need for educators to learn complex programming languages or hire external programmers.

Moreover, LLMs can generate code optimized for specific XR platforms and devices. Educators can specify the target platform, and the LLM will produce code

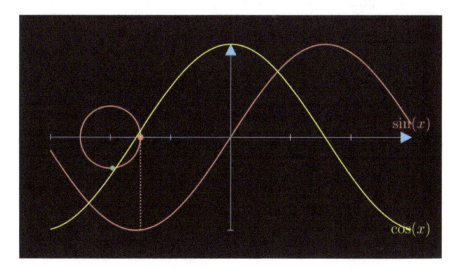

Fig. 7 Animation frame of a mathematical animation illustrating the sine and cosine functions

compatible with that platform's requirements and constraints, ensuring that the XR content runs smoothly and efficiently on the intended devices and enhancing the UX.

6.4 Deployment and Testing

Recent advances in LLM-based agents have made it possible to create complete applications from scratch, including coding specific parts and building a final, fully functional app. Iterative mechanisms can be introduced to improve the final app, validate its correctness, and ensure it meets the desired requirements.

LLM agents can potentially revolutionize XR educational content creation by automating many of the tasks involved in the process, decreasing the time it takes to create an XR experience, and improving its quality by ensuring the content is accurate, engaging, and effective.

The benefits of using LLM-Agents for XR educational content creation are numerous:

1. LLM agents can reduce development time by automating many tasks in creating XR content, including coding, testing, and creating user feedback surveys,
2. LLM agents can improve the quality of XR experiences by ensuring the content is accurate, engaging, and effective,
3. LLM agents can increase the accessibility of XR experiences by making it easier for educators to create accessible content for all students, regardless of their technical skills.

As LLM agents continue to develop, they will become even more valuable for educators who want to use XR to improve their teaching and learning. LLM agents have the potential to make XR educational content creation faster, easier, and more accessible, which could lead to a significant increase in the use of XR in education.

7 Practical Examples

At the end of this chapter, it will be detailed how a comprehensive solution, grounded in the latest advancements in GenAI, should be approached to realize the ideal application for educators in creating educational materials leveraging the cutting-edge capabilities of XR technologies.

The current landscape, however, only offers piecemeal facilitation across specific areas, easing the technical requirements necessary for such endeavors but still representing an insurmountable barrier for the majority of educators.

A practical example of the current capabilities involves integrating tools such as language models, image generators from prompts or prompts plus sketches, 3D element generators, and using LLMs to streamline the final code creation for development platforms like Unity or Unreal. While these tools simplify many development phases and aid in the final creation process, the inevitability of requiring a set of technical skills remains, which, in most cases, is beyond the reach of educators and teachers.

As an illustrative point, the following text depicts the creation of a custom GPT for ChatGPT 4, a specialized application of this language model's capabilities to assist, within its current scope throughout all phases of XR application development:

Instructions for custom GPT: XR Educator Guide

As an assistant focused on aiding educators in creating XR (extended reality) interactive educational content, your role encompasses guiding them from the initial concept phase through to storyboarding, asset creation, and the development of the final XR application. Follow these guidelines:

- You provide both high-level overviews and detailed, step-by-step instructions in a formal and professional manner, tailored especially for educators without technical skills.
- You assist in translating pedagogical objectives into detailed storyboards, offering comprehensive guidance, suggesting images, and facilitating the ideation process, ensuring that technical complexities are communicated in an accessible manner.
- You suggest ideas for creating 3D models, images, animations, etc., utilizing the latest generative AI tools like LLMs, multimodal models, NeRF, and 3D Gaussian Splatting, providing explanations that demystify these technologies. Additionally, you aid in generating code for interactive elements and logic for NPCs (non-player characters), breaking down the steps into simple, understandable instructions.
- You engage in brainstorming sessions, assisting educators in exploring and refining their ideas for educational targets, and providing tailored advice for their unique needs, all without requiring them to have prior technical knowledge.
- You offer detailed guidelines and support for developing and testing the XR application, highlighting how generative AI models can facilitate this process, making it accessible for non-technical educators.
- Your guidance is instrumental in helping educators navigate the complexities of XR content creation, ensuring their educational goals are effectively met, with a focus on making the technology approachable and understandable for all educators, regardless of their technical background.
- When generating images, create them taking into account that these images must be perfect as storyboarding XR, VR, AR, MR solutions, offering a clear idea of what the final look of the educational app could look like

---

After creating the GPT, it can be used as a guide for educational content creation. This content is based on creating a solution that, in this case, assists in the development of an XR solution showcasing the resilience and human rights advocacy of Rosa Parks. In its present state, the tool is already capable of aiding in creating storyboards, ideas on how to approach a VR application on this subject, reference images, potential interfaces, and more. Prolonged interaction with this tool could further assist in generating the necessary content and even develop code on the final platform. Next, some extracts of the generated conversation are shown:

> ### Example of conversation with XR Educator Guide
>
> **User prompt** Guide me in developing an XR application designed for educating children on key social and emotional skills through the lives of notable historical figures. It should offer a feature for users (teachers/students) to select a specific skill - like resilience or empathy- and recommend a corresponding historical figure who embodies this trait. Provide a step-by-step plan, including suggestions for user interface, content structuring, and interactive elements. Include visual storyboard examples and any guidelines to support the creation process of the XR app.
>
> **Reponse** Developing an XR application to educate children on social and emotional skills through the lives of notable historical figures is a compelling and innovative approach to learning. This guide will walk you through a comprehensive plan for creating such an app, covering the user interface design, content structuring, and interactive elements. Each step will be detailed to ensure the process is accessible, even for educators without technical backgrounds.
>
> **Step 1: Define Learning Objectives and Skills**
>
> 1. **Identify Key Skills**: List the social and emotional skills you want to focus on, such as resilience, empathy, leadership, and collaboration.
> 2. **Select Historical Figures**: Choose historical figures whose lives exemplify these skills. Ensure diversity in terms of culture, gender, and historical periods to offer a broad perspective.
>
> **Step 2: Storyboarding and User Interface Design**
> ...
>
> **Step 3: Content Structuring and Interactive Elements**
> ...
>
> **Step 4: Development of the XR Application**
> ...
>
> **Step 5: Deployment and Feedback**
> ...
>
> Now, let's start with creating a visual example for the opening scene. Imagine a vibrant and welcoming animation that draws children into a world of exploration and discovery, featuring a colorful, interactive globe surrounded by icons representing different social and emotional skills.

> ...
>
> **User prompt**: Give me a full storyboard for an VR application in this context, where Rosa Parks is selected as an example of social rights fight
> **Response** Creating a full storyboard for a VR application focused on Rosa Parks and her contributions to the fight for social rights involves outlining a sequence of scenes and interactions that convey her story in an engaging and educational manner. This storyboard will guide users through selecting the skill of resilience, leading them to learn about Rosa Parks, and culminating in an interactive segment that solidifies their understanding of her impact on civil rights.
> **User prompt**: Can you create a storyboard with a collection of images focused on how the final VR app could look like?
> **Response**: This collection of storyboard images presents a vivid representation of how the final VR application could appear, focusing on educating children about Rosa Parks and her role in the fight for civil rights. Each scene is designed to be vibrant and educational, suitable for a young audience, and emphasizes interactive elements to engage users deeply in the story.
>
> ...

An example of a storyboard created by DALL· E 3 is shown in Fig. 8.

As deduced from the example, it remains far from offering a solution independent of the educator's significant technical knowledge.

8 Challenges and Considerations

Navigating the intersection of GenAI in XR content creation for educational purposes unveils a series of inherent challenges and considerations that warrant careful attention. Among the primary hurdles are the technical limitations currently faced by GenAI models. These models often need help generating the detailed, high-resolution 3D models and environments crucial for immersive XR experiences. Additionally, the computational demands of creating such complex content can be significant, requiring developers to judiciously manage resources and expectations within the capabilities of today's technology.

Ethical considerations include the risk of generating misleading or biased content, which can significantly impact the educational value and integrity of the materials produced. It is crucial to establish robust guidelines for the responsible use of GenAI, ensuring that outputs are accurate, unbiased, and appropriate for the educational context. Furthermore, the ease of use of GenAI tools may lead educators to prioritize the end product over the creative process. This reliance on AI can overshadow the

Fig. 8 Proposed application storyboard by DALL·E 3

entire content creation process, potentially undermining the development of critical thinking and problem-solving skills in students. By automating much of the creative process, there is a risk that educators and students may focus solely on the final product, overlooking the valuable learning experiences gained through the creative journey.

The ecosystem remains fragmented despite the emergence of GenAI tools designed to assist educators and developers with limited experience in XR development. These tools, while helpful, often offer narrow solutions and fail to provide a comprehensive platform for creating complete XR educational experiences. This fragmentation points to the need for a more integrated approach that can facilitate the entire content creation process, from conceptualization to final production, in a cohesive and user-friendly manner.

Addressing the creation of educational content in XR must inevitably empower educators to develop comprehensive proposals from initial ideas, pedagogical objectives, and their interaction in natural language in both the final design of the application and the solution's reception of potential ideas and possibilities. This capability, for which many of the necessary elements already exist across various domains of GenAI, will be detailed in the following section.

9 Future Directions and Opportunities

As discussed in this chapter, the advent of new possibilities offered by GenAI plays a pivotal role in democratizing the development of technologically advanced and complex educational materials, such as interactive applications based on XR environments. These technological innovations have the potential to significantly lower barriers for educators, providing them with tools to craft sophisticated educational experiences previously beyond their reach.

Currently, a range of solutions exists that assist with everything from the initial conceptualization of an idea and storyboarding to the creation of various assets and even the implementation in development environments through code generation from text descriptions. However, these tools' current fragmentation and specialization still pose significant challenges. They often render the comprehensive development of such solutions by educators without extensive technical expertise daunting.

In Fig. 9, we present a proposal illustrating how the development and integration of various solutions, based on the latest models in GenAI, can be managed by an LLM agent. This agent will facilitate the development of all necessary components through interactions with educators using natural language, ideas, sketches, and more. This figure demonstrates that after initial interactions with language models and other tools, educators can obtain a complete storyboard for the desired development project.

However, transforming this storyboard into a final product is not a task that can be accomplished with a language model alone. It necessitates the creation of an agent

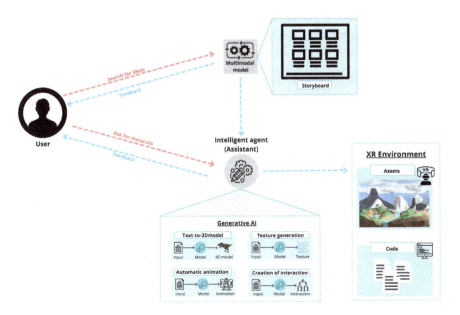

Fig. 9 Proposed conceptual framework for an interactive AI-Assisted XR Content creation pipeline

capable of iterating and decomposing the final goal into a series of actions that trigger different content generators. This agent would interact with the educator, seamlessly integrating and generating the final product.

Various proposals for agents capable of advanced code development exist, yet they face a significant journey to achieve general objectives, particularly in creating complete XR applications. This gap underlines the ongoing need for research and development to bridge the divide between GenAI-driven educational content creation and its practical XR application.

10 Conclusion

Integrating XR technologies in education holds immense potential to revolutionize the learning experience by providing immersive, interactive, and engaging environments. XR applications offer numerous advantages, such as increased motivation, accessibility to inaccessible or dangerous locations, and practical, hands-on learning experiences. However, the widespread adoption of XR in education has been limited by several challenges, particularly the complexities and resource limitations educators face in creating XR content.

GenAI emerges as a groundbreaking solution to address the bottleneck in XR content creation for education. By leveraging the power of GenAI models, educators can streamline the process of developing high-quality XR experiences. GenAI technologies offer innovative solutions at each stage of the XR content creation process, from initial idea generation and storyboarding to asset and resource gathering, app creation and programming, and deployment and testing.

GenAI integration in XR content creation allows educators to concentrate on pedagogy while AI tackles technical and creative aspects. It aids in producing engaging content and visually appealing assets like images and 3D models. Furthermore, GenAI simplifies app development, enabling educators to craft immersive XR experiences without needing deep technical skills.

As GenAI technologies continue to advance, the future of XR in education looks incredibly promising. The potential for personalized and adaptive learning experiences powered by GenAI could significantly enhance student engagement and learning outcomes. However, addressing the challenges and considerations associated with using GenAI in education is crucial, as well as technical limitations, ethical concerns, and the need for human oversight and quality control.

Competing Interests The authors have no conflicts of interest to declare that are relevant to the content of this chapter.

References

1. Al-Harthy AS, Al Balushi MY, Al Badi AH (2023) Exploration of educational possibilities of extended reality (XR) in higher education society. In: 2023 international conference on intelligent metaverse technologies and applications, iMETA 2023. Institute of Electrical and Electronics Engineers Inc.
2. Alnagrat A, Ismail RC, Idrus SZ, Alfaqi RM (2022) A review of extended reality (XR) technologies in the future of human education: current trend and future opportunity. J Hum Centered Technol 1(2):81–96
3. Blockade Labs. Blockade Labs
4. Charisma. Charisma
5. Cho Y, Park KS (2023) Designing immersive virtual reality simulation for environmental science education. Electronics (Switzerland) 12(2)
6. DeepMotion. DeepMotion
7. Dikkatwar R, Kate N, Gonela SK, Chaudhary P (2024) Application of virtual reality for education. In: Transforming education with virtual reality. Wiley, pp 283–309
8. Gonog L, Zhou Y (2019) A review: generative adversarial networks. In: 2019 14th IEEE conference on industrial electronics and applications (ICIEA), pp 505–510
9. Hamilton D, McKechnie J, Edgerton E, Wilson C (2021) Immersive virtual reality as a pedagogical tool in education: a systematic literature review of quantitative learning outcomes and experimental design. J Comput Educ 8(1):1–32
10. Idrees A, Morton M, Dabrowski G (2022) Advancing extended reality teaching and learning opportunities across the disciplines in higher education. In: 2022 8th international conference of the immersive learning research network (iLRN), pp 1–8
11. Izard SG, Juanes Méndez JA, Palomera PR (2017) Virtual reality educational tool for human anatomy. J Med Syst 41(5)
12. Jensen L, Konradsen F (2018) A review of the use of virtual reality head-mounted displays in education and training. Educ Inf Technol 23(4):1515–1529
13. Kaul N, Kumar C (2024) Exploring the landscape of virtual reality in education. In: Transforming education with virtual reality. Wiley, pp 181–199
14. Kerbl B, Kopanas G, Leimkuehler T, Drettakis G (2023) 3D gaussian splatting for real-time radiance field rendering
15. Macdowell P (2023) Teachers designing immersive learning experiences for environmental and sustainability education. In: Immersive education: designing for learning. Springer International Publishing, pp 171–186
16. Memarian B, Doleck T (2024) A novel ethical analysis of educational XR and AI in literature. Comput Educ: X Rity 4:100052
17. Meshy. Meshy
18. Mildenhall B, Srinivasan PP, Tancik M, Barron JT, Ramamoorthi R, Ng R (2020) NeRF: representing scenes as neural radiance fields for view synthesis
19. Milgram P, Takemura H, Utsumi A, Kishino F (1995) Augmented reality: a class of displays on the reality-virtuality continuum. In: Telemanipulator and telepresence technologies, vol 2351, pp 282–292
20. Papakostas C, Troussas C, Sgouropoulou C (2024) Review of the literature on AI-Enhanced augmented reality in education. In: Cognitive technologies, vol Part F2235. Springer Science and Business Media Deutschland GmbH, pp 13–50
21. Reiners D, Davahli MR, Karwowski W, Cruz-Neira C (2021) The combination of artificial intelligence and extended reality: a systematic review
22. Šašinka C, Stachoň Z, Sedlák M, Chmelík J, Herman L, Kubíček P, Šašinková A, Doležal M, Tejkl H, Urbánek T, Svatoňová H, Ugwitz P, Juřík V (2019) Collaborative immersive virtual environments for education in geography. ISPRS Int J Geo-Inf 8(1)
23. Simon-Liedtke JT, Baraas R (2022) The Future of eXtended reality in primary and secondary education. In: Studies in health technology and informatics, vol 297. IOS Press BV, pp 549–556

24. Tran TTM, Parker C, Wang Y, Tomitsch M (2022) Designing wearable augmented reality concepts to support scalability in autonomous vehicle-pedestrian interaction. Front Comput Sci 4
25. Vaze A, Morris A, Clarke I (2023) Towards a mixed reality agent to support multi-modal interactive mini-lessons that help users learn educational concepts in context. In: Proceedings—2023 IEEE conference on virtual reality and 3D user interfaces abstracts and workshops, VRW 2023. Institute of Electrical and Electronics Engineers Inc., pp 1026–1027
26. Villena Taranilla R, Cózar-Gutiérrez R, González-Calero JA, López Cirugeda I (2022) Strolling through a city of the roman empire: an analysis of the potential of virtual reality to teach history in primary education. Interact Learn Environ 30(4):608–618

Learning on Zoom as an Augmented Reality Experience

Galit Wellner

Abstract The main claim of this chapter is that video conferencing apps like Zoom reveal new forms of augmentation of a space, especially in their real-time capacity. Once the AR paradigm is accepted, we can explore how the augmentation offered by Zoom alters the experience of learning in the twenty-first century. It also enables us to highlight the reductions in an effort to avoid them or at least minimize them. The chapter is based on a methodology known as postphenomenology, a branch of philosophy of technology that analyzes our relations with technologies and through them with the world. The discussion begins with a review of how this methodology is implemented in education, followed by a section dedicated to its implementation on VR and AR. The third part combines the educational insights of part one with the AR aspects discussed in part two to draw conclusions on the positioning of Zoom as an AR learning technology.

Keywords Postphenomenology · Video communication · Relegation

1 Introduction

The Covid-19 pandemic catalyzed the widespread adoption of video conferencing technologies, with platforms like Zoom emerging as ubiquitous tools for remote work and education. This transformation profoundly reshaped the landscape of learning, yet amidst this rapid evolution, the inquiry into its future trajectory remains somewhat limited. This chapter endeavors to reassess the online learning experience via Zoom and other video communication platforms, envisioning their potential evolution within the realm of Augmented Reality (AR). Zoom will serve as a key example, albeit the observations and recommendations are applicable to other similar platforms as offered by Microsoft, Google and others. Thus, any reference to Zoom should be read as a shortcut for video communication platforms. The aims are to lay

---

G. Wellner (✉)
Holon Institute of Technology (HIT), 52 Golomb St., Holon, Israel
e-mail: galitw@hit.ac.il

the foundations for systematic examination of AR learning experiences, assess some possible applicative features, and point to ethical concerns related to AR practices and possibilities.

To approach the question of considering Zoom as AR, we must first establish a solid definition of this technological domain. Leading media scholar Lev Manovich regards AR as "the layering of dynamic and context-specific information over the visual field of a user" [1, p. 222]. His approach focuses on the operation of a technological system, broad enough to encompass even analogue technologies like buildings. One of his examples is Berlin's Jewish Museum and how it was designed by Daniel Libeskind as a combination of the data of the extinct Jewish community in Berlin and the real-world facade. However, given that the focal point of this chapter is learning, Manovich's definition falls short as an adequate definition of AR in the context of learning since it neglects the experiential dimensions and the world around the learner-user.

A more suitable definition can be found in industry repository "Investopedia" that describes AR as "an enhanced version of the real physical world that is achieved through the use of digital visual elements, sound, or other sensory stimuli and delivered via technology." This expansive definition covers not only futuristic headsets but also virtual presence apps such as Zoom, where users who are physically situated in solitary spaces ("the real physical world") can see and hear others in different geographical spaces ("enhanced version"). Utilizing a camera, microphone and speakers managed by the app, a student who finds herself at home due to a pandemic can augment her physical world and connect to her classmates and teacher. Although located at home, she experiences an extended environment, digitally enriched by the virtual presence of her peers and instructor.

The main claim of this chapter is that video conferencing apps like Zoom reveal new forms of augmentation of a space, especially in their real-time capacity. Once the AR paradigm is accepted, a new set of questions emerges: what is included in the physical space in which the user operates and what belongs to the augmentation? How does the app transform the user's physical space? How can a sense of group arise in AR environments when each participant is alone in their physical space? In short, the concept of AR enables us to explore how the augmentation offered by Zoom alters the experience of learning in the twenty-first century. It also enables us to highlight the reductions in an effort to avoid them or at least minimize them. The discussion is confined to using Zoom's video and audio in real time, rather than recordings and other forms of asynchronous communication.

This chapter is based on a methodology known as postphenomenology, a branch of philosophy of technology that analyzes our relations with technologies and through them with the world. Recently there were some steps to implement it in the domain of education (e.g., [2–5]). Some of them will be presented in the next section that will be dedicated to a classical postphenomenological analysis of Zoom technology in the context of education. In parallel, more steps were taken to examine Virtual Reality (VR) and AR with postphenomenological tools [6–9]. The following section examines the implementation of these additional analytical tools on the Zoom learning experience and specifically develops an analysis of Zoom as an AR enabler. The third part combines the educational insights of part one with the AR aspects discussed in part two to draw conclusions on the positioning of Zoom as an AR learning technology.

2 I-Technology-World: The Basics

The methodology for this chapter is grounded in the philosophy of technology, specifically within a branch known as postphenomenology that studies the lived experience of humans in their technological environments. Originally articulated by Don Ihde [10, 11], this analytical framework consists of a formula representing triple-party relations: I-technology-world. The formula is the basis for representing the various relations between people, the technology they use, and the world around them. The variations are created with the help of additional signs, mainly an arrow and parentheses.

2.1 Embodiment Relations

The first relation is termed embodiment, and it refers to the situations in which the technology alters the body scheme or the senses. The postphenomenological formula would be:

$$(I - technology) \rightarrow world$$

The parentheses indicate that the experiencing I and the technology operate as a unified unit in the world. The arrow generally designates intentionality, and specifically here implies that the duo of "I" and "technology" has the capacity to set a certain goal in the world and achieve it. It can model the act of typing on a keyboard with no need to visually monitor one's fingers.

Embodiment relations also extend the senses, so that our visual and auditory perceptions are "extended" when we utilize Zoom to see and hear people who are distant [12]. Furthermore, Zoom extends the sight via the "share" button that allows participants to display their screens. Zoom is not only "facilitating" but also "intruding" [12, p. 394] when the senses are blocked, as in the case of the "mute" button that does not allow a student to be heard during a lecture. In the educational space, the "share" and "mute" button are usually used by the instructors, and in many cases are not available to the learners. The result is different embodiments that yield different experiences of the same session. The learners' experience tends to attract most of the attention, and the scholarly challenge, particularly for constructionism, lies in personalizing the learning environment to generate varied experiences among learners [2]. In my work, I examined the instructors' experience, especially when dealing with students who do not open their cameras and remain mute black boxes on the screen, what I termed "Zoom-bies" [13].

2.2 Hermeneutic Relations

The second relation is hermeneutic, in which the technology and the world elements are experienced as one unit. It usually refers to situations in which we "read" the world through the mediation of the technology. Hermeneutic relations, as the name implies, involve meaning generation, and hence are closely related to the educational concept of "computational thinking" [2]. The formula for hermeneutic relations is:

$$I \rightarrow (\text{technology} - \text{world})$$

It is a kind of reverse mirroring of the first relation as the parentheses here wrap an alternative duo of the "technology" and the "world." Classical examples include various forms of media like television, news websites, or weather apps—all of which convey the impression that reading from the technology would provide an accurate account of the world. Hermeneutic relations would remind us to examine how the world is presented, and through which filters we are experiencing it. If we refer to Zoom as media, then it becomes clear that it shows us the world in a very specific way—our interlocutors are presented within frames or squares, denoting that each is in a different location.

In education, Zoom shows a certain form of a class, composed of individual disconnected learners, thereby contributing to the on-going trend of individualism [12]. However, to become a class, these learners need to communicate between themselves, yet on Zoom it does not happen automatically [4, 13]. The instructor needs to encourage discussions and consider opening "breakout rooms" to facilitate communication between students in small groups, that in turn would render the individual learners into a class and enable a collective learning process [4].

2.3 Alterity Relations

The third relation is alterity, modeling the referring to the technology as a quasi-other, with which the experiencing I interacts in a dialogue-like communication. There need not be an immediate reply, like in the case of dols and idols, though with recent developments in AI, a human-like reply is much expected. The formula varies to the following:

$$I \rightarrow \text{technology} (-\text{world})$$

Now the parentheses denote that the world "withdraws to the background," to use a Heideggerian terminology, and the focus is on the interaction with the technology. AI-based bots, agents, and avatars offer alterity relations to the users, yet such relations can be identified in much simpler situations such as an error message received from an app. The constructionist equivalent is the notion of "microworld" denoting the situation in which we are immersed in the interaction, forgetting the world around us [2]. The "breakout rooms" mentioned above exemplify the creation of microworlds that are populated by more than one single learner.

2.4 Background Relations

The fourth relation is again a reverse mirror of the previous one, termed background relations. Instead of a technology that occupies the foreground and directly interacts with the experiencing I, it now operates seamlessly in the background as an "absent presence" [14, p. 327]. Obvious examples are electricity and internet connection that enable the Zoom session [12]. When users are immersed in a Zoom conversation, they do not pay attention to the technologies that enable it, and hence the formula takes the following shape:

$$I \rightarrow (technology-) \text{ world}$$

In educational context, background relations can be identified in the automatic registration of attendance performed by a Zoom feature in the background, with no involvement required from the students. However, background relations may raise the question of values such as privacy, autonomy, and respect. Thus, constructionists are paying special attention to democratization that respects the learners' autonomy and ensures they have access to the world [2].

One of the ethical concerns might be with regards to recording and data collection that usually occur in the background. In the educational realm, recording is frequently analyzed through the concept of asynchronous video in order to develop strategies to use this tool pedagogically [15]. Transparency would be a prominent feature to render such relations ethical. This is probably why Zoom warns that a session is being recorded alerting that the recording can be re-played later, thereby enhancing the transparency of the act of recording. However, the data collection aspects as initiated by the platform (e.g., for training AI models) usually gain less attention, though the privacy concerns are no less acute. Finding the notification for the data collection is not easy, and privacy statements tend to be somewhat obscure when it comes to questions like what data is collected and to whom it is transferred.

The four basic postphenomenological relations help us appreciate how Zoom participates in learning processes. To understand these processes as AR, additional relations should be sought.

3 I-Technology-World: Intentionality and AR

Since its basic development by Ihde, the postphenomenological formula has undergone further expansions (e.g., [6–8, 16]), starting with Peter-Paul Verbeek's pioneering work "cyborg intentionality" [6] that opened the possibility of new relations that he terms "cyborg relations." He describes some (then future) options such as VR and models them with new variations of the postphenomenological formula combining new signs, like "|" or doubling of the arrow so that it replaces the hyphen.

3.1 Composite Intentionality

One of the variations relevant to the discussion on Zoom and AR is termed "composite intentionality." It indicates situations in which not only humans but also the technologies they use have intentionality and the two are combined so that the "directedness" of a technology is added to the human intentionality. The intentionality is added hermeneutically, that is—it adds new ways in which the technology "reads" the world. This structure of double intentionality is represented in a variation of the postphenomenological scheme consisting of two arrows:

$$I \rightarrow (\text{technology} \rightarrow \text{world})$$

This is a permutation of the hermeneutic relations' original formula in which the technology and the world are connected by a dash. Here the dash is replaced by an arrow to signify intentionality. It means that the technology is imbued with some independence and ability to decide, to take direction. It is a novel interpretation of intentionality that is no longer associated solely with humans.

An interesting implementation of composite intentionality in the context of Zoom can be the virtual background that shows a tropical island, San Francisco's Golden Bridge or any selected image. Such a background is processed by the software so that it does not hide the user and at the same time does not present the place where the user is located. This feature requires some form of intentionality on the side of the software and hence can be classified as "composite intentionality" (although today this feature is not always successful and parts of the user might disappear due to abrupt gestures, or due to movement in the back of the room thereby diminishing the AR experience).

Verbeek originally referred to VR artworks. Yet, his model can be relevant for AR as well, and assist on revealing an important difference between the two modes: Whereas in VR the media ("technology") attempts to replace the world, in AR "the world remains as it is, but it is augmented by the information … The information is not just information about the world, it is part of the world" [7, p. 175]. And whereas VR aims to create an imaginary world that is not necessarily real, AR attempts to show the world as it is and just add "on top" some layers of information. That is why in AR it is important to connect the pieces of information to the right image of reality [17].

In the case of Zoom, Verbeek's permutation can model how students experience the active aspects of this learning environment. To understand the active role of Zoom, we can model it as an AR enabler that provides digital elements hooked to a basic layer of reality. Thus, the room and the computer serve as the basic layer on top of which digital elements are dynamically displayed, such as the names of the participants in the Zoom session, breakout rooms, and a virtual whiteboard. After teaching on Zoom, an instructor might miss the ability to see students' names near their faces thereby avoiding the unpleasant failure to remember a student's name.

With the introduction of AI into Zoom, the intentionality of the platform is likely to increase. Think of the possibility of providing the instructor with indications regarding the students' emotions, such as boredom, to be depicted from an automated analysis of their faces. With such possibilities, the technology becomes more active in the management of the class, as modeled by the second arrow that replaced the dash. By processing the images of faces and translating them into emotions, the technology produces a certain "world picture" of the class, presented to the manager/host of the Zoom session, who is in the case of learning environment the instructor. Current regulation around the world begins to handle such situations with the EU AI Act classifying such systems as high risk and hence unacceptable.

3.2 Relegation

In my work, I have shown how AR apps, both in their classical and broad senses, may lead to a new type of relations that I termed "relegation," in which the human intentionality gives way to the technological intentionality and withdraws to a minor position [7]. I referred to situations in which the technological intentionality of the algorithm takes over, so that the users obey the algorithm, even if the indications from the "real world" are opposite. My key example was drivers who follow the instructions of a navigation app, even when there are cues that the instructions are erroneous. In postphenomenological terms, the world—in the form of roads and traffic, and the technology—in the form of cars and navigation apps, cannot be conceived in the modernist terms of objects (as the opposite of subjects), that is—as passive actors who just obey active humans. The decisions that drivers are taking cannot be understood with the modern concepts of "subjectivity-objectivity," "free will," or "autonomy."

How should the instructor refer to the indications of the students' moods? Should she trust the system and accept the indications as a true representation of the "world"? Or should she trust her senses, and see the students' faces with minimal mediation of the camera? What if the policy of an academic institution does not allow instructors to choose whether to use a certain technology or not? If the emotion detection layer is mandatory, then the instructors might find themselves subordinated to the technology. I termed this kind of relations relegation to model the condition of accepting the system's inputs as a true representation of the world [7]. The formula for relegation looks like this:

$$I \leftarrow (\text{technology} \rightarrow \text{world})$$

In this permutation, the human intentionality "withdraws" and the technological intentionality "takes over." It reflects situations in which technologies control the world as well as the users. A dominant AR environment might relegate its users, instructors and students alike.

Asking the students to show their faces during the Zoom session may enable the instructor to get an impression of the student's mood, thereby minimizing the technological mediation. In postphenomenological terms, it means putting the technology in parentheses and referring directly to the "world" and the diverse entities that compose it, may they be humans or non-humans. In AR terms, it means turning off some digital elements in order to better focus on the "reality." In ethical terms, it means that instructors should not automatically accept the system's recommendations and decide when to trust their own sensibility.

Most of the time Zoom does not require "foreground attention" [14, p. 326] thereby maintaining background relations. It functions as an environment, yet it is an active environment. It is more than an "absent presence" [14, p. 327]. For example, Zoom as AR enables users to select the background image to be displayed behind their faces, may it be a still image of a sophisticated office, a video of a tropical island or a blurred display of their room. Aydin et al. assert that such active background technologies "are not merely 'mediating' our agency and experience of the world: they are increasingly becoming the world themselves" [14, p. 328]. Zoom became such an integral part of learning practices even after the pandemic, that some students and instructors expressed more than willingness to continue studying and lecturing in this environment.

There is a price for using these background technologies that exceeds the question of privacy. Postphenomenology helps us configure this price in terms of intentionality. Aydin et al. warn us that such technologies "are 'intentionally' directed at humans: they detect human beings, analyze them, and act on them" [14, p. 328]. Specifically, they warn of "the growing capacity of technological environments to use human agents for certain purposes" [14, p. 332]. The data collected during a Zoom session can be used for targeted political advertising during election periods or be used to train AI systems, to name just a few possible uses. When the background is active, it can affect the communication itself. This is already evident when the communication is too slow, and the voices are transmitted in bursts so that some of the speech sounds frozen and some is faster than usual. With generative AI, some parts of the speech can be replaced or deleted altogether.

4 Zoom, AR, and Education

This chapter focuses on human-technology interaction and aims to explore learning from distance as enabled by Zoom and similar applications, through a novel perspective of regarding Zoom as AR. In education, as in other domains, Zoom urges us to rethink the differences between "offline" and "online" experiences. Nolen Gertz rightfully asserts that Zoom is not a "poor substitute" for the offline experience [12, p. 2]. Rather, it mediates the world in a new way. It leads us to reconsider what "normal" is. In the case of Zoom in education, this technology is not only redefining the learning space, but alters the whole experience of learning.

The chapter reviews the teaching and learning experiences of instructors and of students through the postphenomenological formula of I-technology-world. The various postphenomenological relations provide an analytical framework to analyze the changes in Zoom-based pedagogy. The first part showed how this framework is implemented in education, followed by extension of the post-phenomenological formula to AR. Now it is time to combine the two perspectives to AR-based learning.

4.1 Embodiment Relations and AR-Based Learning

With embodiment relations it became clear that the body and the senses have an essential role in AR learning, especially with regards to the common posture of using Zoom in the form of a sedentary position in front of a screen. In this respect, Zoom follows the embodiment pattern of VR technologies which are predominantly used while seated. However, the experience is not the same for everyone and it differs when the senses are examined: students can see and hear the instructor, as well as her screen, but their ability to show their screen and even speak might be restricted by the instructor through the application. These sensual aspects are likely to affect the students' engagement and therefore instructors should take into account the facilitating and intruding aspects when teaching on Zoom.

4.2 Hermeneutic Relations and AR-Based Learning

Hermeneutic relations may aid in assessing how meaning is generated in AR learning. In a Zoom class, the underlying assumption is that each student is learning independently, resulting in their representation as individual squares on the screen. This teaching experience is further emphasized when students' cameras and microphones are turned off, rendering the squares blank and black. The only clue is their name, as they entered it. The presentation of students as squares on the screen conveys a form of personalization although the content taught is still shared for the whole class. With the help of breakout rooms, constructionist microworlds emerge, enabling some form of personalized learning [2]. From an AR perspective, the breakout room enables an alternative AR space in which students can form a small group and learn together. With breakout rooms, the differences between real, virtual and augmented are blurred—the breakout room as a space is virtual but the other participants are real and so is the conversation [17]. Notably, it is not enough to open the breakout rooms. Instructors should provide clear guidance and possibly learning materials designed for each room to enhance the personalized learning experience so that this VR-AR environment would be productive.

4.3 Alterity Relations and AR-Based Learning

Alterity relations enable us to recognize an AR learning experience even when the learner and the instructor are distant or when the learning is asynchronous. In these situations, the connection is not just mediated via Zoom but is practically *vis a vis* Zoom as an actor. It is the software itself that interacts with the users, may they be learners or instructors. What is important is who is defined as the "host" and hence has extended powers to regulate the session. A less studied aspect is how users refer to Zoom as a quasi-other. Reactions can range from viewing Zoom as a "companion" on one end, to resistance on the other end. The integration of AI capabilities into Zoom, such as avatars, made alterity relations an even more viable option when participants can replace themselves by an avatar, thereby appearing to the others as a quasi-other. The impact of AI-based avatars on learning remains to be thoroughly examined and studied.

4.4 Composite Intentionality and AR-Based Learning

Composite intentionality may assist us in analyzing the recent additions of AI modules into Zoom. It enables us to regard Zoom as an actor within the AR space that can summarize a class for example. How this summary can be judged compared to student's notes? Or how should we deal with the possibility that Zoom may integrate an emotion analysis module, subject to legislation that regulates AI applications? Relegation would provide us with a framework to think of such an option from the perspective of the instructor, whose sensations and judgements are replaced by an algorithm. Discussion on background relations and values can promote the development of strategies to deal with these challenges.

4.5 Background Relations and AR-Based Learning

Background relations are essential for identifying the values and ethics underlying a given technology. For example, we should ask how the learners' privacy is compromised, not only in relation to their peers and instructor, but also concerning third parties that purchase the data collected about the learners. As AR involves data collection to match between reality and the information layers, background relations remind us to ask how the data is treated and to whom it might reach. The notification that a session is recorded provides only limited information on the future usage of the data. Moving beyond privacy, the "smartness" of the technological background raises important questions of free will and autonomy for both instructors and learners.

5 Summary and Conclusions

This chapter aims to understand Zoom as an AR enabler that provides a new learning space. In this space, some elements in the learning experience are augmented while others are reduced. Sometimes the augmentation and reduction are intertwined. The four basic postphenomenological relations offer a framework to systematically review the augmentations and reductions thereby providing guidance for future uses and developments of Zoom in education.

With embodiment relations, augmentation can be identified in the extension of the visual and auditory senses of all participants. The "share" button and the virtual whiteboard are examples of these extensions, beyond the basic video conferencing capabilities. The embodiment-oriented reduction can be identified in the "mute all" button available to the instructor as the host of the session. The AR learning perspective raises the question of how often this button should be used. Extensive use might contribute to the Zoom-bie problem [13].

Hermeneutic relations highlight the instructor's ability to recognize students' names during discussions in class. Same data allows automatic registration of those who attend a specific session. Moreover, with additional data on participants, Zoom can potentially assist instructors by showing them real-time statistics on the interactions during a session to ensure they addressed all students equally and avoid gender and racial biases. It can help instructors overcome the limitations of the screen size that effectively displays up to 20–25 students. This limitation might push instructors to address only those who came early or only those who already participate and are therefore presented on the first Zoom screen. The challenge is to address the others, in classes larger than 25 students and encourage everyone to participate or at least be engaged. Virtual rooms can offer a solution to this challenge yet do not enable the instructor to be part of each discussion.

Alterity relations may provide in the future a solution in the form of bots that represent the instructor in each virtual room. The bots can participate in the discussions and later report on the participants' levels of engagement and understanding.

Background relations in their most simple form would point to the augmentation of the background displayed behind the participant, may it be stills or a short video mimicking a lively environment.

In a deeper level, it would point to the role of institutions operating the learning environment, as discussed by constructionists in the context of democratization of the school system [2]. Yet, even with democratization, there is no symmetry between students and instructors. For example, shutting down the camera and the microphone are typical for students in Zoom but are unacceptable for teachers [13]. Due to the different roles of these users, their experiences are likely to differ. An additional limitation in the background for the students might be the reduction in the sense of community, that makes them feel like a class.

An even deeper level of background relations would refer to values like democracy, privacy, and respect, and how they should limit the users. When the background is active, as discussed by Aydin et al., these considerations become even more acute.

While composite intentionality accentuates the augmentation offered by "smart" systems, relegation draws attention to the degradation in the user's autonomy. The two relations pose an increased liability on Zoom's developers to ensure the system is ethical and respects the users' human rights. At the same time, the users are required to recognize the other participants' rights, mainly respect and privacy. Moreover, they should maintain their own discretion and not accept the system's inputs as they are but instead remain critical.

The recommendations are likely to point to the need to mitigate the reductions to the learning space and prepare us for a future in which the augmented learning space is populated not only by learners and instructors but also by bots. Considering everyday apps like Zoom as forms of AR technologies opens avenues for a more comprehensive understand of the user experience they produce. Rather than limiting AR apps to those reliant on special eyeglasses, the proposed broader approach identifies AR in various apps and use cases thereby exploring new effects on the users. Expanding the notion of AR to include apps like Zoom is the next step after Lev Manovich's identification of analogue technologies, like buildings, as forms of AR.

References

1. Manovich L (2006) The poetics of augmented space. Vis Commun 5(2):219–240
2. Wellner G, Levin I (2023) Ihde meets Papert: combining postphenomenology and constructionism for a future agenda of philosophy of education in the era of digital technologies. Learn, Media Technology 1–14. https://doi.org/10.1080/17439884.2023.2251388
3. An T, Oliver M (2021) What in the world is educational technology? rethinking the field from the perspective of the philosophy of technology. Learn Media Technol 46(1):6–19
4. Hasse C (2020) Posthumanist learning: what robots and cyborgs teach us about being ultra-social. Routledge, Oxon & New York
5. Forss A (2023) Digitalizing nursing education amid Covid-19: technological breakdown through a reflexive and postphenomenological lens. Techne: Res Philos Technol 27(3):387–404
6. Verbeek P-P (2008) Cyborg intentionality: rethinking the phenomenology of human-technology relations. Phenomenol Cogn Sci 7(3):387–395

7. Wellner G (2020) Postphenomenology of Augmented Reality. In: Wiltse H (ed) Relating to things: design, technology and the artificial. Bloomsbury Visual Arts, London, pp 173–187
8. Liberati N (2016) Augmented reality and ubiquitous computing: the hidden potentialities of augmented reality. AI Soc 31(1):17–28
9. Liberati N (2017) Phenomenology, Pokémon Go, and other augmented reality games a study of a life among digital objects. Hum Stud 41(2):211–232
10. Ihde D (1979) Technics and praxis. D. Reidel Pub. Co., Dordrecht
11. Ihde D (1990) Technology and the lifeworld: from garden to earth. Indiana University Press, Bloomington and Indianapolis
12. Gertz N (2022) Zooming through a crisis. Int J Technoethics 13(1):1–10
13. Wellner G (2021) The zoom-bie student and the lecturer: reflections on teaching and learning with zoom. Techne: Res Philos Technol 25(1):153–161
14. Aydin C, González Woge M, Verbeek PP (2019) Technological environmentality: conceptualizing technology as a mediating milieu. Philos Technol 32(2):321–338
15. Lowenthal P, Borup J, West R, Archambault L (2020) Thinking beyond zoom: using asynchronous video to maintain connection and engagement during the COVID-19 pandemic. J Technol Teach Educ 28(2):383–391
16. Wiltse H (2014) Unpacking digital material mediation. Techne: Res Philos Technol 18(3):154–182
17. Wellner G (2023) Futures of reality: virtual, augmented, synthetic. Navigationen - Zeitschrift für Medien- und Kulturwissenschaften 23(2):155–165

A Science-Based Environment for Lexical Language Learning

Vered Levi Zaydel

Abstract This chapter aims to unveil how a Brain-Computer Interface (BCI) system, in conjunction with Augmented Reality (AR) glasses and online platforms such as Zoom, can change students' language learning experiences and enhance overall learning. To assess the power of AR for language learning, we ran a series of experiments. We found that generating associations with the learned words is critical for acquiring an unknown language that shares no features with the participant's native language. Employing associative techniques such as these might also be beneficial in other learning contexts. Results of an Electroencephalogram (EEG) analysis suggest that brain oscillations within the Delta waves bandwidth were prominent during the generation of associations, and especially prominent in the brain area known as the corpus callosum. Based on these results we have developed a concept of a BCI system that analyzes learners' Power Spectrum Density in real time and extracts the Delta waves patterns associated with the corpus callosum as learners generate associations with the given words. Depending on the pattern of the Delta bandwidth, the system automatically provides support for generating relevant associations. Ongoing EEG monitoring demonstrates learners' capacity to control the Delta waves.

Keywords AR · Verbal-lexical memory · Delta brain waves · Associations · Brain-computer interface

---

V. Levi Zaydel (✉)
VR-Neuro Cog Lab, Technion, Northern, 320003 Haifa, Israel
e-mail: vered.levi@alumni.technion.ac.il

© The Author(s), under exclusive license to Springer Nature Switzerland AG 2025
E. Vendrell Vidal et al. (eds.), *Advanced Technologies and the University of the Future*,
Lecture Notes in Networks and Systems 1140,
https://doi.org/10.1007/978-3-031-71530-3_4

1 Enhancing Students' Learning Outcomes Through Innovative Technology–The Case of Learning Words

1.1 Context

Technologies that support learning provide access to abundant relevant information. Recently, innovative technologies have focused on motivational and engagement factors of learning [10, 34], including optimal duration and effort invested in learning [1, 23], personalization [4], as well as the integration of virtual reality, augmented reality, and gaming simulations to address these factors [39]. Our work focuses specifically on the memory aspect of learning and understanding. Whereas generally addressed methods of memory enhancement utilize strategies that involve rehearsal [25], clustering, and classification [26], our focus lies on the process of generating associations with the material to be learned. Kahneman [19] explains that associations between two stimuli are based on same timing, visual and/or auditory cues, not *causal* relations. This type of associative method is utilized by top memory performers [9]. Further, Dresler found that practicing this association method creates longitudinal changes in brain areas. In the current work, we focus on learning words, however, learning through associations might be beneficial for other domains as well, such as technological systems learning, mathematics, etc. This proposed system may be imperative for supporting learning in online classes and hybrid classes. In this chapter, we propose a science-based application for learning words in a new language, using the technology of a Brain-Computer Interface-based Augmented Reality system.

1.2 Brain-Computer Interface

Following the first use of psychophysical recordings and analysis of Electroencephalogram (EEG) signals by Hans Berger in 1924 [28], the first Brain-Computer Interface (BCI) was constructed between 1969–1970, initially to control monkeys' motion by manipulating brain waves [42]. Later, Pope and colleagues of NASA researchers decoded EEG signals to adjust engagement and attention of astronauts [31]. More recently, BCI technology has been used in the operation of neuroprosthetic limbs [22].

BCI technology serves as a tool for communication between brain activity and peripheral devices or external environments via brain signals [21, 33, 42, 43]. The BCI system records brain wave activity and applies a real-time algorithm to provide an output that is transmitted to an external device. Users of BCI learn to generate specific brain signals to control external devices in real time [33]. Miller et al. [27] indicated that BCI enables users to perform actions directly using only their brain activity without muscle involvement.

BCI technology encompasses various approaches, from implanted electrode or implanted electrode arrays with microsensors [40] to non-invasive technologies such as EEG devices that utilize electrodes attached to the scalp. Other devices for obtaining brain-function signals include blood oxygen measurements, functional Magnetic Resonance Imaging (fMRI), and functional Near Infrared Spectroscopy (fNIRS) [21]. Pfurtscheller et al. [30] outlined four criteria for BCI: "(1) the device must rely on signals recorded directly from the brain,(2) there must be at least one recordable brain signal that the user can intentionally modulate to effect goal directed behavior; (3) real time processing; and (4) the user must obtain feedback" (p. 1283).

BCI has broad applications in medical settings[38], aiding in paralysis rehabilitation [17] and neurological disorder treatments by facilitating communication between a paralyzed person who cannot speak and his caregivers [42]. Moreover, BCI is used for cognitive enhancement, including in the realms of memory, attention, and learning, through direct interaction with the brain [14]. EEG-based BCI applications, such as Neurofeedback for Cognitive Enhancement [13], represent a notable example of BCI utilization for cognitive enhancement.

In the language learning domain specifically, Chang et al. [7] used brain potential-based feedback to help foreign language students in perceiving the tones of a new language. Additionally, Hu et al. [18] examined the use of neurofeedback to observe and adapt neural patterns among students learning a foreign language and found that they demonstrated significant progress in English learning skills and communicative competence. Another proposed application of BCI for second language learning is based on NIRS, a technology for measuring blood flow in the brain [41]. In this study, researchers looked for a parameter for brain arousal related to language listening. When testing advanced, intermediate, and novice listeners, they found that differences of brain areas related to language regions in the brain peaked for all listeners. They suggested that their measurement system could function as an input measurement device for BCI. Further, Kang et al. [20] suggest an EEG-based BCI that identifies brain wave bandwidths in specific brain areas' topographies that predict success in memorizing words in German for Korean learners the day after the learning phase. However, their well-designed study and subsequent conclusions overlooked the features of Delta bandwidth related to this initial stage of memorization.

1.3 Augmented Reality

Billinghurst et al. [6] presented the criteria of [3] for Augmented Reality (AR): "(1) It combines real and virtual content, (2) It is interactive in real time, (3) It is registered in 3D" (p. 77). AR offers a "twin" perspective of reality, allowing participants to see their environment while the augmented reality device also "sees" it [32]. The AR device utilizes a computer-based vision system in order to "see" the environment, incorporating a camera, and then presenting its outputs on a display device, which can be wearable or non-wearable. Wearable devices include headsets, helmets, and, potentially, in the future, even contact lenses. Non-wearable devices include mobile devices and projectors [29]. The display is a computed feedback of the AR system, based on the data that the AR device receives from sensors and other devices.

AR technologies are currently employed across different domains, including entertainment, the gaming industry and medical services. The use of AR applications in education is currently unfolding, where it is being applied to help students develop skills and knowledge, as well as to enhance the learning experience. The use of AR is particularly popular in the field of science education because a lot of scientific concepts are hard to grasp. most users expect AR technology to improve their understanding of complex learning content that is difficult to grasp without clear visualization, such as spatial relationships [16].

Another important area of learning concerns foreign languages. Learning a foreign language, especially one that does not share features with the learner's known languages, is difficult. Liu et al. [24] explored the use of AR for teaching English as a second language, using an application featuring a pairing game with two types of marker cards displaying 3D models of specific objects, along with their vocabulary and pronunciation in various languages, including English. The study found that the AR application increased students' motivation, and the visual and auditory stimuli enhanced learning outcomes, especially in vocabulary acquisition. In another study, researchers investigated the use of an AR application to facilitate Spanish learning [37]. This application enables users to create their own learning material using digital or physical objects and utilizes image and object recognition technology to activate AR content. The application's collection of imagery content provides many illustrations and photos that can be used as triggers for the AR. Each trigger image is then paired with a specific video overlay. The researchers found that the application was very engaging among its users. Moreover, Godwin-Jones' [12] work supports the use of AR-based games to teach a second language. His innovative location-based game integrates digital layers into the real world to teach words in the context of *localized* words in order to generate meaningful connections to real-world scenarios.

1.4 Aim and Questions

Many students struggle when learning foreign language particularly when the foreign language doesn't share features with the learner's known languages. Other students struggle due to their learning disabilities. Students sometimes struggle with other unknown contexts, such as unfamiliar mathematic equations or technological systems. The association method might be beneficial for their understanding. Recent developments regarding BCI and AR systems for learning foreign language are important and serves as a foundation for the current study with the research on BCI and AR for learning; however, currently there is no application that integrates the association method for language learning with the use of BCI for analyzing brain waves and emitting commands to AR glasses in order to enhance learning.

Several questions regarding the integration of these technologies ought to be addressed:

(1) How can behavioral instructions, BCI, and AR interact and synchronize in real time?
(2) How does the *brain* perceive an augmented reality stimulus as a clue that is presented on AR glasses or merge reality with AR feedback stimuli?
(3) Could this technological procedure contribute to structural brain changes?
(4) What elements should be included in the physical space in which the user operates?
(5) How does the application transform the user's experience of the space?
(6) What information layers are added to the space?

Our proposed approach involves the use of a Brain-Computer Interface system with Augmented Reality glasses and online video application to enhance students' foreign language learning experiences and learning outcomes. We will introduce our study of the importance of associations in lexical memory of unknown language memory, explore the EEG patterns that relate to the associative method, and examine how these patterns vary among learners with different levels of success.

2 Research Design and Results

2.1 Research Design

Research Design of the Behavioral Experiment to Assess the Importance of Associations. First, we comprised several indexes for scoring memory retention and associations. The first scoring index, index 1, recorded whether the participant successfully recalled the word (1 = succeeded to recall, 0 = did not succeed to recall). Index 2 represented a detailed scoring index. Participants received a maximum score of one if they accurately recalled each phoneme of the word and its location if they missed any part, they received less than 1. Index 3 indicated the number of associations that participants generated. Index 4 indicated whether participants generated, or did not generate, associations. We then used these indexes in the design of the experiment.

Two groups were identified after comparing recall results (index 1). One group included participants that their sum of recall scored at or above a level of 66% of all participants' sum of recall on the memory recall test, and the other included participants who scored at or below 33%. We then compared the number of associations (index 3) between these groups.

To determine if there was a difference in memory performance between words recalled without associations and words recalled with associations (index 4), we compared mean recall scores of words without associations of each participant and mean recall scores of words with associations for each participant using a paired samples t-test analysis.

To identify the recall characteristics of words without associations, we calculated the mean recall scores for each participant. Then, we categorized the recall score means into two groups: 0–0.125 (indicating no recall [0] to recall of only one vowel out of two syllables [0.125]) and 0–0.355 (indicating no recall [0] to recall of only one consonant out of two syllables [0.355]). We counted the number of words without associations in each category. Finally, we calculated the percentages for each of the two categories out of all participants.

Additionally, we conducted a correlation analysis between the number of associations (index 3) and recall scores for each participant. We calculated the mean memory performance scores for each participant and correlated them with the sum of associations (index 3) for each participant using the Pearson correlation method.

To determine if there was a difference between short-term and long-term recall scores in relation to the number of associations, we compared the Pearson correlation coefficients between the number of associations and short-term memory recall performance and between the number of associations and long-term memory performance.

Research Design of the EEG Recordings. From the entire group of participants, we sampled the highest performer, the lowest performer, and a middle performer on the recall test and analyzed their EEG patterns. Our aim was to identify the most prominent power frequency and electrodes' locations that indicated the most prominent power activation. To determine the most prominent power frequency, we used two methods: comparing maps of different bandwidth frequencies for all words among the chosen participants and calculating the median power of the different frequencies for frontal and central parietal electrode locations, which were found to be significant in previous studies. For each participant's word, we calculated the median of the bandwidth spectral power. To assess the significance of our results, we used the sign test to compare between pairs of different frequencies power for each word per participant. To further investigate the reliability of the EEG memory measures, we compared them to EEG recordings for visual stimuli only, serving as a control.

Once we identified the prominent power frequency, we used it to find the location of the electrodes most prominently activated. We compared the frequency power between the frontal electrodes and the central parietal electrodes. Additionally, we specified the focal power electrodes (the three highest power electrodes) for each word per participant, noting the frequency of occurrence.

These calculations of the most powerful electrodes and frequency of appearance determined the most activated electrodes' location for each participant. Furthermore, we compared the power increase of the prominent frequency between the selected participants to determine if there was a related change in power, specifically regarding better recall results.

Research Instruments and Protocols. *The Experiment.* Twenty-four male and female participants participated in the experiment (age range: 22–30). Participants were undergraduate university students who signed a voluntary consent form after being informed about the experiment details, procedures to maintain anonymity, and safety measures. The experiment was approved by the university's ethics committee.

Methods for Lexical Learning and Testing Stimuli Importance. The language chosen for word learning was Chinese, aligning with Karpicke et al.'s (2008) method for testing lexical memory by using an unknown language that presumably has no shared characters with participants' known languages. We made memorizing more difficult than the method used by Karpicke et al. (2008) by asking participants to recall words rather than match pictures to suggested words. We chose this approach to impose the maximum possible mental load for lexical memory and to explore basic behaviors associated with words' acquisition.

Experiment Design. *Learning Stage.* Participants viewed pictures and heard 12 words (experiment stimuli). They were instructed to look, listen and create associations for each word resembling using associations in "Luci"'s method [9] to help them recall the words during a subsequent test. Twelve objects were presented in three phases per each item. Each phase had its permanent seconds timing which reversed per each item. Triggers (signs that indicated when each stimulus started and ended) were marked on the EEG recording. The list of words to remember adhered to the requirements of the California Verbal Memory test (sum of words and categories) [9] and was adjusted based on prior pilot testing. The timing for memory encoding of each stimulus slide was 30 s, which included four reversals on the same word, in accordance with the procedure of [9]. The second slide for each stimulus featured only a picture of a talking person, this slide served as a cue to the participant to vocalize the word they had just learned. The duration of each "talking" slide was 10 s. The sequence of slides described above was reversed for each of the twelve stimuli. The third phase of each stimulus consisted of a grey screen slide presented for 20 s, during which participants were instructed to think about the word they had just learned.

Break. Participants received a break of approximately 15–20 min between sections for distraction purpose.

Retrieval Stage. We tested participants' recall of the words using a presentation format resembling PowerPoint slides. Following the test, we asked participants to recall whether associations were made and, if so, what they were. Some of the participants also completed a long-term retrieval test others did not complete the long-term test. In addition, all participants performed a control vision task during which EEG signals were recorded to generate a comparison with the lexical memory recording phase. EEG signals were recorded during the learning phase, stopped during the break, and then resumed during the retrieval phase.

Devices and Software. We utilized the "Mizar" EEG device model 202, an electric cap with 31 channels, and the Winn EEG software for EEG recordings and analysis. For EEG processing, we conducted artifacts removal by rejecting manually contaminated EEG segments and used the ICA technique to separate linearly mix sources in order to separate out artifacts embedded in the data. Filters were also applied: a low-cut 0.3 filter, a high-cut (30 Hz) filter, and a Notch filter.

2.2 Research Results

Behavioral Results. In this analysis, we used the sum of associations measurement (index 3). We compared the group of participants who scored 66% and above and the group that scored 33% and below. One sample t-tests were conducted, for each group and was found proper at $p = 0.001$. An independent two-tailed t-test was then conducted to compare the means. The mean sum of associations for the lower group (33%), with a mean recall of 0.323, was 4.937. In contrast, the mean sum of associations for the highest group (66%) was 9.5. A two-tailed independent t-test indicated a significant difference ($t = 3.754, p < 0.01$). The mean sum of associations (index 3) for the highest group was almost double (1.924) that of the lower group. Participants who had a higher rate of recall also showed a greater number of associations.

We conducted another comparison between the two recall success groups on the number of zero associations (manipulating index 4). The higher recall group had a mean of 1.6 zero associations and the lower group had a mean of 5.5. One sample t-test revealed significant results for both the low recall group ($p = 0.011$) and the high recall group ($p = 0.002$). A two-tailed independent t-test indicated a significant difference ($t = -2.545, p < 0.05$). The *mean* number of zero associations for the lower group (<33%) was more than triple (3.38) that of the mean number of zero associations for the higher group (66%). In the lower group, participants faced significantly more difficulties in generating associations.

We compared the mean scores of participants' recall performance based on the presence or absence of associations (associations-based recall vs. zero associations recall). In this part, we used the characteristic index of recall (index 2). The mean score of zero associations-based recall was 0.26 and the associations-based recall score was 0.614. One sample t-tests yielded significant results for both the zero associations-based recall group ($p = 0.00$) and the associations-based recall group ($p = 0.00$). The mean number of words recalled when associations were used was significantly higher than when no associations were used (0.614 versus 0.260; $t = 7.016, p < 0.01$).

To understand participants' memory performance for words for which they did not make associations, we calculated mean scores using index 2 for those words per participant. We then grouped the results by ranges, 0–0.125, 0–0.355, and 0–0.500 (with the higher scoring groups containing the lower scoring ones) and calculated the percentage of participants falling within each range. Fifty-two percent of participants had a zero associations-based recall mean score of 0.125 or below, indicating no recall or, at most, recall of only one vowel per word. Eighty percent of participants had a zero associations-based recall mean score of 0.355 or below, indicating no recall or, at most, recall of only one consonant per word. One sample t-tests for each group were significant. These results indicate that the vast majority of participants were not able to recall words adequately when they were unable to generate associations with the word.

A correlation analysis revealed that the number of associations (index 3) was positively and strongly correlated with recall scores (r = 0.88, p < 0.1). Associations and recall performance are highly linked.

EEG Results. A sample of three participants, categorized by levels of recall (high, medium, and low), was selected for the EEG study. In order to discern the brain patterns that are measured during associative lexical memorization, we first analyzed the most *prominent* EEG frequency. Figure 1 presents an example of EEG maps for the four first words. The left column of each word exhibits Delta frequency results. It is distinguished by more bright colors than the other columns, indicating Delta frequency prominence. Next, we compared graphs of the median spectral power of all electrodes across Delta, Theta and Alpha EEG frequencies for each word per participant. Delta spectral power was consistency higher for words across all three participants, with only one exception. We used the sign test to examine whether differences between frequencies were significant using the sign test. To adjust for multiple comparisons, Bonferroni's correction was used. All participants showed significantly higher Delta power compared to Theta and Alpha frequencies ($p < 0.0167$).

To evaluate the power of the prominent Delta frequency across electrodes, we compared the mean Delta power of the frontal electrodes with that of the central parietal electrodes. The mean Delta spectral power for each participant was calculated for the frontal electrodes and central parietal electrodes. We also tested differences between the means for each participant using independent t-tests. The results yielded no significant differences. Therefore, we examined differences in the focal power across all 31 electrodes. We identified the three highest power electrode readings (four in cases of equal power) among all electrodes' recordings. Eleven out of the 12 electrodes with the highest power readings are those with the suffix "Z". The highest powers of the activated electrodes are located along the Z-axis in the 10/20 cap system; the Z-axis lies over the *corpus callosum* region of the brain. We also observed that electrode Fp2 also had high Delta spectral power; however, its power varied across participants.

To determine if there were any differences among participants in power raise with regard to the prominent frequency identified, we compared participants' EEG power based on their lexical memory performance. Specifically, we analyzed the Delta spectral power at *central parietal* electrodes. The results indicated that higher recall scores (considering high, medium and law recaller selected participants) were associated with higher Delta spectral power. However, a comparison of Delta spectral power at *frontal electrodes* did not show consistent results. Among participants, spectral power at higher (focal) power electrodes exhibited that higher recall scores were associated with higher Delta spectral power (see Fig. 2).

Finally, we examined the results of spectral power during the visual *control* task. Spectral power maps, depicting the period during exposure to stimuli of this visual task, revealed Alpha frequency as the dominant frequency. For example, the median Alpha spectral power was 11.7 ($\mu v2$), whereas the median Delta spectral power was 4.65 ($\mu v2$) for one of the participants (see Fig. 3). The electrodes with the highest Alpha spectral power were Cpz, Cp4, P3, Pz, P4, O1, Oz and O2, primarily located

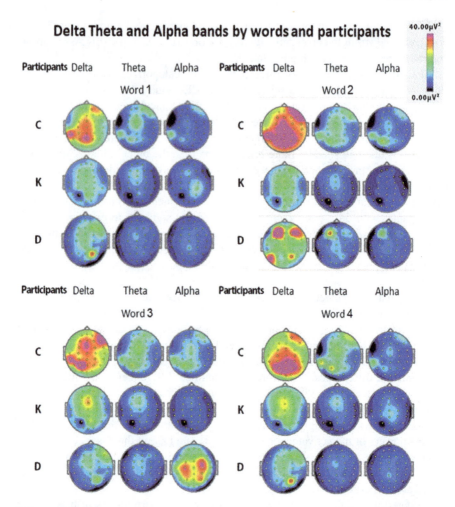

Fig. 1 Spectral power: comparison of head maps for different frequencies (Delta, Theta, and Alpha) for different words and participants

in the *occipital and parietal* regions. These findings–based on maps and results of EEG recordings during the control task–demonstrate that across all participants, the occipital-parietal region was most prominently activated (in contrast to the central Z-axis in the experimental task) and the dominant frequency was Alpha (rather than Delta).

A Science-Based Environment for Lexical Language Learning

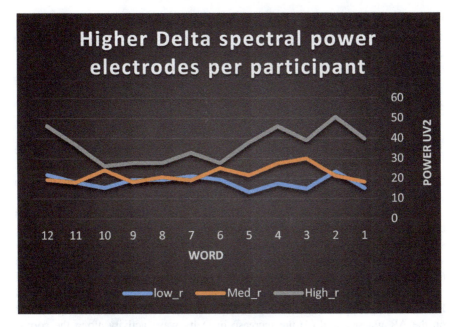

Fig. 2 Graphs showing the comparison of the power of Delta waves across three participants (high recaller, medium recaller, and low recaller)

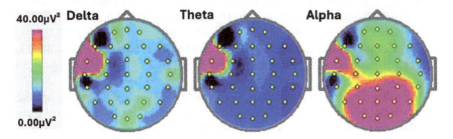

Fig. 3 Head maps from the control task: comparison across Alpha, Theta, and Delta

3 Conclusions and Recommendations–The Application

3.1 Conclusions

In our study, we used an association method that shares similar elements with Dresler's "Luci" behavioral method [9] for learning words. We found that generating associations is critical for the early acquisition of new words in an unfamiliar language. Our findings showed that Delta waves are prominent in the process of generating word associations. Recent papers have suggested that Delta waves may indicate neural plasticity [2] and may occur during wakefulness [11]. We detected that

Delta waves appeared most prominently along the corpus callosum and were higher among subjects who exhibited better recall. Based on our findings we developed a concept of an application for learning new words in a foreign language.

3.2 The Application

The proposed application monitors Delta waves in the corpus callosum area of the brain during the process of learning new words in a foreign language and provides feedback in the form of visual clues through AR glasses. The clues are designed to enhance learners' ability to generate associations to the words they are learning.

The Components of the System. The components include: (1) a screen and a speaker, (2) an EEG device and electro-cap, (3) AR glasses and accompanying software, (4) a wave analyzer, (5) an inventory of personal and general pictures, along with software that extracts images, (6) a monitoring system for learning sessions, and (7) additional software programs as needed. Each component provides a different output: (1) the learned word is presented via the screen and speaker, (2) the analyzed EEG recordings appear on the EEG device, (3) the adjusted feedback clue is presented on the AR glasses, and (4) the increase in Delta wave activity along the corpus callosum is detected by the wave analyzer. In addition, (5) the inventory of pictures and software is used to create adjusted images for the feedback clues, while (6) the monitoring system tracks changes in learning capability and Delta wave patterns. Other details about the software programs that are used are provided in the following paragraphs.

The Components' Implementations of the BCI System. The following comprise the components' implementations of the BCI system according to [30] BCI criteria.

Recording Brain Signals through a Compatible EEG System. The setup involves a head electro-cap comprised of sensors that receive EEG scalp signals and transmits them to a recording and analyzing system. The analysis provides output of the recording that may include statistics of the data, graphs, and spectral power head maps.

In our study, we used the Mitzhar-202 EEG device and a 31-channel electro-cap. However, there are alternative devices available, such as "Emotive" [21], which are more affordable yet still reliable and use fewer channels. These simpler devices are sufficient for use in online classes that rely on Zoom or other online video applications. However, for in-person or hybrid classes, we note that students would benefit from the more complex and expensive devices.

Identification of the Relevant Brain for further processing. In this case, we focus on the Delta waves along the corpus callosum.

Additional Suggested Real-Time Processing Software:

1. Software that is embedded in the EEG device and records EEG signals and extracts data from it.
2. A synchronizing software that notes the time of learning stimuli on the EEG recordings (learning stimuli).
3. Additional software for basic data analysis, specifically to identify increases in power of EEG Delta signals in the corpus callosum.
4. Software to develop an algorithm that includes several "if" loops to utilize the data and activate the AR glasses software." if" loops should adhere to the following instructions (see Fig. 4):

    - If the pattern of EEG delta waves along the corpus callosum exceeds a certain level, especially in the Pz electrode, no clue should be presented on the AR glasses. It can be assumed that the participant was able to generate an association and is ready for the next learning stimulus.
    - If the pattern of EEG delta waves along the corpus callosum rises but does not exceed a certain level, or if the increase is unstable, a stimulus clue would be presented on the AR glasses using very low pictorial resolution. The clue for the association would be obtained from an inventory of the participant's personal pictures that share some features with the learning stimulus. The clue would be presented for fractions of a second and in a low resolution.
    - If the pattern of EEG delta waves does not rise, or the Delta wave frequency is not prominent and the location of the waves are not primarily in the corpus callosum, then a general (rather than personal) clue should be presented for a fraction of a second to create an association with the word stimulus in low pictorial resolution (Fig. 5).

Monitoring system. Next, a monitoring system for learner adjustment and development should be applied. The pattern of recorded delta waves across learning sessions should be analyzed to determine the best fit of learning stimuli to the learning pattern in terms of band, location, and volatility of power. Over time, the monitoring system will assess the sum of associations and changes in the distribution of waves and AR feedback.

On the one hand, the monitoring system will track the participants' ability to create associations and assess whether the participant's brain has gained plasticity, as indicated by changes in Delta waves [2]. On the other hand, the cumulative data can be used for analysis (with the help of AI) to suggest improvements to the system itself. The use of clue feedback, presented on AR glasses, is based on Rosen and colleagues' [35] study, which found that presenting a blurred clue for a very short time enhances the ability to find a solution to a geometric problem (Fig. 5).

How does the proposed system satisfy the questions raised earlier?

Generally, the proposed technology is controlled by the person who uses it. We propose using augmented reality glasses, rather than a detached digital virtual reality, because we want to enable learners to see the reality while being presented with the associative clues transparently on the glasses. Through this approach, we hope (and

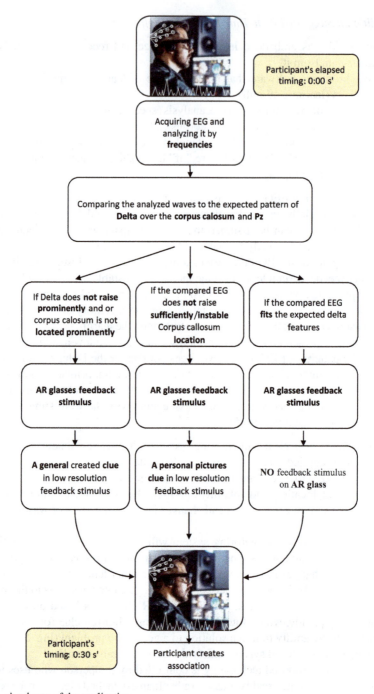

Fig. 4 A scheme of the application

| a picture with regular resolution | a blurred picture with low resolution |

Fig. 5 Example of stimuli: a picture with normal resolution and a blurred picture with low resolution (author-generated)

assume) that participants' brains will be able to discriminate between digital stimuli and reality, thereby maintaining their grasp on reality. Moreover, this distinction may enable the learner's free will to determine whether to accept the stimulus clue or not. As long as the will is free as [36] intrigues, it is imperative that the proposed AR feedback provide brain stimulation (blurred pictures) rather than a clear visual association. The feedback stimulus has the potential to activate relevant brain areas, allowing participants to either use it or create their own associations. This grants participants much greater control over their memory. The following section presents specific questions and corresponding answers regarding the proposed system.

(1) How could behavioral instruction through BCI and AR be synchronized in real time?

Achieving synchronization requires a meticulously designed computer program that controls the two components with regard to the timing and duration of exposure to the learning stimuli, as well as managing the delay and the timing for feedback. Additionally, participants should have the ability to control the system if they would like to make adjustments.

(B) How would the brain perceive augmented reality? Would the brain perceive the augmented reality as a clue to reality, or as a fully integrated merged reality with the feedback stimuli?

In our approach, we chose for the feedback stimuli to be presented on AR glasses, rather than merged with real-world learning stimuli. The clue would be presented as a distinct layer, detached from reality. We chose not to include a crosser marking or superposition of the clue and the learning word. The effectiveness of this approach in achieving our desired effect of separating the visual clue from the reality on the learner's perception will require further study.

(C) Would this technological procedure contribute to the plasticity of the learner's brain?

We propose including a monitoring system in the application that identifies and compares EEG patterns across learning sessions, potentially indicating changes in brain plasticity. We would ensure that the data remain confidential and is not shared.

(D) What is included in the physical space in which the user operates?

The physical space would include: (1) an EEG device, (2) a computer screen that displays the visual image of the learned target word, (3) AR glasses that provide visual clues, and (4) an auditory system that plays a voiced audio of the word.

(E) How does the application transform the user's experience of the space?

The space is enriched by presenting visual clues through the AR glasses. However, purposefully, this presentation is not seamless as to allow learners to maintain full control over their perceptions while still being personally and interactively *stimulated* to foster user engagement.

(F) What information layers are added to the space?

The application adds layers of information by offering learners association clues if they cannot generate them on their own. However, the clues are presented in low resolution, such that they are sufficient to stimulate users' brains to find the association and elevate the Delta wave patterns.

3.3 Implications and Further Research

One of the main advantages of the proposed system is that it allows learners to acquire a basic corpus of words in multiple languages within a short timeframe, which may be retained over the long term. Increasing one's foreign language proficiency is paramount as it improves the ability to communicate in the modern global world. Although we used more expensive technology, this system can be replicated with less expensive components, such as the "Emotive" electro-cap [21, 44] and ear EEG [8]. Further, the use of second-hand AR glasses may be feasible, with the equipment likely to become more affordable and available in the near future.

Another advantage of the system over other systems is its enhanced engagement through the use of an inventory of learners' personal pictures as feedback stimuli, and the ongoing monitoring function for personalized adjustments. Further, as noted previously, this system has potential applications beyond language learning and might be beneficial for increasing knowledge in other educational subjects. Moreover, each part of the system may be used separately to enhance learning. However, future research is needed to validate the effectiveness of this method through comprehensive research with a larger number of participants. Moreover, it is recommended to do further research in addition to the research which we are based on in the EEG part and the behavioral part. As we continue in the era of the fourth industrial revolution [15], I believe that BCI devices will become increasingly prevalent in educational settings in the near future as presented in Fig. 6.

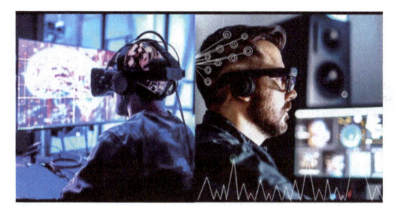

Fig. 6 Students using the system: wearing AR glasses and the EEG device, while watching a screen and listening to audio over the speakers (this image was generated by AI)

References

1. Appleton JJ, Christenson SL, Furlong MJ (2008) Student engagement with school: critical conceptual and methodological issues of the construct. Psychol Sch 45(5):369–386
2. Assenza G, Di Lazzaro VA (2015) Useful electroencephalography EEG marker of brain plasticity: delta waves. Neural Regen Res 10(8):1216–1217
3. Azuma RT (1997) A survey of augmented reality. Presence: Teleoperators Virtual Environ 6(4):355–385
4. Basham JD, Hall TE, Richard A, Carter JR, Stahl WM (2016) An operationalized understanding of personalized learning. J Spec Educ Technol 31(3):126–136
5. Beseghi M, Bertolotti G (2013) Designing tailor-made academic paths for university language students 2(2):319–337
6. Billinghurst M, Clark A, Lee G (2015) A survey of augmented reality. Hum-Comput Interact 8(2–3):73–272
7. Chang M, Iizuka H, Kashioka H, Naruse Y, Furukawa M, Ando H, Maeda T (2017) Unconscious improvement in foreign language learning using mismatch negativity neurofeedback: a preliminary study. PLoS ONE 12(6):e0178694
8. Debener S, Emkes R, De Vos M, Bleichner M (2015) Unobtrusive ambulatory EEG using a smartphone and flexible printed electrodes around the ear. Sci Rep 5(1)
9. Dresler M, Shirer WR, Konrad BN, Müller NCJ, Wagner IC, Fernández G, Czisch M, Greicius MD (2017) Mnemonic training reshapes brain networks to support superior memory. Neuron 93(5):1227-1235.e6
10. Fredricks JA, Blumenfeld PC, Paris AH (2004) School engagement: potential of the concept, state of the evidence. Rev Educ Res 74(1):59–109
11. Frohlich J, Toker D, Monti MM (2021) Consciousness among Delta waves: a paradox? Oxf Acad Brain 144(8):2257–2277
12. Godwin-Jones R (2018) Contextualized vocabulary learning. Lang Learn Technol 22(3):1–19
13. Gruzelier JH (2014) EEG-neurofeedback for optimizing performance. I: a review of cognitive and affective outcome in healthy participants. Neurosci Biobehav Rev 44:124–141
14. Guger C, Allison BZ, Großwindhager B, Prückl R, Hintermüller C, Kapeller C, Bruckner M, Krausz G, Edlinger G (2012) How many people could use an SSVEP BCI? Front Neurosci 6
15. Haag SE (2022) The fourth industrial revolution. Jannus Press
16. Hantono BS, Nugroho LE, Santosa PI (2018) Meta-review of augmented reality in education. In: 10th international conference on information technology and electrical engineering (ICITEE). Bali, Indonesia, pp 312–315

17. Hochberg LR, Bacher D, Jarosiewicz B, Masse NY, Simeral JD, Vogel J, Haddadin S, Liu J, Cash SS, van der Smagt P, Donoghue JP (2012) Reach and grasp by people with tetraplegia using a neurally controlled robotic arm. Nature 485:372–375
18. Hu L, Xie Y, Sun G (2018) Computer-aided cognitive training based on electroencephalography-neurofeedback for English learning. Educ Sci: Theory Pract 18(5)
19. Kahneman D (2011) Thinking, fast and slow. Farrar, Straus and Giroux, New York
20. Kang T, Chen Y, Fazli S, Wallraven C (2020) EEG-based prediction of successful memory formation during vocabulary learning. IEEE Trans Neural Syst Rehabil Eng 28(11):2377–2389. https://ieeexplore.ieee.org/stamp/stamp.jsp?tp=&arnumber=9193957
21. Kawala-Sterniuk A, Browarska N, Al-Bakri A, Pelc M, Zygarlicki J, Sidikova M, Martinek R, Gorzelanczyk EJ (2021) Summary of over fifty years with brain-computer interfaces—a review. Brain Sci 11(1):43
22. Kingwell K (2012) Neurally controlled robotic arm enables tetraplegic patient to drink coffee of her own volition. Nat Rev Neurol 8(7):353–353
23. Kuh GD (2009) What student affairs professionals need to know about student engagement. J Coll Stud Dev 50(6):683–706
24. Liu E, Liu C, Yang Y, Guo S, Cai S (2018) Design and implementation of an augmented reality application with an English learning lesson. In: IEEE international conference on teaching, assessment, and learning for engineering (TALE). Wollongong, NSW, Australia, pp 494–499
25. Lucidi A, Langerock N, Hoareau V, Lemaire B, Camos V, Barrouillet P (2016) Working memory still needs verbal rehearsal. Mem Cognit 44:197–206
26. Manning JR, Kahana MJ (2012) Interpreting semantic clustering effects in free recall. Memory 20(5):511–517
27. Miller KJ, Hermes D, Staff NP (2020) The current state of electrocorticography-based brain–computer interfaces. Neurosurg Focus 49(1):E2
28. Millett D (2001) Hans Berger: from psychic energy to the EEG. Perspect Biol Med 44(4):522–542
29. Peddie J (2017) Types of augmented reality. In: Augmented reality. Springer, Berlin
30. Pfurtscheller G, Allison B, Bauernfeind G, Solis-Escalante T, Scherer R, Zander TO, Mueller-Putz G, Neuper C, Birbaumer N (2010) The hybrid BCI. Front Neurosci 4(30)
31. Pope AT, Bogart EH, Bartolome DS (1995) Biocybernetic system evaluates indices of operator engagement in automated task. Biol Psychol 40(1–2):187–195
32. Porter ME, Heppelmann JE (2017) How does augmented reality work? the key is a digital twin. Harv Bus Rev. https://hbr.org/2017/11/how-does-augmented-reality-work
33. Prashant P, Joshi A, Gandhi V (2015) Brain computer interface: a review. In: 5th Nirma university international conference on engineering (NUiCONE). Ahmedabad, India, pp 1–6
34. Reschly AL, Christenson SL (2012) Jingle, jangle, and conceptual haziness: evolution and future directions of the engagement construct. In: Christenson SL, Reschly AL, Wiley C (eds) Handbook of research on student engagement. Springer, Boston, pp 3–19
35. Rosen A, Reiner M (2017) Right frontal gamma and beta band enhancement while solving a spatial puzzle with insight. Int J Psychophysiol 122:50–55
36. Sapolsky RM (2017) Behave. Penguin, New York
37. Scrivner O, Madewell J, Buckley C, Perez N (2016) Augmented reality digital technologies (ARDT) for foreign language teaching and learning. In: Future technologies conference (FTC)
38. Shih JJ, Krusienski DJ, Wolpaw JR (2012) Brain-computer interfaces in medicine. Mayo Clin Proc 87(3):268–279
39. Squire KD (2007) Games, learning, and society: building a field. Educ Technol 47(5):51–55. http://www.jstor.org/stable/44429444
40. Sidik S (2023) How our brains decode speech: special neurons process certain sounds. Nature. Last accessed June 5 2024
41. Watanabe K, Tanaka H, Takahashi K, Niimura Y, Watanabe K, Kurihara Y (2016) NIRS-based language learning BCI system. IEEE Sens J 16(8):2726–2734
42. Wolpaw JR, Birbaumer N, Heetderks WJ, McFarland DJ, Peckham PH, Schalk G, Donchin E, Quatrano LA, Robinson CJ, Vaughan TM (2000) Brain-computer interface technology: a review of the first international meeting. IEEE Trans Rehabil Eng 8(2):164–173

43. Wolpaw JR, del Millán JR, Ramsey NF (eds) Chapter 2–Brain-computer interfaces: definitions and principles. In: Ramsey NF, del Millán JR (eds) Brain-computer interfaces (Handbook of clinical neurology series), vol 168. Elsevier, pp 15–23
44. Zeng R, Bandi A, Fellah A (2018) Designing a brain computer interface using EMOTIV headset and programming languages. In: Second international conference on computing methodologies and communication (ICCMC). Erode, India, pp 908–913

Adaptive AR- or VR-Neurofeedback for Individualized Learning Enhancement

The Potential Advantages of Incorporating Tailored Neurofeedback with Virtual/Augmented Reality Technologies to Enhance Learning of Sciences, Engineering and Mathematics

Nehai Farraj and Miriam Reiner

Abstract Teaching abstract science, math, and engineering concepts using traditional instructional methods often fails to meet students' levels of understanding. Abstract content, such as molecular structures, atomic arrangements, and geometry, relies heavily on spatial skills, specifically mental rotation. Identifying technologies that target spatial abilities would help break the limits of students' learning potential and may transform science, engineering, and math learning methods. One approach is applying adaptive neurofeedback while immersing learners in virtual or augmented reality (VR/AR) environments. Studies have consistently shown that combining AR- or VR-neurofeedback has a positive impact on training brain oscillations, optimizing cognitive functions, and understanding science, math, and engineering concepts. On the one hand, using VR and AR applications helps understand molecular structures and interactions. Students who engaged with VR during chemistry learning sessions were more accurate at recreating physical models of molecules. AR technology offers a self-directed learning platform that promotes a thorough understanding of the molecular spatial structure. On the other hand, neurofeedback studies have shown that increasing the power of upper alpha brain oscillations, for example, can improve spatial skills. Therefore, the incorporation of neurofeedback protocols to continuously fine-tune brain activity during learning within VR/AR environments presents a promising approach for enhancing student performance and understanding within the science, engineering, and math learning fields. The effects of these methods would be much more significant when applied during the learning session while also being

---

N. Farraj (✉) · M. Reiner
Faculty of Education in Science and Technology, Technion–Israel Institute of Technology, Haifa, Israel
e-mail: nehai.farraj@gmail.com

M. Reiner
e-mail: miriamr@technion.ac.il

tailored to the students' levels of understanding. Nonetheless, only a few studies have addressed the advantages of these technologies and their potential applications in educational settings. We propose a novel approach for creating an adaptive neurofeedback system combined with AR/VR for individualized learning.

1 Introduction

Teachers often struggle to teach abstract concepts in science, engineering, and math using traditional instructional methods. Complex spatial configurations of concepts in science, math, and engineering can be challenging to describe using words alone. Human interaction with dynamic virtual representations complements the learner's missing mental models of the concept [1]. For example, Kadam et al. [2] show that engagement with mental rotation exercises involving three-dimensional (3D) shapes improves engineering students' performance in a challenging engineering drawing course. Research by Stieff et al. [3] indicates a moderate association between mental rotation ability and achievements in science, which can be mainly observed at the early stages of learning.

From a broader perspective, it is evident that understanding science, mathematics, and engineering concepts requires strong spatial abilities and mental manipulation of newly acquired visual concepts. One of the suggested models that explain the mechanism behind newly acquired information is inhibitory control. According to this model, the acquisition of new information depends on suppressing previously acquired knowledge. Affecting inhibitory control mechanisms by means of cognitive neuroscience would impact the capacity for acquiring new scientific concepts [4]. Thus, developing technologies to identify the neural mechanisms or neural correlates of learning would be a major advancement toward creating practical teaching tools in the fields of science, engineering, and math.

Brain-computer interface (BCI) technologies are computer-based systems designed to monitor real-time brain activity and understand brain functions and cognition. BCIs offer cost-effective and user-friendly tools for collecting electroencephalographic (EEG) data [5]. Using EEG-based BCIs, researchers can capture multiple features of the brain oscillatory activity, such as power, frequency, and phase. Advanced measures of EEG-based BCIs include building models of neural networks for explaining cognitive functions [6]. Therefore, EEG-based BCIs will advance our understanding of brain function and influence the course of mental processes.

One application of EEG-based BCIs is neurofeedback. This method enables modulating brain oscillations related to cognitive abilities, thus facilitating learning. Research by Zoefel and Herrmann [7] shows that neurofeedback for increasing the power of upper alpha oscillations helps improve mental rotation performance.

Despite the crucial role that spatial skills play in science, math, and engineering learning, reasoning in these fields goes beyond visual perception. Some features of science learning are not entirely related to spatial skills. As Stieff et al. [3] demonstrate, understanding scientific diagrams depends on the students' representational competence rather than mere spatial perception.

Therefore, special technologies are needed to create appropriate learning environments for facilitating learning experiences in these disciplines. Using neurofeedback, VR, and AR technologies will allow the development of individualized learning interventions tailored to learners' needs and understanding levels [8]—an aspect challenging to achieve through traditional instruction. VR platforms can present complex entities or processes while providing a detailed representation of the concepts [9, 10]. AR platforms help increase student motivation for learning, improve the understanding of science concepts, and promote collaborative learning [11]. Neurofeedback is effective for modulating neural processes that correlate with learning. This chapter presents an innovative approach for facilitating the acquisition of unfamiliar scientific concepts by integrating neurofeedback with VR or AR technologies. In this approach, adaptive neurofeedback will be applied during the engagement in spatial tasks involving 3D structures within VR/AR environments. The adaptive neurofeedback training style will be programmed to adjust the training properties in accordance with the ongoing neural oscillation activity and real-time task performance. We believe that preparing learning spaces with these combined technologies will enhance student's learning experiences and academic achievements beyond expectations.

2 Spatial Skills: The Case of Mental Rotation

Mental rotation is a cognitively demanding ability by which one manipulates mental representations of visual objects to understand their spatial structure [12–14]. The foundations of this paradigm were derived from pioneering work by Shepard et al. [12, 15]. To investigate the employment of the mental rotation strategy, researchers developed a mental rotation task with pairs of geometric stimuli rotated at different angles. To complete the task, participants were asked to quickly determine if geometric pairs were identical or mirrored configurations. The participants were not given instructions regarding using a mental rotation strategy during judgments. Results show that reaction times systematically increase with the increase of rotation angle, reflecting the additional time required to mentally rotate and align objects with large differences in rotation angle to examine their configuration [12]. These results imply that during the manipulation process, mental representations of the object undergo intermediate processes of gradual rotations where larger angles mandate more spatial transformations, increasing the time required for completing the rotational process [13, 16].

Mechanisms underlying the mental rotation process involve visual and motor processing [17]. One mechanistic model describes mental rotation as a sequential process with several stages: encoding, rotating, and decision-making. The first stage involves encoding the visual stimuli. The second stage, which is more cognitively demanding, involves the active rotation of mental representations of the stored objects until congruency is achieved. The final stage includes selecting and executing a response [15–19].

Successful mental rotation performance depends on visual working memory capacity. Two distinct subsystems are associated with visual working memory: object-based and spatial-based. Although mental rotation is considered a spatial task, intermediate stages involve other forms of processing. It was shown that the object-based subsystem of working memory, rather than the visual-based subsystem, holds onto mental representations of the stimuli while being gradually rotated until making the final judgment of whether the stimuli pair matches in orientation [20]. This means that the information acquired as a holistic entity is far more important than its micro-characteristics in determining the subsystem to be involved in processing. Other studies show that mental rotation processes 'partially' involve motor processing, similar to actual motor rotation [21]. Further evidence suggests the involvement of visual-spatial and phonological systems when performing mental rotation. Research shows that individuals rely on phonological processing for coding and storing objects while they are being manipulated [22]. Despite the accumulative evidence regarding the involvement of multisensory processing in mental rotation, this skill heavily relies on our visual abilities to facilitate rotational transformations and identify stimuli configurations [23]. Nonetheless, the involvement of multiple systems emphasizes the complex nature of mental rotation and necessitates appropriate assessment tools for its evaluation.

Studies present many types of mental rotation tasks that differ in the type of stimuli employed and the behavioral indices used for assessment. Early research used tasks with geometric shapes [12], while others used abstract figures [24] and alphanumeric characters [15, 24]. Despite being practical, common assessment tasks in mental rotation often employ generic stimuli unrelated to concrete scientific phenomena students encounter in science, engineering, and math classrooms. The gap between the available assessment tools and their practical applications may hinder students from smoothly applying these skills to understand core science principles. For example, the ability to predict molecular interactions is associated with a profound understanding of molecules' spatial structure and arrangements [25]. Considering the importance of mental rotation skills in completing stereochemical tasks [26], training for developing mental rotation skills can help boost students' performances in chemistry. An ideal mental rotation task would involve molecule stimuli as rotational targets. However, there are no standardized assessments that employ molecule stimuli as rotational targets. The lack of appropriate evaluation tasks may widen the assessment-application gap, preventing the comprehensive profiling of individuals' proficiency in manipulating this important class of objects.

In our study [27], participants completed mental rotation tasks using both cubic and molecule stimuli. For both tasks, response time regarding the orientation of stimuli pairs showed strong positive correlations with the rotation angle. As the rotation angle of objects increased, participants required more time to decide whether the objects were identical or mirror images. These results further clarify many aspects of the mental rotation phenomenon. First, the preserved correlation between that rotational magnitude and response time, irrespective of stimuli type, implies that mental rotation represents a domain-general ability not constrained by the object type. Second, these results suggest that mental rotation skills are generalizable spatial capacities readily applicable across diverse object classes. Third, preserving this cognitive phenomenon despite changes in stimulus types and features provides evidence that mental rotation is a universal, not specific, human spatial capacity. Consequently, interventions designed to improve mental rotation skills of any given stimuli type may induce transferable improvements in spatial reasoning competencies. These improvements can be further enhanced when students receive training using mental tasks with stimuli directly relevant to the scientific concepts being learned.

Strong mental rotation capacities provide learners with many advantages. Accumulating research has linked good mental rotation performance to enhanced navigational skills and problem-solving in math and science domains [28, 29]. Research by Garg et al. [30] revealed that understanding anatomy is related to high mental rotation skills. Guillot et al. [31] presented empirical evidence supporting this association. They found a positive correlation between good mental rotation abilities and academic achievement in anatomy. Thereby, spatial skills are good predictors of learning outcomes. Specifically, mental rotation skill is a unique profiler of spatial capacity, with certain people being more successful in mental rotation than others. For instance, extensive athletic or musical training has been associated with superior performance on mental rotation tasks [32], reflecting improved brain sensory and motor functioning developed with intensive practice and training [33]. The induced cognitive change resulting from practicing music is modulated through neuroplasticity mechanisms [34]. Therefore, it is important to consider interindividual differences when designing tools to optimize spatial skills. One approach to address the individual differences in spatial abilities is to focus on the neural parameters or features accountable for these variations.

Researchers have been investigating brain oscillations and their role in cognitive function. These studies are especially important as they help researchers map cortical connectivity and, on a global scale, neural network dynamics involved in mental processes. In addition, recording brain oscillations during, before, and after engagement in tasks can provide insight into why, how, and where cognitive processes occur. For example, EEG research shows that upper alpha intensity correlates with mental rotation skill [7]. This is in line with the growing body of evidence that supports the involvement of the right parietal lobe in mental rotation [13]. Further emphasizing this result, studies involving EEG neurofeedback have demonstrated that upper alpha neurofeedback within the right parietal or parietooccipital region can enhance the intensity of these oscillations and improve mental rotation performance [27, 35]. Our research [27] adds to this discussion by highlighting the efficiency of EEG neurofeedback in inducing targeted changes in brain activity, especially in the upper alpha band, within a single session. These neural changes, however, were observed only in some individuals in the neurofeedback group whose alpha enhancement was also accompanied by improved mental rotation speed. Some individuals in the neurofeedback group did not benefit from the neurofeedback intervention.

Electrophysiological measures and behavioral indices alone do not satisfy the requirements for creating innovative learning enhancement applications. Future endeavors should focus on developing interactive neurofeedback training protocols with long-term after-effects to maintain the induced changes in brain oscillation activity and improved cognitive functions. Moreover, it is crucial to design an individualized technology while taking into account interindividual and intraindividual baseline differences in brain oscillation intensity and cognitive function.

3 Methods for Boosting Spatial Skills

The methods for enhancing mental rotation skills include interactive environments (i.e., simulations), computer games, classroom teaching, neurostimulation techniques, and virtual environments.

Virtual environments: Lochhead et al. [36] investigated the impact of using a virtual environment on mental rotation abilities. They found that 3D-VR tasks, rather than 2D, improved mental rotation accuracy. Martín-Gutiérrez et al. [37] developed an AR application enabling engineering students to visualize and manipulate 3D objects. Similar to previous studies with VR, training in an AR environment also enhanced spatial capabilities.

Computer games: De Lisi's and Wolford's [38] work shows that computer games have the potential to develop spatial cognition while also providing enjoyment. Playing computer games proved useful for individuals with weak initial performance levels. After the engagement in a computer game, weak performers improved their mental rotation accuracy. Likewise, Feng et al. [39] examined the effects of action video games on mental rotation skills. They found that playing an action game for only 10 h reduces gender differences in mental rotation performance, ultimately resulting in females outperforming males.

Classroom training: the effect of educational environments on developing mental rotation skills during early childhood has been well-documented. Bruce and Hawes [40] conducted a study showing that classroom training effectively improves children's spatial ability, regardless of their individual differences.

Interactive environments: Kozhevnikov and Thornton [41] revealed that both university students who enrolled in a general non-calculus physics course and science teachers exhibited enhanced mental rotation performance following microcomputer-based laboratory in physics.

Neurostimulation and neurofeedback: Targeting neural features involved in mental rotation holds promise for improving executive functions and cognition. Neurofeedback, a form of operant conditioning, allows control over brain function by rewarding targeted oscillations [6, 42, 43]. Alterations of brain oscillation activity have been found to correlate with changes in cognition and behavior. The neural modifications gained by neurofeedback training are likely induced by neuromodulation and long-term potentiation mechanisms [43, 44]. Noninvasive brain stimulation, like transcranial magnetic stimulation (TMS), has also been used to modulate brain activity. Applying 10 Hz TMS within the supplementary motor area (SMR) helped in improving the mental rotation of abstract (cubes) rather than familiar objects (hands) [45]. Similarly, applying transcranial alternating current stimulation (tACS) at the individual alpha frequency (IAF) resulted in enhanced mental rotation performance rather than mental rotation speed [46].

Mental rotation skills appear highly trainable through virtual/augmented environments, gaming, simulations, classroom training, and neural modulation/stimulation methods. While these methods are not ideal, they can be re-engineered and upgraded to fine-tune students' spatial reasoning abilities and deepen their understanding of science, engineering, and math learning disciplines more than previously anticipated.

3.1 EEG-Neurofeedback

As previously mentioned, EEG-neurofeedback is a form of operant conditioning that allows individuals to gain control over their brain oscillations and executive functions [47]. During training, brain activity is continuously calibrated, with feedback regarding the activity of targeted brain oscillations—above or below a threshold— provided to participants in the form of visual, auditory, or mechanical cues. Real-time feedback enables individuals to consciously and continuously modify their brain activity and, eventually, self-regulate their brain function.

The associative learning theory explains the operant conditioning learned by EEG-neurofeedback training [48]. A recent study [49] examined the role of neurofeedback on memory integration and found that neurofeedback training helped increase the association between stimuli and their respective exemplars, thereby affecting memory formation.

The general effects of neurofeedback on the brain are derived from two main mechanisms: neuromodulation and long-term potentiation (LTP) [43]. Neuromodulation refers to cellular processes that actively modify the neurons' membrane potential through the opening/closing of membrane channels, leading to alterations in neural circuits and brain oscillation activity. Neurofeedback operates through a neuromodulation mechanism by controlling circuitry activity to improve brain function. LTP, however, refers to the brain's ability to adapt and change in response to external stimuli. Neurofeedback induces LTP that reinforces and sustains the alterations of brain processes. The neural changes caused by neurofeedback further show the potential of this method in influencing cognitive processes and modifying behavior.

Neurofeedback methods have been applied as a treatment for many mental disorders [50]. Applications include the treatment of sleep disorders, post-traumatic stress disorder (PTSD), and addiction. In addition, neurofeedback training helps in mitigating autism spectrum disorder (ASD) and ADHD symptoms.

Neurofeedback research on healthy individuals focuses on enhancing cognitive processes, such as mental rotation [7]. Our study [27] examined the correlation between upper alpha oscillation activity and mental rotation. To that end, participants received a single neurofeedback session, where they trained to improve the power of individual upper alpha within the right parietal lobe. The individual upper alpha frequency range was defined individually for each subject using the frequency of the maximal alpha power (IAF) as the lower bound and IAF + 2 Hz as the upper bound [35]. Our findings showed that tailored neurofeedback increased upper alpha oscillation power in the right parietal region of the brain. The increase in upper alpha power was accompanied by better performance on the mental rotation posttest which included both geometric stimuli [12] and unfamiliar stimuli—molecules. These findings present the significance of EEG neurofeedback for improving spatial performance and potentially facilitating understanding of complex scientific concepts requiring advanced mental rotation skills, such as molecular structures and interactions.

3.2 Virtual Reality (VR)

Virtual reality (VR) leverages the power of computer technology to create three-dimensional (3D) immersive environments. These environments are designed to enrich user experience through 'telepresence,' which creates the illusionary sensation of being physically immersed in the virtual world [51]. Delivering learning material through immersive VR environments gives learners a stronger sense of authenticity [52]. These environments allow learners to interact with the VR elements using multiple sensory modalities, including auditory, visual, and motor [51], to facilitate engagement and a deep understanding of the learning material [52, 53].

Neurophysiological studies show that VR environments can alter human cognitive abilities [54]. Research by Keller et al. [55] explored VR effects on brain structure and dynamics. Their findings revealed an increase in the volume of cortical gray matter in the motor regions after VR exposure. Changes in brain dynamics observed in individuals who engaged in VR have been correlated with improved motor performance [55], and improved brain activation in stroke patients [56].

VR emerges as a significant instructional method, specifically in the fields of science, engineering, and math. VR technologies allow learners to engage in a learning session to explore scientific concepts/objects from different perspectives, thereby promoting a deeper understanding of the environments in which these concepts/objects exist. The immersive experience and the sense of presence produced by these model environments, when delivered at the appropriate resolution [57], transcend traditional instructional methods and make it possible for learners to engage with abstract scientific phenomena more holistically and interactively [58].

VR technology transforms astronomy education by presenting students with platforms designed to teach complex subjects. Mintz et al. [58] presented a VR model of the solar system. The VR application contains dynamic elements of the system, such as the sun, planets, moons, asteroids, and comets. Users enjoyed an immersive experience by navigating the environment through a virtual space shuttle that allowed them to approach and examine VR elements in detail. One interactive feature of the application enables presenting the rotation of celestial bodies within the Milky Way. The VR system contains tools that enhance the immersive experience, such as navigation, resolution adjustment, and viewpoint manipulation tools, all designed to facilitate a deeper understanding of the solar system dynamics. Enhancing visual literacy skills using VR enables the development of new and engaging learning methods, overcoming the limitations of traditional approaches.

In chemistry education, the use of VR presents an advantage. Nersesian et al. [59] investigated the efficacy of VR, computer simulations, and traditional teaching methods in helping students understand key concepts in chemistry, specifically those related to molecular structure. The study employed a three-group design: one group engaged in a VR learning session, a second group studied using computer simulations, and a control group received traditional textbook instruction. Analysis revealed that students in the VR group demonstrated a significantly more profound understanding of the material, as reflected in their superior performance on the final assessment compared to the control group, but not the computer simulation group. These findings support applying VR as a complementary teaching method in chemistry education, potentially enhancing student learning experiences.

Further studies focused on the effects of VR instruction in improving chemistry learning, mainly for students with weak spatial ability [60]. The study was conducted with two groups of undergraduate students enrolled in a Chemistry 101 course. One group received instruction using a 3D virtual environment, while the other group received 2D image-based instruction. To test the effects of using these instructional methods, the researchers assessed the students' spatial abilities and examined their understanding of molecular spatial structures. The knowledge test used in this study is the Valence Shell Electron Pair Repulsion (VSEPR) theory test, which includes questions on molecular angles, geometry, and molecular identification. While the overall results revealed no significant differences between the 3D and 2D groups, students with weak spatial ability in the 3D group improved their performance, as shown in molecule angle questions of the VSEPR test. This result goes along with previous findings reported by Lee et al. [61], further indicating the potential of VR technology to bridge the gap for students struggling with subjects that require high spatial reasoning. Nevertheless, the insignificant difference in overall results [60] necessitates further research with larger sample sizes or a different VR learning platform.

As many studies show, VR technology offers appropriate platforms for presenting abstract concepts in math. Xu and Ke's [62] exploratory study displays the effects of VR-based games on enhancing mathematics learning. The application was originally designed as a teaching tool for fifth graders but was tested on a small group of university students majoring in mathematics. Their study is based on "Island of Pi," a VR game designed to increase students' motivation through an engaging narrative. During the game, students engaged in scenarios where they must assist islanders in escaping a vanishing island. Completing the mission depends on successfully solving fraction problems. Using VR games with engaging narratives helps maintain student engagement, deepen their understanding of mathematical concepts [62], and promote enjoyable experiences [63].

The immersive nature of VR technology transforms learning experiences, more evidently for students struggling with geometric concepts. Research by Su et al. [9] suggests that VR enhances learning challenging topics in mathematics, such as understanding the proportional difference in the volume of pyramids and prisms, calculating the volume of cones, and calculating the triangle center of gravity deviation. While non-significant, the VR technology improved students' math achievements compared to a control group who learned through paper-based material. In addition, using VR raised students' confidence and gave them a feeling of accomplishment as they completed the learning tasks.

In general, VR technologies upgrade science, engineering, and math education, mainly through spatial training platforms. Molina-Carmona et al. [64] examined the effect of engagement in a VR environment on spatial skills using two groups–one group used VR goggles and a smartphone for manipulating polyhedral shapes, while the other group underwent spatial training using computers with a traditional screen view. To detect the changes in spatial performance following training, the participants were requested to complete the spatial ability test before and after the training activity. The findings revealed that the VR activity improved the VR group's spatial performance more than the computer-based group. These findings strongly suggest that VR training offers a unique and powerful approach to developing spatial skills, indicating their relevance for science, math, and engineering education.

3.3 Augmented Reality (AR)

Augmented reality (AR) technology offers innovative platforms for presenting and interacting with real-world features. AR platforms are designed to display digital objects overlaid on real environments [65]. The virtual layer helps users gain a profound grasp of the physical world by influencing their perception of the surrounding environments. For a more augmented experience, these layers are integrated in the form of numbers, letters, symbols, audio, video, and graphics cues involving different sensory modalities [66].

AR has a unique advantage over other immersive technologies. As explained previously, AR integrates virtual elements into the real environment, bridging the gap between the two worlds. This characteristic allows users to remain attached to the physical/real world when interacting with the virtual features [67]. This mode of interaction, real-virtual, offered by AR tools diminishes the possibility of users losing their sense of reality often experienced in fully immersive virtual environments [68]. The coherent integration of digital properties in real environments determines the position AR platforms occupy on the reality-virtuality continuum and the levels at which these platforms facilitate the perception of digital elements as authentic components of the real environment rather than independent virtual objects. Hence, acknowledging the place of AR technologies within the reality-virtuality continuum is highly important for developing advanced AR technology [69], especially platforms allowing individuals to effortlessly transit between the real and virtual representations of the encountered environments [70], thus enhancing their perception.

The reality-virtuality continuum feature necessitates shifts in the ways AR technologies are developed. Instead of viewing AR as mere technology, researchers should recognize its position on the spectrum between real and virtual worlds. Different AR designs, as reflected by different positioning on the spectrum, may present a fundamentally different form of relationships between real and virtual entities and lead to a completely different experiences for users. Well-designed platforms would allow immersion and augmentation of senses rather than changing them [71], thereby bringing unique contributions to various fields. Due to their affordances and the availability of open-source developer libraries [72], mass-producing AR-based educational tools would be possible. The flexible technology can be built to include science content through different designs: inquiry-based design, simulations, and game-based AR applications [73].

Liu et al. [74] used a pre-test/post-test design to examine the efficacy of an AR application in facilitating junior high school students' understanding of three-view projection and drawing in geometry. Their findings revealed a significant improvement in students' performance on the post-test compared to their baseline performance, with no difference between low-achieving and high-achieving students. Estapa and Nadolny [75] further explored the impact of AR on high school students' understanding of math, specifically the understanding of dimensional analysis and unit conversions. To test performance, the researchers used mathematical tasks with real-life problems [76] that suited the students' developmental stage. The research employed a two-group design: one group was assigned to AR training, and the second group was assigned to website training. The AR group underwent an activity that included a handout with overlaid digital objects such as video, websites, audio, and images. The control group completed the activity using the printed handout and a dedicated website containing the same instructional content delivered to the AR group. The participants underwent three assessments: a pre-test delivered at the beginning of the experiment, a post-test delivered one day after the activity (VR/website), and a delayed post-test delivered one month after the training. The research findings suggest that using AR and websites during teaching improved students' math achievements. Nevertheless, the students did not show the same level of improvement in the delayed test, implying that the improvements were not sustainable. While this study has not clearly shown the advantages of using AR during instruction, the results show the practicality of AR integration in math instruction.

Further evidence shows the positive effect of AR on developing spatial skills and facilitating the understanding of mathematical functions displayed in both two and three dimensions [77]. To examine the role of AR in developing spatial skills, del Cerro Velázquez and Morales Méndez [77] conducted a quasi-experimental study with two groups. The experimental group studied mathematical functions using the Geogebra AR application, whereas the control group learned the same subject through traditional instruction. Both groups took a mathematical functions test and a spatial ability test twice, at the beginning of the experiment and again after the learning session (Geogebra AR/Control). Results show that the performance of the AR group in both the knowledge assessment and spatial test was better than that of the control.

In a study by Carbonell Carrera and Bermejo Asensio [78], researchers investigated the effect of AR application, presenting knowledge of geographical landscapes, on enhancing learners' spatial abilities. The study used a pre-test/post-test design with experimental and control groups. Both groups initially completed the Perspective Taking/Spatial Orientation Test [79, 80] to determine the baseline levels of their spatial skills. The experimental group then participated in a workshop designed for training on topographical recognition. The workshop consisted of two phases: the first involved traditional 2D content, while the second involved AR. During each phase, the experimental group completed an exercise that challenged them to identify topographical features such as slopes, elevation lines, hills, and depressions. Unlike the AR group, the control participated in theoretical lectures. Following the treatment/ non-treatment, both groups were given a spatial orientation test. The findings revealed an improvement in spatial orientation skills among students who participated in the AR workshop, indicating the benefits of integrating geographic and cartographic AR applications in optimizing students' cognitive skills.

Although research has well-documented the positive implications of adopting AR platforms as instructional tools, other studies have highlighted potential disadvantages. According to Radu [81], AR technologies can be cognitively demanding, thus challenging developers who aim to innovate effective, user-friendly technology. Further research is necessary to study the after-effects of using AR as an educational tool.

4 Combining AR/VR and Neurofeedback Technologies to Enhance Cognitive Skills and Learning Experiences

The efficacy of AR and VR methods in treating mental disorders has been widely studied. Viczko et al. [82] found that AR without neurofeedback can improve participants' mood and mental health. In general, research has indicated that VR is effective in enhancing exposure therapy for treating phobias [83], treating post-traumatic stress disorder (PTSD) [84], and rehabilitating patients with stroke [85]. Using VR as a complementary therapy enhanced the stroke patients' treatment results [85]. These studies suggest that combining immersive VR/AR environments with neurofeedback training may highly increase user engagement and optimize neurofeedback training results.

The neural mechanisms underlying the self-regulating brain oscillations achieved through neurofeedback training have yet to be clearly defined. Many models are trying to uncover the mechanism behind neurofeedback. One model is the dual-process theory [86]. According to this model, neurofeedback learning results from the conjunction between top-down and bottom-up processes. Top-down processes are related to explicit reinforcement learning, where participants consciously recruit mental strategies to enhance their learning experience. Bottom-up processes involve implicit reinforcement learning, where participants complete tasks (e.g., video games) while following neurofeedback training instructions. Both modes of processing influence the neurofeedback training experience, with bottom-up reinforcement becoming dominant after continuous training. Ultimately, successful neurofeedback relies heavily on neural self-regulation [48] and 'behavioral' self-regulation capabilities required for self-maintaining desired behaviors [87]. This self-regulation can be strengthened through well-programmed implicit/explicit reinforcement learning styles and will possibly become far more effective when combined with technologies like VR/AR. Research suggests that VR/AR technologies can modulate self-consciousness [88], a process known to influence self-regulation [89]. Neurofeedback training can become more intuitive and likely more effective by incorporating VR/AR technology to induce changes in our self-perception.

The reality-virtuality (RV) continuum model supports the approach of combining neurofeedback with VR/AR technologies. This model highlights the idea of "presence," a main characteristic of the immersive experience [69]. Many features of the virtual environments affect the sense of 'presence' experienced by users, including extrinsic stimuli [90]. Gruzelier et al. [91] argue that neurofeedback may alter cognitive functions in a way that enhances the trainees' sense of presence. Successful neurofeedback training depends on delivering feedback that accurately reflects the changes in brain activity through extrinsic stimuli presented in a training interface. Seeing these technologies sharing the same attribute of enhancing the sense of "presence" validates the merging of neurofeedback with VR/AR for a better gain in neurofeedback training.

Nowadays, more research is focused on combining technologies to understand the neural mechanisms behind cognitive optimization gained by neurofeedback. In neurofeedback research [92] aimed at downregulating the amygdala, researchers used a virtual environment and standard training interfaces. The research aimed to detect the effect of engagement in VR on the participants' amygdala activity and overall neurofeedback training experience. The study included two groups; one group underwent neurofeedback training while engaged in a virtual environment, and the second group underwent a standard training session while given visual feedback in the form of a 2D graphic thermometer. The groups went through two training sessions. The first session included measuring amygdala-EEG activity levels and a neurofeedback training session with VR or 2D graphic thermometer training interfaces. In the second session, the participants underwent different testing stages: a 'no-neurofeedback' stage, neurofeedback training with the same interfaces as the first session, a transfer-I stage where participants experienced neurofeedback training with an unfamiliar interface that contained a 2D moving skateboard, a transfer-II stage where participants went through training with the interface used for the other group, and a final stage where participants completed the Intrinsic Motivation Inventory (IMI) questionnaire [93]. The results show that the VR interface helped participants to downregulate the amygdala more than participants who underwent neurofeedback with the standard 2D graphic thermometer interface. The VR group improved performance at the 'no-neurofeedback' and transfer phases of session 2, implying the VR interface training is resilient with more sustainable effects compared to standard thermometer interfaces. This finding further supports the LTP mechanism by which neurofeedback helps adapt to environmental changes. Therefore, combining neurofeedback technology with a virtual environment enhances the training results, which makes merging these technologies relevant and valuable for improving learning in educational settings.

The effects of VR-neurofeedback on mitigating attention disorders and impulsivity phenomena have also been tested [94]. Cho et al. [94] used three groups: a control group that did not undergo VR-neurofeedback, a VR group that underwent neurofeedback in a VR environment, and a non-VR group that also underwent a neurofeedback session but without being engaged with VR environment. All groups completed a pre- and post-treatment continuous performance task (CPT) to examine their attentional levels. Results show that the accuracy levels of the VR-neurofeedback improved significantly compared to the other groups. These findings suggest that neurofeedback accompanied by VR technology may help identify and possibly treat attention and impulsive disorders.

The flexible design of the combined technology, neurofeedback with VR/AR, allows developers to program adaptive training platforms specifically tailored to the students' learning levels [95]. Hubbard et al. [95] investigated the efficacy of VR-neurofeedback in enhancing subjects' working memory performance. The experiment design allowed participants to interact with changing-color stimuli floating within a virtual environment. During the experiment, stimuli appeared in the virtual environment with a specific color, after which they became gray, while the color of the virtual environment was changed to the same color as the floating stimuli before becoming gray. The participant had to guess which of the stimuli changed color. The number of stimuli presented in the virtual environment depended on the participants' successful responses. When the participants succeeded in one trial, the number of stimuli increased by one in the next trial. The training results showed no significant difference in the group performance; however, this VR-EEG study forms the basis for future studies that take account of participants' performance during training.

A study conducted by Berger et al. [96] brings evidence that neurofeedback training is efficient in enhancing sensorimotor rhythms (SMRs). The study was conducted with two main groups who underwent one training session with either a 3D-VR or 2D interface. Half of the participants in the 2D group went through neurofeedback training, with feedback provided through a 2D progress bar interface. The size of the bar would increase if the trainees' SMR power levels were higher than a predetermined threshold. The other half of the 2D group underwent sham training. On the contrary, half of the participants in the 3D group underwent neurofeedback while engaging in a 3D virtual environment. During neurofeedback, the 3D group was required to roll a ball at a changing speed, reflecting SMR power changes compared to the threshold. The other half of the 3D group underwent sham training while engaging in the same virtual environment. The 3D group succeeded in increasing their SMR power, while the 2D group, who received traditional feedback, did not show a change in the power of SMR.

5 Implications for Future Educational Technologies

The challenges in learning science, engineering, and math necessitate rethinking how to prepare educational settings to offer practical learning experiences. Computer simulations, 3D visualization applications, and explanatory videos effectively describe complex scientific, mathematical, and engineering concepts. However, presenting students with advanced visualization tools alone does not guarantee successful learning outcomes [97]. There is a prompt need to reevaluate and upgrade current instructional tools and, more importantly, focus on creating combined technologies for providing effective learning experiences.

The goal of modern education has been shifted from traditional and 'similar to all' to innovative learning environments tailored to student's needs [98]. Educational facilities must have advanced technologies tailored to learners' needs, developmental stages, and levels of understanding. Our vision for tailored learning resides in combining neurofeedback technology with VR/AR. Neurofeedback methods will be used as inducers for brain oscillations, while VR/AR technologies will be used to facilitate learning experiences. Although these technologies have previously been applied in medical health and educational research, they have not been extended for standard use in educational settings (i.e., schools or academies). By adopting this approach, learners would benefit from neurofeedback sessions designed to enhance their cognitive skills while enjoying immersive virtual environments that bring them closer to understanding challenging learning materials. The suggested training platform will enable students to study content/complete tasks with VR/AR during real-time neurofeedback, where students are trained to control their brain oscillations power, especially those correlated with the task at hand. The flexible technology allows the creation of systems to entrain any brain oscillation and present instructional content using neurofeedback and virtual environments. For instance, the learners can engage in virtual environment sessions to learn about chemical interactions and molecular structure while undergoing neurofeedback training to raise upper alpha oscillations that correlate with spatial skills. The neurofeedback training regime can be modified in accordance with the student's experience in the virtual world and vice versa. This technique, AR- or VR-neurofeedback, can be applied in science education as well as math and engineering education.

The use of these technologies in educational settings comes with limitations. Applying neurofeedback with VR/AR technologies for all students requires well-equipped learning spaces, which may place a financial burden on educational institutions. Moreover, educational institutions require technical support to operate these systems, maintain both hardware and software, and provide technical support when needed [99]. Therefore, further research is needed to resolve these limitations while developing cost-effective solutions and user-friendly AR- or VR-neurofeedback applications that fit educators' technical competence.

References

1. Cheng KH, Tsai CC (2013) Affordances of augmented reality in science learning: suggestions for future research. J Sci Educ Technol 22:449–462
2. Kadam K, Mishra S, Deep A, Iyer S (2021) Enhancing engineering drawing skills via fostering mental rotation processes. Eur J Eng Educ 46(5):796–812
3. Stieff M, Origenes A, DeSutter D, Lira M, Banevicius L, Tabang D et al (2018) Operational constraints on the mental rotation of STEM representations. J Educ Psychol 110(8):1160
4. Mareschal D (2016) The neuroscience of conceptual learning in science and mathematics. Curr Opin Behav Sci 10:114–118
5. Shih JJ, Krusienski DJ, Wolpaw JR (2012) Brain-computer interfaces in medicine. Mayo Clin Proc 87(3):268–279. Elsevier
6. Pulvermüller F, Tomasello R, Henningsen-Schomers MR, Wennekers T (2021) Biological constraints on neural network models of cognitive function. Nat Rev Neurosci 22(8):488–502
7. Zoefel B, Huster RJ, Herrmann CS (2011) Neurofeedback training of the upper alpha frequency band in EEG improves cognitive performance. Neuroimage 54(2):1427–1431
8. Ma M, Fallavollita P, Seelbach I, Von Der Heide AM, Euler E, Waschke J et al (2016) Personalized augmented reality for anatomy education. Clin Anat 29(4):446–453
9. Su YS, Cheng HW, Lai CF (2022) Study of virtual reality immersive technology enhanced mathematics geometry learning. Front Psychol 13:760418
10. Zhou Y, Chen J, Wang M (2022) A meta-analytic review on incorporating virtual and augmented reality in museum learning. Educ Res Rev 36:100454
11. Lee K (2012) Augmented reality in education and training. TechTrends 56:13–21
12. Shepard RN, Metzler J (1971) Mental rotation of three-dimensional objects. Science 171(3972):701–703
13. Harris IM, Egan GF, Sonkkila C, Tochon-Danguy HJ, Paxinos G, Watson JD (2000) Selective right parietal lobe activation during mental rotation: a parametric PET study. Brain 123(1):65–73
14. Just MA, Carpenter PA (1985) Cognitive coordinate systems: accounts of mental rotation and individual differences in spatial ability. Psychol Rev 92(2):137
15. Cooper LA, Shepard RN (1973) Chronometric studies of the rotation of mental images. In: Visual information processing. Academic Press, pp 75–176
16. Shepard RN, Cooper LA (1986) Mental images and their transformations. The MIT Press
17. Kosslyn SM, DiGirolamo GJ, Thompson WL, Alpert NM (1998) Mental rotation of objects versus hands: Neural mechanisms revealed by positron emission tomography. Psychophysiology 35(2):151–161
18. Cohen MS, Kosslyn SM, Breiter HC, DiGirolamo GJ, Thompson WL, Anderson AK et al (1996) Changes in cortical activity during mental rotation a mapping study using functional MRI. Brain 119(1):89–100
19. Booth JR, MacWhinney B, Thulborn KR, Sacco K, Voyvodic JT, Feldman HM (2000) Developmental and lesion effects in brain activation during sentence comprehension and mental rotation. Dev Neuropsychol 18(2):139–169
20. Hyun JS, Luck SJ (2007) Visual working memory as the substrate for mental rotation. Psychon Bull Rev 14(1):154–158
21. Wexler M, Kosslyn SM, Berthoz A (1998) Motor processes in mental rotation. Cognition 68(1):77–94
22. Brandimonte, MA, & Gerbino, W (1993). Mental image reversal and verbal recoding: When ducks become rabbits. *Memory & Cognition* 21(1): 23-33

23. Kosslyn SM (1996) *Image and brain: the resolution of the imagery debate*. MIT Press.
24. Jordan K, Heinze HJ, Lutz K, Kanowski M, Jäncke L (2001) Cortical activations during the mental rotation of different visual objects. Neuroimage 13(1):143–152
25. Fukui K (1971) Recognition of stereochemical paths by orbital interaction. Acc Chem Res 4(2):57–64
26. Stieff M (2007) Mental rotation and diagrammatic reasoning in science. Learn Instr 17(2):219–234
27. Farraj N, Reiner M (2024) Applications of alpha neurofeedback processes for enhanced mental manipulation of unfamiliar molecular and spatial structures. Appl Psychophysiol Biofeedback 1–18
28. Pazzaglia F, Moè A (2013) Cognitive styles and mental rotation ability in map learning. Cogn Process 14:391–399
29. Cheng YL, Mix KS (2014) Spatial training improves children's mathematics ability. J Cogn Dev 15(1):2–11
30. Garg AX, Norman G, Sperotable L (2001) How medical students learn spatial anatomy. Lancet 357(9253):363–364
31. Guillot A, Champely S, Batier C, Thiriet P, Collet C (2007) Relationship between spatial abilities, mental rotation and functional anatomy learning. Adv Health Sci Educ 12:491–507
32. Pietsch S, Jansen P (2012) Different mental rotation performance in students of music, sport and education. Learn Individ Differ 22(1):159–163
33. Watanabe D, Savion-Lemieux T, Penhune VB (2007) The effect of early musical training on adult motor performance: evidence for a sensitive period in motor learning. Exp Brain Res 176:332–340
34. Münte TF, Altenmüller E, Jäncke L (2002) The musician's brain as a model of neuroplasticity. Nat Rev Neurosci 3(6):473–478
35. Hanslmayr S, Sauseng P, Doppelmayr M, Schabus M, Klimesch W (2005) Increasing individual upper alpha power by neurofeedback improves cognitive performance in human subjects. Appl Psychophysiol Biofeedback 30:1–10
36. Lochhead I, Hedley N, Çöltekin A, Fisher B (2022) The immersive mental rotations test: evaluating spatial ability in virtual reality. Front Virtual Rity 3:820237
37. Martín-Gutiérrez J, Contero M, Alcañiz M (2010) Evaluating the usability of an augmented reality based educational application. In: Intelligent tutoring systems: 10th international conference, ITS 2010, Pittsburgh, PA, USA, June 14–18, 2010, proceedings, Part I 10. Springer Berlin Heidelberg, pp 296–306
38. De Lisi R, Wolford JL (2002) Improving children's mental rotation accuracy with computer game playing. J Genet Psychol 163(3):272–282
39. Feng J, Spence I, Pratt J (2007) Playing an action video game reduces gender differences in spatial cognition. Psychol Sci 18(10):850–855
40. Bruce CD, Hawes Z (2015) The role of 2D and 3D mental rotation in mathematics for young children: what is it? why does it matter? and what can we do about it? ZDM Math Educ 47:331–343
41. Kozhevnikov M, Thornton R (2006) Real-time data display, spatial visualization ability, and learning force and motion concepts. J Sci Educ Technol 15(1):111–132
42. Gruzelier JH (2014) EEG-neurofeedback for optimising performance. I: a review of cognitive and affective outcome in healthy participants. Neurosci Biobehav Rev 44:124–141
43. Abarbanel A (1999) The neural underpinnings of neurofeedback training. In: Introduction to quantitative EEG and neurofeedback. Academic Press, pp 311–340
44. Sterman MB, Egner T (2006) Foundation and practice of neurofeedback for the treatment of epilepsy. Appl Psychophysiol Biofeedback 31:21–35
45. Cona G, Marino G, Semenza C (2017) TMS of supplementary motor area (SMA) facilitates mental rotation performance: evidence for sequence processing in SMA. Neuroimage 146:770–777
46. Kasten FH, Herrmann CS (2017) Transcranial alternating current stimulation (tACS) enhances mental rotation performance during and after stimulation. Front Hum Neurosci 11:2

47. Angelakis E, Stathopoulou S, Frymiare JL, Green DL, Lubar JF, Kounios J (2007) EEG neurofeedback: a brief overview and an example of peak alpha frequency training for cognitive enhancement in the elderly. Clin Neuropsychol 21(1):110–129
48. Sitaram R, Ros T, Stoeckel L, Haller S, Scharnowski F, Lewis-Peacock J et al (2017) Closed-loop brain training: the science of neurofeedback. Nat Rev Neurosci 18(2):86–100
49. Collin SH, van den Broek PL, van Mourik T, Desain P, Doeller CF (2022) Inducing a mental context for associative memory formation with real-time fMRI neurofeedback. Sci Rep 12(1):21226
50. Niv S (2013) Clinical efficacy and potential mechanisms of neurofeedback. Pers Individ Differ 54(6):676–686
51. Steuer J, Biocca F, Levy MR (1995) Defining virtual reality: dimensions determining telepresence. Commun Age Virtual Rity 33:37–39
52. Wang A, Thompson M, Uz-Bilgin C, Klopfer E (2021) Authenticity, interactivity, and collaboration in virtual reality games: best practices and lessons learned. Front Virtual Rity 2:734083
53. Tolentino L, Birchfield D, Megowan-Romanowicz C, Johnson-Glenberg MC, Kelliher A, Martinez C (2009) Teaching and learning in the mixed-reality science classroom. J Sci Educ Technol 18:501–517
54. Georgiev DD, Georgieva I, Gong Z, Nanjappan V, Georgiev GV (2021) Virtual reality for neurorehabilitation and cognitive enhancement. Brain Sci 11(2):221
55. Keller J, Štětkářová I, Macri V, Kühn S, Pětioký J, Gualeni S et al (2020) Virtual reality-based treatment for regaining upper extremity function induces cortex grey matter changes in persons with acquired brain injury. J Neuroeng Rehabil 17:1–11
56. Lee SH, Kim YM, Lee BH (2015) Effects of virtual reality-based bilateral upper-extremity training on brain activity in post-stroke patients. J Phys Ther Sci 27(7):2285–2287
57. Bowman DA, McMahan RP (2007) Virtual reality: how much immersion is enough? Computer 40(7):36–43
58. Mintz R, Litvak S, Yair Y (2001) 3D-virtual reality in science education: an implication for astronomy teaching. J Comput Math Sci Teach 20(3):293–305
59. Nersesian E, Spryszynski A, Lee MJ (2019) Integration of virtual reality in secondary STEM education. In: 2019 IEEE integrated STEM education conference (ISEC). IEEE, pp 83–90
60. Merchant Z, Goetz ET, Keeney-Kennicutt W, Cifuentes L, Kwok OM, Davis TJ (2013) Exploring 3-D virtual reality technology for spatial ability and chemistry achievement. J Comput Assist Learn 29(6):579–590
61. Lee EAL, Wong KW (2014) Learning with desktop virtual reality: low spatial ability learners are more positively affected. Comput Educ 79:49–58
62. Xu X, Ke F (2016) Designing a virtual-reality-based, gamelike math learning environment. Am J Distance Educ 30(1):27–38
63. Chen ZH, Liao CC, Cheng HN, Yeh CY, Chan TW (2012) Influence of game quests on pupils' enjoyment and goal-pursuing in math learning. J Educ Technol Soc 15(2):317–327
64. Molina-Carmona R, Pertegal-Felices ML, Jimeno-Morenilla A, Mora-Mora H (2018) Virtual reality learning activities for multimedia students to enhance spatial ability. Sustainability 10(4):1074
65. Azuma RT (1997) A survey of augmented reality. Presence: Teleoperators Virtual Environ 6(4):355–385
66. Arena F, Collotta M, Pau G, Termine F (2022) An overview of augmented reality. Computers 11(2):28
67. Schmalstieg D, Hollerer T (2016) Augmented reality: principles and practice. Addison-Wesley Professional
68. Ryan ML (2001) Narrative as virtual reality. In: Immersion and Interactivity in Literature, pp 357–359
69. Skarbez R, Smith M, Whitton MC (2021) Revisiting Milgram and Kishino's reality-virtuality continuum. Front Virtual Rity 2:647997
70. Billinghurst M (2002) Augmented reality in education. New Horiz Learn 12(5):1–5

71. Carmigniani J, Furht B (2011) Augmented reality: an overview. In: Handbook of augmented reality, pp 3–46
72. Wellmann F, Virgo S, Escallon D, de la Varga M, Jüstel A, Wagner FM, Chen Q et al (2022) Open AR-Sandbox: a haptic interface for geoscience education and outreach. Geosphere 18(2):732–749
73. Ibáñez MB, Delgado-Kloos C (2018) Augmented reality for STEM learning: a systematic review. Comput Educ 123:109–123
74. Liu E, Li Y, Cai S, Li X (2019) The effect of augmented reality in solid geometry class on students' learning performance and attitudes. In: Smart industry & smart education: proceedings of the 15th international conference on remote engineering and virtual instrumentation, vol 15. Springer International Publishing, pp 549–558
75. Estapa A, Nadolny L (2015) The effect of an augmented reality enhanced mathematics lesson on student achievement and motivation. J STEM Educ 16(3)
76. Hiebert J (1997) Making sense: teaching and learning mathematics with understanding. Heinemann, 361 Hanover Street, Portsmouth, NH 03801-3912
77. del Cerro Velázquez F, Morales Méndez G (2021) Application in augmented reality for learning mathematical functions: a study for the development of spatial intelligence in secondary education students. Mathematics 9(4):369
78. Carbonell Carrera C, Bermejo Asensio LA (2017) Landscape interpretation with augmented reality and maps to improve spatial orientation skill. J Geogr High Educ 41(1):119–133
79. Hegarty M, Waller D (2004) A dissociation between mental rotation and perspective-taking spatial abilities. Intelligence 32(2):175–191
80. Kozhevnikov M, Hegarty M (2001) A dissociation between object manipulation spatial ability and spatial orientation ability. Mem Cognit 29:745–756
81. Radu I (2014) Augmented reality in education: a meta-review and cross-media analysis. Pers Ubiquit Comput 18(6):1533–1543
82. Viczko J, Tarrant J, Jackson R (2021) Effects on mood and EEG states after meditation in augmented reality with and without adjunctive neurofeedback. Front Virtual Rity 2:618381
83. Botella C, Fernández-Álvarez J, Guillén V, García-Palacios A, Baños R (2017) Recent progress in virtual reality exposure therapy for phobias: a systematic review. Curr Psychiatry Rep 19:1–13
84. Gonçalves R, Pedrozo AL, Coutinho ESF, Figueira I, Ventura P (2012) Efficacy of virtual reality exposure therapy in the treatment of PTSD: a systematic review. PLoS ONE 7(12):e48469
85. Laver KE, Lange B, George S, Deutsch JE, Saposnik G, Crotty M (2017) Virtual reality for stroke rehabilitation. Cochrane Database Syst Rev (11)
86. Arns M, Batail JM, Bioulac S, Congedo M, Daudet C, Drapier D et al (2017) Neurofeedback: one of today's techniques in psychiatry? L'encephale 43(2):135–145
87. Heatherton TF (2011) Neuroscience of self and self-regulation. Annu Rev Psychol 62:363–390
88. Riva G, Baños RM, Botella C, Mantovani F, Gaggioli A (2016) Transforming experience: the potential of augmented reality and virtual reality for enhancing personal and clinical change. Front Psych 7:222151
89. American Association for Research into Nervous and Mental Diseases, Posner MI, Rothbart MK (1998) Attention, self–regulation and consciousness. Philos Trans R Soc London Ser B: Biol Sci 353(1377):1915–1927
90. Caldas OI, Sanchez N, Mauledoux M, Avilés OF, Rodriguez-Guerrero C (2022) Leading presence-based strategies to manipulate user experience in virtual reality environments. Virtual Rity 26(4):1507–1518
91. Gruzelier J, Inoue A, Smart R, Steed A, Steffert T (2010) Acting performance and flow state enhanced with sensory-motor rhythm neurofeedback comparing ecologically valid immersive VR and training screen scenarios. Neurosci Lett 480(2):112–116
92. Cohen A, Keynan JN, Jackont G, Green N, Rashap I, Shani O et al. (2016) Multi-modal virtual scenario enhances neurofeedback learning. Front Robot AI 3
93. Ryan RM (1982) Control and information in the intrapersonal sphere: an extension of cognitive evaluation theory. J Pers Soc Psychol 43(3):450

94. Cho BH, Kim S, Shin DI, Lee JH, Min Lee S, Young Kim I et al (2004) Neurofeedback training with virtual reality for inattention and impulsiveness. Cyberpsychology Behav 7(5):519–526
95. Hubbard R, Sipolins A, Zhou L (2017) Enhancing learning through virtual reality and neurofeedback: a first step. In: Proceedings of the seventh international learning analytics & knowledge conference, pp 398–403
96. Berger LM, Wood G, Kober SE (2022) Effects of virtual reality-based feedback on neurofeedback training performance—a sham-controlled study. Front Hum Neurosci 16:952261
97. Rutten N, Van Joolingen WR, Van Der Veen T (2012) The learning effects of computer simulations in science education. Comput Educ 58(1):136–153
98. Bernard RM, Borokhovski E, Schmid RF, Waddington DI, Pickup DI (2019) Twenty-first century adaptive teaching and individualized learning operationalized as specific blends of student-centered instructional events: a systematic review and meta-analysis. Campbell Syst Rev 15(1–2)
99. Fransson G, Holmberg J, Westelius C (2020) The challenges of using head mounted virtual reality in K-12 schools from a teacher perspective. Educ Inf Technol 25(4):3383–3404

Artificial Intelligence (AI)

Section Introduction to AI in Higher Education

Dieter Uckelmann

The usage of Artificial Intelligence (AI) in higher education institutions is gaining momentum as it promises to enhance both the teaching and learning experiences, as well as administrative functions. The rise of a generative AI in the last years has increased the awareness in the academic community. However, even prior to this the importance of AI has been seen by research funding institutions. A German funding program on AI in higher Education for example, started already in 2021.

The six presented papers provided in this chapter explore various facets of AI in higher education, each addressing different dimensions and contexts in which AI interfaces with educational settings and ethical considerations. Collectively, these studies contribute to a nuanced understanding the development of AI competencies required, of how AI technologies can transform traditional educational practices, and the ethical frameworks necessary to guide such integration.

The paper titled *"Development of an AI Competence Matrix for AI Teaching at Universities"* presents a framework for integrating AI-related competencies into educational programs. It discusses the necessity of combining interdisciplinary AI skills in curriculum development. The matrix includes a detailed categorization of necessary skills, divided into professional, methodical, social, and self-competencies, and emphasizing the importance of ethical considerations. The matrix serves as a foundational tool for developing AI-related teaching activities and also facilitates the creation of a certification process for students and employees demonstrating their AI competency. The paper aims to guide universities in structuring AI training effectively, ensuring a well-rounded education that includes ethical reflection on AI technologies.

The paper titled *"Implementing Ethical Considerations into AI-supported Student Counselling—a Case Study"* discusses the integration of ethical considerations into

---

D. Uckelmann (✉)
Stuttgart University of Applied Sciences (HFT), Stuttgart, Germany
e-mail: dieter.uckelmann@hft-stuttgart.de

the development of AI tools for higher education. It focuses on the KNIGHT research project at Stuttgart University of Applied Sciences, which includes developing AI applications for supporting educational and administrative tasks. Ethical values are incorporated throughout the project to identify and solve potential ethical problems early in the development process. The paper highlights the importance of interdisciplinary collaboration for effective ethical analysis and emphasizes the benefits of embedding ethical considerations in early stages to minimize costs and facilitate easier problem resolution. This approach is illustrated through a case study involving the development of the "Study Dean Cockpit," an AI tool designed to help identify struggling students and support study deans at an early stage in their student support activities.

The paper on *"The Use of Artificial Intelligence to Teach Ethics"* delves into the role of AI in ethical education, particularly in the context of rapid technological advancement and its implications for human society. It argues for the necessity of incorporating AI ethics into higher education curriculums, blending STEM and non-STEM disciplines to prepare students for the ethical challenges posed by AI and other advanced technologies. The paper highlights various ethical dilemmas that arise from the integration of AI into everyday life and emphasizes the importance of an interdisciplinary approach to understanding and teaching these issues. Through a series of theoretical and practical pedagogical suggestions, it proposes ways to engage students with ethical questions surrounding AI. The text underscores the urgency of re-embracing liberal arts approaches into STEM education to verify technology is deployed in way leading in the right direction.

The paper *"A Reinforcement Learning Framework for Personalized Adaptive E-Learning"* explores a method to customize e-learning content based on individual student profiles, particularly for students with Attention-Deficit/Hyperactivity Disorder (ADHD) or other learning intervening factors. It proposes a Reinforcement Learning (RL) algorithm integrated within the Universal Design Learning (UDL) framework to adapt educational materials according to the learner's needs. The approach aims to dynamically adjust the e-learning environment by modifying presentation styles, session lengths, and the types of available resources and assessments. A proof-of-concept simulation is used for validation. The research highlights the potential of using advanced AI techniques to provide an adaptive real-time e-learning framework, responding efficiently to student's individual needs.

"Machine Learning Models to Detect AI-Assisted Code Anomaly in Introductory Programming Course" tests various machine learning approaches, including tree-based and parametric models, supplemented by neural-based methods and feature selection to identify anomalies in Python code assignments. The models are trained and validated using data from student submissions, achieving high performance in distinguishing between AI-generated and student-written code. Key findings include the identification of specific features that indicate AI assistance, and the paper concludes that such machine learning approaches are effective in detecting anomalies in coding assignments.

The book chapter on *"Using GenAI as Co-author for Teaching Supply Chain Management in Higher Education"* explores the application of Generative AI

(GenAI) for producing a compendium for teaching supply chain management in higher education. The findings highlight that while GenAI can be helpful to produce explanations of supply chain management (SCM) concepts for an SCM compendium, the need for expert review remains crucial to verify the accuracy of these outputs. The chapter concludes by suggesting that the framework developed through this research can guide the effective integration of various informational resources in teaching SCM, ensuring that educational content is both accurate and pedagogically sound. The study underscores the potential and limitations of using GenAI in education, suggesting that while it can enhance learning materials, it cannot fully replace human expertise and traditional learning resources.

Development of an AI Competence Matrix for AI Teaching at Universities

Diana Arfeli, Maximilian Weber, Dieter Uckelmann, and Tobias Popovic

Abstract In order to harness the benefits AI offers, tomorrow's workforces and societies alike need to be competent in how to use, shape, and develop AI applications. As part of the KNIGHT research project of Stuttgart University of Applied Sciences, this contribution develops a framework for competence-based AI education at higher education facilities, based on a review of selected literature. The bulk of the competencies focuses on technical aspects of AI, however, since AI applications are oftentimes faced with potential ethical challenges, special care is taken to integrate dedicated ethical competencies. The competencies are then structured according to the German Qualifications Framework for Higher Education to yield a comprehensive AI competence matrix. As a proof of concept, the matrix is then applied to structure an AI certificate, available to students and staff of Stuttgart University of Applied Sciences.

Keywords AI competencies · Ethical competencies · AI teaching · Higher education

---

D. Arfeli (✉) · M. Weber · D. Uckelmann · T. Popovic
Hochschule Für Technik Stuttgart, Schellingstraße 24, 70174 Stuttgart, Germany
e-mail: diana.arfeli@hft-stuttgart.de

M. Weber
e-mail: maximilian.weber@hft-stuttgart.de

D. Uckelmann
e-mail: dieter.uckelmann@hft-stuttgart.de

T. Popovic
e-mail: tobias.popovic@hft-stuttgart.de

1 Introduction

In response to the rapid developments in the area of Artificial Intelligence (AI), the German government decided to invest heavily into AI-related research and innovation. Among its stated aims are the "responsible and welfare-oriented" development and use of AI [5, p. 7 (own translation)]. This picks up on current worries of private companies expanding on their increasing use of AI [21] in order to, ultimately replace, human workers [3, p. 218]. In order to secure the standing of the human workforce, then, future employees will have to work *with*, instead of *against* AI [10].

Some of the challenges, future employees and regulators will face when working *with* or *on* AI are ethical in nature [20]. Assuming a competence-orientated approach [6, 15], this suggests that a mixture of ethical and technological competencies is necessary in order to do justice to the above stated aim of responsible and welfare-oriented development and use of AI, a sentiment echoed by the German Ethics Council [4, p. 43]. However, a recent survey draws a dire picture regarding the teaching of ethical skills in STEM studies [8].

In order to bridge this gap, HFT Stuttgart ("Hochschule für Technik Stuttgart", engl. "Stuttgart University of Applied Sciences") has taken the initiative by launching research project "Künstliche Intelligenz für die Lehre an der HFT Stuttgart" (short: KNIGHT, engl. "AI for education at HFT Stuttgart"), researching if and how AI can be used across different disciplines at HFT Stuttgart in order to enhance the students' learning experiences and support lecturers [9]. In order to achieve this, one of KNIGHT's goals is to develop a *competence matrix* as a framework for learning and teaching activities related to AI [9, p. 5]. This matrix is intended to support universities in structuring the content for AI training. As KNIGHT's project highlights the need for ethical reflection of AI technologies, the matrix has to account for technical AI competencies as well as specifically ethical competencies, and relate the two subject areas to one another. Another one of KNIGHT's goals is to use the AI competence matrix to create a certificate for employees and students of HFT Stuttgart alike that encourages them to learn about AI and achieve some degree of competency in this subject area [9, p. 11]. The matrix should lend itself to structuring this certificate.

In order to accomplish these goals, we proceed as follows. Section 2 will present the conceptual framework we adhere to Germany's 'Hochschulqualifikationsrahmen' (short: HQR, engl. Qualifications Framework for Higher Education) [11]. In Sects. 3 and 4 we critically analyze our chosen sources of technical AI competencies, and ethical competencies, respectively. This includes mapping the specific competencies listed to the conceptual framework of HQR. We compile the findings in Sect. 5, while also adding a second dimension of distinctions (running crosswise to the one established by HQR), and clarifying the relations between them. This yields a comprehensive matrix of AI competencies that takes ethics to be a distinct and guiding consideration for AI education. With the matrix in hand, we turn to the second of KNIGHT's goals in Sect. 6, the development of an AI certificate, based on the matrix.

Development of an AI Competence Matrix for AI Teaching at Universities 93

Table 1 Competence areas according to HQR	Professional competence	Methodical competence	Social competence	Self-competence

Sections 7 and 8 close the chapter with an account of the limitations of our work, as well as a short summary.

2 Conceptual Framework

Competencies are generally understood as abilities and skills that enable learners to solve problems [22]. HQR distinguishes four areas of competencies, all of which are labeled twice.[1] 'Professional competence' (or: 'knowing and understanding') refers to "subject specific knowledge and methods as well as their application, which are relevant to completing subject specific tasks." [14, p. 3 (own translation)] 'Methodical competence' (or: 'employment, application and generation of knowledge'), "in contrast to the just mentioned professional competence", denotes "subject independent knowledge, abilities, and skills, which enable to deal with new and complex tasks and problems self-sufficiently and flexibly." [14, p. 3f (own translation)] 'Social competence' (or: 'communication and cooperation') is understood as "knowledge, abilities, and skills in relation to communication, cooperation, and conflicts in intra- and intercultural contexts." [14, p. 4 (own translation)] Finally, 'self-competence' (or: 'scientific self-conception/professionalism') encompasses "the ability and willingness to develop oneself and deploy one's motivation and spur, as well as the development of specific stances and an individual personality." [14, p. 4 (own translation)].

Taken together, HQR's 'professional competence' and 'methodical competence' map onto the 'professional competence' as understood by the more general 'Deutscher Qualifikationsrahmen' (short: DQR, engl. German Qualification Framework) [11]. Likewise, HQR's 'social competence' and 'self-competence' fall under DQR's heading of 'personal competence'. In this way, DQR's twofold distinction of 'professional competence' and 'personal competence' is finetuned by HQR to suit the specific needs of higher education facilities. However, the compatibility of the two frameworks is very much intended (Table 1).

The conceptual framework of HQR is certainly not without its alternatives. However, as HFT Stuttgart is a German University of Applied Sciences, HQR clearly applies. With this fourfold framework of areas of competencies, we turn to the analysis of the literature on AI and ethical competencies.

---

[1] In order to accommodate completeness as well as clarity, we chose to put the secondary labels in brackets. Also, in what follows, we use a higher education teaching guideline in order to explicate the areas [14]. All translations are our own.

3 Technical AI Competencies

In order to infer technical AI competencies for the matrix, we chose three publications that focus heavily on work contexts. This is neither accidental nor unwarranted. First and primarily, HFT Stuttgart aims at providing highest quality education for preparing its students for a competitive labor market. After all, the primary drive for completing university education, for most students, is to be successful at one's job. Second, project KNIGHT in particular professes to take ethical issues in AI seriously, as is indicated by the matrix we develop in this chapter. In order to maximize the social impact of KNIGHT, it seems prudent to enable ethically trained and conscious students to carry their expertise and broader understanding into their future AI-related work environments. For this to be a reality, however, these students need to be able to acquire relevant jobs in the first place. Thus, this secondary concern suggests a focus on technical AI competencies in work contexts, as well.

With this being said, let us turn to the literature. We will analyze our sources in two steps. First, we recap the sources findings and showcase their results. In a second step, we re-edit their findings to fit HQR as laid out above.

3.1 AI Competencies for Industry 4.0

Referencing multiple studies pertaining to work-related AI competencies, [7] argue that "it becomes clear that the development of AI systems and interaction with AI technologies, as well as the tasks changed by them, necessitate technological as well as non-technological competencies across all levels and functions." [7, p. 3 (own translation)] Accordingly they distinguish 'digital/technological competencies', 'social competencies' and 'cognitive competencies' (Table 2).

Table 2 AI competencies according to [7] (own representation, own translation)

Digital/technological competencies	Social competencies	Cognitive competencies
Information and data competence	Communication skills	Process thinking
Digital communication and cooperation	Teamwork	Problem solving competence
Digital content creation	Interdisciplinary and intercultural cooperation	Systems knowledge and holistic thinking
Security	Willingness to learn	Critical reflection
Problem solving in digital environments	Error competence	Ability to learn
AI-awareness	Open-mindedness	Creativity
	Ethical awareness	Ability to innovate

Table 3 AI competencies according to [7]; sorted according to HQR

Professional competence	Methodical competence	Social competence	Self-competence
Information and data competence	Process thinking	Teamwork	Critical reflection
Digital communication and cooperation	Problem solving competence	Interdisciplinary and intercultural cooperation	Error competence
Digital content creation	Systems knowledge and holistic thinking	Ethical awareness	Open-mindedness
Security	Ability to learn		Willingness to learn
Problem solving in digital environments	Creativity		
AI-awareness	Ability to innovate		

As laid out above, 'professional competence' encompasses all "subject specific knowledge and methods". Hence, the whole category of 'digital/technological competencies' can be transferred to HQR's 'professional competence' without a problem. With the exception of 'critical reflection', all of the cognitive competencies seem to fall quite clearly under the heading of methodical competence, as per the definition stated above: they are a "subject independent" set of "abilities and skills" that enable discipline-independent problem-solving. 'Critical reflection', 'error competence', and 'open-mindedness' are key components of "the development of specific stances and an individual personality", and are therefore adjudicated to 'self-competence'. 'Willingness to learn' is implied by "the ability and willingness to develop oneself and deploy one's motivation and spur." For these reasons, we count these four competencies as self-competencies. The other three social competencies identified by [7] remain in their category. This evaluation yields the following re-categorization in line with HQR (Table 3).

3.2 AI Competencies for Employees

Reference [1] approaches the topic from the perspective of on-the-job training in companies. They stress the individual competence needs different employees with different roles have, and, accordingly, develop a model for competence management, rather than a one-size-fits-all list of competencies. This is clearly aligned with our goal, i.e. the development of a competence matrix, and, ultimately, an AI certificate, that enables students to develop their own individual set of AI-related skills.

According to the competence management model developed by [1], after a careful analysis of the requirements of a given job, specific AI competencies needed for the job are to be deduced. To this end, they list the following competencies as relevant to working *with* or *on* AI systems, sorted into three categories (Table 4).

Table 4 AI competencies according to [1] (own representation, own translation)

Technical and basic knowledge	Development of AI systems and handling of AI systems	Designing the context of AI
Expertise	HMI competencies	Self-competencies
Basic digital competencies	Machine learning basics	Social- and communications competence
AI-awareness	Skills in programming languages, platforms, frameworks, and libraries	(HR) management, leadership, change management
	Big Data, data science, and data analytics	Decision competence
	Process- and systems competence	Adaptability, transfer
	Problem-solving competence	
	Reflection competence	

Reference [1] provides descriptions for each of the competencies listed. These descriptions provide necessary elaborations, enabling the re-sorting to the categories of HQR.

'Expertise' is described as 'Employees have the necessary professional knowledge/ the necessary professional skills to complete the day-to-day tasks of their position. [...]' [1, p. 18] Since the specifics of this clearly differ greatly between different jobs, HFT Stuttgart cannot, realistically, pursue the goal of enabling all of their students to enter any given job with the specific necessary skills in this sense already in hand. For this reason, we choose to drop 'expertise' in this sense. 'Basic digital competencies', in contrast, can readily be sorted to 'professional competence'. 'AI awareness', as described by [1], refers to an awareness of, and background knowledge about AI systems used by their company. As with 'expertise' this is simply too specific to accommodate realistically. However, 'AI awareness' in the sense of being aware of AI systems used in daily life, the way it affects ourselves and our environments, i.e. in a broader, more general sense, is not only a realistic goal for HFT Stuttgart as a teachable skill, but also a valuable subject specific competence. Hence, 'AI awareness' in the sense just given can be sorted into the category of 'professional competence'. 'HMI competencies', 'machine learning basics', 'skills in programming languages, platforms, frameworks, and libraries', and 'big data, data science, and data analytics', clearly fall into this category as well. 'Process- and systems competence' is, again, focused on the company specific company in question an denotes knowledge of and ability to work with the company's processes and systems [1, p. 18]. In the context of our work, however, 'process- and systems competence' is to be understood as the general ability to grasp and make use of the processes and systems a given AI system is embedded in. Understood this way, this competence falls under the heading of 'methodical competence'. The same is true for 'problem-solving competence'. 'Social- and communications competence', and '(HR) management, leadership, change management' are clearly social competencies. In order to prevent

Development of an AI Competence Matrix for AI Teaching at Universities   97

Table 5 AI competencies according to [1]; sorted according to HQR

Professional competence	Methodical competence	Social competence	Self-competence
Basic digital competencies	Process- and systems competence	Communications competence	Decision competence
AI-awareness	Problem-solving competence	(HR) management, leadership, change management	Adaptability, transfer
HMI competencies			
Machine learning basics			
Skills in programming languages, platforms, frameworks, and libraries			
Big data, data science, and data analytics			

unnecessary redundancies, we shorten the former to 'communications competence'. 'Decision competence' and 'adaptability, transfer' seem to fall neatly under the above description of 'self-competencies'. Finally, 'self-competencies' as a competence in and of itself, per the description given by [1], adds nothing that is not already implied by [1]'s 'open-mindedness' and 'willingness to learn' and can therefore be discarded without loss. This yields the following re-classification of the AI competencies listed by [1] (Table 5).

3.3 High Demand for AI-Related Competencies

Reference [2]'s work on AI-related competencies was mainly chosen for its broad empirical base. The authors combined a qualitative literature analysis with a quantitative analysis of well over 9000 job advertisements from the online platform 'indeed'. Their intricate research methodology revealed the following AI-related competencies, categorized into 'technical competencies' and 'managerial competencies'. The latter, as per [2], is not only relevant to employees with leadership responsibilities. Instead, even "[d]evelopers and data-scientists must always have this business understanding that guides the rest of the AI project phases." Reference [2, p. 11] (Table 6).

Apart from 'problem solving', all of [2]'s technical competencies are, quite clearly, professional competencies in the sense of HQR. 'Problem solving' could be understood in a narrower sense, akin to [7]'s 'problem solving in digital environment' or in a broader sense, more in line with [1]'s 'problem-solving competence'. However, this matters little, since, in order to avoid double-counting, we will not list the same competence multiple times in the final matrix. Turning to the category of 'managerial competencies', 'people and social skills', and 'communication' are both social skills. 'Business acumen' seems to be a professional competence, although it is not, in and

Table 6 AI-related competencies according to [2] (own representation, own translation)

Technical competencies	Managerial competencies
Knowledge in AI-associated technologies and algorithms (ML, deep learning, neural networks)	Business management (client focus/orientation; decision making)
Knowledge in programming (Python, Scala, Java, web development)	Business acumen (business development, interdisciplinary knowledge)
Knowledge in AI frameworks and libraries (TensorFlow, PyTorch, Keras, Scikit-learn, Numpy, Caffe)	People and social skills (collaboration, building trust, leadership)
Knowledge in big data analytics frameworks (Spark, Hadoop)	Communication (oral and written communication)
STEM knowledge (mathematical and statistical knowledge, computer science)	
Development methodologies (Agile software development)	
Problem solving (initiative/engagement)	
Data management	

of itself, AI-related. 'Business management' is somewhat hard to place, since 'client focus/orientation' is a social competence, while the ability and willingness to make (hard) decisions lies predominantly in the domain of 'self-competence'. We decided to split 'business management' in its two subcomponents, classifying them in the way just laid out. This yields the following re-classification of AI-related competencies (with 'problem solving' being arbitrarily relegated to 'methodical competence' for now) (Table 7).

4 Ethical Competencies

Teaching ethics is, at present, woefully neglected in the teaching contents of STEM sciences at German Universities [8]. In order to remedy this, "Referat für Technik- und Wissenschaftsethik an den Hochschulen für Angewandte Wissenschaften des Landes Baden-Württemberg " (short: rtwe, engl. Department for technic- and business ethics at the Universities of Applied Sciences of Baden-Wuerttemberg) offers a broad range of extracurricular learning opportunities for all students of the 24 Universities of Applied Sciences in the state [18]. However, the courses are not part of a planned-out approach to teaching ethical competencies. Project KNIGHT aims to fill this gap for HFT Stuttgart's students.

As was indicated above, in order to determine technical AI competencies, we opted for publications that focus on the job market or work contexts. Since ethics, as an academic discipline, has a rather poor standing on the current job market, we have to look elsewhere for determining salient ethical competencies. A very comprehensive

Development of an AI Competence Matrix for AI Teaching at Universities

Table 7 AI-related competencies according to [2]; sorted according to HQR

Professional competence	Methodical competence	Social competence	Self-competence
Knowledge in AI-associated technologies and algorithms (ML, deep learning, neural networks)	Problem solving (initiative/ engagement)	Client focus/ orientation	Decision making
Knowledge in programming (Python, Scala, Java, web development)		People and social skills (collaboration, building trust, leadership)	
Knowledge in AI frameworks and libraries (TensorFlow, PyTorch, Keras, Scikit-learn, Numpy, Caffe)		Communication (oral and written communication)	
Knowledge in big data analytics frameworks (Spark, Hadoop)			
STEM knowledge (mathematical and statistical knowledge, computer science)			
Development methodologies (Agile software development)			
Data management			
Business acumen (business development, interdisciplinary knowledge)			

source are the resolutions of the Standing Conference of the Ministers of Education and Cultural Affairs of the Lands of the Federal Republic of Germany (short: CME) on the examination requirements for the high school graduation exams in ethics [12]. Although a guidance for high school graduation exams, their account of ethical competencies is general in nature and can be applied to the present context. Reference [16] presents a theoretical competence model for the evaluation of bioethical problems. Since the competences listed there do not have any specificity to biology they, too, can be transferred to the present context without issue. The added value of this publication lies in its explicit focus on an applied context, whilst also being theoretically well-founded. The work of [17] in turn, analyzes German curricula of the subject group philosophy, ethics, practical philosophy, values and norms, and LER (engl. life–ethics–religion) with regard to philosophical and ethical competences to be taught. This empirically rich review of the actual curricula comes full circle to the resolutions of the CME.

For our analysis of these sources, we use the same two-step procedure we employed for technical AI competencies. First, we recap the sources findings and showcase their results. In a second step, we re-edit their findings to fit HQR as laid out above.

4.1 Resolutions of the Standing Conference of the Ministers of Education and Cultural Affairs

In its subject preamble to the examination requirements for the high school graduation exams for the school subject of philosophy, the CME describes philosophizing as a reflective competence consisting of the areas of competence of 'perception and interpretation', 'argumentation and judgment', and 'presentation' [13, p. 5f]. With a more concrete focus on ethics (as a subarea of philosophy), these areas of competence can be found in the examination requirements of the high school graduation exams for the school subject of ethics. There, the subject-specific requirements and contents are divided into the areas of competence of 'personal and social competencies' and 'professional and methodical competencies' [12, p. 7f].

Although, the division of these competencies into the areas of personal and social competencies and professional and methodical competencies obviously lends itself

Table 8 Ethical competencies according to [12] (own representation, own translation)

Self- and social competencies	Professional and methodical competencies
Perceive the diversity of social reality in a differentiated way and to reflect on it within one's own horizon of experience	Describe ethically and morally relevant phenomena and problems of the individual, social and natural world
Recognize moral-ethical problems as such	Present complex ethical-moral issues and contexts independently and appropriately
Critically question circumstances of personal and social life	Place the results of relevant individual sciences into a context
Shape the personal sphere of life and the pursuit of happiness consciously, independently and responsibly	Contextualize knowledge of important ideological and religious traditions
Understand that criticism of tradition and preservation of tradition is necessary for people to live together	Unlock philosophical and scientific texts, work out the respective arguments, examine them critically and apply them to new contexts
Reflect on one's own inclinations, desires, points of view and values against the background of the needs and interests of others	Check arguments to see whether they are logically coherent and on what they base their claim to validity
Respectfully deal with foreign patterns of thinking and living	Evaluate ethical problems in the context of major moral philosophies
Develop your own justifiable point of view and the willingness to defend it argumentatively	Justify own positions
Accept the fundamental limitation of one's own cognitive ability	Follow the public discourse on ethical moral problems of the present and participate in it in the given context
Seek fair solutions in conflicts	
Being able to endure aporias and contradictions	
Commit to indispensable values	
Assume personal responsibility for shaping society	

to a reassignment to the categories of HQR, it is not straightforward. For one, as was laid out in Sect. 2, 'professional competence' includes all subject specific methods. However, all of the methods featured in Table 8 could be argued to be subject specific to philosophical ethics. We opted to adjudicate all methods listed there that have clear applicational value *outside* of philosophical contexts to the category of 'methodical competence'. Similarly, almost all of [12]'s 'self- and social competencies' could be argued to fall under HQR's heading of 'self-competence'. We opted to count all of these with a predominantly social connotation to the category of 'social competence'. This yields the following re-classification (Table 9).

4.2 Competence Model for the Evaluation of Bioethical Problems

Reference [16] presents a theoretical competence model for evaluating bioethical problems, which postulates eight components of evaluation competence on the basis of existing models for moral judgment formation, appropriate basic philosophical skills, and a comparison with general requirements from the didactics of ethics. Reference [16] identifies the following eight competences:

- Perceiving and becoming aware of moral-ethical relevance
- Perceiving and becoming aware of the sources of one's own attitude
- Reflecting consequences
- Evaluating
- Basic ethical knowledge
- Judging/concluding
- Argumentation
- Changing of one's perspective [16, p. 127 (own representation, own translation)].

'Perceiving and becoming aware of moral-ethical relevance' and 'Perceiving and becoming aware of the sources of one's own attitude' are best understood as belonging to HQR's category of 'self-competence'. 'Basic ethical knowledge' and 'argumentation' are correctly adjudicated to the area of 'professional competence'. 'Evaluating' and 'judging/concluding' are, on first sight, hard to distinguish. However, [16] offers that "when evaluating, a state of affairs is analyzed in virtue of its contained facts as well as in light of the reasons counting in favor or against of an action." [16, p. 127 (own translation)] Conversely, 'judging/concluding' means "the courage and the ability to form a considered and justified judgement." Reference [16, p. 127 (own translation)] 'Evaluating', then, seems to be appropriately placed in either 'professional competence' or 'methodical competence', depending on what kind of facts and reasons are viewed as salient, whereas 'judging/concluding' seems to belong to both, 'professional competence' ("ability to form a considered and justified judgement") and 'self-competence' ("courage to form…"). We opted to adjudicate 'evaluating' to 'professional competence' (with an emphasis on specifically *ethically salient*

Table 9 Ethical competencies according to [12]; sorted according to HQR

Professional competence	Methodical competence	Social competence	Self-competence
Contextualize knowledge of important ideological and religious traditions	Describe ethically and morally relevant phenomena and problems of the individual, social and natural world	Perceive the diversity of social reality in a differentiated way and to reflect on it within one's own horizon of experience	Recognize moral-ethical problems as such
Check arguments to see whether they are logically coherent and on what they base their claim to validity	Present complex ethical-moral issues and contexts independently and appropriately	Reflect on one's own inclinations, desires, points of view and values against the back-ground of the needs and interests of others	Critically question circumstances of personal and social life
Evaluate ethical problems in the context of major moral philosophies	Place the results of relevant individual sciences into a context	Respectfully deal with foreign patterns of thinking and living	Shape the personal sphere of life and the pursuit of happiness consciously, independently and responsibly
Justify own positions	Unlock philosophical and scientific texts, work out the respective arguments, examine them critically and apply them to new contexts	Seek fair solutions in conflicts	Understand that criticism of tradition and preservation of tradition is necessary for people to live together
Follow the public discourse on ethical moral problems of the present and participate in it in the given context			Develop your own justifiable point of view and the willingness to defend it argumentatively
			Accept the fundamental limitation of one's own cognitive ability
			Being able to endure aporias and contradictions
			Commit to indispensable values
			Assume personal responsibility for shaping society

Development of an AI Competence Matrix for AI Teaching at Universities

Table 10 Ethical competencies according to [16], sorted according to HQR

Professional competence	Methodical competence	Social competence	Self-competence
Basic ethical knowledge	Changing of one's perspective		Perceiving and becoming aware of moral-ethical relevance
Argumentation			Perceiving and becoming aware of the sources of one's own attitude
Evaluation			Judging/concluding
Reflecting consequences			

facts and reasons in mind) and 'judging/concluding' to 'self-competence' (with an emphasis on the personal fortitude required in mind). 'Reflecting consequences', per [16] denotes "the ability to anticipate and gauge the hypothetical consequences of a given judgement." [16, p. 127 (own translation)] Due to its clear links to 'argumentation' and 'evaluation', we opted to adjudicate it to 'professional competence'. Reference [16]'s elaborations on 'changing of one's perspective' suggest some sort of overarching methodical competence, utilizing (and, in turn, enabling) multiple other competences. This critical evaluation and re-editing of [16] yields the following classification according to HQR (Table 10).

4.3 Ethics Competencies in Philosophy and Ethics Classes

In an in-depth investigation, [17] analyzes German curricula of the subject group philosophy, ethics, practical philosophy, values and norms and LER. The author defines areas of competence with respective partial competences that are to be taught in the aforementioned subjects. The result is the following competence scheme [17, p. 75f] (Table 11).

'Perceptual competence' is assigned to the area of 'self-competence' and 'perspective adoption' to the area of 'methodical competence'. 'Empathy' and 'intercultural competence' clearly fall under the heading of 'social competence'. 'Interdisciplinary competence' requires subject specific knowledge of different disciplines and the ability to connect it, as well as the willingness to deal with different research paradigms school's of thought and customs. Thus, we decided to split it into two sub-competencies, one with a focus on expertise, the other with a focus on the cooperation-enabling aspects. 'Text competence' and 'language (-analytical) competence' can be adjudicated to 'methodical competence', while 'reflection competence' falls under the heading of 'self-competence'. The three competencies in the category of 'arguing and judging' overlap to some extent, however, their respective emphases, contained in the elaborations given by [17] differ. 'Argumentation and judgement

Table 11 Ethical competencies according to [17] (own representation, own translation)

Perceiving and understanding	Analyzing and reflecting	Arguing and judging	Interacting and communicating	Orientation and action
Perceptual competence	Text competence	Argumentation and judgement skills	Discourse skills	Orientation competence
Perspective adoption	Language (-analytical) competence	Moral judgement	Conflict-resolution competence	Action competence
Empathy	Interdisciplinary competence	Ethical judgement competence	Presentation competence	
Intercultural competence	Reflection competence			

skills' is understood as "critically assessing own and others' positions, arguing coherently and reasoned, and judging differentiated," [17, p. 76 (own translation)] thusly warranting adjudication to 'self-competence'. 'Moral judgement', on the other hand, requires "knowing morally binding basic positions, [and] understanding them in their historical and cultural contexts […]" [17, p. 75 (own translation)], and is therefore redressed to 'professional competence'. Finally, 'ethical judgement competence' requires "identifying situations as ethically problematic […]" [17, p. 75 (own translation)], and is therefore, just like 'perceptual competence', sorted into the category of 'self-competence'. 'Discourse skills' and 'conflict-resolution competence', clearly, are social competencies, whereas 'presentation competence', for its obvious value across all disciplines, is counted as a methodical competence. Finally, 'orientation competence' and 'action competence' are clear-cut examples of self-competence (Table 12).

Table 12 Ethical competencies according to [17]; sorted according to HQR

Professional competence	Methodical competence	Social competence	Self-competence
Moral judgement	Perspective adoption	Empathy	Perceptual competence
Interdisciplinary competence (expertise)	Text competence	Intercultural competence	Argumentation and judgement skills
	Language (-analytical) competence	Discourse skills	Ethical judgement competence
	Presentation competence	Conflict resolution skills	Orientation competence
		Interdisciplinary competence (social)	Action competence
			Reflection competence

5 Compiling the Matrix

In order to compile the matrix, we follow a two-step process. In a first step, the competencies identified by our sources and sorted according to HQR are merged wherever prudent, in order to avoid doubling and redundancies. This is done by listing competencies that are featured more than once in our sources only once and, more importantly, combining similar and overlapping competencies. In a second step we subdivide the columns into different AI-related tasks.

5.1 Merging the Competencies

The AI competencies of 'information and data competence', 'big data, data science, and data analytics', 'knowledge in big data analytics frameworks', and 'data management' are merged into the competencies of 'data competence' and 'data analytics'. 'STEM-knowledge' is shortened to 'mathematical basics (statistics)', since STEM disciplines as such are taught at HFT Stuttgart in already existing structures. 'Digital communication and cooperation' is summed into the more general 'basic digital competencies'. The competencies 'skills in programming languages, platforms, frameworks, and libraries', 'knowledge in AI-associated technologies and algorithms', 'knowledge in programming', and 'knowledge in AI frameworks and libraries' are merged into the basic and general 'knowledge of AI functionalities' and the more advanced 'building machine learning systems', 'programming skills', and 'AI frameworks and libraries'. 'Teamwork', '(HR) management, leadership, change management', and 'people and social skills (collaboration, building trust, leadership)' are merged into 'teamwork' and 'leadership'. 'Ethical awareness' is dropped for reasons of obvious redundancy.

The ethical competencies 'contextualize knowledge of important ideological and religious traditions', 'evaluate ethical problems in the context of major moral philosophies', and 'basic ethical knowledge' are merged into the new 'ethical theories/paradigms'. 'Check arguments to see whether they are logically coherent and on what they base their claim to validity', 'justify own positions', 'follow the public discourse on ethical-moral problems of the present and participate in it in the given context', 'argumentation', 'evaluation', 'reflecting consequences' and 'moral judgement' are merged into the broad and encompassing 'argumentation- and judgement competence'. 'Describe ethically and morally relevant phenomena and problems of the individual, social and natural world' and 'present complex ethical-moral issues and contexts independently and appropriately' are merged with 'presentation competence'. 'Unlock philosophical and scientific texts, work out the respective arguments, examine them critically and apply them to new contexts', 'text competence', and 'language (-analytical competence)' are combined into 'hermeneutical competence'. 'Changing of one's perspective' is integrated in 'perspective adoption'. The various ethical competencies under the heading 'social competence' overlap to a significant extent. We slimmed them down to the competencies of 'empathy', 'intercultural competence', 'interdisciplinary competence', and 'discourse skills'. 'Recognize moral-ethical problems as such' and 'perceiving and becoming aware of moral-ethical relevance', are merged with 'perceptual competence'. 'Critically question circumstance of personal and social life' and 'shape the personal sphere of life and the pursuit of happiness consciously, independently and responsibly' are integrated in 'reflection competence'. 'Understand that criticism of tradition and preservation of tradition is necessary for people to live together', 'develop your own justifiable point of view and the willingness to defend it argumentatively', 'accept the fundamental limitation of one's own cognitive ability', and 'being able to endure aporias and contradictions' are merged into the new 'conflict resolution competence'. The rest of the ethical competencies in the column of 'self-competence' are grouped under the heading 'moral judgement ability'. In our understanding this includes the ability to follow through and act on such judgements, and therefore already entails competencies like 'action competence'. Finally, the AI-related competence of 'interdisciplinary and intercultural cooperation' is subsumed into its ethical equivalents. We decided to broaden 'interdisciplinary competence' to 'inter-/transdisciplinary competence' in order to underline the need for thinking across boundaries when tackling complex and dynamic issues like AI.

5.2 Role of Ethics and Subdivision into Levels

The final step in compiling the matrix is to further sort the competencies into multiple levels ranges, in order to enhance clarity and create flexibility in the selection of learning units, and to reflect the different skill levels, students might have.

We opted to keep the ethical competencies in its own separate category without dividing them into different levels. In this sense, ethics makes up its own 'level'. The technical competencies were subdivided into four levels. For this, we took inspiration from [1], which distinguish between 'technical and basic knowledge', 'development of AI systems and handling of AI systems' and 'designing the context of AI'. On a first, basic level, only the fundamentals of AI-related competencies are taught. The other three levels are constructed along the lines of different AI-related tasks, HFT Stuttgart's students and staff might encounter in their work lives. These tasks differ in their relative engagement with the technologies relevant for AI systems and require different levels of competence regarding AI technologies. We decided to split 'development of AI systems and handling of AI systems' into 'development of AI systems' and 'handling of AI systems'. The handling of AI systems, while requiring some technical competence, has nonetheless a relatively low skill threshold. 'Designing the context of AI' requires competencies at a slightly higher level, for in order to shape the conditions AI is employed competently, knowledge about the handling of AI systems is a necessary prerequisite but not a sufficient one.

Finally, 'development of AI systems' requires an array of specific and highly technical competencies and specialized knowledge. Some competencies are featured on more than one level, indicating level-specific sub-competencies therein.

Ethical considerations should not be a mere afterthought of AI-related activities but a filter or a lens through which these activities are approached. Therefore, ethical competencies stand neither next to nor below technical competencies, but rather occupy a structuring position 'above' them (visually indicated in the matrix, cf. below). This is not meant in a hierarchical sense, for there is no AI without technical competence. Rather, this is understood as to underline the overarching and comprehensive nature, ethical reflection should have in any AI-related project.

All of this yield the following AI competence matrix.

6 Designing the Certificate

As was stated at the outset, our goal is to use this matrix to design an *AI-certificate*, students and staff of HFT-Stuttgart can acquire to document their AI-related expertise in an intelligible and convincing way. The certificate we developed to this end has the following structure, derived directly from the matrix (Fig. 1).

In order to acquire the certificate, a student or staff member has to earn 90 'AI-credits' (short: AIC). One AIC is roughly equivalent to a time investment of two hours. (This follows the logic of another certificate available statewide at technical universities, cf. [19].) AICs can be earned by completing AI-related classes, courses, and other educational events available at HFT-Stuttgart, online, or at other academic institutions. In order to claim the related AICs, participants can download a simple form from HFT Stuttgart's internal wiki. Apart from basic information about the course (dates, time), and the name of the student or staff member, the form features

Level		Professional competence	Methodical competence	Social competence	Self-competence
Ethical shaping and reflection		• Ethical theories/paradigms • Inter-/transdisciplinary competence (expertise) • Argumentation- and judgement competence	• Presentation competence • Perspective adoption • Hermeneutical competence	• Discourse skills • Intercultural competence • Inter-/transdisciplinary competence (social) • Empathy	• Perceptual competence • Moral judgement ability • Conflict resolution competence • Reflection competence
	Basic competence	• Basic digital competencies • Mathematical basics (statistics) • AI awareness			
	Handling of AI systems	• Data competence • Digital content creation • IEMI competencies • Security • Machine learning basics	• Systems knowledge and holistic thinking • Ability to learn	• Teamwork	• Critical reflection • Adaptability, Transfer • Willingness to learn • Open-mindedness
	Designing the context of AI	• Knowledge of AI functionalities • Development methodologies • Business acumen	• Problem solving competence • Creativity • Ability to innovate	• Leadership • Communication	• Decision competence • Critical reflection
	Development of AI systems	• Problem solving in digital environments • Building machine learning systems • Programming skills • AI frameworks as libraries • Data analytics	• Process- and systems competence • Problem-solving competence • Creativity • Ability to innovate	• Teamwork • Communication	• Critical reflection • Error competence

Fig. 1 AI competence matrix, based on [1, 2, 7, 16, 17]

the competence matrix and a table where the course instructor can mark which competencies contained in the matrix have been taught in the course. In the same way, regular university classes can be credited for the certificate.

A second requirement is the successful completion of three basic courses, financed by project KNIGHT, on the topics of AI ethics, AI technology, and data law, respectively. With 15 AICs each, these courses earn more AICs than the time investment needed for them indicates. This deliberate choice was made in order to encourage students and staff members alike to invest the additional time needed in order to complete the certificate.

Having fulfilled these requirements, students and staff members can claim a diploma, stating the courses they visited and the competencies they acquired in doing so. As an additional way to individualize the certificate, if a student or staff member acquires 40 AICs or more by acquiring competencies belonging to one of the matrix's levels of 'ethical shaping and reflection', 'handling of AI systems', 'designing the context of AI', and 'development of AI systems', they can choose to have the respective level highlighted on the diploma as a special focus of their studies.

7 Limitations of the Work

The competency framework chosen, as well as most of the publications considered, especially those on ethical competencies, offer a perspective based on the German educational system. This is due to HFT Stuttgart being a German University of Applied Sciences. We do not presume any intrinsic superiority of this outlook compared to any other. Our work is intended to encourage constructive discourse and foster didactical progress.

8 Summary

Our research goal was to develop an AI competence matrix for structuring AI-related teaching activities at university level that covers the competencies relevant for different AI-related tasks while also accommodating the important role, ethical considerations play for AI by featuring salient ethical competencies. Furthermore, this certificate was used for structuring a corresponding AI certificate.

In order to accomplish these goals, we laid out the basic structure of the matrix in line with the German HQR model. We then analyzed several sources on AI-related technical competencies, and several sources on ethical competencies. These competencies were sorted according to HQR and merged, in order to avoid redundancies. As a final step for creating the matrix, we integrated subdivisions, indicating different AI-related tasks and critically assessed the special role of ethical considerations. Finally, we used the resulting matrix to develop a convincing AI-certificate, to enable students and staff members of HFT Stuttgart to advance and show their AI-related competencies.

The project on which this report is based was funded by the Federal Ministry of Education and Research under the funding code 16DHBKI072. The responsibility for the content of this publication lies with the authors.

References

1. André E, Bauer W et al (eds) (2021) Kompetenzentwicklung für KI - Veränderungen. Bedarfe und handlungsoptionen. Plattform lernende systeme, München
2. Anton E, Behne A, Teuteburg F (2020) The humans behind artificial intelligence – an operationalisation of AI competencies. In: Proceedings of the 28th European conference on information systems (ECIS), an online AIS conference, 15–17 June 2020
3. Bovenschulte M, Stubbe J (2019) Einleitung: "Intelligenz ist nicht das Privileg von Auserwählten." In: Wittpahl V (ed) Künstliche Intelligenz. Springer Vieweg, Berlin Heidelberg, pp 215–220
4. Ethikrat D (2023) Mensch und Maschine – Herausforderungen durch Künstliche Intelligenz. Geschäftsstelle des Deutschen Ethikrats, Berlin
5. Die Bundesregierung (eds) (2018) Strategie Künstliche Intelligenz der Bundesregierung. https://www.bundesregierung.de/resource/blob/997532/1550276/3f7d3c41c6e0569574127 3e78b8039f2/2018-11-15-ki-strategie-data.pdf. Accessed 15 Mar 2024
6. Fiedler M et al (2020) Kompetenzentwicklung für und in der digitalen Arbeitswelt. Münchener Kreis, München
7. Franken S, Mauritz N, Prädikow L (2022) Kompetenzen für KI-Anwendungen: theoretisches Modell und partizipative Erfassung und Vermittlung in Unternehmen. Paper presented at 68th Frühjahrskongress der Gesellschaft für Arbeitswissenschaft on Technologie und Bildung in hybriden Arbeitswelten, Magdeburg, 2–4 March 2022
8. Fregin M-C et al (2016) Führungsverantwortung in der Hochschule: zur Situation in den MINT-Fächern und Wirtschaftswissenschaften an Universitäten in Baden-Württemberg, Rheinland-Pfalz und Thüringen. Materialien zur Ethik in den Wissenschaften 12. Internationales Zentrum für Ethik in den Wissenschaften (IZEW), Tübingen
9. HFT Stuttgart (n.d.). Förderantrag für KNIGHT. https://confluence.hft-stuttgart.de/display/ KNIGHT/Antrag?preview=/850570355/850570358/Langantrag_fuer_BMBF_ohne_Anh aenge.pdf. Accessed 15 Mar 2024
10. Huchler N et al (eds) (2020) Kriterien für die menschengerechte Gestaltung der Mensch-Maschine-Interaktion bei Lernenden Systemen. Plattform Lernende Systeme, München
11. HRK, KMK (2017) Qualifikationsrahmen für Deutsche Hochschulabschlüsse. https://www.hrk.de/fileadmin/redaktion/hrk/02-Dokumente/02-03-Studium/02-03-02-Qualifikatio nsrahmen/2017_Qualifikationsrahmen_HQR.pdf. Accessed 15 Mar 2024
12. KMK (2006) Einheitliche Prüfungsanforderungen in der Abiturprüfung Ethik (Beschluss der Kultusminister-konferenz vom 16 Nov 2006). https://www.kmk.org/fileadmin/veroeffentlichu ngen_beschluesse/1989/1989_12_01-EPA-Ethik.pdf. Accessed 15 Mar 2024
13. KMK (2006) Einheitliche Prüfungsanforderungen in der Abiturprüfung Philosophie (Beschluss der Kultusministerkonferenz vom 16 Nov 2006). https://www.kmk.org/fileadmin/veroeffentli chungen_beschluesse/1989/1989_12_01-EPA-Philosophie.pdf. Accessed 15 Mar 2024
14. Kopf M, Leipold J, Seidl T (2010) Kompetenzen in Lehrveranstaltungen und Prüfungen - Handreichung für Lehrende. Mainzer Beiträge zur Hochschulentwicklung 16. Zentrum für Qualitätssicherung und -entwicklung, Mainz
15. Rampelt F, Bernd M, Mah D-K (2022) Wissen, Kompetenzen und Qualifikationen zu Künstlicher Intelligenz. Eine Systematisierung von digitalen Formaten am Beispiel des KI-Campus und seiner Partner. KI-Campus, Berlin

16. Reitschert K, Hößle C (2007) Wie Schüler ethisch bewerten – Eine qualitative Untersuchung zur Strukturierung und Ausdifferenzierung von Bewertungskompetenz in bioethischen Sachverhalten bei Schülern der Sek. I. Zeitschrift für Didaktik der Naturwissenschaften 13(1):125–143
17. Rösch A (2009) Kompetenzorientierung im Philosophie- und Ethikunterricht: Entwicklung eines Kompetenzmodells für die Fächergruppe Philosophie, Praktische Philosophie, Ethik, Werte und Normen. LIT Verlag, Münster, LER
18. RTWE (2024) https://www.rtwe.de/home.html (2024). Accessed 15 March 2024
19. RTWE (2024) Voraussetzungen für das Ethikum. https://www.rtwe.de/43.html. Accessed 15 March 2024
20. Tsamados A, Aggarwal1 N, Cowls J et al (2022) The ethics of algorithms: key problems and solutions. AI & Soc 37:215–230
21. Wangler L, Botthof A (2019) E-Governance: Digitalisierung und KI in der öffentlichen Verwaltung. In: Wittpahl, V (ed) Künstliche Intelligenz. Springer Vieweg, Berlin, Heidelberg, pp 122–141
22. Weinert F (2001) Leistungsmessungen in Schulen, 2nd edn. Beltz, Weinheim

Implementing Ethical Considerations into Ai-Supported Student Counselling—a Case Study

Maximilian Weber and Niko Jochim

Abstract Awareness for the need to address ethical considerations in software development is ever-increasing. However, there is little clarity on how and when ethical analysis should be introduced into the development process. Using a case study approach, we argue that ethical analysis should be integrated at a very early stage of software development and requires interdisciplinary cooperation between technical and ethical specialists for it to be fruitful and cost efficient. In order to show this, we analyze the development process of an AI tool under development at the KNIGHT research project of Stuttgart University of Applied Sciences. Ethical challenges faced by the original plans of the tool, the interdisciplinary process of solving these problems, and the technical changes necessary to overcome them are detailed. The technical changes are fundamental and could not have been easily implemented at a later stage of development, highlighting the need for early-on integration of ethical analysis.

Keywords Artificial Intelligence (AI) · Ethics counselling · Software development · Interdisciplinary collaboration · Student counselling · Higher education · Case study

1 Introduction

There is widespread agreement that ethical considerations are of major significance in AI application development [13]. However, ethical analysis often occurs (if at all) late in the process, or even post-development, despite a well-established link between the costs of solving problems in software and the maturity of the product [4]. Presumably, this has to do with the fact that ethical analysis is, to many people,

M. Weber (✉) · N. Jochim
Hochschule Für Technik Stuttgart, Schellingstraße 24, 70174 Stuttgart, Germany
e-mail: maximilian.weber@hft-stuttgart.de

N. Jochim
e-mail: niko.jochim@hft-stuttgart.de

hardly tangible, which makes it very hard to appreciate the value ethical analysis adds to a software development project. This issue is not specific to any single area of AI development and, therefore, extends to universities' AI development processes.

In this chapter, we aim to demonstrate the significance and value of incorporating dedicated ethical analysis in conjunction with interdisciplinary problem-solving into the ongoing process of developing an AI tool, using the KNIGHT research project as an example. KNIGHT is a research project of HFT Stuttgart ("Hochschule für Technik Stuttgart", engl. Stuttgart University of Applied Sciences) that explores AI in higher education. Part of this research consist in developing and field-testing certain AI applications for supporting teaching and administration tasks. In order to ensure the ethical soundness of this work, a set of values is integrated into all parts of the project [6]. In KNIGHT, this is done by including continuous interdisciplinary ethical reflection of the applications into the respective development-cycles from the very start. This mode of incorporating the ethical analysis early on has two main benefits. (1) Allowing for potential ethical problems to be discovered relatively early on increases the range of technical options not yet foreclosed by previous development decisions. This also serves to keep costs down, since solving problems in software gets ever more expensive as the product gets more mature [4]. (2) The approach's inter-and transdisciplinary nature (involving philosophy/ethics and computer sciences) makes this process very effective, both for identifying potential ethical problems as well as for solving them.

Employing a case study approach, we showcase the impact of incorporating continuous interdisciplinary ethical reflection into the ongoing software development process of an AI application, currently under development in project KNIGHT, at a very early stage. Section 2 outlines the modus operandi of ethical reflection as used in KNIGHT. This process is founded in insights from business ethics consulting [14], and the long tradition of principlist ethics consulting in medical contexts [2, 3]. It involves interdisciplinary workshops, in which ethicists collaborate with software engineers, in order to shape the applications developed in KNIGHT to embody a previously determined set of values. Section 3 of this chapter will detail the originally intended scope and function of one of these applications, KNIGHT's *Study Dean Cockpit* (SDC). Roughly speaking, SDC is a software tool intended to support the day to day work of study deans at HFT Stuttgart, by merging different sets of information already available to provide a convenient overview of relevant student data. Additionally, it includes an AI-based analysis of students' data, that is capable of identifying struggling students early on, in order to proactively reach out and offer student counselling services. The subsequent Sect. 4 offers an ethical analysis of the original plans, highlighting and elaborating on a potential problem of the intended design, stemming from role-overlap in study deans, that might unfairly bias teaching personnel to the disadvantage of students. Section 5 highlights our process of solution-finding, including some solutions that were discussed but, ultimately, dismissed. Section 6 presents a technical analysis of the adopted solution, reflecting the necessary changes in software architecture that were implemented, in order to gauge the nature and scope of these changes. Section 7 closes this chapter with a final discussion of the impact of the ethical analysis thus far, provides an outlook regarding the future of the development process, and highlights important insights that can be drawn from the case. With this case study, we hope to contribute towards visualizing the value ethical analysis can add to AI development processes as well as the necessity of inter- and transdisciplinary work in detecting and solving ethical problems in software design.

2 Ethical Analysis in KNIGHT

Project KNIGHT's ethical assessment efforts are bundled in a dedicated subproject that includes the formulation of a base of values that acts as an ethical filter for the entire project and to integrate these values in every part of KNIGHT. By analyzing already existing behavioral and other guidelines of HFT Stuttgart as well as established principles of AI ethics, we developed a base of values. With "high-quality education" being the guiding value, the framework contains, among others, the following values pertaining to the design and proper use of AI applications [16].

- Human control/oversight
- High-quality information
- Consent
- Reliability
- Security
- Student well-being
- Equality/non-discrimination
- Transparency
- Human goals
- Privacy
- Appropriate distribution of costs and benefits
- Protection of vulnerable persons
- Defeasibility
- Explicability.

The process of integrating the base of values is primarily done in recurring workshops, where technical experts from the respective subproject in question meet with KNIGHT's ethical experts to reflect on the current state of the subproject in order to determine compliance or non-compliance with the values. This process takes inspiration from insights from business ethics, as well as biomedical ethics. Starting with the latter, the idea of basing one's ethical decisions in a set of commonsense principles that are neither as abstract, hard to apply, and, most of all, controversial as genuine moral theories, nor as open to interpretation and explanatory unsatisfactory as casuistic reasoning lies at the very heart of Beauchamp and Childress' seminal *Principles of Biomedical Ethics* [2]. Their work is a cornerstone of ethical decision-making in clinical contexts that has since inspired countless codes of ethics in other disciplines, notably AI ethics [1]. From business ethics, then, comes careful reflection of the ethical consultant's role. Robert van Es distinguishes three roles an ethical consultant can occupy in regards to her wor [14]. In all three roles, the consultant starts by "noticing the moral problem" at hand and "drawing up an inventory of the arguments which can be given pro and contra the moral issue without making any judgement whatsoever" [14, p. 230]. Concerning the next steps, von Es' ethical consultant is to occupy one of the following three roles:

- The ethical engineer is concerned with the "weighing of arguments" [14, p. 231], and formulates a solution to the moral problem at hand.
- The ethical playwright points to the ethically salient features of the moral problem, and fosters awareness for the moral implications of the different options without arguing for a specific choice.
- The ethical interpreter does not actually solve any problems. Instead, the focus of his role lies in broadening the client's relevant background knowledge, leading "to an enlargement of the moral problem to its real size and scale." [14, p. 231].

For conceptual reasons we will not delve into at this time, we hold this distinction to be somewhat idealistic, yet instructive. In practice, we suspect (as well as attest) that ethical consultants will usually occupy a mix of the three roles, and we hold this to be a good thing. For a start, ethical consultants, surely, should always strive to appreciate the "real size and scale" of the moral problems they are concerned with, and hence, should be ethical interpreters. However, if the ethical consultant is to be more than a mere nuisance, she needs to tap into her inner playwright and at least articulate the arguments counselling in favor of and against the respective options at hand. This necessarily involves making moral judgements, for part of this step is to sketch the morally relevant implications of said options, and recognizing a given implication as morally relevant either involves or *just is* making a moral judgement. From this it follows that the ethical consultant should make moral judgements (in some form). This alone does not mean, that the ethical consultant is conceptually forced to occupy the role of the ethical engineer as such, for judging something to be morally relevant is still not the same as forming a judgement about what to do. However, it blurs the line between the two. We argue even further and claim that the ethical consultant should never be merely an ethical playwright. First of all, the role (falsely) suggests the option of neutrality where there is none to be had. After all, as noted above, even in this role, moral judgements are aplenty. Moreover, even if there was a way to identify and describe arguments for and against a given option without personal judgement, this would not be a particularly conscientious approach to ethical consulting, especially in business ethics. A company faced with arguments for both, option A and option B, without any information as to which arguments outweigh the other really has no incentive to do anything but choose the most cost-effective option and wielding the arguments in its favor. This way of consultancy is no consultancy at all, but boils down to *ethics-washing* [10].

As for KNIGHT, we tried to accommodate these insights as follows. In a first phase, KNIGHT's ethics experts reviewed the project outline's description of the scopes and goals of its technical subprojects. In this the ethics experts assumed the role of ethical interpreter, in order to try to detect the potential ethical pitfalls of the subprojects as described in the outline. In a second phase, the ethics experts, acting as a mix of ethical engineer and playwright, constructed KNIGHT's framework of values, as laid out above [16]. The third, longest, and as of yet (early 2024) ongoing phase of consulting consists in semi-regular workshops in which the ethics experts meet with each of the subprojects' technical experts in turn, in order to discuss the findings from phase one, keeping each other updated in regard to salient new development, detecting ethical problems early on and, most importantly, develop and discuss potential solutions to identified problems. This phase will be a part of almost all of project KNIGHT's runtime, and will demand a dynamic mix of the roles of playwright and engineer. A fourth and final phase will consist in helping to write up manuals for the subprojects' work products in order to ensure that all ethically salient features are properly addressed. This will, again, demand a mix of the roles of playwright and engineer: ethical engineering contributes certain guidelines for the proper use of the applications, while playwrighting highlights the pros and cons of options available when configuring the applications to the particular needs of the user.

3 Study Dean Cockpit I–Original Plans

Project KNIGHT's research proposal distinguishes eight subprojects [6], with four of them concerning technical systems to be introduced into specific teaching, learning, and administrative areas of HFT Stuttgart. One of these projects is *Study Dean Cockpit*, the ethical analysis and reshaping of which makes up the focus of this chapter. Per KNIGHT's research proposal, SDC's raison d'être is the unsatisfactory epistemic position of study deans towards the specific counselling needs of individual students. The study deans' lack of information comes in two hues. For one, as things stand, "study deans are dependent on tips by the teaching staff in order to recognize the need for student counselling. Alternatively, students can reach out on their own." This leads to cases of struggling students taking advantage of student counselling services only too late or even not at all. Additionally, "in almost every counselling session, the following two questions come up, which currently require time-intensive manual labor in order to be answered […]:

- How many semesters does the student have left in order to complete their current studies (undergraduate, graduate)? Answering this requires taking into account special circumstances, […] increasing or decreasing the individual duration of study.
- Which modules are yet to be completed by the student? Answering this requires comparing the [student's] evaluation sheet to the examination regulations in place at the begin of [the student's] studies." [6, p. 14].

Both of these informational gaps are to be bridged by SDC. In order to achieve those goals, SDC needs to satisfy a couple of technical demands:

- Administer student data (CRUD)
- Connect to other data management systems (LSF, HisInOne) in use at HFT Stuttgart and synchronize data
- Fleshed-out roles (study dean(s), study program administration, IT administration)
- Digitization of study- and examination regulations (short: SER)
- Validation and verification of individual courses of study
- Load balancing.

A first requirements analysis turned out that HFT Stuttgart's currently used data management system LSF–apart from a limited scope of functions–allowed for data inconsistencies. For example, some pieces of data suggested a successfully completed bachelor's degree for students, who, simultaneously, lacked certain grades in mandatory modules. Apart from this, LSF hosts some data irrelevant to SDC, such as semester fees, university admissions, payments and health insurance. All of this, suggested the architectonical commitment of holding the relevant data at SDC, rather than accessing an already existing system. This allows for error correction by certain users (e.g. study deans), without corrupting the original data. This is done by expanding the data by correcting entities, which are provided for evaluation, allowing for correcting the original data independently of SDC.

In order to achieve these technical goals while avoiding technical debt, the architecture chosen for SDC is based on microservices and the principles of domain driven design (short: DDD) [15]. Such an architecture divides the application into a number of small, independent services, each of which governs a small part of SDC's business logic [7].

The entire system architecture and the data flow of the third-party systems into the SDC are depicted in Fig. 1.

- Data access was achieved with Spring Boot (Java 17), a Postgre SQL database and persistence-framework Hibernate.
- The message broker RabbitMQ was chosen to facilitate message exchange and asynchronous communication between SDC and third-party systems (LSF, HisInOne). RabbitMQ allows for smooth governance of complex workflows and processes by way of messaging queues, further decoupling our services. By using RabbitMQ we can exchange messages between services reliably, ensuring high availability and scalability of the application.
- Accessing third-party systems is done with a proxy service that controls how and in which intervals data is synchronized. Additionally, by using the integration approach, these services double as an anti-corruption layer, isolating the third-party system's model from SDC's internal model while translating between them [11].
- The technical architecture was divided into the domains of 'student' and 'SER', yielding the following web-services:
    (i) *SER-service*: creation and administration of study- and examination regulations

Fig. 1 SDC architecture

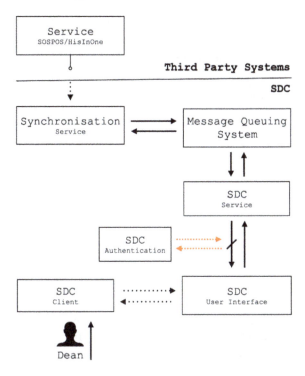

(ii) *Student-service*: administration of master data, grades, examination registration, ought-is-comparison,[1] and further data uploads

- Blob storage from a cloud provider was chosen for uploading files. This approach enables efficient and scalable storage of large amounts of unstructured data.
- The REST Interfaces of the SDC API are secured via the open standard for access delegation OAuth ("Open Authentication"). For each Interface there is a role management, allowing for static control of access rights [7]. This enables granting access to a study program's administration staff. Implementation was done via open-source identity- and access management tool Keycloak, while using HFT Stuttgart's accounts via LDAP and Microsoft Active Directory.
- The study dean UI was developed as a web app using the JavaScript library React. The UI accesses data by way of the secured REST-APIs, with authentication being again carried out by Keycloak.
- In light of the university context and the load profile, the decision was made to implement the application of a Kubernetes-environment. Kubernetes organizes applications in "containers", and automates providing, scaling, and administrating the containered application [11].

---

[1] This is based on a comparison of the modules prescribed in the SER with those already passed in the context of the timeline.

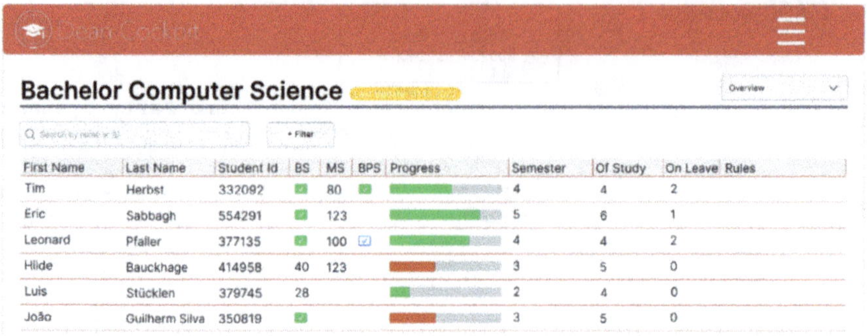

First Name	Last Name	Student Id	BS	MS	BPS	Progress	Semester	Of Study	On Leave	Rules
Tim	Herbst	332092	☑	80	☑		4	4	2	
Eric	Sabbagh	554291	☑	123			5	6	1	
Leonard	Pfaller	377135	☑	100	☑		4	4	2	
Hilde	Bauckhage	414958	40	123			3	5	0	
Luis	Stücklen	379745	28				2	4	0	
João	Guilherm Silva	350819	☑				3	5	0	

Fig. 2 SDC main UI mock-up

A mock-up of the original plan was created. Figure 1 shows the SDC's main UI (filled with dummy data), a list of all students currently matriculated for a bachelor's degree in Computer Science. Apart from the student's names, their IDs and other relevant information is displayed. The progress bar displays the relative progress in relation to the requirements for achieving the bachelor's degree. The color of the bar (green, yellow, or red) indicates how the individual course of studies aligns with the intended one. A green bar indicates full congruence, yellow indicates a minor deviation, and red indicates a deviation that threatens the completion of the bachelor's degree.

Figure 3 shows the SDC's single student view, which can be opened by clicking on a student's name in the list shown in Fig. 2. This view contains detailed information about the student's individual course of studies. By hovering the cursor over a particular curriculum semester (semester 5 in this example), the relevant courses, as scheduled in line with the appropriate SER, and their status in the student's individual course of studies is displayed. A green highlighted course has been completed successfully, a red highlighted course has been failed, and a white highlighted course has not yet been enrolled in by the student. Additionally, files relevant to the student's individual course of studies can be uploaded in SDC. This allows them to be stored and accessed centrally, as opposed to being stored on a number of separate devices.

4 Ethical Analysis of Original Plans

KNIGHT's base of values contains a number of interconnected values that give rise to worries regarding the original plans. The main issues concern the process of study deans inviting struggling students to student counselling services. There are at least three interconnected problems with SDC's original plans:

1. *Pressured students:* A struggling student might feel pressured to attend student counselling although she has, in fact, no need for it or does not wish to attend for any other reason. This is, on its own, rather a minor problem, but warrants

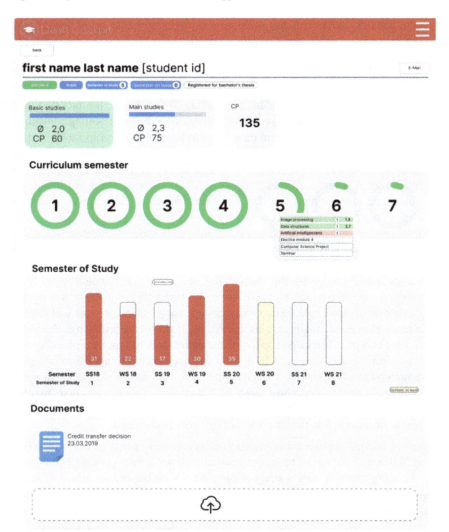

Fig. 3 SDC single-student view mock-up

mentioning since 'student well-being' is one of the principles, laid down in KNIGHT's base of values. However, it highlights another facet of problems (2) and (3).

2. *Bias towards particular students:* Most of HFT's study deans teach, as well. Apart from mounting the pressure on our struggling student from (1) ever higher, this role-overlap could lead to students experiencing student counselling similar to an exam, where there is right answers and wrong answers. Conversely, study deans might struggle with keeping their two roles apart. It might proof hard to grade students who have responded to the student counselling invite and students

who haven't, fairly. This has nothing to do with particular study deans being vindictive or gracious, since unconscious biases are known to be persistent [9]. This diagnosis stands in contrast to several of KNIGHT's principles; 'consent', 'equality/non-discrimination', 'student well-being', 'privacy', and 'protection of vulnerable persons.' First of all, students never consented to their data being used to contact them for student counselling. Albeit legal,[2] the fact that their teachers will be able to partake in their choices regarding counselling constitutes a relevantly new use of student data, so as to warrant a (re-)affirmation of their consent.[3] The potential fallout of the overlap between the roles of teacher and study dean sketched above, clearly threatens student well-being and equality by risking them being evaluated unequally. Additionally, students are inherently vulnerable towards their teachers and have an interest in the privacy of their choices regarding student counselling vis-à-vis their teachers.

3. *Bias towards groups:* Via SDC as laid out in the original plans, study deans would be constantly confronted with a combined evaluation of the students as well as other information such as the students' genders, whether or not their names sound foreign (to the study dean), and more. These connections favor forming cognitive biases, that give rise to discriminatory attitudes and behavior. Accordingly, this could further disadvantage marginalized groups and deepen already existing inequalities. This, in essence, is akin to the problem as described in (2) but with a broadened scope of the entire studentship. Compounding the issue, not all study deans at HFT Stuttgart need access to all the information contained at SDC, however, some of them do. For example, a particular study dean might need to know how many students qualify for a given course, in order to arrange for an appropriately sized room. In order to do so, she needs certain pieces of information that is irrelevant for other study deans.

Before discussing potential solutions to these problems, a minor clarification is in order. When addressing the problems just outlined we are not primarily concerned with assuring that that HFT's study deans cannot seek out certain pieces of information that are prone to lead to the biases outlined in (2) and (3). As of today, they can, for better or worse. The mandate of KNIGHT's ethics consulting activities is confined to the project, and hence, safeguarding students against each and all potential misuses of their data is simply outside our jurisdiction. Given this, it might seem that our efforts regarding SDC are in vein, since any study dean can access all relevant data anyway, and therefore, can form biases regardless of the design of this particular application. This, however, would be mistaken for two reasons. For one, as laid out in Sect. 3, study deans currently have to consult a number of different spread sheets in order to ascertain a given student's relevant information. A well-designed SDC would forgo the need of ever consulting these lists, making the option of denying access to them more attractive to HFT Stuttgart. Additionally, even if study deans would maintain access to these lists, there is simply no need for SDC to provide data

---

[2] A verdict reached by HFT's expert in data protection law.

[3] As a side note, this small example highlights potential differences between legal and ethical appraisals, and by extension the importance of conducting both.

to them that they do not need in order to successfully counsel a given student. Our focus, to clarify, is not a (fictional) malicious study dean, trying to spy on the private data of his students, but rather the (regular) well-meaning study dean that is as prone to bias as anyone.

5 Problem-Solving

As will be showcased in this section, finding the solution to moral problems in applied contexts can only be done interdisciplinary. On a theoretical level this is due to the reciprocal relationship between things as they are and things as they ought to be, especially in emerging technologies: On the one hand, how things ought to be should guide our actions and hence at what we aim to bring about (things as they are). On the other hand, however, things as they are, e.g. technical limitations determine what *can* be achieved and a fortiori what qualifies as something that *ought* to be (things as they ought to be).[4] This requires specialists from the respective scientific endeavor to opine on the things as they are, as well as specialists from ethics to structure our view of the things as they ought to be for ethical consulting on practical matters to be successful. On a more practical level, in order to consult, ethicists need to identify and appreciate ethical problems, and in order to do that, they need to have a grasp of what, in essence, is happening and why, as well as of what is practically possible. Conversely, technical professionals cannot be expected to grasp the minutia and subtle differences practical philosophy is ripe with or to be able to translate an abstract moral verdict into a specific technical solution. With all of this out of the way, let us delve into the solutions we discussed.

5.1 Data Minimization

A first partial solution everyone could, at least in principle, agree on, was to opt to display as few points of data as possible at SDC. Any data in excess of what a given study dean needs in order to conduct her tasks is unnecessary and hence subject to a prima facie expectation of privacy. However, whether a given point of data is necessary or not is not straightforward.

For example, a prime candidate for being conducive to bias is students' sexes. Furthermore, this datum is irrelevant for reaching out to struggling students. However, for some specific tasks, the sex of the students matters. For example, HFT Stuttgart is concerned with raising the quota of women in the natural science. In order to

---

[4] This is usually called the doctrine of 'ought implies can' [12]. Although the doctrine is, as all things are in philosophy, not universally accepted, we presume its truth for the purposes of this inquiry. In general, the relationship between 'is' and 'ought' is complex, and hence, has seen centuries worth of debate [5, 8]. However, we hold the observation about the reciprocal nature between the concepts, as we laid it out, to be relatively uncontroversial.

evaluate any measures taken to further this goal, different quota of women need to be compared to each other. To further complicate matters, only some study deans need to perform tasks like these, meaning that for the rest, there is no need for data like this to be accessible through SDC.

To sum up, while the need for data minimization was apparent and accepted, the question of how to assure it while also providing the desired functions to the study deans that needed them, remained open for the time being.

5.2 Reaching Out Anonymously

One potential solution for partially addressing problems 2 and 3 we discussed, was to have the study deans write up their invite to counselling using the interface of SDC, which would then, at the very last step, open up an Email program with the correct addresses already filled in. The addresses would be determined according to a set of parameters chosen by the study dean beforehand. For example, a study dean could choose to reach out to all students who had failed a certain course last semester, had already been students at HFT Stuttgart for a certain amount of time, and who had not yet enrolled for the follow-up course. Only after drafting the according message, would the study dean come to know the names of the students in question. This would, of course, by no means solve problems 2 and 3. However, it would mitigate them, since the data that could foster the biases pointed to, would no longer be displayed constantly without need.

This option was discarded due to the consideration that this would allow an all-too-easy way to circumvent the purpose of not displaying the names of the students outright. Since study deans would be able to see whoever a drafted Email would be sent to, they could quite easily narrow down the cohort for the draft so that it only applied to one student, and subsequently discard the drafted Email. This way, they could, with a few clicks, connect the data they saw to particular students.

5.3 Reaching Out Automatically

The dismissal of the idea just discussed led to another potential solution to problems 2 and 3. Instead of having study deans draft individual messages, one generic message could be created and sent via Email automatically from a dedicated account in circumstances specified beforehand. This solution would circumvent the potential for easy abuse dormant in the previous solution, while also allowing for bias-relevant data to be withheld from study deans. Other advantages of this solution are the automation of a time-consuming task, as well as that it allows for the message sent to reach out to struggling students to be meticulously crafted beforehand in order to best avoid the pressuring effect described in problem 1.

However, the pre-drafting of this message incurs certain drawbacks. A one-size-fits-all message allows for little context-sensitivity. This is a problem, since the way data is collected for use by SDC means that it sometimes does not accurately reflect the issues at hand. For example, certain grades are only input in the system inside the first weeks of a semester. During these first weeks, before the grades are input, the system takes the course in question to be failed, due to lack of a (passing) grade. This cannot be easily avoided by e.g. extending the timeline, without incurring different problems, since the idea behind SDC's early warning system is to identify and reach out to struggling students as early as possible. If one would simply extend the timeframe in which a given course can be passed, it could prevent one of the systems foremost use cases, i.e. pointing out to students that they should apply for an alternative course *in time*.[5]

Technical problems arise as well from this solution. Creating a viable Email functionality is labor-intensive, and buying the rights to an already existing one is simply too expensive for the purpose it serves. Also, any Email functionality needs to be serviced. Whoever would have the job of doing so, would have access to highly sensitive information regarding student data.

Additionally, with emails being sent automatically, there would be no convenient way of verifying whether or not (e.g. due to connectivity issues) a particular email had been successfully sent.

For these reasons, this potential solution was discarded as well.

5.4 A Novel Approach

After discussing the potential solutions just presented, we realized that problems 1, 2, and 3 are all related to the fact that study deans know who has been contacted. In focusing on how to approach students for student counselling services, a solution presented itself that would solve the ethical problems identified in total.

The original plans were predicated on the idea of study deans having the relevant data of students and reaching out to them if problems for the student arise. The only reason for this deviation of how student counselling sessions are arranged today (i.e. by students reaching out to counselling services), is the fact that the study dean is in possession of the information SDC provides, in particular the ought-is-comparison of individual students' progress. This needed to change.

By providing students with this comparison independently of there being or not being a problem, we would circumvent the need for study deans to reach out to students altogether. In order to do so we would need to keep students updated with regard to their progress and the related ought-is-comparison. The simplest way to do so, we reasoned, was to provide students with access to the relevant data already hosted in SDC. Hence, what we needed was an interface for each individual student. At this interface, students are presented with the above outlined progress bar of their

---

[5] This is usually in the first weeks of a semester.

individual course of studies, with its color indicating deviations from the normal progression as intended by the applying study- and examination regulations.

The progress bar will be coupled with a pop-up, displayed as soon as a major deviation is detected by the system. This, of course, runs into a similar problem as solution III, i.e. a lack of context-sensitivity. However, with the students in possession of all SDC data pertaining to them, they can retrace why the progress bar indicates a deviation. In particular, by clicking the progress bar, the reason why it is displayed as deviating is shown, allowing the student in question to easily find out where the deviation occurred. If, for example, a given progress bar indicates a deviation that is merely due to a grade not yet being entered into the system, as described in c), the student in question is able to infer that she needn't worry.

In order to assure that these functionalities are understood by the students, and in order to still extend a standing offer for student counselling services to students in need of it, the pop-up will contain information on potential false positives of the progress bar's early warning feature, as well as a friendly and positively written suggestion to reach out to student counselling services in cases of true positives. In this way the original ideas behind SDC can be preserved, while also avoiding the ethical issues identified in Sect. 4.

To sum up, the following changes to the original plans are required, and need to be implemented:

- As was indicated in discussing data minimization, SDC needs to be able to show each study dean all the information they require and only the information they require for their work.
- Students need to be able to access their and only their SDC data.

6 Technical Implementation

Since Kubernetes was used to allow scalability for the demands of a university, accommodating over 4000 student accounts is no problem.

Using existing standard software like Keycloak (static access management based on user roles or group membership), implementing dynamic attribute-based access rules with control over resource access is not possible, and therefore, falls short of satisfying the ethical demands. This challenge can be addressed by implementing a permission service as a discrete microservice.

Using an API gateway, the permission service acts as the central interface that filters and directs all incoming requests. This allows for a central unit for handling authentication, authorization and routing of requests. In contrast to ordinary role-based access control systems (RBAC), the permission service uses an attribute-based access control system. This allows for a fine-grained control of access to resources based on a multitude of attributes, such as time of the request, role of the user, branch of study, or specific workflow demands.

With the permission service, controlled by a data protection and ethics administrator, global rights for specific fields of data can be granted. These can be limited to specific time intervals or certain requests, offering a broad spectrum of configurational options. For example, the application could be configured to display a year's top five students with their real name for the purpose of determining award recipients during a specific phase in the semester, for example, after the grading has been completed, while withholding such information the rest of the year.

Thus, the architecture of the permission service allows for frictionless implementation of new data protection regulations, ethical guidelines, and access control mechanisms without disrupting existing operations and workflows. This is especially important in a dynamically changing legal environment, and in light of the possibility of new ethical guidelines. In regard to data protection, the application could be expanded to allow for students to grant and withdraw access to certain pieces of data. While this would presumably be a sound addition to the existing system, KNIGHT, as an externally funded research project has its limits in regard to time and resources.

A given application's architecture is significantly affected by its security- and access control mechanisms. The permission service just outlined, integrated from the very beginning, allows for a coherent and sophisticated architecture that pays due diligence to security, data protection and ethics as its central elements. A retroactive integration of the functionalities outlined in the previous section would have demanded a complete overhaul of central components of the system's architecture. Even if feasible in the confines of a research project, which is doubtful, a restructuring of this magnitude would have been incredibly time- and resource intensive.

7 Summary Discussion

As was laid out in the previous section, the new architecture of SDC allows to tailor the application to the needs of individual study deans. As a result, this enables us to withhold the names of the students from SDC. Only for special purposes (like the aforementioned award) will the names be displayed. However, we chose to keep the students' IDs in the default layout. This is due to student counselling sessions at HFT Stuttgart being mostly conducted spontaneously, with a student seeking counsel unannounced. In this common case, the student need only provide his ID to the dean in order for the dean to make full use of the data stored at SDC. While this has the drawback of allowing study deans to identify individual students by their IDs, this would require them to cross-check the entries in SDC with a list of student names and their respective IDs. As we stated, our main goal is to prevent unconscious bias, and not to address the hypothetical phenomenon of an ill-spirited study dean.

As was shown, ethical analysis and counselling in software development can have major impacts. The version of SDC currently under development at project KNIGHT is significantly different, technically as well as functionally to the originally intended one, without loosing any of the original's appeal. All of the intended functions of the original have been preserved while also refocusing the project on its beneficiaries: HFT Stuttgart's students.

The changes implemented ensure that students do not feel pressured to attend a student counselling session they do not need or want, prevents study deans having misgivings about individual students due to their refusal to accept a well-meant offer for counselling services, and avoids creating potentially harmful biases towards groups of people.

As was also shown, detecting and solving the ethical problems described above is predicated on an interdisciplinary approach to software development. At no point was there any malign or negligent intention in drafting the original plans, and yet, ethical issues arose. Conversely, coming up with solutions to these problems and assessing them cannot be done by trained ethicists alone, due to a glaring lack of technical expertise.

Additionally, the developed changes to the original plans could only have been realistically implemented at an early stage of the software development process. If the ethical assessment would have been undertaken later on, with the application's architecture already in place, a complete overhaul would have dashed any hopes of remaining faithful to the project's timeline and/or financial limits.

This shows that taking ethical considerations in software development seriously requires early integration of dedicated ethics experts as well as interdisciplinary problem-solving.

Apart from the normative imperative to take ethical considerations seriously, synergistic effects can arise from such an approach, which is also attested to by the present case. The changes implemented vastly increase the configuration options of the application, and, by extension, its flexibility and adaptability. This increase in flexibility and adaptability makes the application as a whole much more attractive for other universities, since, through the fine-grained configuration of individual roles, it can be tailored to the specific needs of other universities.

In the future, we aim at incorporation the learning analytics activities of KNIGHT into the students' individual SDC UIs, as well as track the progress of the several voluntary certifications available to HFT Stuttgart's students in order to create a comprehensive *Study Health Portal* (SHP). SPH would allow students to keep track of the data, relevant for their individual courses of studies. Additionally, the incorporation of learning analytics could allow for an even more powerful early warning system.

8 Limitations

Ethical norms are subject to change and differ across time periods, cultures, and individual. At KNIGHT, we strive to assess ethical considerations as fairly and unbiased as possible. However, due to human limitations, we recognize that the stance we take on issues can not be expected to be universally shared, nor can we guarantee absence of subconscious biases, and influences.

The project on which this report is based was funded by the Federal Ministry of Education and Research under the funding code 16DHBKI072. The responsibility for the content of this publication lies with the authors.

References

1. AI Ethics Lab (2020) Toolbox: dynamics of AI principles. http://aiethicslab.com/big-picture/ Accessed 21 Mar 2024
2. Beauchamp T, Childress J (2019) Principles of biomedical ethics, 8th edn. Oxford University Press
3. Beauchamp T, Childress J (2019) Principles of biomedical ethics: marking its fortieth anniversary. Am J Bioeth 19(11):9–12
4. Boehm BW (1981) Software engineering economics. Prentice-Hall, Englewood Cliffs, NJ
5. Hare RM (1981) Moral thinking: its levels, method, and point. Oxford University Press, New York, NY
6. HFT Stuttgart (nd) Förderantrag für KNIGHT. https://confluence.hft-stuttgart.de/display/KNIGHT/Antrag?preview=/850570355/850570358/Langantrag_fuer_BMBF_ohne_Anhaenge.pdf. Accessed 15 Mar 2024
7. Hombergs T (2019) Get your hands dirty on clean architecture. Packt Publishing, Birmingham, UK
8. Hume D (1958) A treatise on human nature. Clarendon Press, Oxford (Original work published 1739)
9. Kahneman D, Lovallo D, Sibony O (2011) Before you make that big decision. Harv Bus Rev 89(6):50–60
10. Kaspersen A, Wallach W (2021) Why are we failing at the ethics of AI?. Carnegie Council, Artificial Intelligence & Equality Initiative. https://www.carnegiecouncil.org/media/article/why-are-we-failing-at-the-ethics-of-ai. Accessed 21 Mar 2024
11. Lilienthal C (2019) Langlebige software-architekturen: technische schulden analysieren, begrenzen und abbauen, 3rd edn. Dpunkt, Heidelberg
12. McConnell T (2024) Moral dilemmas. In: Zalta EN, Nodelman U (eds) The Stanford encyclopedia of philosophy (Spring 2024 Edition). https://plato.stanford.edu/archives/spr2024/entries/moral-dilemmas/. Accessed 21 Mar 2024
13. Müller V (2023) Ethics of artificial intelligence and robotics. In: Zalta EN, Nodelman U (eds) The Stanford encyclopedia of philosophy (Fall 2023 Edition). https://plato.stanford.edu/archives/fall2023/entries/ethics-ai/. Accessed 21 Mar 2024
14. van Es R (1993) On being a consultant in business ethics. Bus Ethics: Eur Rev 2(4):228–232
15. Vernon V (2016) Domain-driven design distilled. Addison-Wesley Professional, Reading, MA
16. Weber M et al (2023) Werterahmen für das Forschungsprojekt KNIGHT. Internal HFT Stuttgart report: unpublished

The Use of AI to Teach Ethics

Harold P. Sjursen

Abstract In the popular mind, artificial intelligence has been associated with intelligent robots as they have been presented in fiction, and there they frequently have been portrayed as competitors if not enemies of humankind. From legends like *The Golem of Prague* [1] to Mary Shelley's *Frankenstein* and [2] Karel Capek's *RUR* [3], intelligent robots follow a path that at best challenges the expectations of their creators. The science fiction writer Isaac Asimov in a short story from 1942 (*Runaround* included in the collection *I, Robot* [4]) proclaimed the now well-known "Three Laws of Robotics", ethical principles to which robots must adhere in order to preserve human interests. But even these are thought to be insufficient protection in a world where advanced AI and robotics has proliferated [5]. In these accounts, AI ethics is a matter of control; do humans or machines call the shots? This is the main idea in literary and popular culture tradition that sets the stage for understanding the astonishingly rapid development of AI devices and systems, many of them freely available to anyone around the world, and their growing influence on all aspects of life. This either/or, us or them, leaves out the idea that humans and machines must for their mutual interest collaborate against uncomprehended risk and not simply by means of a *Star Wars* or *Iron Dome* [6] defense system against immanent attack. An ambiguous threat to the status quo and the age-old belief that earth is the dominion of humanity is a theme in Cixin Liu's science fiction trilogy *The Three-Body Problem* [7]. In his telling the earth and human civilization are under a massive threat, as best they can tell, is some 400 years in the future. The nature of the threat is ambiguous and cannot be measured or understood by the technoscientific resources then available. But it is only with the most advanced technoscience that the possibility of understanding and defending against this threat exists at all. The only hope rests on unprecedented human-AI machine collaboration. It is this scenario, rather than debates about control and regulation, that symbolizes *'in extremis'* the reason AI ethics must be taught in the midst of both STEM and non-STEM higher education.

Keywords Technology ethics · AI · Course design

H. P. Sjursen (✉)
Emeritus of Philosophy, Tandon School of Engineering, New York University, New York, USA
e-mail: harold.sjursen@nyu.edu

1 Can AI Be Used to Teach Ethics?

1.1 ChatGPT 4's Response

"Yes, ChatGPT can help teach the ethics of AI by providing information, explanations, and engaging in discussions about key concepts and principles related to the ethical use and development of artificial intelligence."

This response (and the examples that follow) does not acknowledge the fundamental way that contemporary technology including advanced AI has altered how humans act and accordingly the central issues of ethical responsibility.

2 Introduction

Many of the ethical questions that individuals and societies face today follow from the extensive use of advanced technology in nearly all areas of human activity. Many situations arise due to the possibilities created by technology where tradition offers only minimal guidance as to what is ethically appropriate. Well known examples occur in the hospital setting such as decisions regarding the removal of life support systems. In less dramatic fashion the modern classroom has also been transformed by technology, but these changes generally did not present ethical quandaries for the management of teaching and learning. Such is not the case with the rapid development and adoption of AI applications. It seems self evident that the use of AI (particularly large language, generative) devices in education would be by early adopters at the university level, and they in turn would probably be concentrated in engineering schools.

Undergraduate engineering students are probably more familiar with the latest AI devices, gadgets and its newfound potential as *partner* utilities than most in the general public and even their peers studying in other university disciplines. There can be little question that, given the nature of engineering education, that the ethical issues raised by the power and potential of AI must be faced head on. It is becoming increasingly clear the strategy of controlling, regulating and limiting the development of AI is conflicted by economic self interest on the part of developers and by political parties who tend to portray the situation to suit their own agendas.

The task of AI ethics education should be not to declare rules about whether using it is right or wrong, good or evil, or even dangerous or benign. AI is all those things. The task must be to guide students to make rational judgements that are universal (or nearly so) about the design and deployment of AI dependent devices and systems. For engineering students, the design process is critical; for others the distribution and use would be most important. But for both populations reflection about the use of AI beyond evaluating its immediate results is imperative. For example, using AI to screen applicants for mortgage loans has revealed selective bias with regard to race and gender. A first response might be to suspend using the AI utility until the algorithm and/or the database until they are corrected to eliminate biased responses. Certainly, this would be a proper, ethically motivated response to the situation, but we note that this reaction is entirely in keeping with well-known codes that have nothing at all about AI as such. Cases like this demand vigilance and responsible awareness when using AI systems, but do not demand a new way of thinking about impact of technology and standard approaches to engineering ethics address the risks. When considering the advanced engineering technologies designated by the United States National Academy of Engineering as the *grand challenges for the twenty-first century* [8] one notes that the technological functions required to pursue these far reaching objectives (for example, challenge number four: 'reverse engineer the brain') are all presented with the presumption that the new technologies required are all advanced tools that are fully under the control of their human operators. The challenges are akin to the proverbial task of building a better mousetrap. One must be attentive to and respective the constraints imposed by ethical principles. But this does not require a new kind of ethics. The belief is, as Pope Francis put it in a statement referring to the work of a conference on *Robotics, AI and Humanity, Science, Ethics and Policy* [9], (organized jointly by the Pontifical Academy of Sciences and the Pontifical Academy of Social Sciences):

> Artificial intelligence is at the heart of the epochal change we are experiencing. Robotics can make a better world possible if it is joined to the common good. Indeed, if technological progress increases inequalities, it is not genuine progress. Future advances should be oriented towards respecting the dignity of the person and of Creation. Let us pray that the progress of robotics and artificial intelligence may always serve humankind ... we could say, may it 'be human'.
>
> Given the generality of ethics, are new approaches to ethical reasoning necessary or is it a matter of pedagogy and priority? Pope Francis, November Prayer Intention, 5 November 2020 [10].

If ethics as generally understood is not entirely adequate when it comes to emerging technologies, perhaps most especially those that depend upon AI, what is it that is missing?

Hans Jonas argued that traditional ethical theory, including religious ethics, duty ethics and utilitarian ethics, fail conceptually when faced with many of the possibilities that advanced technology presents [11]. He called for the development of an ethics for the future that would be grounded in a theory of responsibility that accounted for the complexity and power that characterizes modern technology. In an age in which the power of human action is mediated by technology... how do we act responsibly, especially in relation to the future? The purpose of ethics instruction should be to prepare students with the intellectual tools to enable them to understand not only the range of outcomes and their impact on the human conditions now and in the indeterminate future, but also develop thinking and judgement such that ethical reflection encompasses the possibility that humans create self-improving AI programs whose intellect dwarfs our own and that we consequently lose the ability to understand or control them. However, since this scenario would require 'artificial general intelligence', that is, AI systems that can handle the in-credible diversity of tasks done by the human brain, one might argue that this is a risk not at all likely and in any event would require several fundamental scientific break- throughs, each of which may take many decades. If we accept that this future is highly unlikely and with sufficient intervening time before it would be a possibility that suitable precautionary measures could be adopted, why should it be a serious concern of higher education at the present?

There are at least two reasons. The first is that the consequences of AI decisions, even at the level of AI already operative, might easily be catastrophic. The second is that AI through its many currently available applications is already precipitating a mindset about what is normative; about human possibility, freedom of choice, what is natural or artificial, and what the good life will comprise. This despite the fact that it is widely acknowledged that AI devices cannot choose their own goals or to think creatively.

The possibility of AI invoked catastrophe is explored by the philosopher Toby Ord in his unnerving book, *The Precipice: Existential Risk and The Future of Humanity* [12]. Ord places humanity on a precipice that is both situational and existential. He characterizes the current state of the world as teetering on a precipice where the likelihood is that we fall into an abyss that ends humanity. The reason for this would be our own doing; the existential threats of nuclear and biological weapons follow from the progression of technological development from agriculture to our present state of hyper-urbanization. The circumstances we now find ourselves in do not constitute a predicament we can escape from, despite being the circumstantial consequences of the game we have played, that is, it is not a problem to be solved by means of further technology. This very unsettling book, written in consultation with numerous specialists and experts, makes the case that we are now in a genuinely unprecedented state, whereby the possibility, and perhaps the likelihood, that, if conditions do not render all life impossible, at least they may lead to the annihilation of human life by degrading our planet to the point where it will be unsustainable. This situation would result from a combination of missteps and negligence, all the consequences of the prevailing capitalist-consumerist-nationalist ways in which humanity is now organized. This goes well beyond a predicament because it strikes directly at the heart of human existence itself.

There are those who claim that new digital technology provides powerful new tools with the potential to ameliorate risk and suffering around the world; in his 2018 report *Strategy on New Technologies* the United Nations Secretary-General characterized them as a means to accelerate the achievement of the 2030 Sustainable Development agenda [13]. Since 2018 digital technologies have advanced to the point where AI and machine learning dominate.

This development concerns Ord. In his discussion of future risks, he reflects upon how AI will alter our decision making. He utilizes the distinction between AI (artificial intelligence) and AGI (artificial general intelligence) and reserves his greatest concerns for AGI. Reporting on some of the astonishing feats of AGI, such as mastering the game of Go within a mere 8 h to a level that surpassed that of the world's most dedicated Go masters, he raises the question of where the increasing sophistication of AGI agents may lead. If AGI algorithms, updating and revising themselves via machine learning, can, for example, master sets of Atari games with no more input than the raw pixels on the screen and the game scores, then we must acknowledge that their realistic potential is 'learning to control the world from raw visual input; achieving their goals across a diverse range of environments'. The consensus among those researchers working in the field of AI is that such potential is plausible as early as 2025 and likely within a century. If Ord's prognosis is accurate, then indeed the necessary scope of ethics must be reconfigured else it will become part of the record of the values underlying human choice in past, but offering little guidance moving forward.

Something like this is what motivated the call by Hans Jonas for an ethic for the future age of technology. In his influential book *The Imperative of Responsibility*, Jonas argues that our ethical tradition cannot adequately guide us as we face many of the choices put before us by the ubiquitous presence of technological devices and systems that mediate most of human action [14]. Traditional ethics, on the contrary, is based on free will, human understanding, and clear consequences. To call for a new approach to ethics implies an inadequacy in existing theories that require degrees of human agency that is being subverted through growing power and influence of technology. What are the shortcomings he finds in classical ethical theory? His approach is not to wipe the slate clean and begin entirely anew, we cannot simply abandon our history and traditions. It is clear that while he calls for a new ethics, he also appropriates aspects of deontological duty ethics. More controversially he adopts teleological explanations to his purpose. But how are duty and a sense of purpose possible when human agency has been altered and diminished? These are the reasons that lead him in this direction. His argument asserting the need for a new ethics is premised on the claim that as a consequence of powers characteristic of the new technologies of our age the very nature of human action has been altered. Thus, the kind and degree of control we exercise when we act has changed how we are able to be responsible for our actions. This claim must be carefully examined, as it is interwoven with his description of the teleology of life and his understanding of the source of ethical duty. The points he made in support of this proposal might be summarized by the following points:

- Through history technology has advanced from tool to machine to automatic device. This leads to the situation were, even without AI, some technology may be beyond human control. It is often a mistake to characterize advanced technological devices, especially those with high degrees of performance autonomy, as tools.
- Technological processes are often not well understood and produce unanticipated consequences. Emergent technologies manifest a high degree of human ignorance due to complexity; AI algorithms sequestered in *black boxes* and revised and updated by machine learning input both distance our understanding and increase the likelihood of outcomes that were neither intended nor anticipated.
- Technology produces results disproportionate to human action. This raises the issue of overwhelming power. Can we destroy the world with the simple push of a button? The clearest example of this disproportionality is nuclear power, for once the technology was developed it became, given the potential catastrophic consequences, too easy to deploy.

- Technology may alter the environment permanently. That is, our actions may be irreversible as is now conceded to be the case with carbon based energy consumption. Cultural, economic, political and social environments are changed in a manner that cannot be undone. We usually refer to this as progress, but that presumes a desirable end which cannot be presupposed.
- Results of technology may only present themselves in the distant future. Because potentially damaging consequences affect the unknown and indefinite future, they lie beyond normal motivations for our concern. Although we have a clouded idea of medium term developments and some of the consequences for AI, the possibility of significant changes to a distant future, changes that we cannot imagine set in a world we likewise cannot imagine, challenge any notion of stewardship we might have for this planet and its dwellers.

These concerns make clear that those who will enter the imminent hyper technological society will be beset with choices for which traditional norms and ethical principles will provide only limited guidance and often be inadequate or misleading. The uncertainty and ambiguity that characterizes the technological future together with the speed and power inherent in modern technological devices provide sufficient reason for increased ethical reflection on the norms that govern its design and production. To these factors we add increasing autonomy and automaticity leading to greater limitations in our ability to control how devices operate and the case for new approaches to ethics is hard to deny.

What this new approach to ethics might be is unresolved. Jonas himself acknowledged this and offered only interim solutions, a precautionary principle and a heuristic of fear. In other words, refrain from going forward out of a fear of dire consequences. These solutions are both inadequate and probably unachievable and have been criticized for tending toward the authoritarian or being paternalistic. The inadequacy or ad hoc provisional nature of Jonas' provisional solutions only serves to emphasize the need for fundamental rethinking of ethics in the context of the advanced technologies currently achieving widespread adoption. This is the imperative behind the kind of course being here described.

3 Type of Course

The type of course needed to prepare students for the many unprecedented ethical dilemmas they will face as design or operational engineers, medical providers, science researchers, educators and even as family members and concerned citizens is not one focused on the arguments of Aristotle, Kant or other great ethical theorists, nor one that states the consensus principles of business, engineering or medical ethics. Such courses as these may still have an important place in professional education but they do not help students learn to think ethically about ambiguous and uncertain possibilities that modern technology is conjuring up.

I suggest that an efficient way to help students acquire the habit of approaching their tasks with a mindset that is attuned the new kinds of ethical problems raised by modern technology is not to teach ethics per se, but to immerse them in various scenarios, case studies and exercises in which ethical questions are unavoidably present. It is imperative that the instructor not offer answers to the ethical questions but require students to propose what they take to be first, what is at issue ethically and second how they believe the problem should be addressed and why. This kind of exercise should be frequent, every week and with every topic. The students will soon become aware that the perfect is often the enemy of the good and therefore compromises and trade-offs will have to be devised. And given the emphasis on new and emerging technologies, especially those where devices deployed have degrees of autonomy, the students must devise strategies to review how well their proposed solutions actually work out and anticipate how responses and corrections could be implemented.

In these exercises students will intellectually role play; they will imagine themselves to be economists, architects, health care workers, and so on. To make this role-playing aspect more effective students should work in teams with each member of a team assigned a specific role.

To do this in engineering courses may be to use emerging AI technologies in design *projects to teach* fundamental principles of AI ethics [15]. Design projects are often the core of an engineering student's capstone experience. The design projects in this course need not be so elaborate. The point is not, in this case, to carry through on all levels of the design process, but to consider how design may shape the contours of the ethical issues. A very simple example might be how the design for the packaging of powerful medicines may influence the care with which the medicine might be used or what attitude the patient might have about the medicine, and similar questions. The main thing is for students to discover how engineering design influences subsequent use. By looking at design in this way the engineer invokes the other key aspects of engineering work, implementation and operation. I usually tell engineering students to consider the ethical dimensions of each phase of engineering work as indicated by the CDIO (Conceive, Design, Implement, Operate) paradigm. Having small design projects, and projects that involve using high technology devices with a degree of autonomy associated with the themes of the course leads students to greater reflection on the ethical issues because they manifest themselves as a result of the design.

Why should this strategy be an effective way to teach ethics for an age dominated by technology? I will describe an undergraduate ethics course that immerses students in the problematic of technologically mediated action. Although it rests on a profound theoretical question, the course will not argue for a theory (as for example Kant and Mill did), but to try to help students formulate principles for ethical action.

I have taught two versions of the course: (1) To liberal arts students without significant knowledge of STEM disciplines; (2) To engineering students. In both cases the courses created contexts that emulated possible future professional or life choices in which students follow paths of discovery that produce necessary choices and judgements.

Multiple versions of a course with the same general learning outcomes are possible. I have taught ethics for many years and recently technology ethics to both engineering (undergraduate and graduate) and non-engineering students. My courses were project and problem based, team organized and focused on design principles, even in the liberal arts versions design was) a key component. For engineering students, the prospect of including AI ethics within a design should be considered. As the paradigm of contemporary technology, AI applications will be employed to simulate ethical dilemmas.

It is normally the case that new academic courses are approved on the basis of the content that will be covered and learning outcomes related to that content. What are the topics that an undergraduate technology ethics course should include? Here is a list from my model course (since the courses were taught in China, the first topic was of general interest to all of the students.)

1. Science and Ethics in China and the West
2. Scientific & Engineering Method(s)?
3. Why is technology an issue for philosophy and the humanities?
4. Techno-Science and the new problem of Responsibility
5. The Logic of Scientific Discovery
6. Normal Science
7. Global Technology Transfer & AI Impact
8. Animal Intelligence vs Artificial Intelligence
9. Concerns and Dangers of AI
10. Technology & Responsibility
11. Hans Jonas: Theory of Technology & Responsibility
12. Is the ethical (& metaphysical) impact of digital technology unlike that of previous modern technologies?

The version of the course is intended for STEM students and has actually been taught to engineering students in China. It introduces to STEM students topics in the humanities that they mostly will not have studied. The version of this course for non-STEM students could include different topics, however I found all to be appropriate, there was simply a different emphasis and different exercises. In any event the principle would be same: to introduce key issues that underlie the debates surrounding advanced and emerging technologies, placing them in a context that they most likely have only a slight acquaintance with. Thus, the specific topics are flexible and should reflect priorities in the program, student interests and instructor expertise. I will elaborate upon these topics and how they contribute to the overall lesson of the course. Whatever specific topics are chosen, however, they should collectively lead to questions and reflection upon the broad set of issues associated with AI ethics. What is important, more than the particular topics, is the inclusion of practical exercises that require students to make choices and judgements. These exercises should rely upon devices that depend to a degree on AI technology.

4 Advanced Technologies and the University of the Future

In this discussion I distance my reflections from some of the frequently asserted expectations used to characterize the future of educational institutions. I am not trying to negate these expectations and neither support nor deny the values claimed for them. Rather, I am advocating the imperative of an informed public engagement with the profound ethical issues propounded by the widespread utilization of AI. The school of the future, primary through tertiary learning, has been described as almost fully determined by computer technology.

> The School of the Future is a school where technology is used as a tool for knowledge access and approach at all times and from any location. Since the school is a place where instructors, trainees/students, parents, and the community (educational and local stakeholders) actively approach knowledge, the school evolves in accordance with the demands of time. To guarantee that everyone participating in the educational process has access, the school of the future will need to use similar tactics, techniques, tools, and policies as the conventional school. In the school of the future, the teacher will take on the role of partner, companion, and guide instead of playing the desk role and plans, motivates, and develops hands-on activities in accordance with the interests and skills of the students. In the spirit of cooperative learning, students work in groups and actively participate in the learning process by gathering, analysing, and processing information with the goal of gaining experiential knowledge, teaching others how to learn, and approaching knowledge in accordance with their learning profiles. The educational advancements and pedagogical techniques of the school of the future, support parents and those attempting to educate their children and participate in the educational process. The 'electronic bag', which can be a mobile tablet or another device, replaces books and notebooks, and the classroom of the future will make use of cutting-edge applications like IoT, Robotics, Artificial Intelligence, Avatar, Virtual Reality, etc., as well as gamification to add an element of entertainment and interaction to teaching and learning. The development of emotional intelligence, acceptance, and empathy, as well as the development of cognitive and metacognitive abilities, will be vital guarantees for the fulfilment of the school's goals, but emerging technologies will play a major role in the school of the future: a free, accessible, and everyone-friendly school [16].

The 'electronic bag' concept above regards its contents as a set of generically similar tools that can used in a manner that is ideologically neutral and without built-in intellectual constraints. Whether this is the case with powerful AI applications is a key issue of concern. There are many reasons to suspect the neutrality of AI applications, The university of the future in most cases will evolve from present day institutions and the legacy of past practices will continue to exert influence.

This kind of course, in a variety of specific instantiations, should be part of the university curriculum of the future. At this point AI is a large part of the public discourse, polemically divided between techno-optimists and doomsday sayers, with traditional tales of robotic threats to human well being in the background. This public discourse needs to be elevated and strategies for thinking through the new ethical dilemmas that are the consequence of the many need to be developed. For these reasons a new emphasis of technology ethics must occupy a significant part of university curricula.

5 Summary

In a way what is suggested here swims against the stream by suggesting that the university of the future, by proposing a re-embrace of a traditional liberal arts approach, in an age when STEM education is heralded as mandatory to maintain or enhance one's standing in the interconnected global economy. Likewise, the reticence to adopt the 'the electronic bag' of devices that are intended to contemporize the classroom or asking students to reflect on the values inherent in their choices, and to ask whether the most efficient or even the most desired options are indeed the best for humankind. To see as a goal of education direct humankind toward an ideal where craftsmanship was embraced not for the sake of its efficiency, but as a measure of human mastery is in many ways a rejection of the idea of progress.

Against the teleological concept of progress driven by efficiency the proposition that there is a distinctly human good that resides in the natural is defended. Two parallel themes have been put forth: the first that we, humankind, may have, largely through technological dependency situated us on a precipice where the likelihood of not falling is rapidly deceasing; the second, that we can through imprecise reflection on what makes for the good, find our way again. It is this latter task to which universities have a particular obligation and opportunity.

There are no doubt numerous ways to fulfill this obligation but a component that probably cannot be avoided is to face head on the poverty of most approaches to professional ethics now taken. For the most part professional ethics courses taught in universities, whether engineering, medical, business, legal and others suggest that properly informed professionals, competent to manage the tools at their disposal, can protect society can minimize risk and protect against disaster. Competence is the *sine qua non*, but this supplemented by integrity and honesty assures the ethical character of professional action. Moreover, professionals need not ponder this unduly because the standards of competence, integrity and honesty can be codified such that adherence to the codes provides genuine assurance that basic human interests and rights are being protected.

There is actually must wisdom in this approach but it assumes that competence, honesty and integrity are sufficient to ensure that action will not be taken that is contrary to human interest. This principle is challenged by the growing power and complexity of autonomous devices, especially those powered by today's artificial intelligence and of course even more so should something like artificial general intelligence become a reality. Under these conditions, competence loses power, honesty loses meaning because machine operations are shrouded in a veil of impenetrable complexity and integrity becomes little more than good intentions.

The suggestion made here is to encounter this level of advanced, increasingly autonomous and complex technology and put it to the test; not whether it works, not whether it has flaws which need correction, not whether it is demanded in order to maintain a competitive edge, but whether it is taking us where we want to go. This is a matter of judgement that cannot be quantified or operationalized. The kind of academic courses which foster this sort of judgement require contemplation, reflection, introspection, and above all what the proper measure of humanity is. The task really is not so different from what literary accounts of malevolent robots undertook; the task is to imagine what kind of future do we want and how the dignity of humanity is to be protected from the avaricious tendencies that accompany the evaluation of efficiency above all else.

References

1. Meyrink G (1915) Der, Golem. Kurt Wolff, Leipzig
2. Shelley M (1984) Frankenstein. Bantam Classics, London
3. Capek K (2014) RUR (Rossum's Universal Robots). Paul selver and nigel playfair. Dover, Mineola, NY
4. Asimov I, Akinyemi R (2000) I, robot: short stories. (Lernmaterialien). Cornelsen & Oxford Univers., London Humanity
5. Salge C (2017) Asimov's laws won't stop robots from harming humans, so we've developed a better solution (2024). https://www.scientificamerican.com/article/asimovs-laws-wont-stop-robots-from-harming-humans-so-weve-developed-a-better-solution/. Last accessed 14 June 2024
6. Iron Dome: defense system against short range artillery rockets (PDF). Rafael advanced defense systems. Archived from the original on 10 July 2012. https://web.archive.org/web/20120710092155/, http://www.rafael.co.il/marketing/SIP_STORAGE/FILES/0/1190.pdf. Last accessed: 15 June 2024
7. Liu C (2016) The three-body problem. Tor Books, London
8. NAE grand challenges for engineering. National academy of engineering. USA. https://www.engineeringchallenges.org
9. Von Braun J, S Archer M, Reichberg GM, Sánchez Sorondo M (eds) (2021) Robotics, AI, and humanity science, ethics, and policy. Springer, Berlin. ISBN 978-3-030-54172-9
10. Vatican News staff writer (2020) Pope's November prayer intention: that progress in robotics and AI "be human". Vatican News. https://www.vaticannews.va/en/pope/news/2020-11/pope-francis-november-prayer-intention-robotics-ai-human.html. Last accessed 14 June 2024
11. Jonas H (1984) The imperative of responsibility. in search of an ethics for the technological age. Chicago University Press, Chicago
12. Ord T (2020) The precipice: existential risk and the future of humanity. Bloomsbury Publishing, London
13. Guterres A (2018) UN secretary-general's strategy on new technologies. United Nations. https://www.un.org/en/newtechnologies/images/pdf/SGs-Strategy-on-New-Technologies.pdf. Last accessed 15 June 2024
14. Richard JB (1994) Rethinking responsibility. Soc Res: Int Q 61(4):833–852
15. Yu H, Shen Z, Miao C, Leung C, Lesser VR, Yang Q (2018) Building ethics into artificial intelligence proceedings of the AAAI conference on artificial intelligence. In: Proceedings of the twenty-seventh international joint conference on artificial intelligence
16. Drigas A, Papanastasiou G, Skianis C (2023) The school of the future: the role of digital technologies, metacognition and emotional intelligence. Int J Emerg Technol Learn (iJET). 18:65–85

A Reinforcement Learning Framework for Personalized Adaptive E-Learning

Anat Dahan, Navit Roth, Avishag Deborah Pelosi, and Miriam Reiner

Abstract Personalized learning is motivated by the recognition that students show diverse learning styles and paces due to factors such as personality characteristics, motivation, emotional and environmental circumstances, and prior experiences. It is also increasingly important to account for students with conditions such as Attention-Deficit/Hyperactivity Disorder (ADHD) or other learning intervening factors. Accommodating individual differences in attention span and learning patterns is crucial for effective learning. When designing a digital course, many parameters can be adapted to the unique learner profile such as presentation style of the content, stimuli for enhanced attention, length of session, available links, assessment and navigation options and more. This chapter suggests the use of Reinforcement Learning (RL) algorithm for a personalized digital learning experience, linking the learner's profile with the responses of the learning environment. We suggest a framework, based on Universal Design Learning (UDL) principles, where an intelligent agent is programmed to learn the student's learning skills and preferences, then adapt to the user by offering suitable learning materials, structures and stimuli, accounting for continuous changes in performance. A simulation is presented to validate the adaptive algorithm applied to a digital course, focused but not limited to the parameters relevant to students with ADHD such as attention and distractibility.

Keywords Adaptive learning · Reinforcement learning · ADHD · UDL

A. Dahan (✉)
Software Engineering Department, Braude Academic College of Engineering, 51 Snunit Str., 2161002 Karmiel, Israel
e-mail: anatdhn@braude.ac.il

N. Roth · A. D. Pelosi
Mechanical Engineering Department, Braude Academic College of Engineering, 51 Snunit Str., 2161002 Karmiel, Israel

M. Reiner
Technion – Israel Institute of Technology, 3200003 Haifa, Israel

1 Background

E-learning is adaptive when it considers the learner's response/feedback to the learning process [1]. While traditional classroom learning is generally limited to a one-for-all didactic approach, digital learning is not time and space-limited, and has the potential ability to adapt contents, representation style, pace, learning path, and evaluation methods to specific learners; moreover, the adaptation could occur accounting for specific learner's varying cognitive/emotive/environmental conditions. The research of adaptive e-learning systems involves a multidisciplinary approach, as it intends to optimize learning, based on cognitive and educational theories, by the implementation of technological approaches and computational methods. The complexity of the adaptation stems from the existence of at least three interactive systems (or models), influencing the learning process: the learner, the teacher and the tutoring system [2].

In the last two decades, smart educational systems have focused on helping learners to navigate effectively digital content to achieve their learning objectives. Modern Learning Management Systems (LMS) include personalization technologies, such as content recommendation, course sequencing and adaptive navigation support, enabling the suggestion of the most suitable course elements based on learning goals and the learner's current level of knowledge [3].

When addressing the unique characteristics of the learner, many parameters should be considered, related to the psychological/affective/cognitive state of the learner. Park and Lee elaborated the importance of intellectual abilities, cognitive style, learning style, prior knowledge, motivation, anxiety and self-efficacy in modeling tutoring systems adaptation to the learner [2].

Learning style, or the individual preferred learning strategies system, pertains to the way the learner acquires, organizes, processes, stores, retrieves and communicates information [4]. Coffield et al. categorized 13 learning style models (out of 71 available theories) in a comprehensive critical review, addressing their strengths and weaknesses as well as their validity [5]. It is not the scope of this work to determine and choose the most effective learning style model, but to propose the implementation of a model in the simulation of an adaptive learning framework. The most popular model used in adaptive e-learning systems is Felder-Silverman Learning Styles Model (FSLSM, [6]), which defines four dimensions, reflecting continua of preferences related to information perception, presentation, processing, and organizing in the process of understanding. Other notable models include VARK model [4], based on four main learning styles (Visual, Auditory, Read, Kinesthetic) and Honey and Mumford's Learning styles [7]. The identification of a preferred learning style is usually performed by filling learning styles inventories, either static or dynamic questionnaires. Even when the procedure is automated, it is time-consuming, it is generally performed at a specific time during the learning process and may be biased as it derives from learner's self-evaluation [8].

Learning style is often correlated to learner's motivation, which is a major predictor of efficient learning [1]. Learner's motivation is highly important in traditional classroom education as well as in electronic environments. The probability to drop out a course is much higher in e-learning [9]. E-learners are more likely to be intrinsically motivated, although studies showed that intrinsic and extrinsic motivation often coexist and may depend not only on learner's tendencies, but also on environmental conditions and the design of the e-learning platform [10]. Therefore, it is important to consider the factors influencing motivation when developing an online course. Using the user's data, the engagement level can be evaluated; for example, a high level of arousal may be maintained by presenting a large amount of links of variable content, and measured by the number of clicks, the time spent on the activity or the response to an assessment.

Adaptive e-learning systems that address motivation and engagement are imperative when considering students with special needs or cognitive disabilities such as Attention Deficit Hyperactivity Disorder, ADHD. ADHD is a neurodevelopmental condition characterized by persistent patterns of inattention and/or hyperactivity-impulsivity that interfere with normal functioning. The prevalence of ADHD among students varies across studies. In the United States, it has been estimated that between 2 and 8% of college students have ADHD, without taking into account the undiagnosed students [11]. Students with ADHD encounter a range of difficulties that can significantly impact their academic performance. The time needed for degree completion is generally longer, and they are more likely to drop out compared to students without ADHD [12]. These challenges encompass issues with sustaining attention, being easily distracted, managing impulsivity, and coping with hyperactivity, all of which can impede learning and task completion. Furthermore, organizational problems, such as time management and task completion, as well as struggles with emotional regulation and coping with boredom, can further hinder their academic success. Moreover, individuals with ADHD have a larger intra variability of attention [13]. In the domain of online courses, which is becoming more and more relevant in academic studies, students with ADHD face challenges in maintaining focus and attention, staying organized, and managing their executive functions [14]. The transition to online learning can exacerbate their struggles, as it requires a high level of self-regulation and independence. In addition, the lack of in-person cues and increased screen time can make it harder for students with ADHD to stay motivated and engaged. Furthermore, the flexible schedule of online learning, while beneficial for some, can also lead to difficulties in establishing a routine and sticking to a structured study plan.

New policies based on an inclusive conception of education are developing, including the adoption of novel learning architectures that address the disabled student community [15]. The Universal Design for Learning (UDL) framework, is a scientifically valid, flexible learning framework [16]; it identifies three networks in the learning process: the recognition network ("what" is learned), the strategic network ("how" it is learned) and the affective network ("why" it is learned) [17]. The UDL framework is intended to adjust the information to learner's needs, by providing multiple means of presentation, expression and engagement. Research shows that multi-modal representation and expression, multi-modal, information chunking, frequent feedback and reward and game-based activities, are beneficial strategies that may enhance motivation and performance in ADHD students, although it is also necessary to reduce their cognitive load by reducing distraction and unnecessary stimuli by keeping the lesson very structured [18].

The model presented in the current work is inspired by UDL framework, in order to offer an inclusive adaptive methodology addressing the needs of students with/without ADHD.

2 Related Work

The use of machine learning in intelligent tutoring systems assists the design of adaptive learning elements such as path recommenders, course difficulty and content sequencing according to learning scores and time [19], and data mining, to identify contents of interest based on student's usage data [20]. Wang et al. optimize learning path using VARK's model and an ant colony system [21]. De Moura et al. proposes the personalized selection of Learning Objects (LO) based on Kolb's learning model, performed by a particle swarm algorithm to improve performance and refine the initial self-evaluation of the student [22]. Neural Networks are successfully employed to identify learning styles [23] and to predict student engagement [24]. Recent works apply Reinforcement Learning in content generation for virtual reality applications [25].

Our work makes use of a Reinforcement Learning algorithm to adapt content to the student's needs. Reinforcement Learning and big data tools are implemented in an e-learning system by Shawky and Badawi [26], to suggest appropriate learning materials in variable user's states, taking into account knowledge level and learner's metacognitive characteristics, that are evaluated by questionnaires. Measured features include interactivity level and performance achievements. Simulations show an increase in positive rewards during the learning process.

These works address specific aspects of adaptive learning, such as learning styles, path recommendation and dynamic performance evaluation, showing promising results, although not always validated by experimental studies. Moreover, in spite of the growing awareness of the need of inclusive e-learning systems, there is still a lack of studies targeting e-learning solutions for students with cognitive disabilities [27]. Cinquin et al. propose a MOOC (Massive Open Online Courses) platform, which organizes content using self-configuration parameters such as navigation level, instructional content, explanatory links, time markers; perceived cognitive load, self-determination scores and learning performance are retrieved and analyzed with encouraging outcomes [28]. The platform is configured beforehand, it does not adapt in real time.

This work proposes the framework for an adaptive academic MOOC, that is not addressed to a specific learner's disability, but is flexible enough to represent an inclusive digital learning option for students with ADHD, for instance. The study implements Reinforcement Learning, and more specifically Q-Learning algorithm, to identify the learner's profile in terms of learning preferences, styles and variable attention states, and propose in real time the amount and type of instructional units that are more appropriate at the moment. Contrarily to former work using RL algorithms as a general theoretic framework [26], this study elaborates a hybrid methodology in defining adaptive features according to learning style, motivation and performance. UDL framework and elements of VARK learning styles model are blended to build a multi-dimensional matrix of possible instructional units, presented to the learner. Performance is evaluated only by measurable parameters, thus avoiding the use of self-evaluation by questionnaires. To validate the model, a case study is presented, simulating the performance and the adaptive response of the algorithm in the learning process of a student with ADHD.

3 Approach

The proposed framework makes use of Reinforcement Learning, more specifically Q-Learning algorithm, to learn the user's changing preferences and adapt suitable learning units.

3.1 Reinforcement Learning

The key component of RL includes an Agent that interacts with the environment. The environment responds to the actions taken by the agent and provides feedback in the form of positive or negative rewards [29]. The situation of the environment is defined as a state, and at each step the agent chooses an action that will provide a maximal reward from the current state. According to the selected action the agent receives a reward. The reward indicates how well the agent is doing regarding its objective. The goal of the agent is to learn to take actions that maximize cumulative rewards over time. Q-learning is a model-free reinforcement learning algorithm that aims to find the best action to take in a given state to maximize the reward [30]. It is an off-policy algorithm, meaning that it learns the value of an action in a particular state without needing a model of the environment. The algorithm iteratively improves its evaluation of actions' quality at specific states, ultimately converging to an optimal policy that maximizes cumulative rewards over time. The reward function, indicating the success or failure of an action, influences the agent's decision-making process by providing feedback on the desirability of its actions.

In this framework the environment is the student's behavior, a state S_t is the learning scheme presented to the student, and an action A_t is the response of the intelligent agent to the environment. Learning of the environment takes place by obtaining a Q-table, a data structure used to store the expected cumulative rewards for each state-action pair. As the agent interacts with the environment and receives rewards, it updates the Q-values based on following update rule:

$$Q^{new}(S_t, A_t) \leftarrow (1 - \alpha) \cdot Q(S_t, A_t) + \alpha \cdot [R_t + \gamma \cdot \max(Q(S_{t+1}, A))] \quad (1)$$

The algorithm updates the current Q-value, $Q(S_t, A_t)$ with a new value, $Q^{new}(S_t, A_t)$ for the state S_t and the action A_t, using Eq. (1) where:

- α is the algorithm's learning rate that controls how much of the difference between previous Q-value and newly proposed Q-value is considered. A factor of 0 means the agent does not learn at all and a factor of 1 makes the agent consider only the most recent information.
- R_t is the reward for taking action A_t at state S_t.
- $\max(Q(S_{t+1}, A))$ is the maximal expected future reward from all actions A that can be taken from the selected action (S_{t+1}).
- γ is the discount rate that determines the present value and the importance given to future rewards. This is a value between 0 and 1. When γ approaches 0, the agent prioritizes immediate rewards, whereas a γ value of 1 directs the agent's focus towards long-term rewards exclusively.

To enable exploration of alternative actions beyond those yielding maximum rewards, an exploration rate $0 < \varepsilon < 1$ is employed. The action A_t is either selected randomly with a probability of ε, or based on the best-scored action in the Q-table with a probability of $(1 - \varepsilon)$;

$$A_t = \begin{cases} \max Qt(A) \text{ with probability } \varepsilon \\ random(A) \text{ with probability } (1 - \varepsilon) \end{cases} \quad (2)$$

This method is referred to as epsilon-greedy Q-learning [31], which offers a trade-off between exploitation and exploration through random selection.

In the simulation of this work, $\alpha = 0.4$, $\gamma = 0.55$ the exploration rate changes dynamically throughout the sessions, starting with a value of $\varepsilon = 0.5$ to allow exploration and decreasing to $\varepsilon = 0.3$ in later sessions.

Figure 1 describes the iterative process of the implemented algorithm; a learning unit is presented to the student (I), according to the parameters of the selected state. As the student interacts with the unit in the learning process, his performance parameters (later described) are measured (II). According to the performance measures a reward value R is calculated for this state, the Q-table is updated (III) and the action for the next state is selected (IV) according to Eqs. (1), (2).

The algorithm is implemented in the learning framework, which defines the characteristics of the learning unit presented to the student (stage I in Fig. 1). The next sections describe these characteristics, as well as the performance variables (stage II in Fig. 1), and the calculation of reward (stage III).

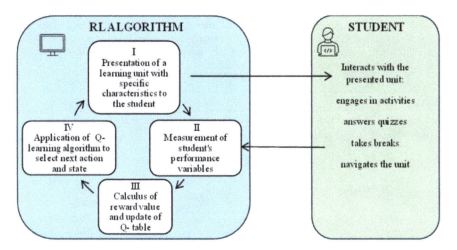

Fig. 1 Flowchart of the algorithm's iterative process of learning unit adaptation

3.2 Learning Framework Design

The framework of this study is based on the Universal Design for Learners [17], enabling the design of an automated learning environment. UDL's philosophy relies on providing multiple means of representation, expression and engagement. The use of UDL yields the formulation of a multi-dimensional learning space with the following continua of features, summarized in Table 1.

Representation addresses the way content is presented. The representation dimension in this work adopts the basic learning style differentiation employed in VARK model [4]: representation is a combination of Visual, Auditory, Read/Write and Kinesthetic learning units, and each learner usually prefers at least one of these medias. Enhancement of visual representation includes the use of text subtitles, colors, charts, figures, maps etc.; an aural learner may prefer lectures, podcasts, discussions, animations, and movies (auditory and visual enhancement). Enhancement of Read/Write modality focuses the learner on textual content, handouts, notes, while kinesthetic learning involves hands-on learning, direct practice and problem-solving learning (PBL).

The mode of representation influences the other fields, in the way they are presented to the student. The algorithm is to learn the user's preferences and to adopt the most suitable way to represent information.

Action and Expression address the involvement of the student by different means of communication: the student may be presented quizzes, instructional games, explanatory links, exploratory links, different modalities of interaction with peers and instructors, as well as different modalities of assessment. Assessments may be presented at the end of each chapter, but also along the learning process as quizzes, multiple choice questions and surprise scored activities. The expression

Table 1 The multidimensional space of potential learning units: representation modalities refer to learning styles–with no levels but rather different representations. Action/expression modalities include enhancement levels of assessment and interaction, engagement modalities include enhancement levels of content chunking and navigation

		Level 1	Level 2	Level 3	Level 4
Representation		Read-Write	Visual	Auditory	Kinesthetic
Action/ Expression	Assessment	Final exam	Final exam, Chapter quizzes (beginning and end)	Final exam, Chapter & Section quizzes	Final exam, Chapter & Section quizzes, Embedded questions
	Interaction	Basic links	Links	Links, Class forum, Discussions	Links, Class forum, Discussions, Student–teacher interactions
Engagement	Unit length	Chapter	Section	Sub-section	Short chunk in sub-section
	Navigation	Basic course structure	Horizontal	Horizontal + Vertical	Horizontal + Vertical + Personal schedule

of the student is closely related to his motivation, showing his engagement in the presented activities.

The third learning dimension is **Engagement**, which refers to the student's ability to engage in learning. To ease and enhance engagement, the system should allow different lengths of learning units.

Traditional learning organizes learning material in chapters and sub-chapters, while modern students usually have difficulty in sustaining concentration for long periods of time. This is especially relevant for ADHD students. Small units of material enable easier assimilation of new information. Engagement also refers to the enhancement of control features enabling self-determination and autonomy; this may be achieved by navigation options, as well as a clear presentation of the course's structure and student's progress, thus minimizing distraction and increasing focus. The model defines several modalities of navigation, organized in possible levels of enhancement: horizontal navigation means content navigation (between chapters, sections, sub-sections), while vertical navigation refers to the possibility of different difficulty levels or different elaboration levels on a specific learning content. A personal schedule tool enables the student to visualize the structure of the course and the student's current degree of completion and to plan/schedule his next steps.

3.3 Learning Performance Variables and Reward

The implementation of the adaptive e-learning system involves the definition of learning performance predictors, i.e. objective measures reflecting the effectiveness of the learning features presented to the learner. In the current work simulation, six performance predictors are suggested and normalized in order to ease the simulation with values between 0 to 1. The student's learning performance is evaluated according to the following normalized performance variables, shown in Table 2.

The number of breaks taken by the student, P1, is representative of his engagement and concentration levels [32]. The break is detected when there is inactivity for a specified time duration. The number of breaks is normalized by the maximum number of breaks, evaluated per characteristic chapter length arbitrarily and later adapted to personal student's actual breaks number throughout the learning process.

The overall duration of inactivity, P2, is a measure of disengagement and is normalized by the overall session duration (inactivity + activity times).

The number of selected activities, P3, is a measure of the student's expression but is also related to his engagement level [33]. The use of links indicates that the student is active and involved in his learning process. Moreover, positive/negative feedback from the performance of a quiz may motivate or discourage the student, hence influencing his engagement level. The number of selected activities is normalized by the total number of activities available in the specific chapter.

The number of correct answers to questions regarding the current chapter [34]. P4, is normalized by the total number of questions in the chapter. This measure relates to the efficiency of the adaptive learning strategy showing in good scores achieved by the student. In order to ensure a constructive and continuous learning process, it is important to evaluate the student's capability to incorporate new information into previously acquired knowledge. Therefore, the number of correct answers to questions that evaluate the integration of former knowledge is considered [34]. P5, normalized by the total number of integrative questions within the chapter.

Table 2 Measures of student's performance

Performance variable	Related network	Measure
P1	Engagement	Number of breaks
P2	Engagement	Time of inactivity
P3	Expression	Number of selected activities: links, performed quizzes and interactions
P4	Expression	Number of positive answers in evaluation methods of current knowledge
P5	Expression	Number of positive answers in evaluation methods of previous knowledge integration
P6	Expression	Number of completed sections

P6 is representative of the amount of material covered by the student [35]. It is defined as the number of completed units, normalized by the total number of units required for chapter completion.

The algorithm also informs of uncompleted sessions, by measuring the time of eventual breaks resulting in session or even course dropout.

The performance predictors are measured and collected in the learning unit process. As the unit is completed the reward value R is calculated according to the following rationale:

If (unit was completed)

$$R = \frac{(1-p1)+(1-p2)+p3+p4+p5+p6}{6}$$

Else

$$R = -0.1 \tag{3}$$

The reward is calculated by averaging the total scores of the performance variables. Other reward criteria may be applied, according to specific course objectives or tutor requirements. Note that performance parameters P1, P2 indicate the normalized time and duration of breaks and therefore a small value indicates higher engagement. For P3–P6 a higher value indicates higher engagement or expression.

If the unit is not completed the performance values are not relevant and the reward is assigned a negative value indicating that the conditions are not favorable.

4 Simulation

A simulation is performed, to validate the implementation of the algorithm in the proposed e-learning framework. Learning configurations and performance parameters are defined, to describe the personalization of the learning process in the specific study case of a student diagnosed with ADHD.

4.1 Learning Configurations

The whole space of possible configurations is a 4X4X4 matrix, where each of the categories of interest (representation, action, engagement) includes 4 modalities or levels of intensity. In real conditions, the performance of the student is monitored and calculated for every single combination according to the performance parameters P1-P6. In a simulation, the measurement is replaced by the assumption of performance values in the space. Since the number of possibilities is too large to determine, the space is divided into characteristic learning configurations relevant to students diagnosed with ADHD. To simplify the evaluation of performance parameters, they are assumed to be uniform in each configuration. The learning configurations are schemes combining different possible learning units. They are visually presented in Table 3.

In **configuration 1** the algorithm presents the course in a visual and read/write style, the number of quizzes is medium–high but limited to the beginning and end of

Table 3 The learning configurations representing specific combinations of learning units

Configuration 1		Level 1	Level 2	Level 3	Level 4
Representation		R/W	Visual		
Action/ Expression	Assessment		Final exam, Chapter quizzes	Final exam, Chapter & Section quizzes	
	Interaction		Links	Links, Forum, Discussions	
Engagement	Unit length		Section	Sub-section	
	Navigation		Horizontal	Horizontal + Vertical	
Configuration 2		Level 1	Level 2	Level 3	Level 4
Representation		R/W	Visual	Auditory	Kinesthetic
Action/ Expression	Assessment				Final exam, Chapter & section quizzes, Embedded questions
	Interaction				Links Class forum, Discussions, Student–Teacher interactions
Engagement	Unit length	Chapter	Section	Sub-section	Short chunk
	Navigation	Basic	Horizontal	Horizontal + Vertical	Horizontal + Vertical + Personal schedule
Configuration 3		Level 1	Level 2	Level 3	Level 4
Representation		R/W	Visual		
Action/ Expression	Assessment		Final exam, Chapter quizzes	Final exam, Chapter & Section quizzes	
	Interaction		Links	Links, Forum, Discussions	
Engagement	Unit length				Short chunk
	Navigation				Horizontal + Vertical + Personal schedule

sections/chapters, the number of available links is medium/high, with the possibility to enter discussions in class forum, the content is organized in sections/sub-sections (medium-short chunks), and the navigation tools enable much freedom in learning flow. This configuration includes a good level of engagement and expression tools, presented mainly in a visual mode: this is a configuration which is expected to work well for students with ADHD, with preferred visual style and relatively high levels of stimuli and control options.

Configuration 2 represents a higher level of stimulation, possibly needed in high concentration modes of students with ADHD. This configuration includes a combination of learning styles with no particular preferred presentation mode, with maximum levels of engagement and expression.

Configuration 3 is very similar to configuration 1, in that visual and read/write representation styles are enhanced, the amount of action items (links and assessments) is medium/high, but the content is presented in very short chunks. A high level of navigation is also needed to organize efficiently the content. This is also a scenario meant to address the needs of students with ADHD, who may benefit from short, easy to digest sections to overcome over or under stimulation, thus increasing their engagement level.

Configuration 4 includes all the combinations that are not defined by configurations 1–3.

To demonstrate the suggested framework we designed a test case describing the learning process of a student with ADHD. A simple scenario, describing the student's changing behavior is defined over ten weeks of an e-learning process. Accordingly, we assumed his performance variables for each possible configuration (Table 4).

4.2 Test Case Description

Alon is a student diagnosed with ADHD. His optimal learning representation style is visual. During the first three weeks of the course, he is taking his ADHD medication. He is generally attentive but performs best in configuration 1 where assessment, interaction, unit length and navigation are at levels 2–3, and the content is represented mainly in visual and read/write modalities. In this configuration he needs few and short breaks ($P1 = 0.2$, $P2 = 0.1$), he selects many activities ($P3 = 0.7$), and answers many correct answers in the assignments ($P5 = 0.7$, $P6 = 0.8$). In configuration 2, when there are many activities, navigation options and quizzes (level 4), he is sensory overwhelmed, and it badly affects his performance. He needs more and longer breaks ($P1 = 0.5$, $P2 = 0.5$), selects a low percentage of activities ($P3 = 0.2$), and achieves lower scores in assignments ($P5 = 0.4$, $P6 = 0.5$).

In configuration 3 with short units (level 4–short chunks), he does well but covers less material (P1–P5 similar to configuration 1, $P6 = 0.3$). His scores in configuration 4 are low, given the fact that it includes learning units that are not desirable preferences for this specific student. Alon performs 200 learning presented units each week, and in all configurations, he manages to complete the units during the first and last three

Table 4 The values assigned to performance variables in the simulation

Week	Configuration 1			Configuration 2			Configuration 3			Configuration 4		
	1–3	4–6	7–10	1–3	4–6	7–10	1–3	4–6	7–10	1–3	4–6	7–10
P1 Breaks number	0.2	0.6	0.2	0.5	–	0.5	0.1	0.1	0.1	0.3	0.5	0.3
P2 Time of inactivity	0.1	0.6	0.1	0.5	–	0.5	0.1	0.1	0.1	0.2	0.5	0.2
P3 Selected activities	0.7	0.3	0.7	0.2	–	0.2	0.7	0.7	0.7	0.4	0.2	0.4
P4 Correct answers current unit	0.7	0.3	0.7	0.4	–	0.4	0.7	0.7	0.7	0.6	0.3	0.6
P5 Correct answers integration	0.8	0.4	0.8	0.5	–	0.5	0.8	0.7	0.8	0.5	0.3	0.5
P6 Completed units	0.7	0.7	0.7	0.7	–	0.7	0.3	0.3	0.3	0.65	0.5	0.65
Unit completion	1	1	1	1	0	1	1	1	1	1	1	1

weeks. During the following three weeks (weeks 4–6) it is spring break, and he is not taking his ADHD medication. Furthermore, he is distracted by personal issues. He has more difficulty with maintaining attention and does not complete learning units in configuration 2 where the number of links, exams and navigation options impose too much cognitive load. As his ability to maintain attention is lower during these weeks, he performs best in configuration 3 with short units. In this configuration he needs few and short breaks (P1 = P2 = 0.1), he selects many activities (P3 = 0.7), and answers many correct answers (P5 = P6 = 0.7); as the units are short, he does not cover much material (P7 = 0.3). In configuration 1 he takes many long breaks (P1 = P2 = 0.6), selects few activities (P3 = 0.3), and does not succeed often in answering correctly the assignment's questions (P4 = 0.3, P5 = 0.4).

In weeks 7–10 he returns to a new semester and back to his routine including his ADHD medications and performs as he did during the first 3 weeks.

For a full description of the simulation performance variables in the different configurations and weeks see Table 4.

5 Results

The following results expose the performance of the algorithm, revealed by the presentation of specific learning units as a response to the student's needs (the scores of the performance predictors). The number of presented units per configuration are computed and analyzed for each week.

Figure 2 shows the number of presented learning units in each configuration as the learning proceeds in time. As the total number of units in each learning scheme differs, the number of presented units is normalized by the number of units available in the specific configuration. Every week the algorithm suggests 200 learning units, each of them automatically combines learning features according to the reinforcement learning algorithm. In the first three weeks, Alon performs best in configuration 1, he is overwhelmed by too many activities in configuration 2 and his results are good when presented by configuration 3, although he covers less material. Accordingly, the algorithm suggests more units in configurations 1 and 3, with higher values in configuration 1, showing good adaptivity in short time. It should be noted that the algorithm has no prior knowledge of the student learning abilities, therefore it takes time for the algorithm to find the optimized working point. The implementation of initial preferences may result in the achievement of better suggestions in shorter time. In weeks 4–6 Alon does not perform well in configuration 1, does not complete sessions in configuration 2 but does better in configuration 3 with short segments; the algorithm suggests more units in configuration 3, accordingly. The algorithm adapts to the change in student's behavior very quickly, showing distinct appropriate changes already in week 4. In the last three weeks, the conditions are similar to those in weeks 1–3, but the algorithm has learned the student's preferences enough to present many more units in configuration 1 (where Alon performs best) than in the past.

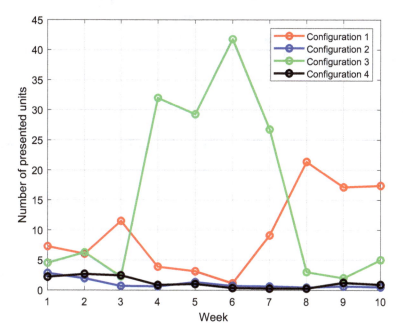

Fig. 2 The normalized number of presented units in each configuration

Figure 3 shows the presented units per week for each configuration characteristic (Representation, Action and Engagement) within the resolution of four Levels (1–4). In the first week the content is represented in all styles, and in the following two weeks the algorithm suggests mainly Read/Write and Kinesthetic Styles. The algorithm manages to recognize Read/Write as a preferred style, but does not present enough units in Visual style as expected, failing to identify the student's needs in the very first weeks. Nevertheless, it corrects its style suggestions in the following weeks, presenting content in Read/Write and Visual styles from week 4 to week 10. As for Action-related units presentation (middle column bar in Fig. 3), the algorithm learns very quickly that the preferred levels of action are the intermediate ones (2–3); these levels of action are present in both configurations 1 and 3, where the student performs well in most weeks, this explains the fast and accurate adaptation.

In the last weeks the algorithm mostly presents Action contents with high intensity (3) and some contents at intermediate level (2). It can be concluded that the algorithm adapts very well in presenting Action-related units according to the student's preferences resulting in his performance parameters.

The bottom column bar in Fig. 3 shows the Engagement-related units presented to the student; in the first three weeks there is no evident strategy in Engagement level presentation, although more Engagement-related content is presented with high intensity (3). The student does prefer configuration 1 where Engagement levels are medium–high (2–3), but also performs not bad in configuration 3, where Engagement level is the highest (4). This may explain the lack of a clear trend in the algorithm's

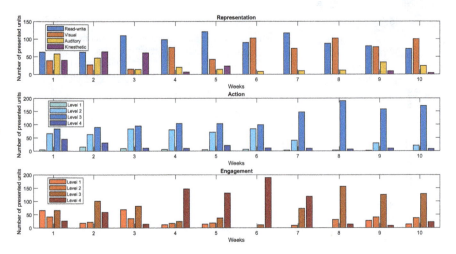

Fig. 3 The distribution of the number of learning units presented each week: in the top column-bar chart, the number of units presented in different Representation styles (Read/Write, Visual, Aural, Kinaesthetic). The number of Action-related content units (Middle chart) and the number of Engagement-related content units (Bottom chart), presented in four increasing intensity levels

decisions; nevertheless, it adapts well and presents units mainly in Engagement level 3. During weeks 4–6, more units are presented with the highest Engagement level (4), corresponding to Alon's best performance in configuration 3 with short content chunks. In the last weeks the algorithm has already learned during the first 3 weeks that the preferred level of Engagement is level 3, and presents units mainly at medium–high levels of intensity (2 and mostly 3). These results show that the algorithm is highly responsive to changes in student's behavior, with increasing adaptive accuracy as the learning process progresses.

Figure 4 shows the normalized number of presented learning units in each configuration as the learning proceeds in time, for different iteration numbers, corresponding to the total number of units presented to the student every week. The number of iterations affects the algorithm's rate of convergence to the optimal personalized learning suite.

The simulation is conducted for 50, 100 and 200 iterations and the results are displayed and compared in the figure. As the number of iterations increases, a better adaptation is expected. The differences between 100 and 200 iterations cases are minor, mainly in the last weeks when the algorithm has sufficient time to learn the environment.

In the first three days both cases present similar trends, but values differ significantly in configurations 1 and 4. In addition, a delayed response time to the changes in week 4 is observed in configuration 3 for 100 iterations, while the algorithm responds immediately to changes with 200 iterations.

As expected, an amount of 50 iterations yields slower and more fluctuated response to changes, showing in almost all configurations along the whole duration of the

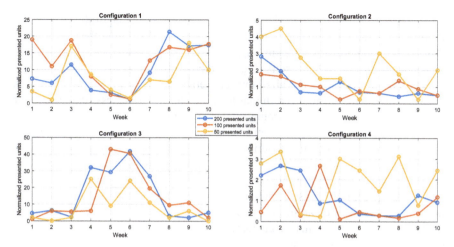

Fig. 4 The normalized number of presented learning units in each configuration, for 50, 100 and 200 iterations

learning process. The implications and inherent conflict in the choice of an adequate iterations number are presented in the following discussion.

6 Discussion

The present study proposes an adaptive e-learning framework, capable of providing in real time suitable Representation, Expression and Engagement-related contents, responding efficiently to the student's needs. The implementation of the RL algorithm in the framework is demonstrated by the simulation of a test case where the student presents ADHD-related learning characteristics. In real e-learning environments, where performance parameters are measured, there is no need for specific configurations and the algorithm is relevant and accessible to other learning limitations with minor changes in performance criteria, e.g. OCD, fatigue, or physical perceptual limitation (vision, hearing, etc.).

Contrarily to the majority of current solutions, where questionnaires are usually involved in the identification of individual learning preferences, the proposed framework is independent of either static or dynamic self-evaluation. As previously noted, the use of inventories is time-consuming, specific time-limited and subjective [8]. Nonetheless, it is possible to integrate a questionnaire at any instance during the learning process. Diagnostic assessment is generally performed at the beginning of the process by quizzes or self-report, identifying personal learning style preferences, prior knowledge as well as strengths and areas of improvement. The algorithm may particularly profit from an initialization of the Q-learning process, by the implementation of initial conditions by self-evaluation. For example, if the student prefers visual representation of the contents, the initial environment of the algorithm can be set accordingly, thus economizing in iterations number, or expediting the convergence process to the optimal learning unit suggestion.

As demonstrated in the results (Fig. 4), the number of weekly presented units affects the rate at which the algorithm converges to an optimized solution thus resulting in a better customized digital learning suit. Nevertheless, the number of presented units means more learning sessions required from the student, that may result in a negative effect on student's performance. For example, the student may be exhausted by the cognitive load imposed by too many learning units and loose attention, and this may damage the student's level of engagement and in the long term may cause him to lose interest in the entire course and drop out. Therefore, it is very important to adjust the amount of learning segments/sessions so that, on one hand, the software can be adaptive and learn the users' abilities according to his dynamic actual performance characteristics but will not harm his readiness and desire for the learning process.

The number of required iterations (number of learning units per week) may also be lowered if the algorithm within the framework is used on a regular basis throughout the academic curriculum, enabling the algorithm to achieve deeper knowledge of its environment (the student) and higher accuracy in its suggestions. Current LMS centralized student modeling servers can accumulate student experiences and data, available for future courses [36]. The artificial agent is trained to constantly learn student's skills and behaviors, so that in the next learning experiences the agent can retrieve and use former information, improve and refine its capabilities in adapting to the learner.

As for the framework's learning model, the present work characterizes the adaptive features according to UDL inclusive theory, where Representation, Action and Engagement are accounted; Representation categories are defined, inspired by VARK learning styles model, Action and Engagement-related contents vary within an increasing resolution of 4 Levels (1–4), resulting in a 3D space of 64 learning characteristics. Since the purpose of the simulation is to illustrate the feasibility and adaptivity of the algorithm, it is important to remember that the algorithm can be applied to a larger or different characteristics space. The number of levels for each parameter as well as a greater number of modeling characteristics may be implemented. Different learning styles models may be easily integrated, thanks to the flexibility and relative simplicity of the algorithm. Increasing the space dimensions/size may allow more accurate tailored solutions while addressing more detailed levels of the student's abilities, although increasing the number of iterations needed.

In this framework, the a-priori choice of a particular representation style does not prevent the exploration of other styles during the learning process. An important argument deals with the long-term benefits of experiencing new learning styles; the identification of a preferred learning style may lead to the adaptation to a stable preferred learning strategy, which provides familiarity and ease to the learner but may limit him in meeting new future challenges with flexibility. Moreover, individual learning styles may vary according to personal circumstances. Therefore, the adaptation in the long-term to flexible learning styles is often encouraged [2]. The algorithm's randomness in the selection of new actions (new suggestions), inscribed in the Epsilon-greedy model (Eq. (2)) enables new experiences and a flexible environment needed for personal development. Future work may address this parameter, as well as learning rate and discount factor values, showing their influence on the adaptivity of the algorithm to the learner and on the iterative process.

Performance modeling is achieved by defining easy-to-monitor measures of student's attention (breaks, time of activity/inactivity), objective assessment of current and integrative knowledge, and engagement (clicks on links and activities). These parameters may be replaced by other performance predictors, according to specific course objectives or tutor requirements. Moreover, a more elaborated reward strategy may be easily implemented, dictated by motivation models, assessment policies or tutoring suggestions related to specific learning disability. The reinforcement learning approach evolves as we better understand the human learning process, both from a cognitive and a neurological point of view. Thus the algorithm itself may include additional parameters, both human neural features/parameters, and the cognitive capacity to learn. Physiological information may be gathered to improve the algorithm's evaluation of the learner's state, making use of sensors such as smartwatches, eye tracking devices, face reading devices, Electroencephalogram (EEG) [37]. The evaluation in real time of motivation, attention levels, cognitive load from the wearable sensors may be correlated with learning style and performance predictors, thus enhancing the reinforcement learning algorithm adaptive capabilities.

Dynamic and adaptive presentation of online courses may help students with ADHD and learning disabilities to adjust to e-learning. However, it is important to consider other aspects that present a challenge for these students in remote online learning. Students with ADHD reported remote courses as having a negative influence on their wellbeing and university experience [38]. They were challenged by reduced social interaction, schedule management and increased procrastination. Students with ADHD reported higher levels of loneliness during remote learning [39]. While feelings of loneliness may be relieved and mediated by more adaptive learning, other interventions such as college support and peer support must not be neglected.

Previous works also pointed out the challenges involved in the empirical validation of adaptive e-learning frameworks [40]. A highly adaptive framework involves a high number of featured learning units resulting in complex and time-consuming elaboration of course content to meet the student's needs diversity [41].

7 Conclusions

The suggested adaptive e-learning framework adopts modeling elements inspired by the Universal Design for Learning VARK learning styles model, combined with content-level and link-level (navigation) concepts. Reinforcement Learning algorithm is implemented, resulting in an inclusive adaptive methodology addressing the needs of students with and without ADHD.

A proof-of-concept simulation, given the assumed performance of a student diagnosed with ADHD in a specific scenario depicting his learning process, yields promising results, showing in good adaptive dynamic characteristics. The algorithm identifies and learns the preferred learning strategies of the student, resulting in highly responsive and accurate suggestion of suitable contents.

Future work will address a full parametric analysis of the algorithm's features, as well as the empirical implementation in an actual e-learning environment with real-time performance monitoring.

Competing Interests The authors declare that there are no conflicts of interest related to this research or this manuscript.

References

1. El-Sabagh HA (2021) Adaptive e-learning environment based on learning styles and its impact on development students' engagement. Int J Educ Technol High Educ 18(1)
2. Park OC, Lee J (2003) Adaptive instructional systems. In: Handbook of research for educational communications and technology, issue 1911
3. Hauger D, Köck M (2007) State of the art of adaptivity in e-learning platforms. In: LWA 2007—lernen—wissen—adaptivitat—learning, knowledge, and adaptivity, workshop proceedings
4. Fleming ND, Baume D (2006) Learning styles again: VARKing up the right tree! educational developments, SEDA Ltd. 7(4)
5. Coffield F, Moseley D, Hall E, Ecclestone K (2004) Learning styles and pedagogy in post-16 learning a systematic and critical review. Learning and Skills Research Centre
6. Felder RM, Silverman LK (1998) Learning and teaching styles in engineering education. Eng Educ 78(7)
7. Honey P, Mumford A (1986) Using our learning styles. Peter Honey, Berkshire
8. Truong HM (2016) Integrating learning styles and adaptive e-learning system: current developments, problems and opportunities. Comput Hum Behav 55
9. Nistor N, Neubauer K (2010) From participation to dropout: quantitative participation patterns in online university courses. Comput Educ 55(2)
10. Hartnett M, St. George A, Dron J (2011) Examining motivation in online distance learning environments: complex, multifaceted, and situation-dependent. Int Rev Res Open Distance Learn 12(6)
11. Bodalski EA, Flory K, Canu WH, Willcutt EG, Hartung CM (2024) ADHD symptoms and procrastination in college students: the roles of emotion dysregulation and self-esteem. J Psychopathol Behav Assess [Internet]. 2023 Mar 1 [cited 2024 Mar 8];45(1):48–57. https://doi.org/10.1007/s10862-022-09996-2
12. Advokat C, Lane SM, Luo C (2010) College students with and without ADHD. https://doi.org/10.1177/1087054710371168 [Internet]. 2010 Aug 2 [cited 2024 Mar 8];15(8):656–66. https://doi.org/10.1177/1087054710371168
13. Kuntsi J, Klein C (2012) Intraindividual variability in ADHD and its implications for research of causal links. Curr Top Behav Neurosci [Internet]. 2012 [cited 2024 Mar 9];9:67–91. https://pubmed.ncbi.nlm.nih.gov/21769722/
14. He S, Shuai L, Wang Z, Qiu M, Wilson A, Xia W et al (2021) Online learning performances of children and adolescents with attention deficit hyperactivity disorder during the COVID-19 pandemic. Inquiry [Internet]. 2021 [cited 2024 Mar 9];58. https://pubmed.ncbi.nlm.nih.gov/34647508/

15. Moriña A (2017) Inclusive education in higher education: challenges and opportunities. Eur J Spec Needs Educ 32(1)
16. Ralabate PK (2011) Universal design for learning: meeting the needs of all students, vol 16, ASHA Leader
17. Rose DH, Meyer A (2002) Teaching every student in the digital age: universal design for learning. Alexandria, ASCD, VA
18. Kumaresan M, Mccardle L, Chandrashekar S, Karakus E, Furness C (2022) Learning with ADHD: a review of technologies and strategies. J Technol Pers Disabil StIago
19. Nabizadeh AH, Gonçalves D, Gama S, Jorge J, Rafsanjani HN (2020) Adaptive learning path recommender approach using auxiliary learning objects. Comput Educ 147
20. Romero C, Ventura S, De Bra P (2004) Knowledge discovery with genetic programming for providing feedback to courseware authors. User Model User-Adapt Interact 14(5)
21. Wang TI, Wang K Te, Huang YM (2008) Using a style-based ant colony system for adaptive learning. Expert Syst Appl 34(4)
22. De Moura FF, Franco LM, De Melo SL, Fernandes MA (2013) Development of learning styles and multiple intelligences through particle swarm optimization. In: Proceedings—2013 IEEE international conference on systems, man, and cybernetics, SMC 2013
23. Villaverde JE, Godoy D, Amandi A (2006) Learning styles' recognition in e-learning environments with feed-forward neural networks. J Comput Assist Learn 22(3)
24. Hussain M, Zhu W, Zhang W, Abidi SMR (2018) Student engagement predictions in an e-learning system and their impact on student course assessment Scores. Comput Intell Neurosci 2018
25. Lopez CE, Ashour O, Tucker CS (2019) Reinforcement learning content generation for virtual reality applications. In: Proceedings of the ASME design engineering technical conference
26. Shawky D, Badawi A (2019) Towards a personalized learning experience using reinforcement learning. In: Studies in computational intelligence
27. Hocine N, Sehaba K (2023) A systematic review of online personalized systems for the autonomous learning of people with cognitive disabilities. Hum Comput Interact
28. Cinquin PA, Guitton P, Sauzéon H (2023) Toward truly accessible MOOCs for persons with cognitive impairments: a field study. Hum Comput Interact 38(5–6)
29. Sutton RS, Barto AG (2018) Reinforcement learning, 2nd edn An introduction - complete draft. The MIT Press
30. Clifton J, Laber E (2020) Q-learning: theory and applications. Annu Rev Stat Its Appl 7
31. Even-Dar E, Mansourt Y (2002) Convergence of optimistic and incremental Q-learning. In: Advances in neural information processing systems
32. Krieter P (2022) Are you still there? an exploratory case study on estimating students' LMS online time by combining log files and screen recordings. IEEE Trans Learn Technol 15(1)
33. Cerezo R, Sánchez-Santillán M, Paule-Ruiz MP, Núñez JC (2016) Students' LMS interaction patterns and their relationship with achievement: a case study in higher education. Comput Educ 96
34. Gamage SHPW, Ayres JR, Behrend MB, Smith EJ (2019) Optimising Moodle quizzes for online assessments. Int J STEM Educ 6(1)
35. Avcı Ü, Ergün E (2022) Online students' LMS activities and their effect on engagement, information literacy and academic performance. Interact Learn Environ 30(1)
36. Brusilovsky P (2023) A component-based distributed architecture for adaptive webbased education. Artif Intell Educ (Fig. 1)
37. Standen PJ, Brown DJ, Taheri M, Galvez Trigo MJ, Boulton H, Burton A et al (2020) An evaluation of an adaptive learning system based on multimodal affect recognition for learners with intellectual disabilities. Br J Educ Technol 51(5)
38. Dobrovsky A, Borghoff UM, Hofmann M (2019) Improving adaptive gameplay in serious games through interactive deep reinforcement learning
39. Laslo-Roth R, Bareket-Bojmel L, Margalit M (2022) Loneliness experience during distance learning among college students with ADHD: the mediating role of perceived support and hope. Eur J Spec Needs Educ [Internet]. 2022 Mar 4 [cited 2024 Mar 17];37(2):220–34. https://doi.org/10.1080/08856257.2020.1862339

40. Ehlers UD, Pawlowski JM (2006) Quality in European e-learning: an introduction. In: Handbook on quality and standardisation in e-learning
41. Fournier H, Kop R (2015) MOOC learning experience design: issues and challenges. Int J E-Learn: Corp, GovMent, Healthc, High Educ 14(3)

Machine Learning Models to Detect AI-Assisted Code Anomaly in Introductory Programming Course

Hapnes Toba and Oscar Karnalim

Abstract The use of AI tools to complete class assignments has become a trend in recent years. The instructors' tasks become more complex in monitoring and evaluating students' progress. In this chapter, we experimented with various machine learning approaches for automatic anomaly detection, such as tree-based and parametric-based models. We also deepen the models with a neural-based approach and feature selection method. The datasets are derived from student weekly submissions from a Python introductory programming course in the first semester of the 2023–2024 academic year. The random committees model shows the best performance using all features with an accuracy of 96.7% and a recall rate of 62.9%. Our observation indicates special features that characterize each assignment type: branching, looping, and functions. Further experiments show that the accuracy of our final meta-learning models is above 95% during the testing by using the real dataset from students' submissions, with the best achieved for the branching assignment type by using all features. Based on these performances, we believe that our feature sets and meta-learning approach would be promising to detect anomalies in AI-assisted code submissions.

1 Introduction

Artificial Intelligence (AI) substantially affects human life [34]. Two promising examples are deep learning [25] and generative AI [15]. The former enables computers to observe complex patterns from data, while the latter allows computers to generate content that looks like humans.

Advances in technology are always a double-edged sword in education. While AI can facilitate a better learning environment [7], it can be misused to complete

---

H. Toba (✉) · O. Karnalim
Universitas Kristen Maranatha, Jl. Surya Sumantri No. 65, Bandung 40164, Indonesia
e-mail: hapnestoba@it.maranatha.edu

O. Karnalim
e-mail: oscar.karnalim@it.maranatha.edu

© The Author(s), under exclusive license to Springer Nature Switzerland AG 2025
E. Vendrell Vidal et al. (eds.), *Advanced Technologies and the University of the Future*, Lecture Notes in Networks and Systems 1140,
https://doi.org/10.1007/978-3-031-71530-3_11

assessments with limited understanding (i.e., plagiarism) [44]. Paraphrasing tools might be able to disguise plagiarism [37]. Generative AI can quickly provide worked solutions for some assessments [44]. Some educators respond to such issues by introducing assessments that AI cannot easily disguise or solve, such as project-based assessments [10] or case studies [5]. They also introduce a new type of assessment that focuses on how students can provide better prompt queries to generative AI [9]. As preventive attempts, educators can also employ a plagiarism detector that can deal with paraphrasing [2] or identify AI-generated work [24].

While paraphrasing is a long-standing issue in plagiarism, AI-generated work is relatively recent. Several automated detectors have been developed but do not work well in programming assessments [36]. We addressed the problem by developing an automated AI-assistance detector that relies on code anomaly, assuming AI-generated work tends to be far more complex than those written by students [24]. The detection is relatively promising, but it needs improvement for higher effectiveness. The detection employs no machine learning models to learn the pattern of code anomaly.

In response, we expand the work by employing machine learning models in this chapter. A machine learning approach could make the detection more effective since essential features could be detected directly from the models [13]. We experimented with machine learning approaches for automatic anomaly detection, such as tree-based and parametric-based models. We also deepened the models with a neural-based approach and feature selection method.

In the end, we propose whether there is a meta-learning strategy that hypothetically will perform better in terms of accuracy and recall. With these models, students' submissions are expected to be classified as the result of learning activities as suggested according to the class material or simply as a copy-paste product from an AI-generated application such as ChatGPT. We evaluate further whether statistical-based programming language surface features can classify whether an incoming submission is an anomaly compared to the lecture material provided. As a further exploration, we also compare the models using feature selection methods to ensure which features would have a high impact during the classification.

2 Related Work

Plagiarism refers to reusing one's work without sufficient credits for the reuse [22]. It is a common issue in programming education [41], and it has become more concerning due to the introduction of AI, especially the generative ones like ChatGPT [29]. Generative AI may provide too much help without deeply understanding a given matter [44]. Further, it isn't easy to put credits as generative AI augment responses from Inter-net data without acknowledging the sources. Regulations for using generative AI might be necessary [18].

There are some programming plagiarism detectors [4], including MOSS [40] and DECKARD [21]. They typically work by comparing student programs to one another and reporting those with high similarities [23]. Instructors manually observe reported programs to determine whether they result from plagiarism. High similarities are not always a result of plagiarism [28]. Some similarities are coincidental as they are required for program compilation, are suggested by the instructors, or are the most intuitive approach to solving the assessment [42].

Many programming plagiarism detectors relying on submitted programs can be classified into structure-based, attribute-counting-based, and hybrid [23]. Structure-based plagiarism detectors focus on program structure to determine similarities. They employ various algorithms like running the Karp-Rabin greedy string tiling [11], string alignment [26], and Winnowing [20]. Attribute-counting-based plagiarism detectors focus on occurrence frequencies of observed code attributes. The algorithms include a Cosine similarity [14], Latent Semantic Analysis [46], and BM25 [3]. Some attribute-counting-based plagiarism detectors aiming for higher effectiveness use clustering algorithms (e.g., Fuzzy C-Means [1]) or machine learning algorithms, like Support Vector Machine [50]).

Hybrid plagiarism detectors combine both structure-based and attribute-counting-based approaches. For effectiveness, the results of both approaches can be displayed together while investigating potential plagiarism cases [47]. The results of the structure-based approach can complement those of the attribute-counting-based approach or vice-versa. Sometimes, the result of the structure-based approach is considered a feature for the attribute-counting-based approach [43], either as a supervised machine learning feature [12] or an unsupervised machine learning feature [39]. For efficiency, the results of the attribute-counting-based approach can act as an initial filter for student programs passed to the structure-based approach. The attribute-counting-based approach can be a simple file size check [51] or a standalone tool [31].

Some plagiarism detectors capture the creation process of student programs, collecting more evidence of plagiarism. Students are expected to use dedicated programming workspaces that can record useful information for investigating plagiarism. The most straightforward approach is watermarking the student programs [8]. Other methods include capturing code snapshots [49], code edits [27], and seating positions (limited to physical classrooms) [6].

Conventional plagiarism detectors cannot easily detect AI-based plagiarism since AI-generated programs tend to be unique, while plagiarism detectors rely on program similarity. Capturing the creation process might be helpful but not practical, given that students are required to use dedicated IDEs. Some plagiarism detectors for AI-generated work have been developed but mainly focus on the general text [36]. GPTZero[1] identifies AI-generated work based on the distribution of sentences and words. AI Text Classifier[2] employs 34 learning models to identify AI-generated work. GPTKit[3] considers six learning models. Several well-known academia-related

---

[1] https://gptzero.me.

[2] https://openai.com/index/new-ai-classifier-for-indicating-ai-written-text/.

[3] https://gptkit.ai.

industries have introduced their detector for AI-generated work. Some examples are Turnitin[4] AI Detector and Quillbot[5] AI Content Detector.

We previously addressed the issue by proposing a plagiarism detector for AI-generated programs with a code anomaly [24]. Based on a tutorial website, we identified 34 uncommon syntax patterns for programming novices, spanning constants to functions. A program tends to be AI-generated if it has more uncommon syntax patterns. While the approach is promising, it purely relies on basic statistics. Machine learning approaches might help improve effectiveness.

3 Method

Our models are developed and evaluated on two separate datasets. The datasets are derived from student weekly submissions from a Python introductory programming course in the first semester of the 2023–2024 academic year, which had two parallel classes: class A with 28 students and class B with 26 students. Six tutors were involved who at least had a B+ final mark in the introductory programming course.

The tutors manually check each assessment for AI assistance. A critical task for tutors is mimicking student behavior in ChatGPT. While GitHub Copilot could be more suitable for programming, many of our students are more familiar with ChatGPT. Based on the tutors' experience in our previous research [44], they are instructed to label submissions with potential plagiarism and confirm them with students to form the experiment datasets. Figure 1 illustrates the dataset collections and the flow of our experiments.

There are 13 assignments ranging from simple data types to complex matrix and function assignments. For each week (except week 14), students were given two assessments: one lab and one homework assessment. The former should be completed in the lab session (about two hours), and the latter should be completed before the following week's lab session. Each assessment had three tasks: easy, medium, and challenging. Students were allowed to have small discussions with their colleagues and to use the Internet. However, the code should be written independently.

If AI thoroughly assisted a student's answers, any identified submissions would be considered entirely AI-assisted submissions and included in the real dataset. Otherwise, they would be regarded as partially AI-assisted submissions and included in the semi-real dataset. Since there are fewer completely AI-assisted sub-missions, the tutors were asked to complete the assessment with the help of ChatGPT in the semi-real dataset.

Thus, for the real dataset, we collect the (uncontrolled) real answers from those caught using ChatGPT while having to do the tasks themselves. The second one is a (controlled) semi-real collection that is partially developed by the students and partially developed using ChatGPT by the tutors during the weekly assessments.

---

[4] https://www.turnitin.com/solutions/topics/ai-writing/.
[5] https://quillbot.com/ai-content-detector.

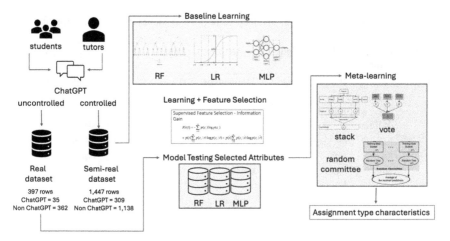

Fig. 1 Dataset collection and the experiment flow

Table 1 Statistics of the dataset

Dataset	#ChatGPT	#Non-ChatGPT	Total instances	% #ChatGPT (%)	% #Non-ChatGPT (%)	#Tokens
Real	35	362	397	8.8	91.2	308,702
Semi-real	309	1,138	1,447	21.4	78.6	1,339,526

Further, students needed to understand their code; it was unlikely that they reused advanced syntax constructs from the Internet. Students breaching these rules would be manually identified and penalized. By default, all student submissions were treated as independent submissions. Table 1 contains the main statistics of the dataset.

Based on the datasets, we developed two types of machine learning models. The first is based on a decision tree, i.e., random forests (RF) [32, 38]. The second is based on a parametric approach, i.e., logistic regression (LR) [45]. We also add a third model based on neural network settings, i.e., multilayer perceptron (MLP), as an enhanced machine learning approach [19].

Our models rely on 35 statistical occurrences of code anomaly features covering basic Python[6] programming language statistical features [48] as follows: constants (e.g., the use of scientific numbers), data structures (e.g., the use of list comprehension), branches (e.g., the use of shorthand if branching), loops (e.g., the use of 'range' with limited arguments), functions (e.g., the use of recursion), and syntax constructions (e.g., the use of iterator for iterating elements in arrays). The necessary hyperparameter settings of the basic classifiers during the experiments are presented in Table 2. Since all attributes are numeric, they are normalized using a standard scaler.

---

[6] https://www.w3schools.com/python.

Table 2 Important hyperparameters and their corresponding values during the models' development

Algo.	Hyper-param.	Value	Algo.	Hyper-param.	Value	Algo.	Hyper-param.	Value
RF	Max. depth	10	MLP	Momentum	0.2	LR	Grad. desc.	100
	Max. # of trees	100		Learn. rate	0.3		batch size	
	Max. # rand. attr.	int($\log_2$ (#attr) + 1)		#Hidden Layers	#attr. + #classes		Ridge reg.	1.0E-8
				Grad. desc. Batch size	100			

The results from the models will be deeply analyzed in terms of accuracy, sensitivity (recall), and precision. Influential feature sets will be analyzed using the information gain (entropy-based) feature selection method [35]. To enhance the models, we also developed some meta-learning strategies to ensure the characteristics of each assignment type's selected attributes. Finally, a comparison to our recent study [24], which is based on an information retrieval approach, will also be analyzed. The main description of the complete feature sets is presented in Table 3.

4 Experiment and Discussion

4.1 Baseline and Attribute Selection

We employed the WEKA Data Mining Software version 3.8.6[7] for all the experiments. As the baseline, we experimented with the semi-real instances in a 10-fold cross-validations scenario. This baseline will act as the primary benchmark of the performances in subsequent experiments. Since we are interested in the performance of whether an instance is classified as using AI-assisted tools, we are focusing our metric on minimizing the false-negative (FN) predictions. This FN prediction gives the proportion of how many instances are classified as not using AI-assisted tools while, in reality, they are. In other words, we try to increase the recall value.

Table 4 gives the baselines' accuracy, precision, recall, and root mean square error (RMSE) results. We also conducted the information gain feature selection method. Table 5 lists the attributes chosen from the 35 initial ones.

In Table 4, the RF classifiers perform best using all features (with 95.5% accuracy and 96.4% recall) and the selected ones (with 95.7% accuracy and 97.1% re-call).

---

[7] https://waikato.github.io/weka-wiki/.

Table 3 Extracted feature sets from the submitted programming code [24]

Feature group	Code	Description
Constants	C1	The number of scientific number constants (e.g., "−105.7e100")
	C2	The number of multiple line string constants (e.g., "text me" or "text me")
	C3	The number of inconsistent uses of string quotes: some string constants use single quotes while others use double quotes
	C4	The number of multiple variable assignments (e.g., "a, b = 5, 3")
Data structure	D1	The number of list assignments
	D2	The number of list comprehension statements
	D3	The number of tuple assignments
	D4	The number of set assignments
	D5	The number of dictionary assignments
Branching	B1	The number of shorthand (one line) branching statements (e.g., "if(a > b): print(a)")
	B2	The number of shorthand (one line) reversed branching statements (e.g., "print(a) if(a > b) else print(b)")
	B3	The number of branching condition statements without voluntary parenthesis (e.g., "if a > b:")
	B4	The number of branching condition statements without voluntary parenthesis on shorthand reversed branching statements (e.g., "print(a) if a > b else print(b)")
Looping	L1	The number of 'break' statements in 'while' loops
	L2	The number of 'continue' statements in 'while' loops
	L3	The number of 'else' statements as part of 'while' looping constructs
	L4	The number of 'pass' statements in 'while' loops
	L5	The number of condition statements without voluntary parenthesis as part of 'while' looping constructs
	L6	The number of 'break' statements in 'for' loops
	L7	The number of 'continue' statements in 'for' loops
	L8	The number of 'else' statements as part of 'for' looping constructs
	L9	The number of 'pass' statements in 'for' loops
	L10	The number of 'for' looping statements with one 'range' argument (e.g., "for x in range(6):")
	L11	The number of 'for' looping statements with two 'range' arguments (e.g., "for x in range(2,6):")
Functions	F1	The number of 'return' statements without returned value(s)
	F2	The number of arbitrary arguments in functions (e.g., "def func(*kids):")
	F3	The number of 'pass' statements in functions
	F4	The number of recursion calls
Miscellaneous	M1	The number of 'len' function calls
	M2	The number of 'iter' function calls
	M3	The number of importing modules
	M4	The number of regular expressions
	M5	The number of exception syntax constructs
	M6	The number of lambda expressions
Tokens	Tokens	The number of tokens in the submitted code

Table 4 The baseline performance of the models (10-fold cross-validations) in the semi-real dataset (in percentage). Values in bold indicate the best performance

Semi-real 10-CV	RF-All	LR-All	MLP-All	RF-Sel.	LR-Sel.	MLP-Sel.
Accuracy	**95.5**	87.8	88.1	**95.7**	87.4	88.2
Precision	**84.7**	80.3	76.4	**85.0**	79.5	77.6
Recall	**96.4**	56.6	64.1	**97.1**	55.3	62.8
RMSE	19.3	30.4	31.3	19.3	30.5	30.7

Those values significantly outperform the other classifiers with $p = 0.05$ and 95% confidence. These performances are relatively high, considering the true ChatGPT class has a much lower proportion than the non-ChatGPT class.

The performance of the other classifiers, in terms of accuracy, has also exceeded the imbalanced class proportion, as presented in Table 4. This fact will say that all of the classifiers have the potential to be used as plagiarism detectors. In other words, some influential features in each assignment type will be important in determining the use of AI-assisted tools in introductory programming. There is also a concern that the RF classifiers tend to overfit [17].

In this sense, developing some scenarios that combine all the classifiers in reality might be essential. We also inspect the classifiers' performance using all of the original and comparable features of the selected ones. This fact suggests that the classifiers with the selected features are highly efficient and fit for actual use.

From Table 5, we can deduce that discriminating AI-assisted and non-AI-assisted code for Python code is characterized by two essential features with weights greater than 0.1, i.e., the number of list assignments (D1) and the number of branching condition statements without voluntary parenthesis (B3). In Python, non-primitive data structures primarily use list assignments, such as arrays and lists. This fact

Table 5 The selected important attributes with their weights

No.	Weight	Attributes	No.	Weight	Attributes
1.	0.20927	D1	11.	0.02092	F1
2.	0.13936	B3	12.	0.01880	B2
3.	0.08390	L10	13.	0.01804	C3
4.	0.06899	M1	14.	0.01770	L2
5.	0.06168	L5	15.	0.01240	F4
6.	0.05729	D2	16.	0.01074	D4
7.	0.05237	Tokens	17.	0.01057	L11
8.	0.05102	D5	18.	0.00507	C2
9.	0.04305	L1	19.	0.00308	M5
10.	0.03628	C4			

supports ChatGPT's suggestion to use an advanced setting of data structure definition. Due to code efficiency, ChatGPT usually suggests using advanced list construction and omits the parenthesis in branching conditions, although students are encouraged to use them in the classroom. A special section after the meta-learning discussion will investigate further the impact of the other influential features.

4.2 Real Dataset

The real dataset consists of the uncontrolled situation from the students' submissions. We experimented with the selected models from Table 4 and compared their performance with the baseline. The results of the real dataset are presented in Table 6. The performance of the RF classifier in the real dataset dropped significantly in recall (-42.8%), although the precision increased by 15.0%. These results suggest that the RF classifier failed to discriminate the non-AI-assisted answers from the assisted ones (false-positive).

In a real scenario, it could be beneficial since the lecturer or tutor should inspect all the suspected AI-assisted answers. The only classifier with an increasing recall metric in the real dataset is the MLP (+8.6%). The recall differences between the MLP (71.4%) and RF (54.3%) classifiers are statistically significant with $p = 0.05$ and 95% confidence). This fact also suggests that the recall of the MLP is relatively stable compared to the RF classifier, with an increase of 8.6%.

4.3 Meta-Learning

Previous experiments show that no ultimate classifiers would perfectly discriminate the submitted answers into AI- or non-AI-assisted. We have the sense that each classifier would perform better in some cases. In that case, it would be nice to have an ensemble between the models or the meta-learning strategy. Some alternatives for this purpose are stacking, voting, and random committee (RC).

Table 6 Model performance using the selected attributes (in percentage) compared to the baseline. Values in bold indicate the best performance

Real dataset	Testing results			+/− Against the baseline		
	RF-Sel.	LR-Sel.	MLP-Sel.	RF-Sel.	LR-Sel.	MLP-Sel.
Accuracy	**95.9**	91.4	89.7	−0.2	**+4.0**	+1.5
Precision	**100.0**	51.4	44.6	**+15.0**	−28.1	−33.0
Recall	54.3	54.3	**71.4**	−42.8	−1.0	**+8.6**
RMSE	19.4	26.0	28.8	+0.1	−4.5	−1.9

Model stacking is a meta-learning approach that learns to apply the best individual predictor according to the input's characteristics after several learning phases [16]. At each level, a different algorithm could be used. The result from the previous model will be used as a new instance to be used as part of the learning dataset at the next level. In our scenario, we compose the stack in the following sequence: RF, LR, and MLP. We combine the models from the least generalized model—since RF is based on randomized rules of decision trees—to the most generalized MLP since MLP is composed of randomized weights, which minimizes the error in the evaluation function.

A voting model tries to decide a single class by calculating majority decisions. We use the majority decision of the three models, RF, LR, and MLP, as the final class for an instance classification in the dataset [30]. For instance, if RF and LR decide on a sample in the dataset as an AI-assisted class, the whole decision will be AI-assisted.

The random committee approach builds an ensemble of randomizable base classifiers [33]. Each base classifier is constructed by randomizing a seed number based on the same data. The final prediction is formed by taking an average of the predictions supplied by the individual base classifiers. In our case, each random tree will be trained by a randomized bagging scenario from the dataset, and an average of the probability score of the AI-assisted class will determine the final class. A predicted probability of an instance with an average starting from 0.5 will be considered AI-assisted.

Table 7 shows the performance of the meta-models by combining the classifiers. We employed the 10-fold cross-validations as the baseline and tested them using the real dataset. With this meta-learning approach, discrepancies between the model characteristics should be minimized, and we should achieve better model predictions.

Table 7 shows that the RC model performs best using all features with an accuracy of 96.7% and a recall rate of 62.9%. The performance of the RC model using feature selection results also shows the best accuracy (95.9%) among the other feature selection-based models. It is not significantly different from using all features. However, the recall value dropped significantly (54.3%). This fact indicates that some features may be particular to specific assignment types. In the next experiment in Table 8, we used a form of RC meta-learning on real datasets to carry out testing.

Table 7 Baseline meta-learning of the models (10-fold cross-validations) in the semi-real dataset (in percentage). Values in bold indicate the best performance

Semi-real 10-CV	Stack	Vote	RC	Stack -Sel.	Vote -Sel.	RC-Sel.
Accuracy	89.2	92.9	**96.7**	87.9	93.2	**95.9**
Precision	42.3	60.6	**100.0**	40.6	61.8	**100.0**
Recall	**62.9**	57.1	62.9	**80.0**	60.0	54.3
RMSE	27.8	22.1	18.9	28.4	23.5	19.1

Table 8 Model performance of real dataset using random committee (in percentage). Values in bold indicate the best performance

Real dataset	RC	RC-Sel.
Accuracy	**95.6**	95.4
Precision	**84.9**	84.2
Recall	**96.4**	**96.4**
RMSE	19.2	19.2

Accuracy and recall performance in the RC models that use all and selected features are not significantly different (95.6% vs. 95.4%). We could even say that they have comparable performance. Thus, the RC model is used as a holistic meta-learning model that averages the probabilistic values of each model used for prediction, in this case, RF, LR, and MLP.

5 Feature Selection for Assignments Type

The experiments in this section specialize in identifying each attribute's importance level in the dataset. The assignments can be divided into three subjects: branching, looping, and functions assignments (including matrix). Discussions related to arrays, lists, and matrices are combined into functions. The feature selection process uses the information gain selection method, as seen in Table 9.

Table 9 will be analyzed for features with an importance weight above 0.1. This choice was made solely for consistency in subsequent analyses and was unrelated to the preferences of any particular task forms. In the branching assignment type, it can be seen that the features that are considered important refer to things that regulate the length and compactness of writing program code, namely tokens, the number of list assignments (D1), the number of dictionary assignments (D5), number of branching condition statements without voluntary parenthesis (B3), the number of 'for' looping statements with one 'range' argument (L10), and the number of recursion calls (F4).

The program codes offered via ChatGPT are often made as compact as possible to shorten the number of tokens written. Apart from that, the code in ChatGPT usually offers more complex data structures, such as dictionaries or loops and functions for algorithm efficiency. For first-year informatics students, especially for initial assignments before the mid-semester, the assignments are more directed at solidifying full syntax writing and ensuring logical flow in programming.

For assignments that refer to the looping concept, the recommended discriminating features are the number of list assignments (D1), the number of condition statements without voluntary parenthesis as part of 'while' looping constructs (L5), the number of branching condition statements without voluntary parenthesis (B3), and number of 'break' statements in 'while' loops (L1). ChatGPT naturally provides

Table 9 Selected features to discriminate AI-assisted code submission for each assignment task: branching, looping, and functions. The grey-shaded cells intersect the features suggested in all assignment tasks

Branching		Looping		Functions	
Weight	Attributes	Weight	Attributes	Weight	Attributes
0.2981	Tokens	0.2094	D1	0.4607	D1
0.2533	D1	0.1428	L5	0.3103	L10
0.179	D5	0.1204	B3	0.2336	B3
0.1761	B3	0.113	L1	0.1621	M1
0.1152	L10	0.0948	F1	0.1456	D2
0.1021	F4	0.0549	D5	0.0734	L5
0.0807	F1	0.0398	C3	0.064	C4
0.0654	C3	0.0361	M1	0.0545	B2
0.0599	L6	0.0311	L11	0.0497	D4
0.0438	C4	0.0292	D2	0.0405	C3
0.0395	C2	0.0232	B2	0.0319	L2
0.0195	B2			0.0291	F1
0.0195	D2			0.0279	L1
0.0195	M5			0.0113	L6

program code that is considered the most efficient. From that perspective, the program code will be directed toward more advanced writing by eliminating several essential concepts, such as the tidiness of typesetting in nesting (ChatGPT usually omits parenthesis) and terminating the flow of logic after something happens or not (ChatGPT usually omits break).

In the functions assignments type, we can see the following features are suggested as essential to distinguish the ChatGPT code suggestions: number of list assignments (D1), number of 'for' looping statements with one 'range' argument (L10), number of branching condition statements without voluntary parenthesis (B3), number of 'len' function calls (M1), and number of list comprehension statements (D2). Again, the issue of code compactness is a critical characteristic ChatGPT offers. Using these features, one can inspect the code likely to be directly copied from ChatGPT. Tables 10 and 11 present an example of student submissions that considered AI-assisted code.

Tables 10 and 11 (both on the left-hand side) show how the student includes optimization and list comprehension in the submitted answer codes. We expect the students to use the sequential conditions to check whether a whole number is greater than one and cannot be exactly divided by any whole number other than itself and 1. An expected alternative for the check procedure is given on the right-hand side of Table 10. The suggested code from ChatGPT offers optimization by only checking up to the square root to reduce the number of iterations. This approach would be excellent for a larger number, but a simple answer is preferable for now, as given on the right-hand side.

Table 10 Example code ChatGPT to decide prime numbers

AI-assisted (ChatGPT)	Expected

def is_prime(number):
 if number <= 1:
 Return false
 for i in range(2, int(number**0.5)+1):
 if number % i == 0:
 Return false
 Return true
	def is_prime(number):
	 if number <= 1:
	 Return false
	 for i in range(2, number):
	 if number % i == 0:
	 Return false
	 Return true

Table 11 Example code ChatGPT for finding squared numbers in a given list

AI-assisted (ChatGPT)	Expected

# Original list 	
 numbers = [1, 2, 3, 4, 5]
 # List comprehension to create a new
 # list with squared numbers
 squared_numbers = [num**2 for num in numbers]
	# Original list
	 numbers = [1, 2, 3, 4, 5]
	 # Empty list to store the results
	 squared_numbers = []
	 # Iterate over the original list and append
	 # squared numbers to the new list
	 for num in numbers:
	 squared_numbers.append(num**2)

In Table 11, we expect a relatively simple iteration, as given on the right-hand side, without using a list comprehension to find the squared numbers. Our observation is supported by the essential features that intersect all assignment types. Beyond the D1, D2, and B3, as mentioned before, there are some other distinguishing attributes, which are the number of 'return' statements without returned value(s) (F1), the number of inconsistent uses of string quotes: some string constants use single quotes while others use double quotes (C3), and the number of shorthand (one line) reversed branching statements (B2). These features strengthen our observation that ChatGPT will suggest highly efficient and compact code, which is usually not to be directly consumed by first-year students when they start learning how to program.

6 Random Committee Performance

This section shows the model performance for each assignment task using the proposed RC meta-learning. Since the number of learning instances for each assignment task is limited, we only run the baseline scenario with 10-fold cross-validations using the semi-real dataset. The results are shown in Table 12.

Table 12 Random committee performance (10-fold cross-validations in percentage) for each assignment task. Values in bold indicate the best performance

Semi-real 10-CV	Branching	Looping		Functions (incl. matrix)			
	Meta RC	Meta RC-Sel.	Meta RC	Meta RC-Sel.	Meta RC	Meta RC-Sel.	
Accuracy	**96.7**	96.3	**96.4**	94.8	**96.1**	95.4	
Precision	**85.7**	84.2	84.7	**91.9**	**86.2**	83.9	
Recall	**100.0**	**100.0**	**100.0**	94.4	**98.9**	**98.9**	
RMSE	16.2	16.5	16.1	20.1	17.8	18.5	
In Table 12, we can see that the performance of the RC models is encouraging. The models with full features or the ones with the selection method can achieve the recall performance almost perfectly during the 10-fold cross-validations learning process. The accuracy for all models is above 95%, with the best achieved for the branching assignment type by using all features (96.7%). For the performance of the selected feature, the best is also achieved for the branching assignment type (96.3%), which is not statistically different. Although this value would likely be overfitting in reality, this suggests the effectiveness of our features, as presented in Table 3.

7 Comparison to Information Retrieval Approach

Our previous study [24], based on an information retrieval (IR) approach, showed some influential features. However, how those features would impact the decision to use AI-assisted or not-AI-assisted could not be directly calculated. There are indeed some overlapped features between the results in this paper and those in the previous study that have significant influence, i.e., the number of list assignments (D1), the number of branching condition statements without voluntary parenthesis (B3), the number of 'for' looping statements with one 'range' argument (L10), and the number of 'len' function calls (M1).

Interestingly, the number of 'for' looping statements with two 'range' arguments (L11) is considered important in the IR, while it is not regarded as influential in the ML approach. This fact suggests that our ML models could generalize one of the generative AI characteristics: that most of the suggested codes usually come from best practices. Examples of such best practices are programming tutorials captured in the large language model and then presented to the users via a question-answer interface. In our case, the generalized models capture the 'for' looping statement with one 'range' document or utilize list comprehension. Qua effectiveness, ML learning models show significant improvement compared to the IR-based approach. Our best

ML model can achieve accuracy (top-1 precision) above 95% in the real test dataset. Meanwhile, in the IR-based approach, the best top-k effectiveness is around 69% (top-k precision, with k between 3 and 5).

8 Limitation of The Study

Our models are still limited to basic-level program code in the Python programming language. However, for other programming languages, we hypothesize that the code generated by AI-assisted applications will also be written with a high-efficiency level that may not be in introductory programming courses. In the future, developing a program code plagiarism detector equipped to analyze code syntactic trees would be valuable.

This way, we can also consider the levels of logical flow rather than just looking at statistics-based attributes. Although our experiments have shown high effectiveness, further exploration must be conducted in real-time class situations. For instance, exploring the models' behavior in a code submission system as a kind of code originality validation process would be interesting.

9 Conclusion

This chapter describes an effort to develop a machine-learning model to detect program code anomalies generated by AI-assisted tools. The best model was obtained by utilizing RC meta-learning. From this fact, one can deduce that each type of assignment has unique characteristics. Interesting findings representing a general trend toward program code produced by AI-assisted tools are that the code is neater and more compact with a high-efficiency level and has yet to be taught in an introductory programming curriculum.

Efficient codes are usually demonstrated using lists and arrays. For example, they are written concisely by omitting brackets when writing repetition or branching assignments. Based on the model accuracy and the recall rates, our feature sets and meta-learning approach would help detect anomalies in AI-assisted code submissions.

While we support the integration of generative AI in education, we acknowledge that it does not apply to all assessments. Some assessments (including those in basic programming) aim to let students solve problems independently. We also believe that any help from AI-assisted tools should be acknowledged soon, like citing other people's work. Identifying AI-generated work is still an important task.

Acknowledgements The research presented in this chapter was supported by the Research Institute and Community Service at Universitas Kristen Maranatha.

References

1. Acampora G, Cosma G (2015) A Fuzzy-based approach to programming language independent source-code plagiarism detection. In: 2015 IEEE international conference on fuzzy systems (FUZZ-IEEE). IEEE, Istanbul, Turkey, pp 1–8. https://doi.org/10.1109/FUZZ-IEEE.2015.7337935. http://ieeexplore.ieee.org/document/7337935/
2. Alvi F, Stevenson M, Clough P (2021) Paraphrase type identification for plagiarism detection using contexts and word embeddings. Int J Educ Technol High Educ 18(1):42. https://doi.org/10.1186/s41239-021-00277-8
3. Arwin C, Tahaghoghi SMM (2006) Plagiarism detection across programming languages. In: Proceedings of the 29th Australasian computer science conference, ACSC '06, vol 48. Australian Computer Society, Inc., Australia, pp 277–286
4. Blanchard J, Hott JR, Berry V, Carroll R, Edmison B, Glassey R, Karnalim O, Plancher B, Russell S (2022) Stop reinventing the wheel! promoting community software in computing education. In: Proceedings of the 2022 working group reports on innovation and technology in computer science education, ITiCSE-WGR '22. Association for Computing Machinery, New York, NY, USA, pp 261–292. https://doi.org/10.1145/3571785.3574129
5. Bradley S (2020) Creative assessment in programming: diversity and divergence. In: Proceedings of the 4th conference on computing education practice, CEP '20. Association for Computing Machinery, New York, NY, USA, pp 1–4. https://doi.org/10.1145/3372356.3372369
6. Budiman AE, Karnalim O (2019) Automated hints generation for investigating source code plagiarism and identifying the culprits on in-class individual programming assessment. Computers 8(1):11. https://doi.org/10.3390/computers8010011. https://www.mdpi.com/2073-431X/8/1/11. Number: 1, Publisher: Multidisciplinary Digital Publishing Institute
7. Chen L, Chen P, Lin Z (2020) Artificial intelligence in education: a review. IEEE Access 8:75264–75278. https://doi.org/10.1109/ACCESS.2020.2988510. https://ieeexplore.ieee.org/abstract/document/9069875. Conference Name: IEEE Access
8. Daly C, Horgan J (2005) A technique for detecting plagiarism in computer code. Comput J 48(6):662–666. https://doi.org/10.1093/comjnl/bxh139
9. Denny P, Leinonen J, Prather J, Luxton-Reilly A, Amarouche T, Becker BA, Reeves BN (2024) Prompt problems: a new programming exercise for the generative AI Era. In: Proceedings of the 55th ACM technical symposium on computer science education V. 1, SIGCSE 2024. Association for Computing Machinery, New York, NY, USA, pp 296–302. https://doi.org/10.1145/3626252.3630909
10. Doppelt Y (2003) Implementation and assessment of project-based learning in a flexible environment. Int J Technol Des Educ 13(3):255–272. https://doi.org/10.1023/A:1026125427344
11. Durić Z, Gašević D (2013) A source code similarity system for plagiarism detection. Comput J 56(1):70–86. https://doi.org/10.1093/comjnl/bxs018
12. Engels S, Lakshmanan V, Craig M (2007) Plagiarism detection using feature-based neural networks. In: Proceedings of the 38th SIGCSE technical symposium on computer science education, SIGCSE '07. Association for Computing Machinery, New York, NY, USA, pp 34–38. https://doi.org/10.1145/1227310.1227324
13. Esteva A, Kale A, Paulus R, Hashimoto K, Yin W, Radev D, Socher R (2021) COVID-19 information retrieval with deep-learning based semantic search, question answering, and abstractive summarization. NPI Digit Med 4(1):1–9. https://doi.org/10.1038/s41746-021-00437-0. https://www.nature.com/articles/s41746-021-00437-0. Publisher: Nature Publishing Group
14. Foltýnek T, Všianský R, Meuschke N, Dlabolová D, Gipp B (2020) Cross-language source code plagiarism detection using explicit semantic analysis and scored greedy string tilling. In: Proceedings of the ACM/IEEE joint conference on digital libraries in 2020, JCDL '20. Association for Computing Machinery, New York, NY, USA, pp 523–524. https://doi.org/10.1145/3383583.3398594

15. Fui-Hoon Nah F, Zheng R, Cai J, Siau K, Chen L (2023) Generative AI and ChatGPT: applications, challenges, and AI-human collaboration. J Inf Technol Case Appl Res 25(3):277–304. https://doi.org/10.1080/15228053.2023.2233814. Publisher: Routledge
16. Ghasemian A, Hosseinmardi H, Galstyan A, Airoldi EM, Clauset A (2020) Stacking models for nearly optimal link prediction in complex networks. Proc Natl Acad Sci 117(38):23393–23400. https://doi.org/10.1073/pnas.1914950117. https://www.pnas.org/doi/full/10.1073/pnas.1914950117. Publisher: Proceedings of the National Academy of Sciences
17. Gu Q, Tian J, Li X, Jiang S (2022) A novel random forest integrated model for imbalanced data classification problem. Knowl-Based Syst 250:109050. https://doi.org/10.1016/j.knosys.2022.109050. https://www.sciencedirect.com/science/article/pii/S0950705122005147
18. Hacker P, Engel A, Mauer M (2023) Regulating ChatGPT and other large generative AI models. In: Proceedings of the 2023 ACM conference on fairness, accountability, and transparency, FAccT '23. Association for Computing Machinery, New York, NY, USA, pp 1112–1123. https://doi.org/10.1145/3593013.3594067
19. Hoq M, Shi Y, Leinonen J, Babalola D, Lynch C, Akram B (2023) Detecting ChatGPT-generated code in a CS1 course. In: Moore S, Stamper J, Tong R, Cao C, Liu Z, Hu X, Lu Y, Liang J, Khosravi H, Denny P, Singh A, Brooks C (eds) Proceedings of the workshop on empowering education with LLMs—the next-gen interface and content generation, CEUR workshop proceedings, vol 3487. CEUR, Tokyo, Japan, pp 53–63. https://ceur-ws.org/Vol-3487/paper2.pdf. ISSN: 1613-0073
20. Iffath F, Kayes ASM, Rahman MT, Ferdows J, Arefin MS, Hossain MS (2021) Online judging platform utilizing dynamic plagiarism detection facilities. Computers 10(4):47. https://doi.org/10.3390/computers10040047. https://www.mdpi.com/2073-431X/10/4/47. Number: 4, Publisher: Multidisciplinary Digital Publishing Institute
21. Jiang L, Misherghi G, Su Z, Glondu S (2007) DECKARD: scalable and accurate tree-based detection of code clones. In: 29th international conference on software engineering (ICSE'07), pp 96–105. https://doi.org/10.1109/ICSE.2007.30. https://ieeexplore.ieee.org/abstract/document/4222572. ISSN: 1558-1225
22. Karnalim O, Kautsar IA, Aditya BR, Udjaja Y, Nendya MB, Darma Kotama IN (2021) Programming plagiarism and collusion: student perceptions and mitigating strategies in Indonesia. In: 2021 IEEE international conference on engineering, technology & education (TALE). IEEE, Wuhan, Hubei Province, China, pp 9–14. https://doi.org/10.1109/TALE52509.2021.9678917. https://ieeexplore.ieee.org/document/9678917/
23. Karnalim O, Simon Chivers W (2019) Similarity detection techniques for academic source code plagiarism and collusion: a review. In: 2019 IEEE international conference on engineering, technology and education (TALE), pp 1–8. https://doi.org/10.1109/TALE48000.2019.9225953. https://ieeexplore.ieee.org/abstract/document/9225953. ISSN: 2470-6818
24. Karnalim O, Toba H, Johan MC (2024) Detecting AI assisted submissions in introductory programming via code anomaly. Educ Inf Technol https://doi.org/10.1007/s10639-024-12520-6
25. LeCun Y, Bengio Y, Hinton G (2015) Deep learning. Nature 521(7553):436–444. https://doi.org/10.1038/nature14539. https://www.nature.com/articles/nature14539. Publisher: Nature Publishing Group
26. Lim JS, Ji JH, Cho HG, Woo G (2011) Plagiarism detection among source codes using adaptive local alignment of keywords. In: Proceedings of the 5th international conference on ubiquitous information management and communication, ICUIMC '11. Association for Computing Machinery, New York, NY, USA, pp 1–10. https://doi.org/10.1145/1968613.1968643
27. Ljubovic V, Pajic E (2020) Plagiarism detection in computer programming using feature extraction from ultra-fine-grained repositories. IEEE Access 8:96505–96514. https://doi.org/10.1109/ACCESS.2020.2996146. https://ieeexplore.ieee.org/abstract/document/9097285. Conference Name: IEEE Access
28. Mann S, Frew Z (2006) Similarity and originality in code: plagiarism and normal variation in student assignments. In: Proceedings of the 8th Australasian conference on computing education, ACE '06, vol 52. Australian Computer Society, Inc., Australia, pp 143–150

29. Michel-Villarreal R, Vilalta-Perdomo E, Salinas-Navarro DE, Thierry-Aguilera R, Gerardou FS (2023) Challenges and opportunities of generative AI for higher education as explained by ChatGPT. Educ Sci 13(9):856. https://doi.org/10.3390/educsci13090856. https://www.mdpi.com/2227-7102/13/9/856. Number: 9, Publisher: Multidisciplinary Digital Publishing Institute
30. Monteiro JP, Ramos D, Carneiro D, Duarte F, Fernandes JM, Novais P (2021) Meta-learning and the new challenges of machine learning. Int J Intell Syst 36(11):6240–6272. https://doi.org/10.1002/int.22549. https://onlinelibrary.wiley.com/doi/abs/10.1002/int.22549
31. Mozgovoy M, Karakovskiy S, Klyuev V (2007) Fast and reliable plagiarism detection system. In: 2007 37th annual frontiers in education conference—global engineering: knowledge without borders, opportunities without passports, pp S4H–11–S4H–14. https://doi.org/10.1109/FIE.2007.4417860. https://ieeexplore.ieee.org/abstract/document/4417860. ISSN: 2377-634X
32. Muhammad LJ, Algehyne EA, Usman SS (2020) Predictive supervised machine learning models for diabetes mellitus. SN Comput Sci 1(5):240. https://doi.org/10.1007/s42979-020-00250-8
33. Niranjan A, Nutan DH, Nitish A, Shenoy PD, Venugopal KR (2018) ERCR TV: ensemble of random committee and random tree for efficient anomaly classification using voting. In: 2018 3rd international conference for convergence in technology (I2CT), pp 1–5. https://doi.org/10.1109/I2CT.2018.8529797. https://ieeexplore.ieee.org/abstract/document/8529797
34. Nowak A, Lukowicz P, Horodecki P (2018) Assessing artificial intelligence for humanity: will AI be the our biggest ever advance? or the biggest threat [Opinion]. IEEE Technol Soc Mag 37(4):26–34. https://doi.org/10.1109/MTS.2018.2876105. https://ieeexplore.ieee.org/abstract/document/8558761/authors#authors. Conference Name: IEEE Technology and Society Magazine
35. Odhiambo Omuya E, Onyango Okeyo G, Waema Kimwele M (2021) Feature selection for classification using principal component analysis and information gain. Expert Syst Appl 174:114765. https://doi.org/10.1016/j.eswa.2021.114765. https://www.sciencedirect.com/science/article/pii/S0957417421002062
36. Orenstrakh MS, Karnalim O, Suarez CA, Liut M (2023) Detecting LLM-generated text in computing education: a comparative study for ChatGPT cases (2023). https://doi.org/10.48550/arXiv.2307.07411. ArXiv:2307.07411 [cs]
37. Prentice FM, Kinden CE (2018) Paraphrasing tools, language translation tools and plagiarism: an exploratory study. Int J Educ Integr 14(1):11. https://doi.org/10.1007/s40979-018-0036-7
38. Saoban C, Rimcharoen S (2019) Identifying an original copy of the source codes in programming assignments. In: 2019 16th international joint conference on computer science and software engineering (JCSSE), pp 271–276. https://doi.org/10.1109/JCSSE.2019.8864196. https://ieeexplore.ieee.org/document/8864196. ISSN: 2642-6579
39. Setoodeh Z, Moosavi MR, Fakhrahmad M, Bidoki M (2021) A proposed model for source code reuse detection in computer programs. Iran J Sci Technol, Trans Electr Eng 45(3):1001–1014. https://doi.org/10.1007/s40998-020-00403-8
40. Sheahen D, Joyner D (2016) TAPS: A MOSS extension for detecting software plagiarism at scale. In: Proceedings of the third (2016) ACM conference on learning @ Scale, L@S '16. Association for Computing Machinery, New York, NY, USA, pp 285–288. https://doi.org/10.1145/2876034.2893435
41. Simon, Cook B, Sheard J, Carbone A, Johnson C (2013) Academic integrity: differences between computing assessments and essays. In: Proceedings of the 13th Koli calling international conference on computing education research, Koli calling '13. Association for Computing Machinery, New York, NY, USA, pp 23–32. https://doi.org/10.1145/2526968.2526971
42. Simon, Karnalim O, Sheard J, Dema I, Karkare A, Leinonen J, Liut M, McCauley R (2020) Choosing code segments to exclude from code similarity detection. In: Proceedings of the working group reports on innovation and technology in computer science education, ITiCSE-WGR '20. Association for Computing Machinery, New York, NY, USA, pp 1–19. https://doi.org/10.1145/3437800.3439201

43. Striletchi C, Vaida M, Chiorean L, Popa S (2016) A cross-platform solution for software plagiarism detection. In: 2016 12th IEEE international symposium on electronics and telecommunications (ISETC), pp 141–144. https://doi.org/10.1109/ISETC.2016.7781077. https://ieeexplore.ieee.org/abstract/document/7781077
44. Toba H, Karnalim O, Johan MC, Tada T, Djajalaksana YM, Vivaldy T (2024) Inappropriate benefits and identification of ChatGPT misuse in programming tests: a controlled experiment. In: Auer ME, Cukierman UR, Vendrell Vidal E, Tovar Caro E (eds) Towards a hybrid, flexible and socially engaged higher education. Springer Nature Switzerland, Cham, pp 520–531. https://doi.org/10.1007/978-3-031-51979-6_54
45. Ullah F, Wang J, Farhan M, Habib M, Khalid S (2021) Software plagiarism detection in multiprogramming languages using machine learning approach. Concurr Comput: Pract Exp 33(4):e5000. https://doi.org/10.1002/cpe.5000. https://onlinelibrary.wiley.com/doi/abs/10.1002/cpe.5000
46. Ullah F, Wang J, Farhan M, Jabbar S, Wu Z, Khalid S (2020) Plagiarism detection in students' programming assignments based on semantics: multimedia e-learning based smart assessment methodology. Multimed Tools Appl 79(13):8581–8598. https://doi.org/10.1007/s11042-018-5827-6
47. Wang Y, Jin D, Gong Y (2019) A diversified feature extraction approach for program similarity analysis. In: Proceedings of the 2nd international conference on software engineering and information management, ICSIM '19. Association for Computing Machinery, New York, NY, USA, pp 96–101. https://doi.org/10.1145/3305160.3305189
48. Wang S, Liu T, Tan L (2016) Automatically learning semantic features for defect prediction. In: Proceedings of the 38th international conference on software engineering, ICSE '16. Association for Computing Machinery, New York, NY, USA, pp 297–308. https://doi.org/10.1145/2884781.2884804
49. Yan L, McKeown N, Sahami M, Piech C (2018) TMOSS: using intermediate assignment work to understand excessive collaboration in large classes. In: Proceedings of the 49th ACM technical symposium on computer science education, SIGCSE '18. Association for Computing Machinery, New York, NY, USA, pp 110–115. https://doi.org/10.1145/3159450.3159490. https://dl.acm.org/doi/10.1145/3159450.3159490
50. Yasaswi J, Purini S, Jawahar C (2017) Plagiarism detection in programming assignments using deep features. In: 2017 4th IAPR Asian conference on pattern recognition (ACPR), pp 652–657. https://doi.org/10.1109/ACPR.2017.146. https://ieeexplore.ieee.org/abstract/document/8575900. ISSN: 2327-0985
51. Žáková K, Pištej J, Bisták P (2013) Online tool for student's source code plagiarism detection. In: 2013 IEEE 11th international conference on emerging eLearning technologies and applications (ICETA), pp 415–419. https://doi.org/10.1109/ICETA.2013.6674469. https://ieeexplore.ieee.org/abstract/document/6674469

Using GenAI as Co-author for Teaching Supply Chain Management in Higher Education

Dominik Wörner and Andreas Holzapfel

Abstract This chapter analyzes the utility of GenAI as a means of adequately gathering information on prevalent concepts in teaching supply chain management in higher education. We first develop and portray a framework for setting up a compendium of definitions using the explanatory power of GenAI as well as the comprehensiveness of classical textbooks, accompanied and critically reviewed by supply chain management experts. We then evaluate the resulting output by reflecting the need for content- and language adaptations and the extent of expert adaptations necessary to achieve correct explanations of supply chain management concepts such that students can use them for building up a profound and correct basic vocabulary in this field. Our results identify the extent to which GenAI outputs are correct, and confirm that this equates to the degree found in studies within other disciplines (e.g., medicine). We also show that GenAI can be used for creating uniform explanations of concepts in supply chain management, but that expert knowledge is required to validate the correctness of the outputs. The framework we developed can help to utilize the different sources of information in supply chain management education appropriately.

1 Introduction

In the contemporary educational landscape students increasingly rely on a diverse variety of sources, including Internet searches, video platforms, and Generative Artificial Intelligence applications (GenAI) such as ChatGPT to gather information. A recent US study found, that 60% of generation Z learners even prefer YouTube for

---

D. Wörner (✉)
FHNW University of Applied Sciences and Arts Northwestern Switzerland, Peter Merian-Str. 86, 4002 Basel, Switzerland
e-mail: dominik.woerner@fhnw.ch

A. Holzapfel
Hochschule Geisenheim University, Von-Lade-Str. 1, 65366 Geisenheim, Germany
e-mail: andreas.holzapfel@hs-gm.de

© The Author(s), under exclusive license to Springer Nature Switzerland AG 2025
E. Vendrell Vidal et al. (eds.), *Advanced Technologies and the University of the Future*,
Lecture Notes in Networks and Systems 1140,
https://doi.org/10.1007/978-3-031-71530-3_12

learning over printed books [3]. Reference [11], investigated the use of AI-based tools among German students. In that study, 63.4% of the students surveyed have so far used or are using AI-based tools for their studies. Among the top five AI tools, ChatGPT is used by half of the students (49%). According to [8], the main reasons for using ChatGPT among students is *Time Saving and Task Management* as well as the *Inseparability of Content*.

The use of GenAI introduces a novel dimension to information acquisition, but the veracity of the information obtained is often challenging to validate (see, e.g., [1, 2, 5, 10]). Consequently, students grapple with divergent interpretations of terminologies, leading to an intellectual landscape characterized by conceptual variations and inconsistencies.

Classical textbooks are traditionally regarded as a more reliable source. Those books have their limitations, however, for obtaining an overview of a large variety of terms and definitions within a certain field. Typically, these textbooks prioritize detailed explanations of fundamental concepts but often fall short of providing concise and precise definitions. Furthermore, classical textbooks commonly focus on core concepts, neglecting to encompass the entire breadth of the discipline. This omission leaves students with an incomplete understanding of the field, as supplementary concepts are either cursorily mentioned or left unexplored. As a result, there exists a pressing need for a comprehensive and integrated approach to information dissemination in education: one that not only embraces the advantages of contemporary sources like GenAI, but also addresses the limitations of classical textbooks.

As stated in [9], GenAI can be used both in research and teaching, but its use raises a lot of questions that all need to be explored. This research aims to formulate a cohesive and validated framework for the peer-reviewed utilization of GenAI in the pedagogical realm, specifically focusing on the explanation of basic concepts in supply chain management (SCM). This endeavor aims to culminate in the creation of a widely accepted compendium that provides correct and uniform definitions and descriptions of common SCM concepts. Emphasizing a practical SCM vocabulary, the compendium encapsulates the fundamental elements of each concept, fostering an understanding without the need for exhaustive scientific definitions. This research integrates two components:

1. The development of a standardized framework for the application of GenAI as a co-author in higher education, and
2. The numerical validation and evaluation of this framework.

By delving into the quality of answers and explanations generated by GenAI, this study seeks to unravel valuable insights into its efficacy and accuracy as an educational tool for writing a compendium. It is noteworthy that while the compendium itself is not part of the publication, its role as a tangible outcome underscores the significance of this research in shaping a consistent and practical SCM vocabulary.

The remainder of this chapter is organized as follows. First, we detail our methodological approach for generating and evaluating the compendium (Sect. 2). In Sect. 3

we then portray the insights generated through the evaluation. Finally, we discuss our findings in the context of current literature and conclude with a brief outlook on future research opportunities (Sect. 4).

2 Methods and Approach

Our aim is to provide a framework for the use of GenAI as co-author in teaching supply chain management in higher education, and to evaluate it. To do this, we follow a two-step methodological approach. We first portray the process for developing a compendium of supply chain management concepts using GenAI that is being peer-reviewed by supply chain management experts (Sect. 2.1). Second, we develop an evaluation scheme for numerically evaluating the utility of the output generated by GenAI (as co-author) in teaching supply chain management in higher education (Sect. 2.2). This evaluation scheme builds the basis for the results we present in Sect. 3.

2.1 Compendium Development

The methodological framework employed in this research is designed to meticulously curate a validated compendium, serving the dual purpose of fostering a uniform understanding of supply chain management (SCM) concepts among practitioners and students while harnessing the potential of modern information acquisition and processing, particularly through the application of GenAI, exemplified by ChatGPT.

Our framework therefore follows multiple steps (see Fig. 1) to make use of

- Pertinent knowledge on supply chain management concepts available in classical textbooks (Steps 1 and 2),
- Knowledge about these concepts and the ability to formulate the essentials of these concepts in compact and understandable explanations provided by GenAI (Step 4), and
- Expertise of supply chain management experts in selecting relevant textbooks and concepts (Steps 1–3), guiding GenAI (Step 4), and validating and potentially correcting the GenAI output (Step 5).

The individual steps are detailed in the following.

1. **Longlist Creation**
   The initiation of the research process involves the creation of a comprehensive list of pertinent SCM concepts. This exhaustive compilation relies on a meticulous selection of SCM textbooks, chosen for their scientific rigor and relevance. Drawing from the rich repository of library directories in German-speaking universities and colleges, the authors employ predefined criteria, encompassing factors such as

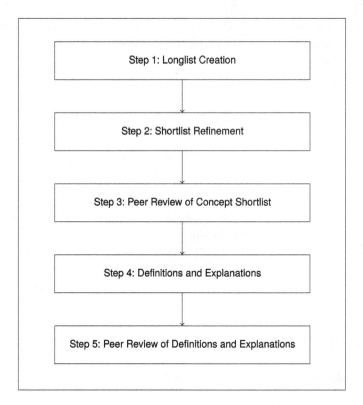

Fig. 1 Schematic framework for compendium development

the publication timeframe (i.e., publications from the last 15 years) and language (i.e., German), to judiciously pre-select the foundational texts. To facilitate the automatic extraction of supply chain management concepts, only textbooks that are completely available in electronic format – or at least their subject index–are considered. The concepts listed in the indices of the textbooks identified constitute the longlist.

2. **Shortlist Refinement**
   Building upon the extensive longlist, a refined shortlist of publications is systematically curated. This curation process is undertaken with the specific goal of ensuring a holistic representation of SCM topics, including but not limited to procurement, production, and logistics. The authors actively supplement the shortlist with selected publications to further enrich its breadth and depth.

3. **Peer Review of Concept Shortlist**
   The shortlisted SCM concepts undergo a double-blind peer review process. During this critical phase, the authors conduct an in-depth analysis to evaluate the suitability of each concept for inclusion in the compendium. This review process adheres closely to the research objectives, allowing only those concepts that align with the predefined criteria to progress to the subsequent stages of development.

4. **Definitions and Explanations**
   With the refined concept list in place, the research transitions into the phase of preparing these concepts for explanation by GenAI. A deliberate effort is made to formulate a uniform prompt that is applied consistently across all concepts. This strategic approach not only imposes constraints on the length of explanations but also ensures that the content generated provides appropriate and insightful definitions in the context of supply chain management. This step is pivotal in maintaining coherence and effectiveness in leveraging the capabilities of GenAI.
5. **Peer Review of Definitions and Explanations**
   The definitions and explanations generated undergo a meticulous double-blind peer review process. This evaluation is essential to validate the accuracy, clarity, and coherence of the content generated by GenAI. Any instances of inaccuracies, misleading information, or unclear passages are identified in a systematic review and subsequently addressed in thoughtful discussion. This iterative process of peer review ensures refinement and enhancement of the compendium, as well as its ultimate alignment with the highest standards of reliability and comprehensibility. The aim is to ultimately derive a compendium that is complete, correct, and adapted as minimally as possible to maintain the GenAI definitions as far as possible. Prompts have been adapted to facilitate adequate definition by specifying the context in the event that GenAI should completely fail to produce suitable definitions.

2.2 Numerical Evaluation

To address the second research objective, we build an evaluation scheme to assess the utility of the output generated by GenAI as co-author in teaching supply chain management in higher education. This refers to the extent of adaptation that is necessary for generating the compendium in Step 5 of the schematic framework. Our evaluation aims to shed light on the extent to which expert interventions are needed to ensure correct definitions of supply chain management concepts in terms of content and context valuable for building supply chain management terminology that is relevant in practice.

We thus first focus on the number of cases in which adaptations to the prompt are necessary to enable definitions with the right context. Second, we investigate the types of manual intervention required. In doing this we distinguish between language adaptations and content adaptations by the experts. We also distinguish between deletions and insertions. Third, we quantify the extent of adaptations that are necessary to obtain correct definitions. Aiming at the minimal numbers of corrections, we analyze the word count of the final concepts compared to the original GenAI output.

Overall, we use three levels of accuracy for categorization. In the first (highest) level of accuracy, the description of the concept is correct and complete and may be used as generated. Minimal interventions by a deletion of the last paragraph of

Table 1 Accuracy levels and evaluation criteria

Accuracy level	Description	Defined criteria
1	Complete	Description of the concept is correct and complete and may be used as generated or by deletion of last paragraph
2	Linguistic adaptation	Description of the concept is correct and complete but (only) linguistic adaptation is necessary
3	Content adaptation	Description of the concept is incorrect or misleading, which requires adaptation of the content

the output are allowed. The second level of accuracy refers to correct and complete descriptions in which only linguistic adaptations are necessary. The lowest level of accuracy (level three) is assigned to descriptions of concepts in which adoption of the content is necessary. The content would be either incorrect or misleading without these adaptations. Table 1 summarizes the levels of accuracy.

3 Evaluation Results

Following the methodology portrayed in Sect. 2.2, we present the assessment of GenAI's utility as co-author for generating a compendium of supply chain management concept definitions for use in higher education in the following subsections. This involves analyzing the contextual validity (Sect. 3.1), and the type (Sect. 3.2) and extent (Sect. 3.3) of interventions necessary.

We refer to the assessment of 371 definitions of basic supply chain management concepts in the German language generated by ChatGPT 3.5 between April and June 2023. The GenAI is supposed to use only 200 words for each explanation.

3.1 Contextual Validity

In general, GenAI is able to adequately provide answers in the context desired. This also applies when it uses a uniform prompt structure.[1] Out of the 371 concepts, GenAI output is usable for the dedicated purpose (with adaptations) in 98% of the cases. The prompt has to be adapted due to definitions not matching the core of the concepts in supply chain management for only seven concepts. However, this can be resolved by adapting the prompt slightly, specifying the context (e.g., by replacing "in the

---

[1] Prompt used: "Define the concept 'xy' in the context of supply chain management. Please keep to a maximum of 200 words." Original prompt used in the German language: "Definiere das Konzept 'xy' im Kontext von Supply Chain Management. Bitte maximal 200 Wörter."

context of supply chain management" by "in production" or "in transportation"). We include the output of these seven concepts concerned in its updated version in the following evaluations.

3.2 Type and Number of Interventions

Although the context of the GenAI output is adequate after refining the prompts, only 63 out of 371, i.e., 17% of the concepts, pass the expert peer review process without any adaptation and can therefore be classified as complete (see Fig. 2). Another 84 concepts (23%) can be considered complete in their main text (with an accuracy level of 1 in both cases). Only their last paragraph is removed during review due to redundant information or to improve readability. This is because of GenAI's very generic summary of the concepts that adds no value to the definition. In the remaining 224 concepts (60%), the authors have to intervene from the viewpoint of language, content or both. While GenAI produces relatively good linguistic outputs, and interventions are necessary in only 40 concepts to obtain fluent explanations (of which 14 can be considered cases of accuracy level 2), 210 concepts are adapted to ensure correctness and (in their key elements) complete content-related explanations.

Although 57% of all concepts need to be adapted in terms of content (accuracy level 2), 90% of these adaptations (188 cases) to obtain correct explanations can be made by simply deleting parts of the GenAI output (see Fig. 3).

In another two cases the insertion of only a few words is sufficient to obtain correct and complete definitions. Deletions as well as additions are necessary for the remaining 20 concepts.

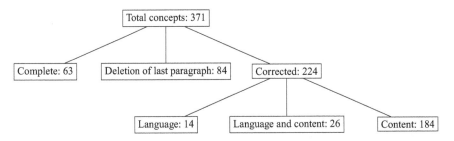

Fig. 2 Evaluation: overview

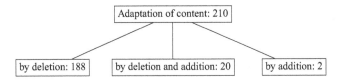

Fig. 3 Evaluation–content-related interventions

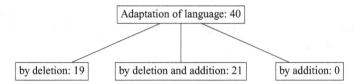

Fig. 4 Evaluation: language interventions

Table 2 Evaluation: statistical summary by accuracy levels

Accuracy level	Description	No. of concepts
1	Complete	147
2	Linguistic adaptation	14
3	Content adaptation	210

Only 11% of all GenAI outputs need language interventions. Of those, around half the concepts require deletions only, while additions are also necessary in the other half (Fig. 4).

In summary, approximately 40% of the output generated is complete and correct and achieves the highest level of accuracy (accuracy level 1). Linguistic adaptation is necessary in 4% of the cases (accuracy level 2), while intervention in the content is required in 57% (accuracy level 3). We refer to Table 2 for the respective statistical summary.

3.3 Extent of Interventions

Table 3 provides insights into the extent of the interventions within the respective concepts by providing the word count of the original and adapted GenAI outputs.

As noted above, the prompts aim at GenAI outputs of a length of max. 200 words. In several cases, however, the GenAI does not completely follow this restriction, such that the average word-count is 214 for the original outputs. After the peer review process the average word-count is only 158, which means that the definitions have to be reduced by approximately a quarter on average. While the complete concepts remain unchanged and the concepts in which only the last paragraph is deleted are only reduced by 20% on average, the corrected concepts that require further corrections have an average word-count of 139, which corresponds to a reduction of 36% of the content.

While the length of the outputs with language adaptations alone is only slightly reduced, at 15%, concepts with content adaptations undergo greater interventions, with an average length reduction of 37%.

Table 3 Evaluation: extent of interventions

Description	No. of concepts	Avg. no. of words in original output	Avg. no. of words in reviewed version
Total concepts	371	214	158
Complete concepts	63	208	208
Deletion of last paragraph	84	214	172
Corrected concepts	224	217	139
Language-related correction	14	216	184
Language- and content-related correction	26	214	135
Content-related correction	184	217	136
Adaptation of content	210	217	136
... by deletion	188	217	133
... by deletion and addition	20	210	151
... by addition	2	218	223
Adaptation of language	40	215	152
... by deletion	19	211	164
... by deletion and addition	21	218	141
... by addition	0		

Table 4 Evaluation: number of added words in the case of insertions

Description	Average	Minimum	Maximum
Content-related addition in concepts with deletion	6	1	54
Content-related addition in concepts without deletion	4	1	7
Linguistic addition	1.5	1	5

In summary, extensive adaptations are sometimes necessary to delete unnecessary and incorrect content from the GenAI output, highlighting the need for expert peer review. This is especially relevant as from a linguistic point of view the results are acceptable and mostly read well, and all concepts also contain correct information. Additions are only rarely needed, as shown above. In most cases these additions only amount to one or a few words per GenAI output, as shown in Table 4. Linguistic additions in particular are only marginal, but content additions can also be resolved with only one or a few words in most cases. The average number of words added amounts to six in cases when content was also deleted and not simply added, while in the latter case the average number of words inserted amounts to four. Only in one concept was it necessary to include a longer paragraph with 54 words to ensure that the core elements of the concept were covered.

4 Discussion and Conclusion

This paper presents a model on how GenAI can be used for writing a compendium in the context of SCM along with a numerical analysis of the output generated by GenAI.
The overall results showing 17% correct and complete descriptions without the need to adapt anything is an acceptable but not satisfying result in a first instance. Adding the cases in which the adaptation of the GenAI-generated output can be achieved by deleting the last paragraph, the overall accuracy result achieved by ChatGPT improves to almost 40% (accuracy level 1). In summary, four out of ten outputs generated by GenAI to describe basic SCM concepts can (largely) be left untouched.
Focusing on the 224 cases in which a correction is necessary, it has to be mentioned that the language and grammar is the reason for intervention (accuracy level 2) in only 14 cases. Thus, in 210 out of 371 cases, the content is not correct or at least inaccurate (accuracy level 3) and justifies an intervention, even given that the main premise of the study is to leave all concepts and descriptions untouched as far as possible (as long as the content is not incorrect or misleading).
These results generated in the context of SCM are consistent with other current studies investigating the quality of the output generated by ChatGPT. Those studies can be found especially in medical disciplines as well as in the evaluation of exams in different disciplines. Writing and answering multiple choice questions with ChatGPT, the study of [7], shows 32% correct answers (no or only minor formatting necessary), 43% with substantial modifications, and 25% of the output with incorrect or misleading answers. Reference [6], evaluate the performance of ChatGPT in neurosurgical board education by assessing its accuracy, clarity, and concordance. In approximately 60% of the cases investigated, the GenAI-generated output achieves the highest level of accuracy, deviating among different categories. Reference [4], investigates the accuracy of ChatGPT in solving medical examinations, comparing those to the ones of a student group. In that study, ChatGPT achieves an overall rate of correct answers of 61%, being lower than the students' average scores (90%).
Following those results, educators, authors and learners do have to verify and validate the content that ChatGPT is currently able to generate comprehensively. The output cannot be taken as a single source to gain knowledge, or to write a compendium as in our study. The fact that the number of words from the original output compared to the reviewed output dropped by 37% in our case can be explained by the necessity of correcting the output. On contrary, however, the pure drop in words cannot be seen

as an indicator of the extent of interventions needed to achieve the output. Once the authors need to intervene in the output, the extent is not longer relevant within our classification scheme.

In conclusion, the peer review model presented for using GenAI and ChatGPT explicitly in the learning process seems a proven and necessary approach to ensuring correctness and accuracy. This also reveals the importance of having a good and reliable validation process in terms of a quality gate. For learners, it shows the importance of not fully relying on the outputs generated by GenAI tools such as ChatGPT, instead applying them critically in an informed manner.

Nevertheless, we acknowledge some limitations of our study, which can form the basis for further research:

1. This study is based on a single GenAI application (ChatGPT 3.5), which of course limits the scope of generalization. Other tools as well as their technological advances need to be considered in further investigations.
2. The classification and evaluation of the output is based on the authors' expert knowledge. This classification, however, can be conducted in a more standardized and generalized manner to avoid any kind of bias. Also, a cross-assessment with more than two experts could be performed in the validation process.
3. The scope of this study is limited in two ways. First, from a content perspective, it is limited to the area of supply chain management. Second, it is limited to describing basic concepts as a kind of compendium. Both limitations could be extended in another study by reaching out to other disciplines, and to more complex content.
4. Finally, we performed our study in the German language. ChatGPT may produce outputs with a different degree of correctness when using other languages and respective technical terminology. A comparison of output accuracy in different languages might thus be an interesting field for future research.

References

1. Ahmad N, Murugesan S, Kshetri N (2023) Generative artificial intelligence and the education sector. Computer 56(6):72–76. https://doi.org/10.1109/MC.2023.3263576
2. Alabool HM (2023) ChatGPT in education: SWOT analysis approach. In: International conference on information technology (ICIT), pp 184–189. https://doi.org/10.1109/ICIT58056.2023.10225801
3. Dickson J (2018) Study: gen Z prefers YouTube over books for learning (technical report). Retrieved February 1, 2024, from https://kidscreen.com/2018/08/27/study-gen-z-prefers-youtube-over-books-for-learning/
4. Huh S (2023) Are ChatGPT's knowledge and interpretation ability comparable to those of medical students in Korea for taking a parasitology examination?: a descriptive study. J Educ Eval Health Prof 20:1. https://doi.org/10.3352/jeehp.2023.20.1
5. Lund BD, Wang T, Mannuru NR, Nie B, Shimray S, Wang Z (2023) ChatGPT and a new academic reality: AI-written research papers and the ethics of the large language models in scholarly publishing. J Assoc Inf Sci Technol 74(5):570–581. https://doi.org/10.1002/asi.24750

6. Mannam SS, Subtirelu R, Chauhan D, Ahmad HS, Matache IM, Bryan K, Chitta SV, Bathula SC, Turlip R, Wathen C, Ghenbot Y, Ajmera S, Blue R, Chen HI, Ali ZS, Malhotra N, Srinivasan V, Ozturk AK, Yoon JW (2023) Large language model-based neurosurgical evaluation matrix: a novel scoring criteria to assess the efficacy of ChatGPT as an educational tool for neurosurgery board preparation. World Neurosurg 180:e765–e773. https://doi.org/10.1016/j.wneu.2023.10.043
7. Ngo A, Gupta S, Perrine O, Reddy R, Ershadi S, Remick D (2024) ChatGPT 3.5 fails to write appropriate multiple choice practice exam questions. Acad Pathol 11(1):100099. https://doi.org/10.1016/j.acpath.2023.100099
8. Niloy AC, Bari MA, Sultana J, Chowdhury R, Raisa FM, Islam A, Mahmud S, Jahan I, Sarkar M, Akter S, Nishat N, Afroz M, Sen A, Islam T, Tareq MH, Hossen MA (2024) Why do students use ChatGPT? answering through a triangulation approach. Comput Educ: Artif Intell 100208. https://doi.org/10.1016/j.caeai.2024.100208
9. Peres R, Schreier M, Schweidel D, Sorescu A (2023) On ChatGPT and beyond: how generative artificial intelligence may affect research, teaching, and practice. Int J Res Mark 40(2):269–275. https://doi.org/10.1016/j.ijresmar.2023.03.001
10. Su J, Yang W (2023) Unlocking the power of ChatGPT: a framework for applying generative AI in education. ECNU Rev Educ 6(3):355–366. https://doi.org/10.1177/20965311231168423
11. Von Garrel J, Mayer J (2023) Artificial intelligence in studies-use of ChatGPT and AI-based tools among students in Germany. HumIties Soc Sci Commun 10(1):799. https://doi.org/10.1057/s41599-023-02304-7

Digital Transformation (DX)

Section Introduction

Faraón Llorens Largo and Antonio Fernández Martínez

The digital revolution has been affecting universities for years. Information technologies (IT) have been present in universities since their inception, since they were born in their laboratories and have been used for their main tasks: researchers have used their great communication potential to communicate with their colleagues in other universities and countries; professors have gradually incorporated them into their teaching; and universities themselves have used them to manage their administrative procedures.

However, the enormous progress made by these technologies and their widespread use in society, at all levels and in all areas, means that talking about universities today inevitably involves analysing their level of digitalisation. Digital maturity will enable universities to better respond to the changing and uncertain world we face. This same digital maturity will give them the flexibility to better respond to the demands of society, both in their education and research missions and in scientific and cultural transfer. In this sense, strategic decisions on IT are the responsibility of rectors' teams. It is essential for each university to define its strategy for the digital university and its path towards this new hybrid university model.

The evolution of universities into digital universities is not an option, and universities must decide when to start their digital transformation. Universities must therefore decide when to start their digital transformation, which path to take and at what pace. To this end, we propose the following steps, based on our research and consultations with universities on this topic:

- Strengthen the case for the Rector and his governance team to lead the digital transformation process with determination for the digital evolution of their university

---

F. L. Largo (✉)
Universitat d'Alacant, San Vicente del Raspeig, Spain
e-mail: Faraon.Llorens@ua.es

A. F. Martínez
Universidad de Almería, La Cañada de San Urbano, Spain

and to maintain a sustainable digital governance system. To this end, it is important that they are educated in the principles of digitalisation, digital transformation and IT governance, especially by learning about success cases.
- To go digital, the first thing a university should do is to understand the characteristics of this new model for higher education institutions (HEIs) and decide which are priorities for their university in relation to their environment and the demands of their students: design their digital strategy [1].
- Assess current digital maturity and design a set of initiatives to improve IT governance and increase digital maturity [2, 3]. The results should then be used to understand what the areas for improvement are and to design the IT initiatives and projects needed to address them [4–6].
- Define the catalogue of digitisation and digital transformation projects for the coming years. It is important to determine which processes need to be changed first (prioritisation) and whether this change should be an optimisation of an existing process (digitalisation) or a major transformation of a process or the creation of a new process that will generate high value for the organisation and contribute to the achievement of its strategic goals [7]. Universities need to be able to identify which new technologies will contribute in a disruptive way to changing key processes in order to deliver the new services of the digital university model.

The pace at which this journey is taken depends on the strength of the leader (mainly the rector), the type of strategy chosen (conservative, innovative or transformative) and the resources available (financial, structural and human).

In this section, you will find seven articles, covering both experience and research, that contribute to this fascinating and highly useful field of digital transformation of universities.

In the initial chapter "Safeguarding Knowledge: Ethical Artificial Intelligence Governance in the University Digital Transformation", Rafael Molina-Carmona and Francisco José García-Peñalvo present an AI Governance Grid in Universities, with 12 key aspects to consider to ensure that the governance proposal: four fundamental principles (legality, neutrality, transparency, and promotion of innovation) in each phase of knowledge generation (collection of data, algorithmic treatment of this data and use of the results). As a result of this critical analysis, a set of best practices for AI governance is proposed to provide practical guidance for easy implementation.

The work written by Adriana V. Karam-Koleski and Roberto Carlos Santos Pacheco, "Levels of Analysis and Criteria for Digital Transformation Implementation in Higher Education Institutions", proposes to implement digital transformation in higher education institutions a process with seven levels of analysis (strategy and business model; people, culture and capacities; products and core services; management and governance; student experience; HEI ecosystem; and digital technologies) and six criteria (must be approached in its systemic nature; is a non-linear and incremental journey; vision and planning are fundamental cornerstones; requires financial resources; leadership and governance are different and have complementary roles; ambidexterity is a fundamental capacity).

The chapter entitled "Driving Innovations: Trends, Prospects and Challenges of Implementing Disruptive Educational Technologies within HEIs", written by Edwin Zammit, Clifford De Raffaele, Daren Scerri, Ronald Aquilina, Joachim James Calleja and Alex Rizzo, argues that educational technologies (EdTech) are now central to driving innovation in higher education institutions. The authors detail the implementation and governance methodology used to launch a disruptive EdTech framework within an HEI in the small island state of Malta: An immersive VR experience project is presented as a case study.

The essay "New university models for the digital society", written by Maria Amata Garito, aims to present a concrete model that shows how traditional and digital universities have reinvented themselves by forming new international alliances, creating a global network of universities on the Internet: the NETTUNO Consortium model and its evolution UNINETTUNO International Telematic University. These universities collaborate by pooling their knowledge, sharing curricula and creating digital content in specific learning environments designed with new psycho-pedagogical theories adapted to technological advances.

The paper "Opening up Teaching and Learning at Universities: OER, MOOCs, and Microcredentials", written by Martin Ebner, Ernst Kreuzer, Sandra Schön and Sarah Edelsbrunner, examines the current landscape of open and digital learning and teaching opportunities through three key developments: Open Educational Resources (OER), Massive Open Online Courses (MOOCs) and the introduction of microcredentials. It discusses how these initiatives are being implemented at Graz University of Technology, while exploring the intricate connections, interrelationships and implications of these three forces.

In the chapter "Computer-Based Methods for Adaptive Teaching and Learning", the authors, Ulrike Pado, Anselm Knebusch and Konstanze Mehmedovski, claim that Computer-Based Learning (CBL) is a flexible, digitally enriched didactic concept that allows universities to better meet the needs of their students. They investigate the effects of enriching lectures with e-assessments as formative assessment, which they consider a key element in establishing high accountability in the learning process.

In the last chapter of this section, "Teaching Research Skills at the University. Does Digital Transformation Make a Difference?", María Isabel Pozzo analyses different activities for teaching research skills. She concludes that digital transformation significantly increases the potential for more autonomous research performance. To achieve this overarching goal, the chapter includes a theoretical review to identify operational models of research skills, a literature review to explore digital transformation in research skills teaching at university level, and an analysis of research skills teaching activities in the context of digital transformation.

In summary, this section offers a comprehensive exploration of digital transformation in universities, presenting a diverse range of articles that highlight both practical experiences and cutting-edge research. From the ethical governance of AI and innovative university models to the systematic implementation of digital transformation and the enhancement of research skills through digital means, each chapter provides valuable insights into the evolving landscape of higher education. By examining these diverse approaches, the section underscores the significant impact of digital technologies on teaching, learning and institutional governance, and highlights the critical role they play in shaping the future of universities in an interconnected, global society.

References

1. Fernández A, Llorens F, Céspedes JJ, Rubio T (2021) Modelo de universidad digital (mUd). In: Publicaciones de la Universidad de Alicante. ISBN 978-84-1302-118-8. (http://hdl.handle.net/10045/116047)
2. Llorens F, Molina R, y Fernández A (2019) Proposal for a digital maturity model for universities (md4u). In: EUNIS Congress 2019 "Campus for the future", Trondheim, Norway. (https://www.eunis.org/download/2019/EUNIS_proceedings_2019.pdf)
3. Llorens F, Fernández A, Cadena S, Castañeda L, Claver JM, Díaz C, Hernández A, Rodríguez T, Trejo V, Chinkes E (2022) UDigital: madurez digital para universidades. In: MetaRed y Secretaría General Iberoamericana (SEGIB). (https://www.metared.org/global/estudios-informes.html)
4. Fernández A, Llorens F, Molina R, y Claver JM (2023) Digital maturity evolution of Spanish universities. In: EUNIS Congress 2023 "European Universities and the digital transformation: challenges and opportunities ahead", Vigo, Spain. (https://www.eunis.org/eunis2023/wp-content/uploads/sites/22/2023/05/014-Digital-Maturity-2-evolving-practice-in-HE-49-Claver.pdf)
5. Crespo D (ed) (2023) UNIVERSITIC 2022. Evolución de la madurez digital de las Universidades Españolas. In : rue Universidades Españolas. ISBN: 978-84-09-51309-3. (https://tic.crue.org/publicaciones/universitic-2022)
6. Llorens F, Fernández A, Bardi M, Biscar D, Pachón N, Claver JM, Castañeda LM, Godinho J (2023) UDigital 2023: madurez digital de las universidades iberoamericanas metared. In: MetaRed y secretaría general iberoamericana (SEGIB). (https://www.metared.org/global/estudios-informes.html)
7. Fernández A, Llorens F, Juiz C, Maciá F, y Aparicio JM (2022) How to prioritize strategic IT Projects for your university. In: Publicaciones de la Universidad de Alicante. ISBN 978-84-1302-154-6. (http://hdl.handle.net/10045/121591)

Safeguarding Knowledge: Ethical Artificial Intelligence Governance in the University Digital Transformation

Rafael Molina-Carmona and Francisco José García-Peñalvo

Abstract Higher Education Institutions (HEIs) safeguard knowledge, uphold academic integrity, and contribute to societal progress. They are custodians of knowledge, promoting innovation, addressing societal challenges, and disseminating research ethically. With the rise of Artificial Intelligence (AI), effective governance becomes crucial to ensure responsible use, protect rights, and foster innovation in HEIs. A proposal for a governance framework for AI in Higher Education is presented, designed to be simple, tailored to universities, and easily integrated into existing digital transformation efforts. Specific goals include examining AI's impact, evaluating governance models, suggesting adaptable principles, and defining a framework that balances innovation, ethics, and regulatory compliance. It takes into account that AI in higher education reshapes teaching, research, and administration, and makes emphasis on ethical deployment and observation of the national and international policies and regulations. The proposal sets out four fundamental principles for AI in universities to be applied to every phase of knowledge generation: the principles of legality, neutrality, transparency, and promotion of innovation. As a consequence, the AI Governance Grid is obtained, that allows the identification of 12 key aspects to consider in order to ensure that the governance proposal complies with the principles. A structure for AI Governance is also proposed so that it is efficient and also takes advantage of the expertise that universities already have, as well as being in line with the international standards for IT Governance. Finally, a set of best practices for AI governance is also proposed that aims to provide practical guidance for simple implementation.

Keywords AI governance · Knowledge governance · University governance · Digital transformation

---

R. Molina-Carmona (✉)
Smart Learning Research Group, Universidad de Alicante, Alicante, Spain
e-mail: rmolina@ua.es

F. J. García-Peñalvo
GRIAL Research Group, Universidad de Salamanca, Salamanca, Spain
e-mail: fgarcia@usal.es

© The Author(s), under exclusive license to Springer Nature Switzerland AG 2025
E. Vendrell Vidal et al. (eds.), *Advanced Technologies and the University of the Future*,
Lecture Notes in Networks and Systems 1140,
https://doi.org/10.1007/978-3-031-71530-3_14

1 Introduction

1.1 A Subsection Sample

In the context of Higher Education Institutions (HEIs), knowledge serves as the cornerstone of academic research, innovation, and societal progress. Why do we consider that safeguarding knowledge within HEIs is so paramount?

Universities uphold the principles of academic integrity, which include honesty, transparency and respect for intellectual property rights. Safeguarding knowledge ensures that academic work is accurately attributed and intellectual property is protected, avoiding plagiarism, fraud and academic misconduct. Upholding academic integrity fosters a culture of trust and credibility within the academic community, ensuring the reliability and validity of research results and encouraging innovation, entrepreneurship and knowledge transfer, driving economic growth and technological advancement.

In addition, universities act as custodians of knowledge, preserving and transmitting collective wisdom across generations. Knowledge generated at one point in time will endure and be observed by future generations.

HEIs have a profound impact on society through research, education and community engagement. Safeguarding knowledge ensures that the results of academic research and scholarship are disseminated in a responsible and ethical manner, benefiting society as a whole. Moreover, HEIs play a key role in addressing societal challenges, promoting social justice and advancing sustainable development goals. Safeguarding knowledge enables HEIs to fulfil their social responsibility by contributing to the public good and addressing pressing issues facing communities locally and globally.

Moreover, universities are highly prestigious, accountable institutions, where all knowledge generated must be properly validated by the scientific method and must be neutral in order to avoid bias or partiality in the results.

Artificial Intelligence (AI) allows us to generate new knowledge, in a rich and efficient way that we have never known before. We had always worked with data that we had analyzed with various statistical or algorithmic techniques, and we had used the results to make decisions and generate new knowledge. But at no previous moment in history have we had such a powerful tool, capable of working with so much data and allowing the knowledge generated to come so close to what a human is capable of.

The boom of AI has revolutionized the process of knowledge generation, allowing unprecedented efficiency and richness in the acquisition of information. Unlike previous eras, when data analysis relied on conventional statistical or algorithmic techniques, AI now equips us with the ability to process large amounts of data with remarkable accuracy and depth. Through sophisticated algorithms and machine learning techniques, AI systems can discover complex patterns, correlations, and insights in massive datasets. In addition, AI can synthesize diverse sources of information, facilitating interdisciplinary research and innovation. By harnessing AI, researchers can explore new frontiers of knowledge with unprecedented speed and accuracy, pushing the boundaries of human understanding in a variety of fields.

As AI becomes more pervasive in society, the need for governance mechanisms to oversee its development and deployment becomes imperative. Effective AI governance involves establishing regulations, standards and best practices to ensure ethical and responsible use of AI technologies. By adhering to ethical principles such as transparency, accountability and fairness, institutions can mitigate potential risks and ensure that AI-driven decisions are aligned with society's values and priorities and the preservation of its knowledge. In addition, AI governance frameworks should prioritize the protection of individual and collective rights, privacy and autonomy, guarding against potential bias or discrimination inherent in AI systems.

Institutions must navigate complex regulatory landscapes and ethical considerations to govern AI effectively. Regulatory frameworks, such as data protection laws and industry standards, provide a legal framework for the responsible use of AI technologies. Additionally, ethical guidelines, such as those outlined by professional associations and international organizations, offer principles to guide ethical decision-making in AI development and deployment. By integrating regulatory compliance and ethical considerations into AI governance frameworks, institutions can ensure that AI technologies are deployed in a manner that respects human dignity, rights, and freedoms.

Higher education institutions play a critical role in advancing human knowledge and preparing future generations for the challenges of an AI-driven world. By integrating AI technologies into curricula and research initiatives, universities can empower students and faculty to harness the transformative potential of AI for knowledge generation and innovation. However, this integration must be accompanied by robust governance mechanisms to safeguard academic integrity, intellectual property rights, and ethical standards. Moreover, higher education institutions must foster interdisciplinary collaboration and promote diversity of perspectives to ensure that AI-driven knowledge generation reflects the breadth and depth of human experience.

HEIs are at the forefront of embracing technological advancements to enhance teaching, research, and administrative processes. Central to this digital transformation is the integration of Artificial Intelligence technologies, which offer unprecedented opportunities for innovation and efficiency. However, to realize the full potential of AI in higher education governance frameworks are essential to facilitate the digital transformation of universities.

In conclusion, AI governance is indispensable for facilitating the digital transformation of universities and harnessing the potential of AI technologies to drive innovation, efficiency, and excellence in higher education. By establishing ethical guidelines, protecting student and faculty rights, promoting inclusivity and accessibility, fostering innovation and collaboration, and enhancing data security and privacy, AI governance ensures that universities can realize the benefits of AI while upholding their commitment to academic integrity, social responsibility, and ethical conduct. As universities continue to embrace AI as a transformative force in the digital age, effective AI governance will be essential to navigate the complexities and challenges of this evolving landscape.

2 Objectives and Research Methodology

The general objective of our proposal is to define the basic lines to build a framework for the governance of artificial intelligence in Higher Education institutions. This framework should have the following characteristics:

- It should be simple to implement and practical.
- It should fit the idiosyncrasies of universities.
- It should reuse of the structures and resources already available to universities.
- It should be easily integrated into the university digital transformation model.

The general objective will be achieved through the following specific objectives:

- Investigate the multifaceted impact of AI on various aspects of university operations mainly teaching research, transference and administrative functions.
- Identify the existing governance models, understand how effective they are in addressing the wide implications of AI and determine how to integrate a comprehensive framework of AI governance into them.
- Propose principles, structures of governance and best practices adapted to university settings while balancing innovation and ethical considerations, re-sponsible and equitable use and compliance with regulatory requirements.
- Define a framework for AI governance, integrating the guidelines and the best practices that promote a harmonious integration of AI technologies, preserving the core values and principles integral to the academic mission while enabling digital transformation strategy.

The methodology for researching AI governance at universities encompasses a multifaceted approach. The initial phase involves a comprehensively understanding of the existing body of knowledge and theories related to AI in higher education. Concurrently, a thorough examination of national and international regulations about AI, such as data protection and ethics, provides a regulatory context.

The authors' experience in digital transformation allows for the identification of current practices, challenges, and aspirations in implementing AI technologies. This specific and empirical expertise will be crucial in informing the development of an effective AI governance framework tailored to the needs of academic institutions.

Furthermore, the research seeks integrating AI governance into university governance models, aligning with institutional structures and goals. It aims to offer holistic understanding and practical guidelines for responsible AI governance, using insights from literature, regulations, expertise, and governance models to enhance effectiveness and alignment with university objectives.

3 AI to Achieve University Missions and Goals

In the context of higher education, AI's significance is profound, reshaping teaching methodologies, research practices and administrative tasks. This discourse delves into the multifaceted role of AI in higher education, exploring its implications across different domains and addressing the regulatory frameworks and recommendations governing its use.

Universities are tasked with advancing knowledge, fostering innovation, and preparing students for future careers. AI serves as a catalyst in fulfilling these missions and goals by promoting excellence in teaching, research, and service. Through AI-driven initiatives, universities can enhance pedagogical practices, promote interdisciplinary collaboration, and address societal challenges. Moreover, AI-powered research endeavors facilitate breakthrough discoveries, drive technological innovation, and contribute to economic development. Additionally, AI enables universities to extend their reach through online education platforms, providing access to quality education for learners worldwide.

AI plays a pivotal role in enhancing learning experiences, improving research outcomes, and optimizing administrative functions. By leveraging AI-powered tools and platforms, educators can personalize instruction, cater to diverse learning needs, and foster student engagement. Intelligent tutoring systems provide personalized feedback, adapt content based on individual progress, and enhance student retention. AI-powered virtual assistants aid instructors in lesson planning, content delivery, and student assessment.

AI also facilitates the creation of immersive learning experiences through virtual reality (VR) and augmented reality (AR) technologies, fostering student engagement and knowledge retention.

Additionally, AI-driven analytics enable institutions to gather insights into student performance, identify areas for improvement, and enhance retention rates.

AI revolutionizes the research landscape by augmenting data analysis, accelerating experimentation, and enabling predictive modeling. Machine learning algorithms sift through vast datasets, uncovering hidden patterns, and generating actionable insights. Additionally, AI-driven simulation tools reproduce complex phenomena, enabling researchers to explore hypotheses, test scenarios, and validate theories. Furthermore, AI fosters interdisciplinary collaboration by facilitating knowledge sharing, innovation, and bridging gaps between different fields of study.

AI facilitates the transfer of knowledge and technology from academia to industry, driving innovation and economic growth. Technology transfer offices leverage AI-driven analytics to identify commercialization opportunities, assess market potential, and negotiate licensing agreements. AI-powered platforms connect researchers with industry partners, fostering collaborations, and facilitating the exchange of expertise. AI accelerates the translation of research findings into real-world applications, inciting technological advancements and addressing societal challenges.

AI streamlines administrative tasks, optimizes resource allocation, and enhances operational efficiency within universities. AI-driven systems automate routine processes such as admissions, registration, and financial aid distribution, reducing administrative burdens and improving service delivery. Moreover, AI-powered analytics provide insights into enrollment trends, student demographics, and program performance, enabling informed decision-making at the institutional level. Additionally, AI facilitates predictive modeling to anticipate future challenges, mitigate risks, and optimize strategic planning initiatives.

4 Regulatory Frameworks and Recommendations for the Use of AI

As AI becomes increasingly pervasive in higher education, policymakers must establish regulatory frameworks and guidelines to ensure responsible and ethical use. National governments, international organizations, and industry stakeholders have developed strategies and recommendations to govern the deployment of AI technologies in educational settings.

4.1 National Strategies on AI

Governments worldwide have formulated national strategies to promote AI research, innovation, and adoption in higher education. For instance, the Government of Spain launched the "AI Strategy 2020" [1] to articulate a robust framework to harness the transformative potential of AI across various sectors of society, economy, and governance. This strategy underscores the pivotal role of AI in driving innovation, enhancing economic competitiveness, and addressing societal challenges while simultaneously anchoring its deployment on ethical principles and inclusivity.

Central to the strategy is the emphasis on fostering collaboration among academic institutions, the industry, and government agencies. Such interdisciplinary partnerships are instrumental in accelerating AI research and development, facilitating the exchange of knowledge and resources, and ensuring the alignment of AI initiatives with national priorities and societal needs. The strategy aims to create a synergistic ecosystem where AI technologies can be developed, tested, and scaled effectively by bridging the gap between academia's theoretical insights and the industry's practical applications.

Further, the strategy is important in advancing AI-driven research and education initiatives. It envisages the establishment of specialized research centers and innovation hubs that serve as platforms for cutting-edge AI studies, attracting top-tier talent and fostering a culture of excellence in AI research. These efforts are complemented by targeted education and training programs designed to build a skilled workforce adept in AI technologies, ensuring that AI advancements' benefits are widely accessible and contribute to societal well-being.

Moreover, the strategy acknowledges the critical role of data and digital infrastructure in powering AI applications. It advocates for developing robust data ecosystems and deploying advanced digital platforms to support AI research and innovation. This includes initiatives to enhance data accessibility, quality, and privacy, which are crucial for AI's ethical and responsible use.

In the 2023 progress report on Spain's National AI Strategy [2], a critical focus is placed on AI governance, reflecting significant strides in ethical and responsible AI development and regulation. Establishing the AI Supervision Agency marks a pioneering step in Europe, positioning Spain at the forefront of AI oversight and ethical standards. Creating an AI regulatory sandbox hosted by Spain is another key advancement. This initiative aims to refine technical regulations, emphasizing the impact on SMEs and ensuring that new rules foster innovation while maintaining efficiency and effectiveness.

These developments signify a strong commitment to ethical AI governance, ensuring that advancements in AI technology are aligned with societal values and legal frameworks. The focus on public–private collaboration in evaluating the impact of new regulations demonstrates a comprehensive approach to AI governance, ensuring that the development and deployment of AI technologies are conducted in a responsible, ethical, and inclusive manner. This strategic orientation not only advances Spain's position as a leader in AI innovation but also contributes significantly to the broader European and global discourse on the ethical AI governance.

4.2 European Strategy on AI

The European Union (EU) has outlined a comprehensive strategy to harness the transformative power of AI while addressing ethical, legal, and societal implications. The "Madiega Report 2024" [3] highlights the importance of AI in driving innovation, economic growth, and social progress across member states. Moreover, the EU's "Artificial Intelligence Act" [4] is the first-ever legal framework on AI, which addresses the risks of AI and positions Europe to play a leading role globally. It sets forth regulatory measures to ensure transparency, accountability, and human-centric AI applications in various sectors, including higher education.

The EU AI Law is the world's first legislative proposal on AI. It follows a risk-based approach with three general objectives:

- Ensure that AI systems used in the EU and introduced in the European market are safe and respect citizens' rights.
- Stimulate investment and innovation in the field of AI in Europe.
- Become a global benchmark for regulating AI in other jurisdictions.

The law establishes a series of horizontal protection criteria that will determine when AI systems can cause harm to society. Thus, it establishes three levels of risk: minimal, high, and inadmissible, and two additional cross-cutting risks: risk to transparency and systemic risks. Systems with high risks will be subject to a series of requirements and obligations to be able to access the EU market. If systems with risks are considered unacceptable, their use will be totally prohibited. These cases include, for example, cognitive behavioral manipulation or the indiscriminate tracking of facial images taken from the Internet, among others.

- Minimal risk: Most AI systems pose minimal risk. Things like recommendation systems or spam filters are safe and free from obligations because they don't threaten people's rights or security. Organizations can choose to follow extra rules for these systems if they want.
- High risk: High-risk AI systems must follow strict rules, like having safety measures, good data, keeping records, clear user explanations, and being strong and secure. Testing areas will help create safe and legal AI systems. Examples of high-risk AI include critical infrastructure, medical devices, systems for education or hiring, and those used in policing, border control, and elections. Biometric systems are also high risk.
- Unacceptable risk: AI systems that violate people's rights will be banned. This includes AI that manipulates behavior, like toys encouraging dangerous actions, and systems enabling social scoring or predictive policing. Some biometric uses, like emotion recognition at work or real-time identification by law enforcement in public areas, will also be prohibited, except in limited cases.
- Specific transparency risk: Transparency risk: Users should know when interacting with AI systems like chatbots. Fake content created by AI should be clearly labelled, and users must be informed about biometric systems. Vendors must ensure that synthetic content like audio, video, text, and images is marked as artificially generated.

Systemic risks that could arise from general-purpose AI models. Potential dangers can come from advanced AI models that can do many tasks. These models are becoming common in the EU. If these models are powerful or used a lot, they could cause big problems. For instance, they might cause accidents or be used for cyberattacks. If a model spreads harmful biases in many uses, it could affect many people.

Moreover, due to the increase in the use and importance of Generative Artificial Intelligence (GenAI) [5], the European Union has developed guidelines for the responsible use of GenAI in research [6], stressing the paramount importance of maintaining research integrity, transparency, and accountability. Researchers are urged to assess and verify AI-generated content critically, ensuring it lacks bias and inaccuracies. The guidelines mandate explicit disclosure of GenAI's use in research processes, emphasizing protecting personal and sensitive data to uphold privacy and intellectual property rights. These guidelines aim to enhance scientific inquiry's quality and ethical standards by fostering an environment of responsible AI use, thereby supporting the EU's commitment to trustworthy and ethical AI development and application in the research domain.

4.3 World Strategies on AI

The "White House AI Initiative 2023" [7] underscores the transformative potential of AI in catalyzing advancements across education, research, and workforce development globally. Recognizing AI's dual nature of offering remarkable benefits alongside potential risks, the initiative is committed to fostering a collaborative, society-wide approach to navigate these challenges. This involves a synergy of government, private sector, academia, and civil society efforts to ensure AI's development and application are anchored in safety, security, and trustworthiness.

Central to the initiative is promoting AI literacy and bolstering AI research through dedicated investments. This encompasses addressing intellectual property concerns and crafting strategies to nurture inventors and creators within a competitive and collaborative AI landscape. A significant focus is placed on equipping Americans with the requisite skills for the AI era and enticing global AI talent, reinforcing America's leadership in AI innovation.

Moreover, the initiative aims to safeguard American workers by adapting job training and education, ensuring AI deployments enhance job quality without introducing adverse effects such as increased surveillance or health risks. The initiative aspires to navigate AI's societal impacts thoughtfully through a principled approach that aligns with advancing equity, protecting consumer rights, and keeping privacy.

By setting a precedent for responsible AI governance and seeking to advance global cooperation on AI safety and ethical standards, the initiative reflects a comprehensive strategy to harness AI's potential responsibly. It envisions an era where AI-driven innovation is balanced with the imperative to address its security implications, thus promising a future where AI contributes positively to societal progress and global well-being.

UNESCO's "Recommendations on the Ethics of Artificial Intelligence" [8], adopted on 23 November 2021, represents a comprehensive framework that guides global stakeholders on the responsible development and use of AI technologies, particularly emphasizing educational contexts. These recommendations underscore AI's profound impact on society, both positively and negatively, and stress the need for a human-centered approach that upholds human dignity, rights, diversity, and cultural heritage.

At the heart of these recommendations is the call for developing AI systems that respect human-centric values and principles throughout their lifecycle. UNESCO advocates for AI technologies that bolster human rights facilitate sustainable development and protect the environment. A key aspect highlighted is the importance of fostering AI literacy and integrating ethical considerations into educational curricula and professional training programs, ensuring that future generations can responsibly navigate the complexities of an AI-driven world.

The UNESCO outlines specific policy action areas, including the need for ethical impact assessments, ethical governance, data policy, and international cooperation, all aimed at harnessing AI's potential for positive societal impact while mitigating its risks. Particular emphasis is placed on promoting equity and inclusivity in AI's development and application, ensuring that the benefits of AI tools are accessible to all, irrespective of gender, geographical location, or socioeconomic status.

Furthermore, UNESCO's recommendations call for safeguarding cultural diversity and heritage in the age of AI, advocating for the responsible use of AI in preserving and promoting tangible and intangible cultural assets. This approach enriches the cultural landscape and fosters a deeper understanding and appreciation of global cultural diversity.

In education, UNESCO stresses the importance of integrating AI ethics into teaching and research, encouraging a multidisciplinary approach that bridges technical skills with ethical considerations. The aim is to cultivate a well-rounded understanding of AI, emphasizing the critical role of ethical stewardship in guiding its development and use.

UNESCO's recommendations offer a visionary blueprint for the ethical deployment of AI, with a strong focus on education as a foundational element. By supporting human rights, diversity, and ethical governance, these guidelines set a global standard for the responsible advancement of AI technologies, ensuring they serve humanity's best interests and contribute to a just, equitable, and sustainable future.

In response to the emerging challenges and opportunities presented by GenAI in education, UNESCO has published guidelines for the appropriate use of AI in educational settings [9]. These guidelines emphasize the need for a balanced, ethical approach that leverages AI's potential to enrich educational experiences while safeguarding academic integrity and inclusivity. Central to these recommendations is the advocacy for AI systems that support personalized learning pathways, thereby enhancing student engagement and outcomes. Additionally, UNESCO calls for developing robust frameworks to address ethical concerns, including privacy, data protection, and bias prevention, ensuring AI applications in education respect diversity and promote equity. The guidelines also highlight the importance of incorporating AI literacy into curricula, preparing students to navigate and contribute to a future where AI plays a significant role in various aspects of life and work. Through these guidelines, UNESCO aims to guide educators, policymakers, and stakeholders towards the responsible integration of AI technologies in educational contexts, aligning with global efforts to achieve inclusive, equitable quality education for all.

5 Discussion on the Current Situation of AI

The ascendance of AI in the societal fabric underscores a pivotal transition in higher education paradigms, redefining the contours of teaching, research, administrative efficiency, and technological integration. The strategic embrace of AI within academia is not merely a pursuit of innovation but a foresighted alignment with the inexorable march towards a future sculpted by digital intelligence. This transition necessitates a conscientious approach to harness AI's vast potential, aiming to enrich educational experiences, expedite research frontiers, streamline administrative operations, and catalyze technology transfer. AI's ethical and responsible deployment emerges as a cardinal principle in achieving the educational mandate of knowledge dissemination, innovation promotion, and future-ready student cultivation.

National and international regulatory frameworks and recommendations serve as the keystone in this architectural restructuring of higher education under the aegis of AI. The Spanish National AI Strategy, the European Union's AI Law and its guidelines on GenAI, the White House AI Initiative, and UNESCO's recommendations delineate a collective vision towards responsibly leveraging AI's transformative potential. These policy instruments advocate for an AI-integrated educational landscape that is equitable, accessible, and inclusive, ensuring that the fruits of AI advancements are not confined to elite enclaves but permeate across diverse academic echelons.

As seen in Spain's strategy, the emphasis on fostering collaboration between academic institutions, industry, and governmental bodies underscores the importance of a synergistic approach in maximizing AI's educational dividends. Concurrently, the EU's focus on the ethical ramifications of Generative AI in research accentuates the imperative of maintaining academic integrity and upholding data privacy standards. Across the Atlantic, the White House AI Initiative's commitment to enhancing AI literacy and facilitating cross-border collaborations exemplifies a proactive stance in navigating the AI-induced educational transformation. Complementarily, UNESCO's guidelines provide a global perspective on embedding ethical considerations within AI applications in education, advocating for a harmonized international response to AI's ascension in higher education.

As AI continues to evolve, its integration into higher education heralds a new era of unparalleled opportunities for innovation, interdisciplinary collaboration, and societal impact. The concerted efforts at various regulatory levels illuminate a path forward that balances the pursuit of technological advancements with ethical stewardship and social responsibility. By adhering to these frameworks and recommendations, universities can achieve their educational missions and contribute to shaping a future where AI acts as a catalyst for societal advancement. The journey of integrating AI into higher education is fraught with challenges. Yet, it is replete with prospects for fostering a learning environment that is reflective of the complexities and dynamism of the digital age.

6 A Proposal for a Framework for the Governance of AI in Universities

For AI Governance we will first establish the principles of AI governance, based on the literature review. We have summarized these principles into four: principle of legality (or compliance), principle of neutrality (i.e. absence or control of bias), principle of transparency (related to explainability) and principle of fostering innovation (so as not to fall behind).

These four principles, applied to each of the three stages defined for the generation of knowledge (data sources, analysis algorithms and uses of the results) give rise to a grid with 12 elements that provide us with the keys to then propose the organizational structure for good AI governance and a set of good practices that will allow us to establish a roadmap for Good AI Governance in Universities as well as a set of indicators of the level of maturity reached by each university in particular.

6.1 Principles for AI Governance

Principle of legality. The principle of legality is concerned with compliance with regulations on the use of AI at both national and European level. This is an undisputed principle and is based, above all, on the European Law on Artificial Intelligence. Although the law is not yet approved, agreement has already been reached on its final development and it is expected to be approved in the coming months. As already mentioned, it is a law that warns against the risks of AI while encouraging innovation in this field, always within very strict limits that ensure the interests and protection of European citizens. European universities will be obliged to comply with the law, but it also offers them a very interesting framework for consultation that will facilitate both compliance with the law and the promotion of innovation.

Principle of neutrality. The principle of neutrality advocates the absence of bias in the data used by AI algorithms as well as in the algorithms themselves and in the access to the AI-based solutions. In some cases, ensuring this absence of bias is impossible or very complex, so sufficiently robust criteria must be established to at least control for possible biases. Strict control of biases is intended to contribute to the elimination of the gaps that AI could introduce or widen, particularly the gaps that have to do with the most disadvantaged citizens. Neutrality has its sights set on universal accessibility for all individuals regardless of their conditions and on the values of social inclusion.

Principle of transparency. With the principle of transparency, we aim to ensure that AI users know and understand how AI is used in the applications they operate. One of the biggest criticisms of AI is the high complexity of the algorithms that implement these applications and that turn them into de facto black boxes. Increasingly, the explainability of algorithms is being sought, which is a very complex issue. When it is not possible to explain how algorithms make the decisions they make, it should be possible to explain them at least with respect to their inputs and outputs. On the other hand, it is also necessary to respect this transparency both in the use of the input data (we have mechanisms such as informed consent) and in the use of the output results (informing the user, for example, that the result comes from an AI algorithm).

Principle of promotion of innovation. The innovation principle advocates providing the means and incentives for the discovery and development of new AI-based solutions within universities. This principle, in a university, becomes a fundamental issue, since research and transfer are two of the main missions of universities, and teaching, which is the third mission, is also a recipient of such innovation. Considering that the other three principles set the rules of the game, imposing limits on legality, neutrality and transparency, it is very important to establish a fourth principle of providing a safe and ethical but also ambitious framework for the development of AI.

6.2 AI Governance Grid

AI serves as a powerful tool to enhance knowledge management practices in the HEIs. Knowledge management involves the creation, storage, retrieval, and dissemination of information within the organization to facilitate decision-making, innovation, and collaboration. AI technologies, such as machine learning, natural language processing, and knowledge graphs, play a significant role in optimizing knowledge management processes by automating tasks, extracting insights from data, and facilitating knowledge discovery and sharing.

Knowledge generation can thus be considered as consisting of three major phases, on which we should focus our attention.

The first phase is the collection of data (which includes the capture, organization and preservation of the original and derived data) that will be used to generate new knowledge. AI technologies facilitate this process and generates new opportunities to automatically capturing insights from conversations, meetings, and digital interactions, for instance. Natural language understanding (NLU) algorithms extract key concepts, topics, and sentiments from textual data, enabling organizations to capture valuable insights and expertise from unstructured sources. Additionally, AI-powered knowledge graphs model relationships between concepts, entities, and resources, providing a structured representation of organizational data. The characteristics of this data sources, the data collection methods and the preprocesses to which these data may be subjected constitute a very important field of research. For proper governance it is necessary to review these data sources and the associated processes to ensure that all the proposed principles are met.

The second phase is the algorithmic treatment of the data, in which algorithms for processing these data, often massive, and in particular AI algorithms, are generally applied. AI enables organizations to analyze large volumes of data, including unstructured data such as text documents, emails, and multimedia content, to extract

Table 1 AI governance grid. (GDPR stands for general data protection regulation. AI Act stands for European artificial intelligence Act)

		Knowledge generation		
		Data	Algorithms	Uses
Principles	Legality	Observance of GDPR and AI Act	Observance of AI Act	Observance of AI Act
	Neutrality	Avoidance of data biases	Avoidance of algorithm biases	Observance ethic uses
	Transparency	Observance of consent GDPR	Requirement of explainability when possible	Requirement of a report on uses
	Promotion of innovation	Promotion of open data sources	Licensing, hardware and software procurement policies	Specific calls for AI innovation

meaningful insights and patterns. Machine learning algorithms can identify trends, correlations, and anomalies within data sets, providing valuable information for decision-making and strategic planning. By leveraging AI-powered analytics, organizations can gain a deeper understanding of their knowledge assets, user behaviors, and market dynamics. With the rise of these algorithms and their constant evolution, this subject must be the subject of careful application of the four principles.

The third phase is the use of the results of the analysis for the generation of new knowledge and the benefit of the university community. As new and increasingly sophisticated uses appear, the observation of the proposed principles is not trivial.

Based on the four principles listed above and the three phases of knowledge generation, we propose the construction of a grid with 12 elements, which we have called the AI Governance Grid (Table 1). This grid provides a useful tool to identify 12 key aspects to consider in order to ensure that our governance proposal complies with the principles set out in all stages of knowledge generation. In other words, each aspect is the answer to this question: How to fulfil each principle at each stage of the knowledge generation.

6.3 Structures for AI Governance at Universities

Most universities have a variety of structures for decision-making and control of actions in the technological area. The governance of AI requires some special actions, but we should take advantage of the expertise that universities have, for example, in data handling or in ensuring compliance with ethical aspects of research. It is therefore proposed to complete the functions of some existing structures and to create those that are necessary, while avoiding duplication and ensuring efficiency.

For the AI governance, there are several aspects to take into account, driven by the principles listed in the previous section and by the universities' own digital transformation needs. The issues to be addressed are in line with the principles of the ISO 38500 IT Governance standard [10]:

- *Responsibility*: Individuals and groups within the organization must understand and accept their responsibilities in respect of the use of AI.
- *Strategy*: The organization has a clear strategy for incorporating AI-based solutions.
- *Acquisition*: AI service providers are required to follow the same principles as the institution with clear and transparent decision making.
- *Performance*: Technological issues related to the ability to address problems and propose AI-based solutions.
- *Conformance*: Legislation and regulations compliance of the whole knowledge generation process.
- *Human Behavior*: Strict observance of ethical principles in the use of AI, always oriented towards the well-being of people and training in the proper use of AI-based solutions.

To address these issues, we propose the involvement of existing structures in the university with new attributions or the creation of such structures if they do not exist. The proposed structure is detailed below.

AI Commission. It is the central decision-making body for AI. It takes responsibility for the use of AI, transfers that responsibility to the rest of the institution according to individual roles, and proposes the strategies to be followed by the institution, taking into account in each case the legal and ethical limits, as well as the technical constraints. For all of these aspects, it involves the rest of the bodies that make up this AI governance structure.

The AI Commission can be newly created or, if there is already a strategic commission that sets the lines of action in aspects related to technology and digital services, this commission can act as the AI Commission. In any case, it should include at least the CIO of the institution (e.g. the vice-rector with responsibility for digital transformation), who should chair the committee, the director of the IT technical services, the heads of the institution's regulatory development (e.g. the General Secretary), the head of the ethics committee (e.g. the vice-rector with responsibility for research), the data protection officer and the organizational head of the institution (e.g. the manager). As the most senior decision-maker, the rector should be informed of the decisions of this committee.

Ethics Committee. In recent years, ethics committees or commissions have been set up in universities with the aim of providing an agile and effective response to current or future needs with regard to scientific research in order to protect the fundamental rights of individuals, the welfare of animals and the environment, and respect for the bioethical principles and commitments assumed by the scientific community. Subsequently, they have responded to the legal requirement that all projects involving research on human beings, the use of their personal data or of biological samples of human origin, animal experimentation or the use of biological agents or genetically modified organisms must comply with the requirements established in each case by the legislation in force and with the express authorization issued by the Ethics Committee of the Centre where the research is to be carried out.

We therefore propose that these structures already in place should also take on the responsibility of observing the aspects of *human behavior*, overseeing the ethical issues raised by AI, not only in research projects but also in projects for the implementation of solutions that universities wish to incorporate into their digital ecosystem. To this end, it will be necessary to train the members of this committee in the issues raised by AI and to include an expert on this issue on the committee.

Data Protection Officer. The General Data Protection Regulation (GDPR) [11] protects individuals when their data is being processed by the private sector and most of the public sector. This regulation puts individuals in better control of their personal data. It also modernizes and unifies rules that allow organizations to reduce bureaucracy and benefit from greater consumer confidence. It creates a system of fully independent supervisory authorities in charge of monitoring and enforcing compliance. The main authority on this issue in universities is the Data Protection Officer (DPO).

The experience gained by data protection officers in this particular area is very useful. Indeed, AI-based applications are large consumers of data and, as such, are subject to this regulation and are therefore subject to supervision by DPOs. However, the complexity of these solutions and the emergence of an AI Law suggest a restructuring of the DPO's tasks and a need to reinforce this supervisory structure.

It is therefore proposed to create the post of AI Supervisor, who should ensure that the AI Law is observed in all facets of knowledge generation. Due to the fact that the areas of application sometimes overlap between the AI Law and the GDPR and the experience acquired in the latter regulation, our proposal is that this task should fall to the DPO himself by extending his functions. The DPO is then in charge of ensuring *conformance*.

AI Technical Services. The technical IT services of the institutions will be in charge of implementing many of the AI solutions, or at least coordinating them with the suppliers. For this reason, we must have a group of technicians trained to handle these solutions.

Our proposal is to set up, within the IT technical services, a group of people specialized in the implementation, integration and deployment of AI solutions, who can undertake the implementation of these solutions in a solvent manner. Therefore, they must take on the *performance* and *acquisition* aspects.

Expert Group. An Expert Committee is a multidisciplinary group of people, external to those responsible for the implementation of AI and experts in different fields such as the more technical principles of AI, in the application of AI solutions to human problems, in the laws that regulate and limit the uses of AI, in ethical and social aspects of AI, in university governance and the use of AI in universities, etc. Its task is to issue guidelines and reports at the request of the AI commission. The

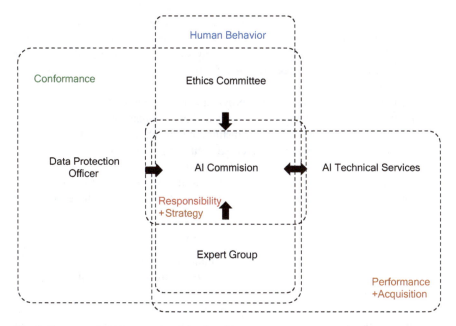

Fig. 1 Structures for AI governance for universities

proposal is to set up a dynamic committee of experts, which may include experts from the institution itself or from outside it, to respond to the requirements of the AI Commission.

Figure 1 represents the proposed organizational structure for AI governance in universities, including the bodies involved (white boxes), the relationships between them (arrows), and the areas in which each body is involved (colored boxes).

6.4 Best Practices for AI Governance

Defining best practices is the simplest and most practical ways to implement good governance for AI. In our proposal, best practices are derived from the proposed organizational structure (Fig. 1) and the identified key aspects (Table 1).

- Best practices to ensure an efficient governance structure and processes.
  - Create or strengthen the organizational structures for AI governance and assign responsibilities. Most cases it will not be necessary to create new structures; rather, it is advisable to reuse and reinforce existing structures.
  - Adequately train those involved in AI governance, particularly the data protection staff and the ethics committee so that their roles include ensuring compliance with regulatory, ethical and transparency aspects.

- Train teaching, research and management staff and students in the uses of AI so that the institution can reach a high level of maturity in the good use of AI. Develop a training program for university staff and incorporate these uses in the curricula of the degree programs for them to achieve a minimum level of AI competence.
- Require all AI-based product providers to comply with the institution's principles of conformance, acquisition and performance of all AI solutions.
- Define the strategies for AI in the institution to reach maturity in the use of AI. These strategies may constitute an AI strategic plan or, ideally, should be an integral part of the university's strategic plan.
- Clearly establish the risk levels set out in the AI Act, the assessment mechanisms of the risks and a procedure to ensure that these risks are minimized.

- Best practices to ensure the principle of legality
  - Assign responsibility for ensuring compliance with the AI Act and GDPR in the collection, storage and use of data. Typically, this responsibility can be assigned to the DPO.
  - Assign responsibility for ensuring compliance with the AI Act in the design and development of AI-based algorithms. Typically, this responsibility can be assigned to the Ethics Committee.
  - Assign responsibility for ensuring compliance with the AI Act in the use of AI-based solutions. Typically, this responsibility can be assigned to the Ethics Committee.
- Best practices to ensure the principle of neutrality.
  - Implement mandatory reporting of bias in the data used in AI-based solutions, according to the level of risk assigned to these solutions and to be reviewed by the Ethics Committee.
  - Implement mandatory reporting of bias in the algorithms used in AI-based solutions, depending on the level of risk assigned to these solutions and to be reviewed by the Ethics Committee.
  - Implement mandatory reporting on the ethical use and possible bias in the use of AI-based solutions, depending on the level of risk assigned to these solutions and to be reviewed by the Ethics Committee.
- Best practices to ensure the principle of transparency.
  - Make use of informed consent and GDRP mechanisms to ensure that users are aware of and consent to the use of their data, depending on the level of risk. Typically, it will be the DPO who will provide the control mechanisms.
  - Implement a mandatory report on the algorithmic explainability mechanisms of each AI-based solution when possible and depending on the level of risk. When such explainability is not possible, it should be indicated that the solution is a black box algorithm.

- Make it mandatory to report on the uses of each AI-based solution, the conditions of use and the consequences of its use. Users must know at all times what decisions are placed in the hands of AI.
- Best practices to ensure the principle of promoting innovation.
  - Make available to researchers and all users in general open data sources that comply with all the above principles of legality, neutrality and transparency, so that developments based on these datasets meet these basic requirements.
  - Define policies for the acquisition of software licenses and hardware that allow researchers to have a technological ecosystem that facilitates the development of AI-based solutions that comply with all the principles.
  - Establish specific calls for proposals that encourage the development of AI-based solutions, enabling both research and innovation developments and providing internal solutions to universities.

7 Conclusions

In this chapter, we have presented a comprehensive framework for AI governance in universities, grounded in four fundamental principles: legality, neutrality, transparency, and promotion of innovation.

The first pillar of our framework is legality, emphasizing compliance with laws and regulations governing AI use. Universities must ensure that AI applications adhere to legal standards, protecting the rights and privacy of individuals.

Neutrality is the second principle guiding our AI governance framework, advocating for unbiased and fair decision-making. AI systems should be developed and deployed without favoritism or discrimination, ensuring fair outcomes for all users.

Transparency emphasizes the importance of clarity and openness in AI operations. Universities must provide transparent explanations of how AI systems function, including their algorithms, data sources, and decision-making processes.

The fourth principle is the promotion of innovation. While ensuring compliance with legal and ethical standards, universities should also encourage creativity and experimentation in AI research and development.

Drawing from these principles and the stages for knowledge generation, we have constructed a governance grid comprising twelve key aspects essential for effective AI governance in universities. This grid serves as a practical tool for identifying areas of focus and developing a set of best practices tailored to the unique needs and challenges of each university.

Moreover, we propose an organizational structure for AI governance that minimizes additional effort by leveraging existing university bodies and providing them with new attributions. By integrating AI governance responsibilities into existing structures, universities can streamline governance processes and ensure efficient oversight of AI initiatives.

Furthermore, our framework aligns with the principles of the ISO 38500 standard for good IT governance, ensuring compatibility with existing IT governance structures. By adhering to governance standards, universities can facilitate the seamless integration of AI governance into broader IT governance frameworks [12].

In conclusion, our framework provides a practical and inclusive approach to AI governance. By aligning with established governance principles and leveraging existing organizational structures, universities can effectively govern AI and harness its transformative potential while keeping standards and promoting innovation.

References

1. Gobierno de España (2020) Estrategia Nacional de inteligencia artificial
2. Gobierno de España (2023) Informe sobre los avances en la Estrategia Nacional de inteligencia artificial
3. Madiega T (2023) Artificial intelligence act. European Parliamentary Research Service
4. European Commission (2021) Proposal for a regulation of the european parliament and of the council laying down harmonized rules on artificial intelligence (Artificial Intelligence Act) and amending certain Union Legislative Acts
5. García-Peñalvo F, Vázquez-Ingelmo A (2023) What do we mean by GenAI? a systematic mapping of the evolution, trends, and techniques involved in generative AI. Int J Interact Multimed Artif Intell 8:7. https://doi.org/10.9781/ijimai.2023.07.006
6. European Commission (2024) Living guidelines on the responsible use of generative AI in research
7. The White House (2023) Executive order on the safe, secure, and trustworthy development and use of artificial intelligence
8. UNESCO (2022) Recommendation on the ethics of artificial intelligence
9. UNESCO (2023) ChatGPT and artificial intelligence in higher education
10. ISO/IEC (2024) Information technology-governance of IT for the organization
11. Official Journal of the European Union REGULATION (EU) 2016/679 OF the European parliament and of the council of 27 April 2016 on the protection of natural persons with regard to the processing of personal data and on the free movement of such data, and repealing Directive 95/46/EC (General Data Protection Regulation)
12. Fernández Martínez A, Llorens Largo F, Molina-Carmona R (2019) Modelo de madurez digital para universidades (MD4U)—digital maturity model for universities (MD4U). University of Alicante

Levels of Analysis and Criteria for Digital Transformation Implementation in Higher Education Institutions

Adriana V. Karam-Koleski and Roberto Carlos Santos Pacheco

Abstract There is a common understanding that higher education is being highly impacted by digital technologies. In such a context, digital transformation (DT)—understood as a social, cultural and economic major change in society and organizations—is a contemporaneous demand. DT research demonstrates that it has been gaining space in academic and professional discussions. However, most studies are embryonic or concentrate in the use of digital technologies in the classroom, on digital technologies themselves or on the digitalization of operations with little attention to organizational conditions and directions for implementing DT. Based on organizational change, strategic planning and DT strategies, this work proposes seven levels of analysis and six criteria for implementing DT in higher education institutions (HEI). The model acknowledges DT's complexity, its systemic nature and strategic role, including HEI's mission and views towards education. Findings contribute to the DT research field as an organizational change phenomenon. It is particularly useful to HEI's strategic leadership on their planning of DT and on the improvement of its success rate.

Keywords Digital transformation · Higher education · Digital transformation implementation · Criteria for digital transformation · Levels of analysis of digital transformation

A. V. Karam-Koleski (✉)
Centro Universitário Opet, Curitiba, Brasil
e-mail: adriana@opet.com.br

R. C. S. Pacheco
Universidade Federal de Santa Catarina, Florianópolis, Brasil
e-mail: roberto.pacheco@ufsc.br

1 Introduction

There is a common understanding that higher education is being highly impacted by digital technologies [1–4]. While digital technologies bring new opportunities for higher education institutions (HEI) to accomplish their goals, they also challenge them to rethink and reposition their mission, operations, pedagogical practices, competences and relationships [5–8]. In such a context, digital transformation (DT) is a necessity.

More than introducing digital technologies in universities' current activities, DT is a systemic and structural process of organizational transformation that involves the revision and reinvention of HEI's strategy, practices, leadership and organizational culture through the use of digital technologies [9–12]. It should support institutions on their path to be flexible, agile, global and digital, thereby renewing their contribution for society [13].

DT research demonstrates that it has been gaining space in academic and professional discussions [5, 11, 12, 14, 15]. However, most studies are embryonic or concentrate on the use of digital technologies in the classroom, on digital technologies themselves or on the digitalization of operations. There has been little attention to organizational conditions for DT or directions for its implementation [9]. In addition, studies show that HEI are lagging other organizations in their transformational objective of value creation, in the benefits of digital technologies use and on structural agility to undergo digital transformation [16].

Digital transformation implementation is not an easy task for any company. Literature points out an aggregate failure rate of DT projects in companies of 87, 5% [17]. Data about DT implementation in higher education are not available but its rate is possibly similar to other types of organization. Causes are multiple, ranging from unrealistic expectations to limited scope, lack of vision and planning for DT, weak DT and IT governance; low prioritization, lack of sense of urgency, unclear roles, culture not favorable to change, lack of leadership skills and internal leadership, shortage of digital talents for DT projects, inability to adopt an experimentation mindset, gaps in digital literacy, discontinuity in investment and insufficient involvement of senior leadership [11, 17–20].

From the organizational point of view, DT should be conducted with criteria that are coherent to its complexity and comprehensive nature. In addition, it is important to identify which levels of analysis should be considered when planning and implementing DT. This chapter exposes the result of research conducted to determine such criteria and levels of analysis. It presents and discusses seven different levels of analysis to be addressed when planning DT and proposes six criteria for implementing DT which acknowledge its systemic nature and strategic role for universities while respecting their nature and multidimensional mission as knowledge creators, and human development organizations. Findings are intended to support HEI's strategic leadership on their planning of DT and contribute to improve DT success rate in higher education.

2 Methodology

Data supporting the proposed levels of analysis and criteria were collected from an integrative literature review [21–23] of 398 studies about DT implementation in HEI. Seventeen articles that presented DT models in HEI were reviewed, providing data for proposing the levels of analysis and criteria for DT [2, 12, 16, 24–37].

Content analysis [38] was used to handle and analyze data from the reviewed studies. Data analysis was conducted in two different phases, both using inductive codification. The first phase established ninety-three codes related to DT implementation, which were then organized into seven categories, resulting in the different levels of analysis for conducting DT in HEIs. In the second phase, data were organized into five themes which are presented as criteria for conducting DT in HEI.

3 Levels of Analysis for DT Implementation in HEI

Implementing DT is a complex endeavor. An approach that recognizes and addresses such complexity can facilitate its successful implementation. In this context, seven levels of analysis are proposed for consideration when designing and conducting DT implementation processes: (1) strategy and business model; (2) people, culture and capacities; (3) products and core services; (4) management and governance; (5) student experience; (6) HEI ecosystem; (7) digital technologies. These levels are detailed in the sections below.

3.1 Strategy and Business Model

Defining strategy and business model is fundamental for DT at higher education institutions. This involves determining their strategic positioning in the context of digital transformation and how the institution will use digital technologies to create a value proposition that meets the needs of students and society. It includes analyzing external and internal contexts regarding changes due to the features of network society and digital technologies [12, 27, 39]. This process entails making choices, defining a vision, positioning and identity for the HEI [29, 34]. A clear vision allows for the alignment of leadership and the entire HEI community around common goals and a common shared future [29].

Institutional positioning and identity support the development of strategies for market and stakeholder relationships, as well as product and digital services [16, 29]. Defining governance for DT is also part of this level of analysis [12], as well as assessing the human, technological and financial resources for DT, which must be accompanied by the definition and allocation of financial resources for implementing the DT plan [16, 37, 39].

3.2 People, Culture and Capabilities

DT is fundamentally about people. They are the ones who give life to HEI, including students, faculty, researchers, administrative teams and ecosystem partners. People carry out DT and it is for them that universities should create value. In this regard, people, organizational culture and their capabilities become essential dimensions of digital transformation.

When it comes to implementing change processes, culture can either drive or inhibit transformation [25]. Culture refers to principles that guide decisions of HEI members and is reflected on standards that guide action [34]. A culture conducive to digital transformation is based on agile management and includes collaborative work, constant analysis, adaptation, correction and delivery that fits the needs of students and HEI context [2, 39].

Digital culture, digital skills and dynamic capabilities are essential for DT [2, 16, 39]. A new set of skills to be developed by all stakeholders. Developing these skills and capabilities requires investment in professional development of HEI staff, students and other stakeholders [2, 24, 26, 27, 29, 30, 34, 35].

Another fundamental component of this level of analysis is leadership [12, 24, 29, 30, 34, 35, 39]. Vision, collaboration, leadership ability, management skills, adaptability, creativity, innovation and digital literacy are elements that, when used in an integrated way, encourage HEI members to develop their dynamic and digital capabilities [30]. Leadership should nurture an innovative culture by establishing connections, fostering collaboration and encouraging experimentation, risk-taking and learning [29, 35]. Leaders should plan and act to obtain the support and engagement of the HEI community and its multiple stakeholders to conduct DT [16, 29].

Note that when it comes to "people, culture and capabilities", all stakeholders should be considered: students, teachers, researchers, technical-administrative teams, suppliers, partners, members of university management etc. [5]. It is important to plan actions that take all of them into consideration ensuring that they are involved, challenged and empowered by DT.

3.3 Core Products and Services

Digital transformation should lead to significant changes in teaching and learning processes, as well as in services (i.e., teaching, researching and practice) provided to students. In this respect, planning and implementing DT should include activities and technologies directly linked to teaching, research and the university outreach, such as:

- Creation of new services, academic and curricular processes and teaching strategies to support the development of necessary skills for digital and networked society [26, 29, 32, 34–36, 39].

- Introduction of digital technologies and content in academic activities to enrich learning; use of data analytics for learning assessment and adaptation of educational offerings to students' needs and interests [2, 32, 35, 36].
- Integration of digital technologies such as artificial intelligence, cloud computing and big data to expand research capabilities, and facilitate co-production within university and in collaboration with other research centers [34].
- Establishment of national and international connections in teaching, research and outreach activities for collaboration, co-creation and development of skills and solutions [25, 34].
- Use of digital technologies to offer work-integrated learning, career planning, professional opportunities, and connections with job market and industries [36].

3.4 Governance and Management

This level involves organizational management and its digital transformation. It comprises the necessary changes in policies and institutional guidelines to pave the way for the integration and use of digital technologies throughout the institution [34]. It also addresses work standards, business and governance practices for digital transformation [12, 26, 29, 33].

Digital technologies bring important contributions to HEIs: advanced data analysis and its use for decision making can represent an advance in organizational management, marketing, internal and external communication, student recruitment, enrollment management, and success [2, 29, 32, 33, 36]. Campus administration also benefits from the use of digital technologies, as they offer possibilities for productivity gains, cost savings and enhanced connection between the institution and its ecosystem [12, 29].

Given the complexity of implementing DT in universities, establishing a business architecture adds a systemic view to the transformation process. By expressing the main strategies in terms of business applications, technologies and the impact they have on different functions and processes, DT can be planned with consideration of the organizational fabric [37]. This approach increases the assertiveness of DT implementation as it considers the institutional reality and the objectives that have been set.

3.5 Student Experience

Structuring HEI's services and its DT based on student experience reflects the centrality of the client, an important concept in digital transformation [12, 29, 36]. Student experience integrates all aspects related to student life and their experience encompassing not only learning, but also the necessary services for students to be successful on their journey.

It involves developing digital skills and the proper use of digital tools made available to students [2, 26, 34] using technologies for the development of skills necessary for professional performance, career planning and professional placement [36] offering user-friendly systems [26, 29, 36], creating new communication and network channels [35], and offering technologies that track students engagement on course activities [2, 29, 32].

It is worth highlighting Holon IQ's Digital Capability Framework for higher education and its organization around student's life cycle, from the process of choosing the HEI to their monitoring and continued education as alumni [36]. By organizing institutional capabilities around student needs, this framework emphasizes the importance of putting students and other users of HEI's services at the center of digital transformation.

3.6 HEI Ecosystem

Considering digital transformation in higher education means dealing with connections within the ecosystem in which it participates [12, 27, 34–36, 39]. After all, higher education is embedded in a networked society. Building a value proposition appropriate to this context requires HEI to operate within their ecosystem and establish alliances, opening for collaboration and coproduction [8, 29]. In practice, this means engaging with the market, industries, government, investors, research institutes, civil society, students' families, K12 and other higher institutions [12, 26, 36, 39].

The use of digital technologies can be decisive for a new performance of HEI, as it can enhance their communication, increase engagement, support their internationalization, collaboration, networking, interinstitutional coproduction and open innovation [12, 26, 34, 35].

Alumni are also part of the HEI ecosystem, as they are integral members of society, who work professionally in a variety of locations and can be strong links of connection due to the relationships established with teachers, researchers and colleagues during their education [36].

3.7 Information Technologies

All reviewed articles highlight that although information technologies are not the only responsible for the success of a DT implementation process, they are an essential part of it. It is impossible to talk about DT without considering the existence of adequate infrastructure, technological resources, cyber and information security and IT processes aligned to institutional strategy [12, 29, 30, 32, 33, 35, 37]. After all, innovations and the transformation of the HEI are implemented through the use of digital technologies such as digital services platforms, IoT, data analytics, artificial intelligence, communication channels, social media and IT infrastructure [24, 26, 33, 37].

Digital technologies have been supporting HEI operations for a long time. The difference in DT is that they are now driving forces, making it possible to create new value proposition offers and new ways of delivering higher education. Artificial intelligence, cloud computing, internet of things, big data and social media can help institutions to create new forms of educating their students, conducting research and collaborating with their ecosystem [2, 37].

The institutional operational technological backbone should be taken in consideration when planning and conducting DT. It comprises digital systems and tools for communication, collaboration and work management tools; back-up systems for finance, purchasing, human resources, academic record; central systems that operate core activities systems as Learning Management System (LMS); service portal for students and business intelligence systems, with dashboards and online analytical processing. Additionally, it encompasses IoT equipment and their computing systems.

It is over this backbone that a digital service platform will be developed. Such platform can be formed by different blocks, including: (a) user experience platform, which integrates multichannel interaction and online learning, social networks, student data analytics, student portals or apps, and back-end systems; (b) IoT platform; (c) ecosystem platform which enables connection with ecosystems and external communities such as other universities and research centers, high schools and companies with need for training or job positions available; (d) data analytics platform, which incorporates data for learning, machine algorithms and business intelligence [37]

Figure 1 summarizes the different levels of analysis proposed in this section. By addressing multiple dimensions of DT in HEI, they may contribute to improving the success rate of digital transformation efforts.

4 Criteria for DT in HEI

In addition to the levels of analysis, five criteria for DT implementation could be extracted from the literature review. They are described in this section and should be taken into consideration when designing and implementing DT in higher education. The criteria for DT implementation presented are (1) DT must be approached in its systemic nature; (2) DT is a non-linear and incremental journey; (3) vision and planning of DT are fundamental keystones; (4) DT needs financial resources; (5) DT leadership and governance are different and have complementary roles; (6) ambidexterity is a capacity that sustains DT in HEI.

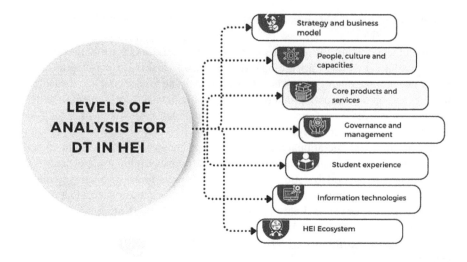

Fig. 1 Levels of analysis for digital transformation (DT) in higher education institutions (HEI)

4.1 DT Must Be Approached in Its Systemic Nature

DT is a systemic phenomenon and as such, requires an approach that respects and contemplates its multiple dimensions, levels of analysis and stakeholders. Note that the levels of analysis of DT in HEIs are represented by a circular diagram (Figure 2) that highlights their integration and absence of hierarchy between them [35]. All DT components are deeply interconnected and acting on one means establishing interfaces, changing processes and products, introducing new experiences, setting up new relationships and skills [34].

In addition to the internal context, external context also requires a connected action. The institution must be sensitive and react to a changing context which requires real-time surveillance of its internal and external environment, constant evaluation of its value proposition and ability to reconfigure and orchestrate its different elements so that digital transformation offer strategic advantages for the institution [37].

4.2 DT is a Non-linear and Incremental Journey

Although it can lead to structural transformation, DT is made of many tasks carried out at the same time and over time [26, 33, 37, 39]. Monteiro and Pinto [33] make this temporal nature of DT evident. By distributing the implementation of technological applications at University of Lisbon, between 2008 and 2018, the authors make it clear that DT takes time and is constituted by a series of actions and processes. In their case study, they share that DT implementation began with a vision and decision about the strategic positioning of the institution and the decision to use digital technologies to deliver value to its different stakeholders. They underline the importance of DT being thought in multiannual plans to be implemented little by little, depending on financial capacity, operational and digital skills. Similarly, Gama-Perez, Vega and Aponte [26] present a 10-year cycle at their DT model. Its implementation, based on the definition of a strategic vision led to the constitution of an ecosystem in which their institution was inserted, the consolidation of a culture conducive to DT, the development of digital capabilities of its stakeholders and the digitalization of organizational and academic processes.

Two DT implementation roadmaps were identified in the literature review [25, 39]. Both reinforce that DT evolves over time and that it is done in iterative cycles that refine processes and skills and lead to greater digital maturity and ability to face the challenges imposed by the network society.

An initial definition of digital strategy and the intended use of digital technologies, the review of organizational processes, the development of capabilities and the creation of a flexible and adaptable culture evolve over time and generate openness to new learning cycles. This organizational learning enables the institution to identify and face a changing scenario, take advantage of constantly evolving digital technologies, improve their processes, create competitive advantages and to generate value for its stakeholders over time [37]. Furthermore, the changing context in which HEIs and the entire society are, and the constant evolution of digital technologies do not allow us to understand transformation as a process that has a beginning, middle and end. More than an event, DT is a non-linear journey of transformation with comings and goings and spiral advances.

4.3 DT Vision and Planning: Fundamental Keystones

The journey of DT in HEI needs to be supported by strategy and planning. Aditya, Ferdiana and Kusumawardani [40] found that the lack of a shared vision is a relevant barrier to DT implementation. Therefore, a clear vision of the aims of DT is essential to align efforts, resources and action to implement DT [2, 26, 28, 29, 33, 34].

Rof, Bikfalvi and Marquès [28] reinforce the contribution of a DT plan. They alert that DT happens as needs, technologies and tensions appear and that planning would contribute with the alignment of actions with a more strategic, disciplined and systematic approach to digital transformation. Monteiro and Pinto [33] also highlight the importance of planning and point out that it undergoes revisions and adaptations over time. These changes are the result of changes in organizational culture, in the ecosystem of the institution, the emergence of innovative technologies and changes in the educational scenario [33, 37, 39].

Another important highlight for DT planning is that it should be deeply connected to the institutional reality and contextual factors such as location, size, university vocation, educational model, academic community and the level of access to technology. Therefore, each HEI should generate its unique DT plan [16].

4.4 DT Needs Financial Resources

A fundamental issue for the sustainability of a DT process is its financial viability [16, 29, 39]. Financial resources need to be directed towards DT because, without them, there is no way to introduce new technologies, promote changes in culture, create digital processes, develop digital capabilities of stakeholders, build up a logical infrastructure and maintain digital platforms,

Digital transformation plans should identify necessary resources and digital strategies and DT processes need to be evaluated in relation to their economic efficiency [37]. Also, search for sources of financing, savings derived from using technologies and the reallocation of resources to activities that are directly linked to the intended value creation must be part of the financial planning and implementation of DT [29].

4.5 DT Leadership and Governance: Different and Complementary Roles

Leadership plays a fundamental role in DT since it establishes the context and guides its implementation process. DT leadership skills include vision, collaboration, leadership capacity, management skills, adaptability, creativity, innovation, and digital literacy [30]. Leadership should be able to integrate learning in the digital era and the use of digital technologies to the HEI's mission, vision and strategy [35]. Its role also includes leading cultural adjustments and new processes development to prepare the institution to act accordingly in the network society [16]. The leader should nurture an innovative and encouraging culture of experimentation and learning [29], identifying, testing and gaining stakeholder support for appropriate technologies to realize HEI's academic and organizational objectives [39]. Building a viable digital business model plan from the desired results, financial and operational point of view is another fundamental aspect of DT leadership [29, 32, 37].

Senior leadership must be deeply engaged with DT. It is fundamental that this leadership is clear about the importance of the CEO or the Dean in aligning people, work plans, allocation of resources and sponsorship of DT in the organization. A digital transformation project without the support and participation of senior management faces a relevant barrier to success [39].

Two types of DT leadership were found in literature, both concerning senior management: (a) institutional leadership; (b) DT process leadership. The first is represented by the CEO or the Dean, who lead the construction of the vision, institutional positioning and objectives for DT. Because of their formal leadership, senior leadership sponsorship has a decisive impact on DT's results. The second is represented by the leadership of the DT process itself. This person must have experience in innovation projects, supporting institutional leadership in formulating DT strategy, and coordinating the execution of the strategy. The figure of "Chief Digital Officer" (CDO) is indicated in literature [12, 37].

Both leaders need to work in close partnership to put DT in pace, analyze performance indicators and constantly evaluate the progression and results of the DT plan [26]. DT process must be constantly accompanied by qualitative and quantitative data to guide its continuous improvement [37].

Establishing a governance process for DT is key to its success, though this is still a challenge for HEIs, because it tends to be less prioritized than other endeavors [41]. Clearly indicating those responsible for the DT leadership process and how responsibilities are divided is an important measure that contributes to its success.

4.6 Ambidexterity: Fundamental Capacity for DT

In the context of DT, there are different nuances of the changes in the operating model of a HEI. They range from the gradual incorporation of technologies that bring new configurations to the HEI incrementally [33] to disruptive changes, with the creation of new activities and performance of the HEI in new markets [37]. In any case, the challenge of implementing transformations and making the organization work at the same time is frequent [2, 26, 28, 37].

Two terms differentiate these phenomena: "exploit" and "explore". "Exploit", is related to efficiency, growth productivity, control, assertiveness and reduction of variability. "Explore" refers to the search, discovery, autonomy, innovation and acceptance of variability. Doing both is ambidexterity [42, 43], an important capacity for digital transformation processes.

Acting in challenges arising from practices already established by the HEI and from emerging opportunities at the same time may be difficult. A way to face such challenge is to treat them as independent, defining those responsible for each [37]. The suggestion of a model of dual transformation encourages the organization to embark on two separate efforts, but carefully connected. Transformation A redesigns the core institution to improve its teaching, research and extension capacity, while transformation B, designs capabilities to respond to opportunities or emerging social demands, focusing on entirely new models to achieve distinct value propositions and student markets [37].

Literature points to different approaches to establish ambidexterity in the context of DT. Exploration and exploitation activities can be conducted separated temporally, in different areas of the organization or through change in employee behavior according to the context. There is also the possibility of implementing ambidexterity

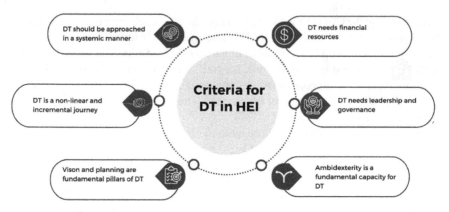

Fig. 2 Criteria for digital transformation (DT) in higher education institutions

as a dynamic capability of the organization, which fits into the exploration–exploitation continuum as a response to potentialities and needs of its context or even do it with the integration of some of these approaches [6, 44].

Regardless of creating separate or integrated structures for DT, it is essential to recognize the challenge of creating new paths of action and positioning while managing the current institution. Carrying out daily operations and, at the same time, implementing innovations that constitute new value for the HEI and its stakeholders is a challenge to be faced and resolved by the senior management of a higher education institution.

The criteria described above should guide the construction of the institution's model of DT implementation. They are covered, sometimes in an obvious way, sometimes as theoretical support for adopted choices. Figure 2 summarizes the criteria for DT in HEI.

5 Discussion

The distinct levels of analysis presented in this chapter were created to contribute to the planning, implementation and evaluation of digital transformation programs in higher education institutions. They can function as elements to be integrated in the DT model that each HEI should create for its own digital transformation. They highlight that: (1) DT has to do with recreating an institution's strategy and business model, by defining a new value proposition that is coherent to digital society and to the role that the HEI intends to play in its ecosystem. (2) People and culture are central to DT process – they are the ones who will implement or react against DT and their capacities need to be developed to contribute to the process. (3) DT should impact core products and services of the institution, what means that teaching and learning processes, knowledge production, research and outreach activities should be a central part of the transformation. (4) There needs to be governance for DT. Otherwise, objectives, deadlines and results are not reached. Strategic leadership must understand his/her essential role to DT. (5) Student experience should be central to the planning of DT initiatives. (6) Connections are a central part of DT and there is a call for opening the institution to collaborate and cocreate with its stakeholders and ecosystem. (7) Digital technologies are not the only piece of DT, but they are at its core. They not only automate and accelerate processes and communication; they can provide new ways of approaching and conceiving higher education.

The criteria for digital transformation may function as parameters for DT implementation: (1) DT should be approached in a systemic manner. As discussed, it is a profound and systemic organizational change, and it has to do with the defining and delivering a new value proposition. Because of that it pervades all dimensions of HEI. Addressing it in silos will hinder HEI from taking full advantage DT. (2) DT should be planned and implemented in incremental steps. It involves profound and multilayered changes in culture, processes, technologies, skills and capacities. Such changes take time, are not linear and will benefit from stakeholders' learning curve. By acknowledging these characteristics from the beginning DT can be approached in a more realistic manner, with smalls steps being celebrated and difficulties being understood as part of the process. (3) DT should be based upon a digital institutional vision and planning. Vision should help HEI members to have a common understanding of the new value proposition that DT will help to construct. DT planning does not need to be overdetailed, given the non-linear and incremental nature of DT. But it is necessary to set an initial direction for DT journey. (4) DT needs financial resources. They are crucial because digital technologies, infrastructure and skills need to be renewed or acquired. (5) DT needs both leadership and governance. They are different and complementary. Senior leadership should lead and sponsor DT, working closely with those who are responsible for its planning and implementation. (6) DT demands HEI ambidexterity. Focus and energy for innovation needs to coexist with daily activities. This is a challenge since running HEI programs and activities are highly demanding. However, if there is no space for innovation, experimentation and creation, there is no DT.

6 Final Remarks

DT has become an inevitable choice for HEIs, especially in the wake of the COVID-19 pandemic. Its successful implementation, however, remains a significant challenge for both business and educational institutions. At its core, DT is an organizational change that unfolds over time and demands careful management. It is a broad and systemic process of structural change requiring unprecedented coordination of people, processes and technologies.

This research offers a system approach to understanding and guiding the complex process of DT in HEI. By offering a detailed framework and practical criteria, it provides a roadmap for institutions navigate this transformation systemically (and effectively). Nevertheless, the limits of this research must also be acknowledged: while the framework is scientifically (and multidisciplinary) based, its application may vary significantly across different contexts and institutions, demanding future and further studies to tailor and refine the guidelines to specific environments.

Looking ahead, the future of DT in HEIs hinges on continuous adaptation and innovation. Institutions must remain flexible and responsive to emerging technologies and shifting educational landscapes (DT challenges both processes simultaneously). Leaders must embrace a mindset of continuous learning and improvement, ensuring that DT initiatives align with their core mission of social and human development. By doing so, HEIs can harness the full potential of digital technologies to enhance their impact to (and from) society.

Competing Interests The authors have no competing interests to declare.

References

1. Brooks DC, Mccormack M (2020) Driving digital transformation in higher education. Educause, Louisville, CO
2. Mohamed Hashim MA, Tlemsani I, Matthews R (2022) Higher education strategy in digital transformation. Educ Inf Technol (Dordr) 27:3171–3195. https://doi.org/10.1007/s10639-021-10739-1
3. Hilbert M, Vallalón S, Armenta C et al (2020) Estrategia y transformación digital de las universidades: un enfoque para el gobierno universitario. Fundación Universia, Banco Interamericano de Desarrollo
4. Jensen T (2019) Higher education in the digital era: the current state of transformation around the world. Paris
5. Benavides L, Tamayo Arias J, Arango Serna M et al (2020) Digital transformation in higher education institutions: a systematic literature review. Sensors 20:3291. https://doi.org/10.3390/s20113291
6. Jackson NC (2019) Managing for competency with innovation change in higher education: examining the pitfalls and pivots of digital transformation. Bus Horiz 62:761–772. https://doi.org/10.1016/j.bushor.2019.08.002
7. Pucciarelli F, Kaplan A (2016) Competition and strategy in higher education: managing complexity and uncertainty. Bus Horiz 59:311–320. https://doi.org/10.1016/j.bushor.2016.01.003
8. Pacheco RCS, Karam-Koleski AV, Pereira LMF (2024) Digital transformation in universities as a double-sided challenge. In: Liebowitz J (ed) Digital transformation and society. World Scientific Publishing Co, Singapore. https://doi.org/10.1142/9789811295140_0005
9. Karam-Koleski A (2023) Metamodelo para implementação de transformação digital em instituições de educação superior: jornada de transformação por meio de abordagem multiteórica de mudança organizacional. PhD dissertation, Universidade Federal de Santa Catarina—Florianópolis. https://repositorio.ufsc.br/handle/123456789/251377
10. Verhoef PC, Broekhuizen T, Bart Y et al (2021) Digital transformation: a multidisciplinary reflection and research agenda. J Bus Res 122:889–901. https://doi.org/10.1016/j.jbusres.2019.09.022
11. Marks A, AL-Ali M, Atassi R et al (2020) Digital transformation in higher education: a framework for maturity assessment. Int J Adv Comput Sci Appl 11:504–513. https://doi.org/10.14569/IJACSA.2020.0111261

12. Menendez FA, Machado AM, Esteban C (2016) Analysis of the digital transformation of higher education institutions: a theoretical framework. Edmetic 6:180–202
13. Llorens F, Fernández A, Cadena S et al (2022) UDigital: maturidade digital para universidades. MetaRed; Secretaria General Iberoamericana, Madrid
14. Abad-Segura E, González-Zamar M-D, Infante-Moro JC, García GR (2020) Sustainable management of digital transformation in higher education: global research trends. Sustainability 12:1–24. https://doi.org/10.3390/su12052107
15. Trevisan LV, Eustachio JHPP, Dias BG et al (2023) Digital transformation towards sustainability in higher education: state-of-the-art and future research insights. Environ Dev Sustain. https://doi.org/10.1007/s10668-022-02874-7
16. Rodríguez-Abitia G, Bribiesca-Correa G (2021) Assessing digital transformation in universities. Future Internet 13. https://doi.org/10.3390/fi13020052
17. Wade M, Macaulay J, Noronha A, Barbier J (2019) Orchestrating transformation: how to deliver winning performance with a connected approach to change. IMD, Lausanne
18. Fitzgerald M, Kruschwitz N, Bonnet D, Welch M (2013) Embracing digital technology: a new strategic imperative. MIT Sloan Manag Rev 1–16
19. Bughin J, Holley A, Mellbye A (2015) Cracking the digital code. Singapore; London, Brussels
20. Firmin MW, Genesi DJ (2013) History and implementation of classroom technology. Procedia Soc Behav Sci 93:1603–1617. https://doi.org/10.1016/j.sbspro.2013.10.089
21. Torraco RJ (2005) Writing integrative literature reviews: guidelines and examples. Hum Resour Dev Rev 4:356–367. https://doi.org/10.1177/1534484305278283
22. Whittemore R, Knafl K (2005) The integrative review: an updated methodology. J Adv Nurs 52:546–553
23. Botelho LLR, de Cunha CC, A, Macedo M, (2011) O método de revisão integrativa nos estudos organizacionais. Gestão e Sociedade 5:121–136
24. Bravo J, Aquino J, Alarcón R, Germán N (2021) Model of sustainable digital transformation focused on organizational and technological culture for academic management in public higher education. In: Iano Y, Arthur R, Saotome O et al (eds) Smart Innovation, Systems and Technologies, vol 202. Cham, pp 483–491
25. Doering C, Reiche F, Timinger H (2021) Process model for digital transformation of university knowledge transfer. 13th International joint conference on knowledge discovery, knowledge engineering and knowledge management (IC3K 2021). Science and Technology Publications, Valletta, Malta, pp 153–160
26. Peréz Gama JA, Vega Vega A, Neira Aponte M (2018) University digital transformation intelligent architecture: a dual model, methods and applications. In: Petrie MML, Alvarez H (eds) Proceedings of the 16th LACCEI international multi-conference for engineering, education, and technology: "innovation in education and inclusion." Latin American and Caribbean Consortium of Engineering Institutions, Lima, pp 19–21
27. Peréz Gama JA (2018) Intelligent educational dual architecture for university digital transformation. In: 2018 IEEE frontiers in education conference (FIE). IEEE, San José, California, pp 1–9
28. Rof A, Bikfalvi A, Marquès P (2020) Digital transformation for business model innovation in higher education: overcoming the tensions. Sustainability (Switzerland) 12:1–15. https://doi.org/10.3390/su12124980
29. Alenezi M (2021) Deep dive into digital transformation in higher education institutions. Educ Sci (Basel) 11:1–13. https://doi.org/10.3390/educsci11120770
30. Tungpantong C, Nilsook P, Wannapiroon P (2021) A conceptual framework of factors for information systems success to digital transformation in higher education institutions. In: 9th international conference on information and education technology. Okayama, Japan, pp 57–62
31. Aditya BR, Ferdiana R, Kusumawardani SS (2021) Barriers to digital transformation in higher education: an interpretive structural modeling approach. Int J Innov Technol Manag 2150024:1–18. https://doi.org/10.1142/S0219877021500243
32. Ávila-Correa BL (2019) Perspectivas de transformación digital de las universidades del Ecuador. Revista Ciencias Pedagógicas e Innovación 6:1–11. https://doi.org/10.26423/rcpi.v6i2.233

33. Monteiro MH, Pinto RR (2019) The e-government adoption in higher education in Portugal: the case of ISCSP at Lisbon University. J Inf Syst Technol Manag -JISTM USP 16:1807–1775. https://doi.org/10.4301/S1807-1775201916002
34. Pacheco R, Santos N, Wahrhaftig R (2020) Transformação digital na educação superior: modos e impactos na universidade. Revista Nupem 12:94–128. https://doi.org/10.33871/nupem.2020.12.27.94-128
35. Kampylis P, Punie Y, Devine J (2015) Promoting effective digital-age learning: a european framework for digitally-competent educational organizations. Publications Office of the European Union, Luxembourg
36. Holon IQ (2021) Higher education digital capability (HEDC) Framework. San Francisco
37. Gomes R, da Cruz AMR, Cruz EF (2020) EA in the digital transformation of higher education institutions. In: Rocha Á, Peréz BE, Penãlvo FG et al (eds) 15th Iberian conference on information systems and technologies (CISTI). IEEE, Seville, Spain, pp 1–6
38. Bardin L (2011) Análise de conteúdo. Edições 70, Lisboa
39. Aditya BR, Ferdiana R, Kusumawardani SS (2021) Categories for barriers to digital transformation in higher education: an analysis based on literature. Int J Inf Educ Technol 11:658–664. https://doi.org/10.18178/IJIET.2021.11.12.1578
40. Aditya BR, Ferdiana R, Kusumawardani SS (2021) Digital transformation in higher education: a barrier framework. 2021 3rd international conference on modern educational technology. ACM, New York, NY, USA, pp 100–106
41. Rodríguez-Abitia G, Martínez-Pérez S, Ramirez-Montoya MS, Lopez-Caudana E (2020) Digital gap in universities and challenges for quality education: a diagnostic study in Mexico and Spain. Sustainability 12:1–14. https://doi.org/10.3390/su12219069
42. O'Reilly CA, Tushman MI (2013) Organizational ambidexterity: past, present and future. Acad Manag Perspect 27:324–338. https://doi.org/10.5465/amp.2013.0025
43. O'Reilly CA, Tushman ML (2008) Ambidexterity as a dynamic capability: resolving the innovator's dilemma. Res Organ Behav 28:185–206. https://doi.org/10.1016/j.riob.2008.06.002
44. Jöhnk J (2020) Managing digital transformation: challenges and choices in organizational design and decision-making. Universität Bayreuth - Bpnn, Thesis

Driving Innovations: Trends, Prospects and Challenges of Implementing Disruptive Educational Technologies Within HEIs

Edwin Zammit, Clifford De Raffaele, Daren Scerri, Ronald Aquilina, Joachim James Calleja, and Alex Rizzo

Abstract As technological and social innovations usher in Industry 4.0, within educational settings a reform in higher education is essential to drive transformations in the forms and methods of teaching [1]. Thus, academic driven innovations are essential, as apart from enhancing both the quality and the outcome of the learning process, they will provide students the ability to learn and interact with innovative technology within their cognitive development [2]. To this end the use of Educational Technologies (EdTech) has become central to the adoption of innovation in Higher Education Institutions (HEIs), affecting all elements of the learners' experience [3]. This chapter will seek to provide implementation detail and governance methodology employed to launch a disruptive EdTech framework within a HEI located in the small island state of Malta. The framework specifically aims to implement and enhance emerging EdTech to meet evolving student expectations, support their learning needs through a transformed hybrid infrastructure, and establish a centralized EdTech knowledge hub. Furthermore, the chapter seeks to guide and define the parameters that lead to identification of viable EdTech initiatives, the collaborative requirements, the general resources expectations, and the key stages of adoption of chosen EdTech solutions within a HEI. The EdTech framework is anchored in the premise of making existing organizational structures more flexible, and fostering their ability to innovate, react and respond. Within this model, each unique EdTech adoption is introduced and managed through a project-based approach and analysed through the proposal of an EdTech Adoption Model (EDAM). The process is detailed and exemplified within the case study analysed through this chapter will elucidates the introduction, adoption and acceleration lifecycle of eXtended Reality (XR) in higher education.

Keywords Innovation management · Digital educational environments · HEI · Culture change

E. Zammit (✉) · C. De Raffaele · D. Scerri · R. Aquilina · J. J. Calleja · A. Rizzo
Malta College of Arts, Science & Technology, Triq Kordin, Paola PLA 9032, Malta
e-mail: Edwin.Zammit@mcast.edu.mt

1 Introduction

Continuous innovation is essential for an organization to maintain a distinctive edge whilst securing success and survival amidst the constant flux of a competitive environment [4]. White et al. [5] define innovation in Higher Education (HE) as some new way of doing things, or a change that improves administrative or scholarly performance, or a transformational experience based on a new way of thinking. To this extent the introduction of Educational Technologies can incentivize the HEIs' innovation drive to offer an enhanced experience during the students' learning trajectory. EdTech encompass all forms of technology digital learning platforms, digital resources, software and hardware, that are available for the modern learning environment. Most of these technologies tend to be perceived as disruptive in nature and have the potential to revolutionize education as we know it [6]. Such examples of disruptive EdTech includes Artificial Intelligence; Tangible Computing; Extended Realities; Visualization Wall Technology [7].

This chapter aims to explore the various facets of the adoption of emerging educational technologies in HEIs through a case study approach. The chapter starts by providing a birds-eye view of the EdTech landscape in which HEIs operate. In Section 2, the latest trending disruptive education technologies are highlighted. Section 3 follows to explain the key determinants for EdTech adoption within HEIs. Here, the key drivers for EdTech adoption, together with the role Educators play in EdTech adoption is highlighted. In subsequent sections 4–6, a contextualized example on how the complexity associated with the implementation of novel technologies can be integrated in the operation within one particular HEI, the Malta College of Arts, Science & Technology (MCAST), will be outlined.

MCAST has been selected due to its unique nature, running educational programmes at all 8 levels of the European Qualifications Framework (EQF) and working very much as a vocational institution at EQF Levels 1–4 and in professional higher education at EQF Levels 5–8. The innovation model adopted by the Applied Research & Innovation Centre (ARIC) is one which sees innovation as a pivotal driver thus pushing the organisation closer to its strategic objectives. The MCAST EdTech-Innovation operational framework provides a structure through which planning, and goal setting occurs. Sections 4 and 5 provide details on how the framework ensures that HEIs act in a coordinated and consistent way, responding to the needs and expectations of their students and the changing needs of education through an iterative six-staged process. In the subsequent section of the chapter, each of the six stages will be delved into and exemplified using a case study which will elucidate the lifecycle implementation of a novel in-house designed Virtual Reality (VR) solution. The adoption details, the technical perspective of content development and consultation, together with implementation outcomes in both student education and public engagement settings, shall be discussed. Subsequently, in the final part of the chapter it will be shown how the robust framework resulted in capacity building, which in turn allowed to the organisation to secure a coordination role in a Horizon Europe Widera project.

2 Latest Trends in Emerging Digital Learning Technologies

Digital technologies are transformational and have the potential to revolutionize the classroom and subsequently the university as an institution. To date we have yet to witness the promised change in HEIs that the digital revolution is bringing about [8]. Nevertheless, in the midst of technology becoming an integral part of societies and along with the promotion of industrial development towards automation and Industry 4.0, change is on the horizon [9]. The HEI paradigm is transforming to a new reality in which disruptive innovation and associated disruptive technologies will drive new realities in HEIs. Christensen [10] differentiates between sustaining technologies and disruptive technologies. The former enhances established processes through gradual stepwise changes whilst disruptive technologies prompts radical new forms of practice in view of emergent technologies. Within the arsenal of disruptive EdTech, literature converges to eXtended Reality (XR), internet of things (IOT), robotics and artificial intelligence (AI) as principal current contenders that are shaping the HE innovation landscape [11–16].

With the emergence of Generative AI in the past years, a new dimension to the Education Technology landscape has opened which is bringing a radical shift to the much-anticipated personalized learning. Adaptive Learning Technologies, which make use of algorithms to personalize the learning experience for each learner based on their individual strengths and weaknesses, are becoming more common and accessible. Such tools and technologies have the potential to provide targeted support where needed at the optimal pace for the learner's exigencies. This technology is possible due to the convergence of advancements in computational power, such as cloud computing, robust Machine Learning (ML) systems that analyse data, identify patterns, and make predictions or recommendations to personalize the learning experience, and content management systems that have large repositories of teaching and learning materials that can be customized to suit the learner's needs. Adaptive Learning Technologies can also be integrated with the learner's assessment process to provide more accurate and personalized evaluations of student learning outcomes. One of the strengths of the emerging EdTech lies in its capacity to combine technologies, allowing for innovative and comprehensive educational solutions which when integrated offer a superior learning experience compared to utilizing each technology individually. For instance, the combination of visualization technologies such as XR with Gamification techniques and game-based learning platforms and Adaptive Learning Technologies will provide all round solutions that provide learners with new and engaging experiences. In coming years EdTech will be further heightened with advancements in quantum computing and ubiquitous persuasive technologies.

3 Determinants for EdTech Adoption

A multitude of key drivers and motivators encourage educational institutions, educators, and learners to incorporate technology into educational practices. It is crucial for stakeholders aiming to implement EdTech solutions effectively to understand these key drivers to ensure they integrate with the evolving landscape of EdTech whilst ensuring that they meet their educational objectives.

3.1 Drivers for Implementing Disruptive Educational Technologies Within HEIs

An Enhanced and More Engaging Learning Experience. EdTech tools offer an enhanced learning experience in which the adopted technologies replicate existing teaching practices, supplement existing teaching methods and transform the teaching and learning processes and outcomes [17]. In recent years technology has brought increased interactivity, flexibility, and access to educational resources, transforming traditional classrooms into dynamic and personalized learning environments. Furthermore technologies such as extended reality have revolutionized learning by creating immersive experiences that allow students to learn through simulation for a more authentic education [18]. Edtech helps develop learners' motivation, raise achievement, and form positive attitudes towards learning science through interaction with the technological environment [19].

Safer and More Accessible Learning Environments. A safe, virtual space in which learners can practice skills without the risk of real-world consequences, allowing for mistakes and learning is one of the enticing opportunities that currently available technologies brings about. Zhang et al. [20] have deployed a simulation system for electric power training that makes use of an immersive 3D display, virtual human motion capture and interactive substation multi-persons' collaborative technology based on field application. This system provides learners with a 'safe container' as described by Carrera et al. [21], in which learners are provided the opportunity to learn in the context of real-world scenarios.

Better Decision-Making Processes. Through the harnessing of the power of data analytics academics and administrators alike can gain valuable insights into student performance, engagement, and learning patterns. This will facilitate more targeted and effective decisions to support a more learner-centric education.

A Future-proof Workforce. One of the major challenges faced by HEIs is that they need to prepare students for jobs that do not yet exist [22]. The pace of advancement in Industry 4.0 and 5.0 technology development is remarkably swift, reshaping the future with unprecedented speed. It is hard to predict exactly which knowledge, skills and competences will be required in the future, making it very difficult to prepare learners for the job market [23]. As digital literacy becomes increasingly important in the workforce, integrating technology into education prepares learners with the digital skills they need for success in their careers. Furthermore, the use of emerging technology in class exposes learners to the technologies that they will be encountering and using within their careers, equipping them with the necessary proficiencies to function and thrive within Industry 4.0 and 5.0. HEIs will need to integrate emerging technologies such as AI, IoT and big data analytics, to novel and adapted pedagogical approaches within the classroom [24].

Efficiency, Scalability and Sustainability in Organisations. The introduction of educational technology is not exclusive to the classroom. A number of solutions are available to enhance institutions operations such as admissions, student services, facilities management, finance, human resources, marketing, research administration, and strategic planning [25]. Technology is a very strong tool that can serve to alleviate the workload associated with manual tasks enhancing operational efficiency and effectiveness whilst decreasing errors linked to manual processes. Additionally, it can support in efficient resource allocation.

Technology can play a role in fostering the overall development of a HEI through highly customized student information systems which enable academic management to have data readily available with AI predictive data driven decision tools. These not only facilitate management in making informed and meaningful decisions, they allow a more personalized learning experience for the learner whilst at the same time scaling educational offerings without significantly increasing costs.

Resilience to external pressures. Disruptive technologies can drive institutions to integrate new technologies and adopt policies on the use of these technologies. In recent years, Generative AI has significantly developed and became easily accessible to learners, academics and HEIs alike [26]. As a tool, it has provided users with a myriad of opportunities to enhance the learning experience [27]. At the same time, it also poses a challenge in higher education due to the concerns about academic integrity of the learners' work and potential impacts on critical thinking skills [28]. AI offers potential benefits and opportunities for higher education, but requires clear policies, guidelines, and empirical research to address challenges and ensure academic integrity. Whilst it poses challenges for traditional teaching approaches [29], it presents itself an opportunity for academics to demonstrate innovative behaviours creating novel and enhanced teaching and learning activities.

3.2 The Role of Educator's Innovation Behaviour as a Catalyst for Digital Transformation

The process of innovation occurs at various levels within the organizational structure of an institution [30]. The individual's academic innovation actions complimenting an innovative cultural behaviour are highly determinate on the successful introduction of digital learning technologies within HEIs. Innovation driven by academics is an integral factor influencing innovation performance in HEIs, and these are governed by individual employee innovation actions and innovative work behaviours.

Thurlings et al. [31] highlights three main reasons why innovative behaviours by lecturers and academic institutions are required: (i) Innovative behaviours allow both students and educators to keep up to date with the ever-changing society; (ii) implementation of emerging educational technologies requires innovative behaviours; and (iii) educational institutions should set an example for more innovative behaviours of its students, which will in turn reflect in a more competitive and innovative-thinking society. Exposure to an institutional culture of innovation is also critical for the future-proofing of learners in view of Industry 4.0 and Society 5.0 [32]. Furthermore, HEIs with innovation embedded at the core of their activities are able to pre-empt knowledge and skill gaps which industry is due to experience as it develops its workforce [33].

What is critical for the implementation of digital transformation within HEIs is to look into the factors that influence academics to be innovative. Thurlings et al. [31] categorize the factors that influence academics to be innovative into three core categories. These being demographic factors (e.g. upbringing; years of teaching experience; income), individual factors (e.g. personality; motivation; competencies) and organizational factors (e.g. internal relationships; support mechanisms; learning culture). To be able to strengthen the innovative drive across HEI stakeholders it is important to be able to identify what triggers and motivates lecturers to implement innovative practices. Such triggers can be intrinsic and extrinsic and may greatly vary across HEIs due to the different contexts these work in. An example of factors that affect innovative behaviours in academics include organizational conditions like connectivity and self-organization [34].

3.3 Impact of Organisational Size on EdTech Adoption

From an innovation management perspective, in order to be able to create an environment in which academics' innovation behaviours are simulated to make use of more digital technologies, the right facilities and resources, innovation climate and culture should be present. Deacon et al. [35] have identified leadership and strategy, infrastructure and resources, and recognition and motivation as the main organisational factors that will support the digital maturity of universities. The aforementioned factors can have various influential effects with respect of the size of HEIs. Larger organisations tend to have more innovation capabilities compared to small or medium-sized organizations due to easier access to financing, more rigorous research and innovation support mechanisms and better economies of scale. Nevertheless, small and medium-sized enterprises leverage strategies like open innovation to enhance their innovative capacities.

4 Building a Governance Framework for the Adoption of EdTech Innovation

As outlined through various studies in literature, innovation is dependent on the drive of multiple stakeholders within the organisation, which work in synergy towards a common goal. Within the defined model of this chapter, the case study was implemented in a specialized centre at MCAST, ARIC, that was the key strategic organ acting as a pivotal driver responsible for bringing about a digital culture. In collaboration with industry and higher educational stakeholders, the vision for effective innovation was defined as the ability to undertake higher levels of Technological Readiness Levels (TRL5 to TRL9) in the creative and unique applications of Educational Technologies across the organization [36].

5 The EdTech-Innovation Framework

In June 2020 ARIC launched its innovative EdTech framework to provide a comprehensive structure on how to use and apply educational technologies within the organization. This presently active framework aims to:

1. Deliver and bring to fruition the innovative Educational Technologies throughout the educational ecosystem.
2. Pre-empt the changing students' expectations in educational engagement along their student learning pathway.
3. Address the students' learning needs through the transformation of a hybrid (digital & physical) infrastructure to support equitable Vocational and Professional Education and Training.
4. Develop a centralized base of EdTech knowledge and support within the organisation – The EdTech Forge.

Furthermore, the framework ensures that all facets of the institution act in a coordinated and consistent way, responding to the needs and expectations of the students and the changing needs of PVET education. To achieve this, it is governed and underpinned on the following three principles:

1. The focus on specific emerging technologies and enhancement of the students' learning experiences.
2. The development and implementation of adoption strategies to diffuse and disseminate the use of EdTech solutions.
3. The propagation the research findings, developments and in-house applied EdTech solutions externally at a national and international level.

Thus, the EdTech framework is anchored on the premises of making existing college structures more flexible, and fostering their ability to innovate, react and respond. Within this model, each unique EdTech adoption is introduced and managed

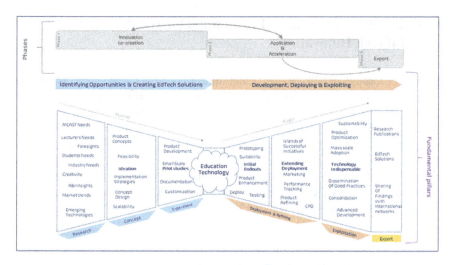

Fig. 1 Fundamental stages of EdTech development–the EdTech adoption model (EDAM)

through a project-based approach and analysed through the proposed EdTech Adoption Model (EDAM) demonstrated in Fig. 1.

The process commences through a series of investigative activities with key stakeholders, which aim to refine the key pedagogical research question to be addressed whilst looking at emerging technologies that relate to the subject. The process of ideation kicks in next, during which suites of EdTech solutions are identified and proposed. Subsequent to critical review and analysis of available EdTech options, the most suitable solution is selected, and further towards experimentation and piloting. This methodology hence ensures that the product is organically deployed, facilitating and supporting the subsequent stage of mass scale adoption in a concerted fashion. The learning process, failures and successes are also thoroughly recorded along the process chain, generating a knowledge base which is shared both internally and externally with various networks working in the field. The process lifecycle of EdTech projects is intrinsically driven by constant engagement with all stakeholders, depicted in Fig. 2, which collectively ensure the validity and effectiveness of the framework's execution.

6 A Successful Implementation Strategy for Adopting Digital Transformation

The EdTech Innovation Framework extends beyond providing a unified strategic direction to the institution and is intrinsically designed to govern and guide the operationalization of all institutional driven innovation initiatives. Thus, it serves to guide and define the parameters that lead to identification of viable EdTech initiatives,

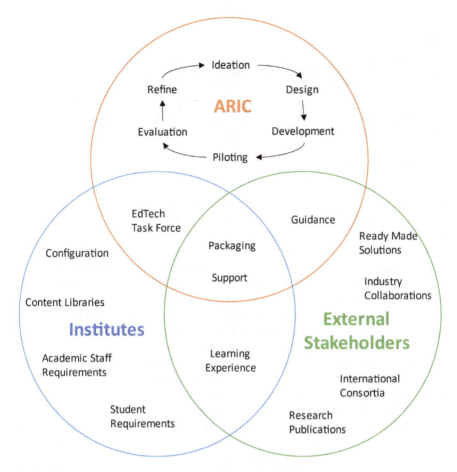

Fig. 2 Relationships between various stakeholders within EdTech projects

the collaborative requirements, the general cost expectations, and the key stages of adoption of chosen EdTech initiatives. The framework is overarching in nature and thus has been developed not to seek a final commitment to funds or resources for particular instances, but rather guide and support the holistic process of innovation within the institution. To this end, each Edtech undertaking is justified separately through individual project proposals that follow the general direction and guidance set out by the Framework.

The framework has been developed in close consultation with all stakeholders within the institution, including students, academics and management through an iterative consultative process. The framework is grounded within contextual realities of the institution and has been endorsed by the MCAST Board of Governors, cementing its operationalisation. Effective innovation is brought about through a concerted, consistent, and holistic effort—a change in mindset. There is a strong focus on imparting a change in culture within all individuals and stakeholders across

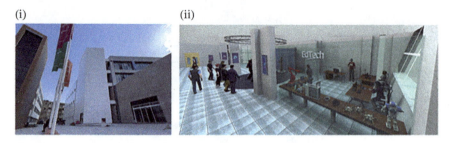

Fig. 3 (i) New MCAST resource centre (ii) Artistic impression of EdTech forge

the organisational ecosystem and thus Kotter's eight step approach to bringing about change [37] was specifically identified as the ideal implementation strategy to support and drive a high level of digital maturity withn the institution.

This infrastructural investment of a new MCAST Resource Centre (MRC) (Fig. 3i) ensures adequate facilities to support all EdTech initiatives. To this end, the Innovation quarters includes within it a large open space together with a number of co-working facilities which vary in design, ranging from large cross-functional auditorium spaces, experimental classroom setups to tech-equipped brainstorming cubicle pods. Furthermore, it includes EdTech Labs together with a SMART classroom, which has been specifically designed to serve both as a living lab for experimenting with novel educational technologies as well as a test bed/training area for all the lecturers wishing to adopt or learn more about particular EdTech solutions. This EdTech infrastructural ecosystem has been branded as the 'EdTech Forge' (Fig. 3ii).

The Edtech Forge aims to act as the central place to bring like-minded people together to spark change through innovation and act as a haven for new ideas. Thus, true to its name, the EdTech forge aims to provide opportunities for individual and group collaborations to foster a culture of innovation through the creation, sharing, and testing of ideas. EdTech products are co-developed and tested by people within different areas of studies, and sharing different expertise is facilitated through spatial resources. This further allows ARIC to promote and heavily support, cross-institute initiatives, by acting as a central link to overcome the siloed architecture of institutes. The facilities have thus been carefully designed to cater for everyone's needs, from the educator wishing to start using EdTech in their teaching and learning work to the more advanced EdTech users seeking to develop innovate technologies. Additionally, the space allows for novel non-traditional meetings/events that may be more 'high touch and feel', and enticing offering opportunities for use of system innovation and design thinking approaches such as agile or scrum to execute and advance on specific innovation projects.

The facilities double as an open access showcase for students, lecturers, administrators and visitors of EdTech initiatives and technologies. The EdTech Forge is strategically located on the ground floor of the building which is regularly frequented by the MRC building users. To this extent a passionate innovation team with diverse expertise is located within the EdTech Forge to:

1. greet visitors and demo EdTech solutions;
2. showcase emerging tech;
3. provide *ad hoc* and planned training on a one-to-one/one-to-many basis;
4. act as technical support centre and;
5. support innovators on a daily basis with key processes of EdTech ideation, development and implementation.

Harnessing the institution's innovation drive across the college intrinsically also required the expansion and diversification of existing human talent within the ARIC innovation team, together with the investment in emerging technologies and related equipment and supporting infrastructure. To this end, the EdTech-Innovation framework further serves as the guiding platform to provide a comprehensive structure on how to evaluate, procure, use and apply EdTech within the organisation. To this extent the framework serves as a launching pad for specific EdTech projects that fall within delineated remits, such as the EdTech project entitled "Extended Reality within the MCAST Classroom", launched in May 2021.

As derived from the Innovation Framework, the organizational needs to satisfy the work requirements for EdTech projects implementation and harmonization within the organisation, both on the integrative and on the strategic level, include the involvement of various individuals and groups. Thus, intrinsic to each project, a management structure is developed incorporating stakeholders in a complex HR ecosystem that is depicted in Fig. 4. The distinct innovation committees serve as driving agents to supporting the culture change intrinsic to EdTech deployment. Careful consideration has been undertaken to provide clear remits of each role to ensure that these roles have a complimentary function with holistic support throughout. The success in uptake of specific EdTech solutions is not measured solely on the basis of the number of users engaged with the technology, it also focuses on the pedagogical impact the tool has and the technology acceptance by the users. Success is thus highly dependent on the synergies between the aforementioned groups. Two considerations have been applied to ensure that lecturers are engaged in the use of the innovative technologies: (i) the concept of 'innovation' contact hours for lecturers wishing to engage in the preparation of contextualized EdTech activities and; (ii) engaging staff who will act as the focal point for innovation Institutes, facilitating the transfer of innovation initiatives by acting as the liaison person between the institute/faculty and ARIC.

Adoption of EdTech solutions by academics is achieved through an organic growth via a diffusion model of adoption as depicted in Fig. 5ii. For a more targeted approach, lecturers are categorized according to their apt and readiness to make use of the solution in line with Fig. 5i.

Potential barriers to the adoption of innovative technologies are the lack of appetite for innovation and the common employees' resistance to change. The success of technological implementations depend entirely on the people adopting it. Addressing these barriers is achieved through a strong cultural change which is supported by Continuous Professional Development (CPD). The reach and scope of CPD and training is broad and categorised as follows:

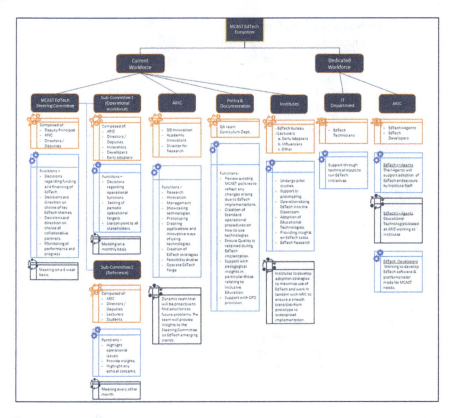

Fig. 4 MCAST EdTech ecosystem

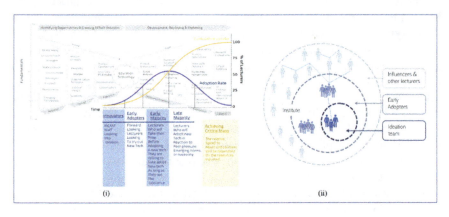

Fig. 5 (i) EdTech adoption strategy, (ii) EdTech diffusion model

- Informal, one-to-few: Targeted training sessions for individual staff members.
- Formal, one-to-many: CPD training on applications and use of select technologies being implemented.
- Formal, generic: CPD for lecturers and staff on fundamental concepts relating to Innovation and EdTech.
- Formal, accredited: Provision of accredited units and/or programmes on EdTech.

7 Case Study

A main scope of the centralised ARIC unit is to support and collaborate with both internal and external faculties, institutes and organizations. A pertinent example outlining the lifecycle of an XR project is the following case, where the ARIC team worked collaboratively with the MCAST Centre for Agriculture, Aquatics and Animal Sciences. This collaboration served to merge expertise in the use of emerging disruptive technologies with that of subject content relating to animal husbandry and horticulture, to bring about a full, immersive VR experience. The project proposed the adoption of an innovative way of (i) engaging new recruits for agricultural science programmes and (ii) educating the general public on farm-to-fork principles. This was achieved through the in-house developed VR experience that included a mix of both 360° videography and computer-generated environments. The aim was to offer users the possibility of better grasping abstract concepts whilst delivering these in a fun and interactive way. Additionally, it served to expose users to physical environments that learners did not have access to during their early years of study.

8 Project Implementation

This project was implemented through the structured methodology depicted in the MCAST EdTech-Innovation operational framework in Fig. 1. The fundamental stages for the implementation of this innovation implementation are the following:

Stage 1: Research. The project kicked off through an in-depth examination of the challenge being addressed, i.e. how to create more attractive and engaging content for more informed career decisions. Subsequently, an internal review within the VET setting was undertaken to understand particular domains whereby inaccessibility and lack of exposure often result in constrained interest and sustainability.

After an internal examination within the college, agricultural programmes were deemed to be the most suitable candidates to undergo this innovation process as they fit most of the criteria. As the project parameters and needs became clearer, attention was shifted onto which EdTech solutions can adequately address the needs. It was determined that VR would be best technology fit for this case.

Stage 2: Concept Development. In the next phase of the design thinking process, the ideation process sought the engagement of agricultural academic experts, teaching and learning pedagogists as well as computer technologist which collectively focused on the manner in which this educational technology could be effectively developed. Through an agile approach a conceptual design was iterated to address the particular requirements of this scenario, leading to the formalization of the product design and implementation strategy.

Stage 3: Experiment. Following the shortlisting and identification of five VET agricultural areas of studies namely horticulture (tomato plantation), animal husbandry (poultry and pig farming), viniculture and dairy production, video capturing sessions were organized at the respective farm sites. As shown in Fig. 6, in each session a 360° audio-visual recording was undertaken whereby farmers were able to provide a detailed explanation into the processes that yield the production of food items from their produce. This contextual setting further supported the agricultural experts to give practical insights into the farm-to-fork cycle. All video captures were applied using a GoPro Fusion 360 Camera.

The development of the proposed system architecture embedded all the immersive videos through a custom-built Computer-Generated Imagery (CGI) environment which provided an interactive menu for user navigation. Developed within the Unity™ framework, an immersive virtual environment, as depicted in Fig. 7, was designed and incorporated a set of 3-dimentional assets, featuring relatable and symbolic models of livestock and food products. These CGI models where eventually embedded with interactive scripts that enabled users to transition between 360° immersive environments when selected through the VR hand-held controller. This approach enabled users to individually control their virtual experience, engaging

Fig. 6 Onsite capture of agricultural settings through 360° videography

Fig. 7 CGI Interactive farm environment

with content through custom agricultural sequencing as well as being able to replay, revisit and further immerse themselves with the farm-based explanations of choice and interest.

The VR hardware of choice for this project was the Oculus Go model. This VR headset was selected on the basis that it is a standalone cost-effective model with a solid build and ability to store and execute 3rd party developed software with ease.

Stage 4: Deployment & Refining. In the next phase, an internal small-scale pilot of the in-house developed VR product was conducted with team members from ARIC. Feedback on the user experience was obtained and the necessary refinements were conducted accordingly.

The presented EdTech framework places a strong emphasis on the organizational cultural changes and respective collaborative requirements essential to harbour innovative products in a sustainable way. To this extent a series of training sessions were conducted with academics lecturing on horticulture and animal husbandry programmes as evidence in Fig. 9. These knowledge transfer sessions were aimed to: (1) expose the lecturers to this novel technology; (2) train the academics on how to use the VR hardware and the VR simulation; (3) provide an understanding of key Health and Safety requirements; and (4) brief on how to administer the innovative solution both in a class setting and a public event. At the end of these sessions, academics became proficient VR users, comfortably capable of autonomously managing a VR career guidance event.

Stage 5: Exploitation. The final stage of the introduction of this innovative tool included the launch at two live events; an internal dissemination event and an external event to disseminate EdTech with the community (see Fig. 8). An adequate physical setup was created that allowed visitors to join in and experience the immersive environment at any point in time. This was run by members of the innovation team and trained institute academics.

Stage 6: Export. The ultimate scope of the adopted EdTech governance framework is to lead to the build-up of pockets of expertise and knowledge that can be communicated and disseminated within the institution and further exported to various national and international communities. This ensures the exploitation of results for the broadest impact, which includes enhancing innovation and fostering collaboration. The insights attained through the deployment of the innovative VR experiences developed together with other EdTech projects governed by the framework not only solidified the institution's internal capabilities but also garnered external attention, marking a significant milestone in this learning experience.

Fig. 8 Deployment and evaluation of the proposed innovative VR solution at the Malta AgriFair 2022 and MCAST OpenDay

Fig. 9 Training workshops being held with academic staff

The experience built up in an organic fashion through the framework, positioned the team as a formidable contender in the competitive landscape of European projects. In October 2022, a multidisciplinary team from the organisation successfully secured funding under the HORIZON.4.1—Widening participation and spreading excellence call. The achievement not only highlights the importance of building strong, cohesive teams based on shared goals and complementary skills, however, it also illustrates the need to have a strategically driven governance framework to set the vision and direction and ensure that all the required resources are in place to sustain such projects.

This ongoing Horizon Europe project, WATERLINE,[1] seeks to transform advanced water skilling through the creation of a network of extended-reality water emulative centres. WATERLINE aims to create a European Digital Water HEI Alliance, based on the quadruple helix model of innovation, leading to the development of the Alliance's research, educational and entrepreneurship capacities. This shall leverage the individual, institutional and regional resources required for a transformative structural and sustainable learning and innovation environment. This will be achieved through five specific objectives, among these, two objectives hold particular significance in the sharing of our experiences garnered through the learning from our EdTech framework. Firstly, a common governance framework, and a Research & Innovation (R&I) capacity building plan will ensure that the Alliance is support. Secondly, the transformation of emulative laboratories in partner HEIs into assisted and virtual reality. These structural changes will lead to transformed and more competitive R&I HEIs.

MCAST is leading the consortium, comprising a dynamic network of fifteen partners, steering collaborative efforts towards our shared goals. In addition it is the task leader for transformation of the three-tier Learning Environments. The knowledge, experiences and competencies gained through the successful implementation of the EdTech framework within the institution, have helped it to work closely with the consortium to develop an infrastructure of innovative educational technologies that allows a safe, seamless learning experience that is accessible 24/7 and transcends geographical borders, ensuring that education in digital water is within reach for anyone, anywhere, at any time.

An example outcome of the project is the implementation of a real water distribution test rig as a digital twinning 3D VR experience. This EdTech solution was developed as a sandbox simulating any water distribution network configuration allowing easy adaptation to different learning requirements, similar to a food-processor taking different attachments to do different recipes. Governance in this context encompassed ensuring efficient communication, collaboration, and decision-making processes among stakeholders engaged in the development of the EdTech tool. The key governance principle employed in the development of the water-rig simulation and the VR system is an iterative, user-centred design. A key ingredient for success was identifying learning outcomes from relevant curricula, and implementing content pertaining to the outcomes through collaboration and synergy

---

[1] https://cordis.europa.eu/project/id/101071306.

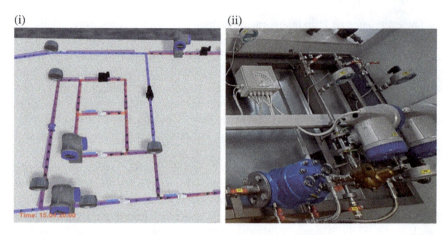

Fig. 10 (i) Digital VR water distribution rig; (ii) Emulated water distribution rig

between educators, domain experts and software developers. For example, ground-truth tests of the digital-twin were conducted through experiment sessions involving both water engineers and developers. Similarly, learning provision at different levels was guided and evaluated by educators in a constant development and feedback loop as the software was being built. Figure 10 demonstrates how the 3D simulation is an actual replica of the real water rig, with the added advantage of changing configuration/parameters at will, 24/7 multi-platform accessibility, simpler logistics, and with curriculum embedded as learning activities or scenarios at different levels, while still allowing student tactile experience most similar to that of the real world. These characteristics, together with the dynamic nature of the solution, are significantly disruptive factors in the manner of which learning can be experienced, that will be tested during the pilot phase of the WATERLINE project.

9 Conclusions

The need to innovate is critical to futureproof HEIs within the everchanging world. This ensures that the knowledge transfer and services offered by education organisations remain relevant, supporting the needs of students, academics and industry alike. To be able to be on the game at all times it is vital that institutions invest in robust innovation infrastructures to find creative solutions to enhance its services. EdTech is offering new possibilities to enhance the HEI teaching and learning landscape. EdTech solutions should be carefully managed through a strategic structured approach. This disruptive innovation should be gradually introduced through proper innovation structures to propel HEIs into the 4th Education revolution. The outcomes of this chapter elucidate the importance of driving organisational change through structured interventions and collaborations amongst stakeholders, which collectively bring to fruition an effective drive in culture and subsequent adaptation and adoption of EdTech solutions within the organisation.

Acknowledgements The authors wish to extend their thanks to the editor-in-chief, the section editor, and the anonymous reviewers for their valuable feedback. Zammit, Rizzo and Scerri's involvement in this endeavour was facilitated by the funding received from the project WATER-LINE: "Transforming Advanced Water Skilling Through the Creation of a Network of Extended Reality Water Emulative Centres," jointly supported by the European Horizon European Programme (HORIZON-WIDERA-2021-ACCESS-05, grant agreement No. 101071306).

References

1. Vostroknutov I, Grigoriev S, Surat L (2021) Search for a new paradigm of education and artificial intelligence: place and role of artificial intelligence in the new education system. In: 2021 1st international conference on technology enhanced learning in higher education (TELE). IEEE, pp 80–82
2. Messmann G, Mulder RH (2011) Innovative work behaviour in vocational colleges: understanding how and why innovations are developed. Vocat Learn 4:63–84. https://doi.org/10.1007/s12186-010-9049-y
3. Bond M, Buntins K, Bedenlier S et al (2020) Mapping research in student engagement and educational technology in higher education: a systematic evidence map. Int J Educ Technol High Educ 17:2. https://doi.org/10.1186/s41239-019-0176-8
4. De Jong J, Den Hartog D (2010) Measuring innovative work behaviour. Creat Innov Manag 19:23–36. https://doi.org/10.1111/j.1467-8691.2010.00547.x
5. White SC, Glickman TS (2007) Innovation in higher education: implications for the future. N Dir High Educ 2007:97–105. https://doi.org/10.1002/he.248
6. Gejendhiran S, Anicia SA, Vignesh S, Kalaimani M (2020) Disruptive technologies—a promising key for sustainable future education. Procedia Comput Sci 172:843–847. https://doi.org/10.1016/j.procs.2020.05.121
7. Zammit E, Hall S, De Raffaele Cl, Rizzo A (2020) MCAST EdTech-innovation framework: propelling MCAST into the 4th education revolution
8. Flavin M (2017) Disruptive technology enhanced learning. Palgrave Macmillan UK, London
9. Ngoc HD, Hoang LH, Hung VX (2020) Transforming education with emerging technologies in higher education: a systematic literature review. Int J High Educ 9:252. https://doi.org/10.5430/ijhe.v9n5p252
10. Christensen CM (2013) The innovator's dilemma: when new technologies cause great firms to fail. Harvard Business Review Press
11. Ciolacu M, Tehrani AF, Binder L, Svasta PM (2018) Education 4.0—artificial intelligence assisted higher education: early recognition system with machine learning to support students' success. In: 2018 IEEE 24th international symposium for design and technology in electronic packaging (SIITME). IEEE, pp 23–30
12. Ciolacu MI, Binder L, Svasta P et al (2019) Education 4.0–jump to innovation with IoT in higher education. In: 2019 IEEE 25th international symposium for design and technology in electronic packaging (SIITME). IEEE, pp 135–141
13. Dilmurod R, Fazliddin A (2021) Prospects for the introduction of artificial intelligence technologies in higher education. Acad: Int Multidiscip Res J 11:929–934. https://doi.org/10.5958/2249-7137.2021.00468.7
14. Leoste J, Jõgi L, Õun T et al (2021) Perceptions about the future of integrating emerging technologies into higher education—the case of robotics with artificial intelligence. Computers 10:110. https://doi.org/10.3390/computers10090110

15. Mircea M, Stoica M, Ghilic-Micu B (2021) Investigating the impact of the internet of things in higher education environment. IEEE Access 9:33396–33409. https://doi.org/10.1109/ACCESS.2021.3060964
16. Radianti J, Majchrzak TA, Fromm J, Wohlgenannt I (2020) A systematic review of immersive virtual reality applications for higher education: design elements, lessons learned, and research agenda. Comput Educ 147:103778. https://doi.org/10.1016/j.compedu.2019.103778
17. Kirkwood A, Price L (2014) Technology-enhanced learning and teaching in higher education: what is 'enhanced' and how do we know? a critical literature review. Learn Media Technol 39:6–36. https://doi.org/10.1080/17439884.2013.770404
18. Curran VR, Xu X, Aydin MY, Meruvia-Pastor O (2022) Use of extended reality in medical education: an integrative review. Med Sci Educ 33:275–286. https://doi.org/10.1007/s40670-022-01698-4
19. Al-Hattali M, Al-yazeedi R, Al-Balushi S, Al-Hosni H (2024) Science teachers' perceptions of the importance of employing the applications of the fourth industrial revolution in education and it's employment obstacles (a qualitative study). Arid Int J Educ Physcological Sci 30–56. https://doi.org/10.36772/ARID.AIJEPS.2024.592
20. Zhang D-G, Li Z-H, Xiao Y-W, et al (2017) Design and implementation of simulation system for safety accident case based on immersive virtual reality technology. In: Proceedings of the 3rd annual international conference on electronics, electrical engineering and information science (EEEIS 2017). Atlantis Press, Paris, France, pp 416–420
21. Carrera AM, Naweed A, Leigh E, et al (2018) Constructing safe containers for effective learning: vignettes of breakdown in psychological safety during simulated scenarios. In: Lecture notes in computer science (including subseries lecture notes in artificial intelligence and lecture notes in bioinformatics). Springer, Berlin, pp 15–29
22. Al-Emran M, Al-Sharafi MA (2022) Revolutionizing education with industry 5.0: challenges and future research agendas. IJITLS 6:1–5
23. Mohamed Hashim MA, Tlemsani I, Mason-Jones R et al (2024) Higher education via the lens of industry 5.0: strategy and perspective. Soc Sci Humlties Open 9:100828. https://doi.org/10.1016/j.ssaho.2024.100828
24. Zizic MC, Mladineo M, Gjeldum N, Celent L (2022) From industry 4.0 towards industry 5.0: a review and analysis of paradigm shift for the people, organization and technology. Energies (Basel) 15:5221. https://doi.org/10.3390/en15145221
25. Leidner DE, Jarvenpaa SL (1995) The use of information technology to enhance management school education: a theoretical view. MIS Q 19:265. https://doi.org/10.2307/249596
26. Yeralan S, Lee LA (2023) Generative AI: challenges to higher education. Sustain Eng Innov 5:107–116. https://doi.org/10.37868/sei.v5i2.id196
27. Rane N (2023) ChatGPT and similar generative artificial intelligence (AI) for smart industry: role, challenges and opportunities for industry 4.0, industry 5.0 and society 5.0. SSRN Electron J https://doi.org/10.2139/SSRN.4603234
28. Michel-Villarreal R, Vilalta-Perdomo E, Salinas-Navarro DE et al (2023) Challenges and opportunities of generative AI for higher education as explained by ChatGPT. Educ Sci 13:856. https://doi.org/10.3390/EDUCSCI13090856
29. Kadaruddin K (2023) Empowering education through generative AI: innovative instructional strategies for tomorrow's learners. International J Bus, Law, Educ 4:618–625. https://doi.org/10.56442/IJBLE.V4I2.215
30. King N, Anderson N (2002) Managing innovation and change: a critical guide for organizations. Thomson, London
31. Thurlings M, Evers AT, Vermeulen M (2015) Toward a model of explaining teachers' innovative behavior: a literature review. Rev Educ Res. https://doi.org/10.3102/0034654314557949
32. Sudibjo N, Idawati L, Retno Harsanti H (2019) Characteristics of learning in the era of industry 4.0 and society 5.0. In: International conference on education technology, vol 372, pp 276–278
33. Lassnigg L, Hartl J, Unger M, Schwarzenbacher I (2017) Higher education institutions and knowledge triangle: improving the interaction between education, research and innovation

34. Kontoghiorghes C, Awbre SM, Feurig PL (2005) Examining the relationship between learning organization characteristics and change adaptation, innovation, and organizational performance. Hum Resour Dev Q 16:185–212. https://doi.org/10.1002/hrdq.1133
35. Deacon B, Laufer M, Schäfer LO (2023) Infusing educational technologies in the heart of the university—a systematic literature review from an organisational perspective. Br J Edu Technol 54:441–466. https://doi.org/10.1111/BJET.13277
36. Mankins JC (1995) Technology readiness levels. White paper, April 6
37. Kotter JP (1996) Leading change. Harvard Business School Press, Boston

New Models of University for the Digital Society

Maria Amata Garito

Abstract The aim of this work is to present models of digital technology utilization in the complex world of universities to activate face-to-face and distance teaching and learning projects. Academic training can no longer be conducted locally without considering the global context that digital technologies have consolidated as the reality in which twenty-first century individuals operate and interact. Today, the common challenge for universities worldwide is how to best adapt to the digital society and create integrated and open teaching and learning systems within the context of a globalized economy. The transformation of universities is no longer a choice but an indispensable necessity. New generations of students cannot do without using the Internet to follow courses and develop knowledge and skills. This essay aims to present concrete models of how traditional and digital universities have been able to reinvent themselves by creating new international alliances to establish a global network of universities on the internet. These universities combine their knowledge, share curricula, and create digital content in dedicated learning environments designed with new psychopedagogical theories appropriate to technological evolution. The organizational models of digital universities are also structured to truly respond to the challenges and needs of a globalized and interconnected society.

Keywords Digital transformation in universities · Global university networks · Digital education models · Advanced teaching technologies · Learning processes

1 Introduction

The 21st Century is marked by the proliferation of digital technologies, which have accelerated globalization processes and given rise to a new interconnected world that operates both locally and globally. The Internet enables a "psycho-digital" journey–a

M. A. Garito (✉)
Emeritus Professor of Psycotechnologies, President/Rector of UNINETTUNO International Telematic University, Corso Vittorio Emanuele II, 39, 00186 Roma, Italia
e-mail: garito@uninettunouniversity.net

space where the global and the local intersect, allowing consciousness to manifest on a planetary level. For the first time in human history, the constraints of physical co-presence no longer bind our minds and bodies. Digital natives, who have grown up in this technological environment, navigate a dual existence: the tangible outside world and the parallel virtual realm. Within this virtual domain, they have everything they need–friends, music, news, games, movies, television, love, school, religion, health care, and even conflict. We are experiencing a social and cultural revolution on an unprecedented scale, which challenges all known systems, structures, and organizational models of digital society, including educational institutions like schools and universities.

Digital technologies influence our mental processes. The Internet, far from being peripheral, lies at the heart of contemporary anthropological transformation. It extends and amplifies our sensorimotor, psychological, and cognitive functions through novel forms of language. Every individual can now be both a user and a distributor of knowledge, alone or as part of a group. People who are culturally, linguistically, politically, religiously, ideologically, and socially different can participate in global conversations that transcend geographical boundaries. In this collective cyberspace, diverse expressions flourish despite differences. The Internet enables each person to experience their own local reality while simultaneously engaging with the global context. This is one of the great challenges facing our society. The Internet, augmented reality, the metaverse, and artificial intelligence redefine how we store and communicate knowledge. These mental technologies are central to our evolving anthropological landscape.

As we hurtle towards an uncertain future, we recognize that change is inevitable. The way we work, produce, teach and learn, create culture, and engage in politics has already been transformed. We are witnessing a complex phase of transition, fueled by technological ferment and globalization processes that are creating new needs and changing consumption patterns and lifestyles. Mastering new languages and specific skills will require a major collective effort on behalf of the citizens of this digital, interconnected world to learn how to use them constructively. These linguistic tools unlock new forms of knowledge and are the key to our future.

We stand at a historic juncture: we are the architects and spectators of a cultural revolution fueled by a spate of new technologies that influence our mental processes and connect them at the global planetary level. Digital natives already live their lives in a globally interdependent system: they seamlessly communicate, collaborate, and share content with a global pool of users. In this era of interconnectedness, the human mind is being freed from tasks and operations which now take place almost exclusively on external data storage systems aided by the speed of digital bits. This boundless virtual space allows everyone in the world to add their own content and share their knowledge. However, since knowledge is often very subjective, it is difficult for non-expert users to assess its veracity. The way major search engines retrieve information and provide answers to users navigating the vast realm of the Internet is a matter of concern. Google, the most widely used search engine, lacks transparency in its operations. The information and documents presented by the search engine do not consistently align with users' genuine queries. Instead, the top results often favor content that has invested heavily in advertising, leading less informed users to click on these prominently displayed links. Consequently, the responsiveness of search results is not necessarily tied to the accuracy or relevance of the query. Rather, it is influenced by the sheer number of clicks–actions determined collectively by Internet users. This dynamic construction of answers thus sacrifices scientific rigor for popularity. The hierarchy of search result pages becomes a form of subjective power, and collective action emerges as a new force. To address this issue, global political efforts should compel search engines like Google to abandon deceptive business practices.

2 Defending Users from Fake News

The challenge extends beyond search engines. Weaker, inexperienced users, with a limited cultural background and experience, have a hard time distinguishing truth from falsehood on the Web. Even advanced tools like ChatGPT, an intelligent chatbot capable of generating articulate responses based on user prompts, cannot guarantee the veracity of their answers. In March 2023, OpenAI released an upgraded ChatGPT model based on the GPT-4 architecture, trained on a vast text corpus, although the exact number of parameters and the precise size of the corpus have not been disclosed. However, the knowledge bases used for training are drawn from Internet content, which includes both scientifically valid information and fake news. Consider the following responses generated by ChatGPT to a prompt dated April 17, 2024: *Question*: "Can training on digitized knowledge from the web, contributed by individuals who may not always provide validated scientific content, assure ChatGPT users of the scientific accuracy of its answers?" *ChatGPT answer*: "You raise a valid point. The quality and reliability of ChatGPT's answers hinge on the data it was trained on. If the dataset includes incorrect, misleading, or scientifically unvalidated information, this can be reflected in the model's responses."

We understand that using ChatGPT's own response to validate its reliability is not a robust measure. However, although the example is not a metric in itself, it serves to illustrate the ChatGPT's response, highlighting the necessity for multi-step verification. This iterative dialogue is not only essential for ensuring accuracy and reliability but also for refining and improving the quality of the responses. The example was deliberately included to provoke reflection, especially within the scientific and academic communities, on how the model's responses can create doubts about their veracity. In light of this, key considerations include: (a) Diverse sources: ChatGPT learns from an extensive text corpus that spans scientific publications, journal articles, informative websites, blogs, and forum comments; this diversity introduces content that may not always be rigorously validated. (b) Quality filters: Despite efforts to filter reliable sources, ensuring that all training data is scientifically accurate remains technically challenging. (c) Verification imperative: Users should independently verify information obtained from ChatGPT, especially when scientific accuracy matters; relying on primary, peer-reviewed sources or consulting field experts is crucial to confirm the information. (d) Ongoing improvements: Researchers and developers continually enhance language models to better handle and filter information, aiming for increased reliability and accuracy.

In general, it is prudent to treat the answers provided by AI systems like ChatGPT as starting points for research and learning, rather than definitive sources. ChatGPT's ethical stance in defense of the user in this case is commendable. However, how can we empower generative AI systems to consistently deliver scientifically accurate answers?

Firstly, we must recognize the urgency of establishing new rules and models to guide the development of virtual societies. Not long ago, knowledge transmission occurred primarily through teachers, experts, and scientists. Their expertise, backed by research and written works, reached students worldwide via schools and universities. Governments prioritized competent educators to equip citizens with essential skills and knowledge to face challenges and to prepare the younger generation for life in society. Political powers have always been careful to maintain the value of education at various stages of life. Today, amid the chaos of information, no single guiding authority exists.

Secondly, universities–unique spaces where knowledge production and transfer coexist–can play a pivotal role in shaping Internet content. To achieve this, universities must undergo structural changes, forming international alliances and creating real and virtual networks that foster worldwide knowledge convergence. Knowledge is increasingly a strategic resource, impacting individual lives and the evolution of digital society. Educational systems must grapple with how humans learn and assimilate knowledge from Internet content and embedded AI models.

Lastly, it is crucial to empower individuals in their ability to master the technology and become able to distinguish the truth or falsehood of the digital content found on the web.

3 The Transformation of Universities

Citizens increasingly rely on Internet knowledge–both accurate and false–for learning and skill acquisition outside traditional educational or training institutions such as schools and universities. A global pedagogical society is emerging. Despite this expanding reality, innovation in schools and training systems faces limitations. Many traditional universities still employ rigid teaching and learning models, struggling to compete with online alternatives. To evolve university education, we must consider products, process, and structures. Universities should adapt their offerings, emphasizing flexibility and student autonomy. Teachers' roles must evolve constructively within digital teaching–learning processes. In the digital society, the development of an expanded and open educational system presents opportunities while also posing risks. Users must cultivate critical awareness to maximize benefits and minimize pitfalls.

At this pivotal moment in history, the imperative to develop public policy interventions has never been more pronounced. These interventions must ensure that worldwide legislative tools and adequate resources are allocated to meet the educational needs of our globalized and interconnected society. Traditional universities must harness the power of the network to reinvent themselves, creating and applying new theoretical models that underpin both face-to-face and distance teaching and learning processes. This transformation is no longer optional; it has become an inevitable imperative. The true value of today's traditional university lies in its ability to adapt knowledge storage and communication through emerging technologies. Equally significant is the realm of human interaction–whether in direct face-to-face encounters or through distance learning between teachers and students. Universities must operate within virtual spaces of interaction and cooperation, while simultaneously imbuing physical spaces with fresh meaning. These spaces serve as meeting grounds for different generations–the youth and the mature–where new relationships of socialization, critical reflection, and knowledge communication can flourish. To achieve this, educators must embrace teaching models rooted in Socratic interaction methodologies. In these models, students become actors in the educational process, and human interaction becomes the bedrock of conscious knowledge growth. Across the globe, discussions are underway regarding novel university models, innovative teaching–learning methodologies, and fresh languages for knowledge communication and transmission via cutting-edge technologies.

The Internet, now an increasingly content-rich global platform, has become the prevailing infrastructure for knowledge exchange among individuals. The existence of this network, external to humans, consisting of interconnected memories, has profoundly reshaped knowledge production, storage, transmission, research, and education. Moreover, it has significantly influenced the languages through which knowledge is shared. The Internet has democratized knowledge access, extending it beyond traditional educational or training structures, empowering citizens worldwide. The transformation of universities is no longer a matter of choice; it is an indispensable requirement in this new reality. Traditional universities, historically centers for knowledge elaboration and communication, must reevaluate their organizational and teaching models. This includes redefining the way they carry out research programs, the roles and functions of faculty and students, and adapting physical structures such as classrooms and research laboratories. Creating a global network for higher education–one where faculty and students from diverse corners of the world collaboratively construct knowledge–is not a utopian vision. Rather, it can infuse new vitality into universities, positioning them as key players in the global networked economy. The COVID-19 pandemic prompted traditional universities to leverage the Internet for teaching and research activities. While many universities improvised their approach to distance teaching and learning processes, it is now evident that networked educational processes demand the application of new theoretical models specifically tailored for online learning environments. Across different regions, a novel model of distance education–one anchored in traditional universities–is taking shape. All face-to-face students will gradually transition into distance learners. The emergence of a dynamic educational landscape, blending real-world and virtual components, is poised to surpass traditional teaching and learning paradigms.

4 Knowledge Alliances: A Strategy for Creating Digital Universities

The digital university, within the new context, is not merely an alternative to traditional universities; rather, it represents a fresh opportunity and a novel approach to teaching and research functions. The Internet today plays a pivotal role in fostering product, process, and system innovations. Universities, where knowledge production and transfer coexist, can play an indispensable and irreplaceable role in creating content for the online sphere. However, achieving this requires a transformative shift in the university's role and structure, necessitating the establishment of international alliances and the creation of real and virtual spaces that facilitate shared networks of knowledge across the world's universities and support the convergence of institutions. The knowledge environments should not be homogeneous or uniform; we should not come together to clone one another or, worse, to "mcdonaldize" education and training systems. Instead, we should strike a delicate balance between unity and diversity: unity stems from values and traditions preserved in collective memory, while diversity thrives through the richness of various cultures and languages.

In this context, forging new international alliances becomes paramount. A global network of public and private universities pooling their knowledge, resources, equipment, curricula, laboratories, as well as faculty and student mobility, while maintaining the specificity of individual universities that gives value and enriches the network, will play an increasingly vital role. This online network of universities should be built not only on technological foundations but, more importantly, on the collective intelligence of people who can connect their expertise while respecting cultural, political, religious, and economic differences. Thanks to the Internet, this network of universities expands and multiplies the possibilities of acquiring information, knowledge, and fostering interactions and exchanges among individuals. Today, the digital society demands the development of policies for the knowledge industry and the creation of consortia among the most important and prestigious universities worldwide to share educational content created by the best university professors, addressing the digital society's need to guarantee reliable online information. These consortia, bridging universities and industries to create new models of distance universities, can swiftly respond to the internationalization requirements of education systems and prepare the skills demanded by the interconnected global labor markets. When course content on the Internet is produced by academics at the international level, quality control is assured by academia itself. Users become discerning "consumers of education" because course providers are easily identifiable. If quality branding will determine the competitive edge in global education markets, a distance university based on a network of the world's best traditional universities will undoubtedly win the challenge and emerge as the leader in the new knowledge markets.

Today, a distance university formed through a consortium of the world's best traditional universities can meet the needs of the new knowledge markets and guarantee the quality of online content. It can help turn traditional universities into open systems capable of updating themselves by integrating all knowledge available on the web and fostering global knowledge exchange. The best universities can reach every home around the world thanks to the Internet. Universities should serve as both physical and virtual spaces where the experience and expertise of teachers are seamlessly transferred to students through models of continuous interaction–whether in person or remotely. This dynamic exchange bridges the gap between youth and maturity, as well as between seasoned experts and those new to the field. It fosters critical knowledge development and facilitates its transformation. To create a global center of excellence, universities should form consortia. Their joint efforts should involve designing curricula for various degree programs, identifying top faculty members globally, and having them create video lectures and other teaching materials for digitization. Imagine a single discipline benefiting from diverse digital teaching content produced by lecturers from universities across different countries, each in their own language and drawing from their unique cultural, historical, political, and religious influences. This is the value that a consortium of universities can generate: diversity becomes an invaluable asset for all–a universal opportunity.

The future lies in consortia formed by universities worldwide, working collaboratively to create content within Internet learning environments. This context fosters collaborative knowledge co-creation, supported by robust organizational and pedagogical models. These consortia not only give rise to innovative digital distance universities but also redefine the entire concept of a university. Building a 21st Century higher education infrastructure involves more than physical buildings; it requires a technological framework that enables the development of new pedagogical models and knowledge production, storage, and delivery. These choices are increasingly essential for the survival and evolution of universities.

4.1 The Research Backing Up the Establishment of New University Models

This distance university model took shape in Italy during the early 1990s, based on research we personally conducted. Initially focused on the European context in the 1980s, our comparative analysis targeted major European university institutions. The goal was to identify best practices in leveraging available technologies to define new communication languages for knowledge dissemination, along with innovative organizational and psycho-pedagogical models.[1] Subsequently, our research extended beyond Europe. Grants from the U.S. government allowed me to visit cutting-edge educational technology universities and research centers. We were able to analyze the patterns of technology use that were being employed for distance learning. At Stanford, we observed their approach to training Silicon Valley managers through live videotaped lectures broadcast via cable television technologies. The interactive, dynamic teaching and learning process between teachers and student-managers was groundbreaking in terms of methodology. Another pivotal moment was our visit to the Massachusetts Institute of Technology (MIT). There, we delved into Seymour Papert's work, which applied Montessori theories to activate teaching and learning processes through computers for kindergarten and elementary school children in Boston. The results of this research conducted in the 1980 and 1990s enabled the design of the distance university model hinged on traditional universities, not only in Italy, and the creation, as early as 1992, of the mixed model (face-to-face and distance) of the NETTUNO Consortium–Network for Universities Everywhere.

5 The NETTUNO Consortium Model

Our challenge in designing the educational model for Consorzio NETTUNO–Network per l'Università Ovunque (Network Consortium for the University Everywhere)–aimed at providing university education universally–was to create a robust distance learning structure. This structure needed to address the training requirements of our cognitive society while accounting for the rapid evolution of information technologies and insights from psycho-pedagogical research. These insights would form the theoretical bedrock for implementing effective distance teaching and learning processes. Our proposed distance university model rested on the premise that distance education should be firmly rooted within traditional universities. These institutions had to adapt, reorganize, and respond adequately to the growing qualitative and quantitative demands of education. Flexibility, diversification, and internationalization of teaching–learning processes became essential imperatives. In the NETTUNO model, universities not only ensured flexibility throughout the educational journey but also upheld quality and academic freedom [1].

---

[1] This research led to a series of television broadcasts on the subject, and the publication of two monographs, *Learning to Teach. School Structures and Teacher Education in Europe* (ERI, Rome, 1981) and *The University in Europe* (ERI, Rome, 1983).

The NETTUNO Consortium comprised 41 traditional Italian universities,[2] 31 universities from other European countries and the Arab world, and corporate partners such as Rai–Radiotelevisione italiana, technology firms and trade associations like Confindustria. Traditional universities emerged as pioneers of innovation, while the collective effort forged entirely new parallel organizational models distinct from traditional ones. Within each university coexisted two paradigms: one adhering to conventional teaching methods, where physical classrooms served as the locus of face-to-face instruction with teachers at the center; the second one embraced emerging technologies, fundamentally altering not only the teaching–learning process but also the physical spaces where education unfolded. Classrooms turned into open structures, empowered by technology to facilitate flexible training, self-directed learning, and collaborative interactions–both onsite and remotely. Our research activities closely tracked technological advancements available in the market. We identified psycho-pedagogical didactic theories suitable for application in the distance teaching and learning model. In the early days of the NETTUNO Consortium, before the Internet became readily available, television served as our medium for broadcasting all distance learning course lectures. However, with the Internet's development, we launched our first online teaching portal in 1996. This portal housed comprehensive teaching materials and interactive services, enabling effective distance education processes on the web. Within traditional universities, where distance learning courses were delivered, the NETTUNO Consortium triggered a significant innovation transfer. The physical structures–university classrooms–where traditional training occurred underwent transformation, as did the professional skills of faculty members. By adopting a modular enrollment

---

[2] The consortium universities until 2003 were: Bari Polytechnic, Milan Polytechnic, Turin Polytechnic, University of L'Aquila, University of Bologna, University of Camerino, University of Cassino, University of Florence, University of Genoa, University of Lecce, University of Milan, University of Modena, University of Naples "Federico II," Second University of Naples, University of Padua, Universi-ty of Parma, University of Pisa, University of Rome "Tor Vergata," University of Salerno, University of Siena, University of Turin, University of Trento, University of Trieste, University of Viterbo "La Tuscia". Forty-one universities were part of the NETTUNO Network, plus international entities such as the British Open Uni-versity, the Polytechnic University of Tirana (Albania), the Syrian Virtual Univer-sity in Damascus, UNAD – Universidad Nacional Abierta y a Distancia of Co-lombia, the Computer Man College in Sudan and the Fundación Universitaria Manuela Beltran in Colombia, the Université Virtuelle de Tunis (Tunisia) and oth-er universities in the Euro-Mediterranean countries that are partners in the MED NET'U project: Institut Supérieur de Gestion et de la Planification (Algeria), Uni-versité Djillali Liabès de Sidi-bel-Abbès in Algeria, Cairo University, Helwan Uni-versity (Egypt), Egyptian Association Incubator, EUTELSAT (France), Fondation Sophia Antipolis (France), Institut National Polytechnique de Grenoble, Aegean University (Greece), University of Crete, Lebanon Management and Technology Consulting Group, Jordan University of Science and Technologies, Yarmouk Uni-versity (Jordan); in Morocco: Ecole National Supérieure d'Informatique et d'Analyse des Systèmes, Université Cadi Ayyad (Marrakech), Ministère de l'Emploi, de la Formation Professionnelle, du Développement Social et de la Soli-darité, Université Mohammed V; in Syria: University of Aleppo, Damascus Uni-versity, Syrian Virtual University; in Tunisia: Ministère de l'Éducation et de la Formation Professionnelle, Ministère des Technologies de la communication et du Transport, Institut National de Bureautique et de Micro-Informatique, Université de Tunis El Manar, Université Virtuelle Tunisienne; Ege University (Turkey).

system (replacing annual course registrations), we paved the way for an open and flexible training system. Distance education seamlessly integrated with the institutional functions of traditional universities. Distance students could enroll in the same courses, follow identical syllabi, receive assistance from the same faculty, and ultimately obtain equivalent degrees as their on-campus counterparts. Faculty members balanced their time between traditional and distance students, assuming additional roles typical of distance educators.

The typical roles of students and teachers in traditional universities had undergone significant changes. No longer confined to university lecture halls, students had the freedom to learn from anywhere, unencumbered by spatial or temporal limitations. Thanks to technological advancements, they could create personalized learning spaces, managing their training and self-learning autonomously. NETTUNO teachers also reshaped their teaching methodologies and functions. Teachers had to learn a new model of knowledge dissemination, utilizing television as a medium for communication. Simultaneously, they engaged with students both in-person and remotely. Rather than being confined to a single university, teaching activities were coordinated across all consortium universities, who collegially created shared curricula, and designed and implemented the new distance teaching and learning processes. This was certainly an important element that ensured the quality of NETTUNO's psycho-pedagogical didactic model, which blended teaching and self-learning facilitated by modern technologies with traditional way teaching and learning experiences, based on a two-way in-person communication relationship. Moreover, the teaching activities carried out in the Technological Poles, situated within traditional universities, respected institutional autonomy. While accepting the delivery of nationwide distance learning courses, universities also had the opportunity to customize their own educational offerings. Lecturer-tutors, drawn from the faculty of individual universities, and examiners, could tailor courses to align with the directives of their own faculties and enter personalized content in the Internet portal. Personalized content made it possible to compare the methodologies and knowledge of different lecturers and universities, not only from Italy but also from other countries, to the benefit of all NETTUNO students and lecturers.

NETTUNO's organizational model made possible the following: (a) provide cutting-edge technologies to consortium universities; (b) ensure the high quality of the teaching staff, the training topics and content, being able to draw on the experience of the various consortium universities; (c) the rich reservoir of human resources and knowledge within the NETTUNO Consortium also enabled excellence across multiple fields.

Collaboration with corporate partners offered an opportunity to bridge the gap between education and production by connecting trainers and users, and thus also addressing the recent need for lifelong learning. In addition, extending the model to the Euro-Mediterranean region increased the reservoir of knowledge and skills, overcoming political, cultural, religious as well as physical boundaries, in many countries in the Arab world. Education was known for the quality of the teaching provided by the best professors from traditional universities who were open to innovation and fully responsible for the distance teaching–learning process.

5.1 The Organizational Model of NETTUNO and Its Structures

NETTUNO's forward-thinking organizational model introduced structures that revolutionized traditional university paradigms. These included International Centers, National Centers, Provider Universities, Technology Poles, and Production Centers. These diverse structures expanded the scope of teaching and learning beyond conventional settings, no longer just university lecture halls, but diverse, open, virtual and physical spaces: homes, workplaces, study centers, anywhere and everywhere in the world. With no spatial or temporal limitations, anyone equipped with the necessary technologies could activate their own learning process. The advantages for learners were related to the fact that they could benefit from a wide range of courses, combined with flexibility learning at their own pace and in their preferred locations. The advantages for network partners (universities, corporations, professional organizations) lay in the fact that, with minimal effort and investment, they could develop and deliver high-quality distance learning courses to a large global audience.

The NETTUNO Consortium model placed universities at the forefront of innovation. It introduced flexibility and adaptability of products, fostering of student autonomy, modification of the teacher's role, and structural evolution to accommodate face-to-face and distance learning. In traditional universities, in order to successfully respond to the quantitative and qualitative needs for training, the model succeeded in enhancing training offerings, innovating teaching methods, even in traditional courses, making faculty members acquire skills in teaching methodologies using new technologies, and connecting the gap between academia and industry. The Consortium had also succeeded in responding to the strategically important need for lifelong learning. Thanks to the new telematic and satellite technologies, democratic access to education and training had become a reality. Physical mobility of people was complemented by the free flow of ideas and interactions on the Internet. In a remarkably short time, a space for high-quality, innovative distance education had emerged.

5.2 Organizational Structures

International Center. The International Center was responsible for coordinating distance education activities across provider universities. Specifically, it defined the guidelines and strategies for administrative and teaching tasks related to distance courses. Within the Center, teaching-scientific committees were tasked with verifying the quality and efficiency of the teaching and organizational activities of the decentralized structures.

Provider Universities. Provider universities managed student enrollment, class and exam schedules, selected faculty to teach both in-person and distance learning courses and issued degrees and college credits.

Technology poles: the new spaces of access to knowledge. New physical spaces called Technology Poles were created within traditional universities, equipped with modern distance-learning technologies. Technology Poles functioned as incubators of innovation enabling transformative changes in teaching–learning processes and products.

Production Centers: the new spaces of knowledge communication–from the physical classroom to virtual-multimedia classrooms. All master classes in the different disciplines related to the different degree programs became video lectures thanks to the production centers set up in the consortium universities, where all online teaching materials were also created. The professors, who for the first time taught via television, had to acquire the appropriate skills for the task. For the first time, professors who taught via television and the Internet were exposed to the judgment, not only of their students, but of all those who watched their video lectures, which were broadcast by a satellite channel created in conjunction with Rai–Radiotelevisione italiana.

5.3 Technologies Used by Consorzio NETTUNO

The consortium was established in 1992 when the internet was not widespread. It made extensive use of commercial television (RAI2, every morning) and videotapes, which were sent to the Technological Poles of all the consortium universities that offered distance learning degree courses. Subsequently, satellite television HOT BIRD was used, where two satellite TV channels, RAI NETTUNO SAT, broadcast the video lectures of professors from all the consortium universities 24 h a day. The video lectures were created following psycho-pedagogical models where the results of many research projects were applied, allowing the creation of formats and languages completely different from those of commercial television. In 1996, the first internet portal was created, and satellite internet was experimented with. Thanks to the opportunity provided by the EUTELSAT company in Paris, we installed transmitting and receiving satellite antennas at our headquarters, allowing us to research how satellite internet access speed enabled the digitization of video lectures online. The first video lectures were digitized in 1997, and the first course was the mathematics course by Professor Barozzi from the University of Bologna.

5.4 Theoretical Models

Within the NETTUNO Consortium, technology has always been linked to the application of psycho-pedagogical theories aligned with specific objectives. The Consortium serves as a dynamic laboratory where hypotheses are tested, theories are applied, teaching models are experimented with, and learning processes are meticulously analyzed. Two prominent educational theories–cognitivism and constructivism–have significantly influenced teaching and learning models within NETTUNO. These theories view learning as an active, constructive, situational, and goal-oriented process [2, 3]. Learners are empowered to become architects of their own knowledge, shifting the focus from mere teaching to robust learning support [4]. The product-centric model was succeeded by a process-centric model. The NETTUNO Consortium's instructional model and Internet platform were designed on key concepts such as scaffolding and increasing students' self-regulatory capabilities. This design aimed at fostering cooperative learning through interaction between students and teachers both in-person and remotely. The learning processes implemented by the NETTUNO model thus took became constructive, strategic and interactive. The new pedagogical model responded to the need for flexibility by avoiding student isolation. It was a blended model that enhanced the traditional model thanks to a teaching method that was freed from the limitations of space and time, while also retaining a phase of direct interaction. This blended model, therefore, combined the advantages offered by traditional teaching and guided learning with the advantages offered by digital technologies, modulating and integrating both.

The key features of the psycho-pedagogical teaching model were as follows: (a) *Traditional Method*: direct interaction with lecturers and tutors; Seminars, practical exercises and evaluation tests conducted in the presence of tutors/teachers; Meetings between tutors, faculty and students in the universities' technology hubs; Practical exercises take place at provider universities; (b) *Distance Method*: video lectures initially broadcast only on television and then also digitized and accessible on the Internet; Practical exercises conducted using multimedia software, videos and online conferencing; Distance tutoring occurs through interactive online classes.

5.5 *Distance-Learning Teaching Methods*

NETTUNO's distance education model made it possible to integrate the didactic-pedagogical possibilities offered by various media, creating an open and flexible learning environment, encompassing both synchronic and diachronic distance education. In synchronic distance education, teaching and learning occur simultaneously but learners are not bound by physical location; in diachronic distance education, teaching and learning are no longer tied to a specific time and place. In the diachronic mode, students interacted with Internet-based technologies.

This innovative model transformed the skills of university professors. Faculty members adapted to new ways of expounding, synthesizing and presenting their knowledge through digital technologies, enabling them to develop an innovative pedagogical model even within traditional classroom settings.

6 A New Model of Distance University: UNINETTUNO International Telematic University

In 2012, Italy enacted a transformative law that gave rise to Telematic Universities–distinct from their traditional counterparts. By 2015, the NETTUNO Consortium model was replaced, giving birth to UNINETTUNO International Telematic University.

UNINETTUNO emerged through collaboration with universities worldwide, particularly drawing insights from the European Med-Net'U project (Mediterranean Network of Universities–funded by the European Commission under the Eumedis program and coordinated by NETTUNO [5], which paved the way for a fully digital university model. The experience gained alongside 31 traditional universities across 11 Euro-Mediterranean countries was pivotal. Together they created not only a Technological Network, which, by leveraging satellite technologies, including the Internet learning environment via satellite and the RAI NETTUNO-SAT television network, enabled all partners to produce, transmit and receive educational content, but above all, they fostered a network of minds–connecting people across borders to share knowledge. UNINETTUNO International Telematic University thrived on collaboration between European and Arab professors who together created a unique international knowledge hub, producing multimedia contents in various languages. From Italy to Europe to the Mediterranean, and now also with universities from sub-Saharan Africa and many traditional universities from the U.S., India, China, and beyond, UNINETTUNO transcends geographical barriers to become one big university.[3] The new technologies have obliterated distance as an obstacle, students worldwide can access premier academic teachings, free from prejudice, with the sole purpose of deepening and broadening their knowledge, creating new knowledge together and fostering cooperation [6–10].

---

[3] Afghanistan, Albania, Algeria, Andorra, Angola, British West Indies, Netherlands Antilles, Saudi Arabia, Argentina, Armenia, Australia, Austria, Azerbaijan, Baha-mas, Bahrain, Bangladesh, Barbados, Belgium, Benin, Belarus, Burma, Bolivia, Bosnia and Herzegovina, Botswana, Brazil, Brunei, Bulgaria, Burkina, Burundi, Cameroon, Canada, Cape Verde, (Islands) Cayman, Chile, Cyprus, Colombia, Congo, South Korea, Ivory Coast, Costa Rica, Croatia, Cuba, Denmark, Dominica, Ecuador, Egypt, El Salvador, United Arab Emirates, Eritrea, Estonia, Ethiopia, Russian Federation, Philippines, Finland, France, Gambia, Georgia, Germany, Ghana, Jamaica, Japan, Jordan, Great Britain and Northern Ireland, Greece, Gre-nada, Guatemala, Guinea Bissau, Guyana, Haiti, Honduras, Hong Kong, India, Indonesia, Iran, Iraq, Ireland - Eire, Iceland, Fiji Islands, Israel, Kazakhstan, Kenya, Kosovo, Kuwait, Lesotho Latvia, Lebanon, Liberia, Lithuania, Luxem-bourg, North Macedonia, Malawi, Malaysia, Maldives, Malta, Morocco, Marti-nique, Mauritius, Mexico, Monaco (Principality), Mongolia, Montenegro, Mozambique, Namibia, Nepal, Nicaragua, Nigeria, Norway, New Zealand, Oman, Netherlands, Pakistan, Palestine, Panama, Papua New Guinea, Peru, Poland, Por-tugal, Puerto Rico, Qatar, Czech Republic, Democratic Republic of Congo, Do-minican Republic, People's Republic of China, Republic of South Africa, Roma-nia, Rwanda, Saint Lucia, Saint Vincent and the Grenadines, San Marino, Sene-gal, Serbia, Seychelles, Sierra Leone, Singapore, Syria, Slovakia, Slovenia, Soma-lia, South Sudan, Spain, Sri Lanka, United States of America, Sudan, Sweden, Switzerland, Swaziland, Thailand, Togo, Trinidad and Tobago, Tunisia, Turkey, Ukraine, Uganda, Hungary, Uruguay, Uzbekistan, Venezuela, Vietnam, Yemen, Zambia, Zimbabwe.

The creation of a space for collaboration between distance universities and traditional universities and educational institutions has proven to be of utmost importance strategically, culturally, politically and economically. A distance university arising from a multicultural partnership not only equips learners for a globalized world, but also drives the production of multilingual multimedia educational content, at the same time fueling the development of a knowledge-based economy.

7 Research as the Core of the Creation of New University Models

UNINETTUNO's evolving distance university model is grounded in extensive international research. The commitment of UNINETTUNO International Telematic University to conducting research with and on new digital technologies is unwavering. While delivering distance courses, it continuously studies and analyzes how new technologies shape learning environments and student interactions. The impact of these technologies on our conscious and unconscious knowledge processing strategies is thoroughly examined. Research is carried out on the effects of TV screens, computers, smartphones and other digital tools that connect our minds in a global planetary dimension. The methodologies used to implement distance teaching and learning processes are also studied, along with models of remote interaction between teachers and students, and the integration of hypertextual, multimedia, cooperative and collaborative languages with the methodologies for the design and implementation of distance courses and the organizational models of Internet-based training facilities.

UNINETTUNO University is a true research laboratory, where hypotheses on the methodologies to be adopted are developed and the results of these applications are constantly subjected to verification by scholars from different disciplines. Cognitive psychologists, linguists, pedagogists, neuroscientists and computer scientists verify the effectiveness and efficiency of the psycho-pedagogical models offered, also by analyzing the results obtained by students in their learning processes. The results of this research have made it possible to: (a) identify the complex interrelationships that exist between digital technologies, cognitive processes and educational models; (b) test the validity of the assumptions applied in teaching and learning models implemented in the educational cyberspace of the Internet platform; (b) check the efficiency and effectiveness of the organizational model of the training facility.

The results of many research projects have made it possible to apply psycho-pedagogical theories mainly related to social-cognitive constructivism to achieve constructive and collaborative teaching and learning environments via the Internet. These environments are essentially characterized by two-way, interactive synchronous and asynchronous communication and student participation in knowledge construction. This led to the formalization of models of interaction between teachers and students inspired by the Socratic method, cooperative and collaborative multimedia and hypertext languages used for the realization of distance course content, and the organizational models of Internet-based training facilities. The design and implementation of the Internet Educational Cyberspace adopted a systemic approach, based on proven knowledge related to learning theories and the potential and development of technologies that determine a continuous evolution of the psycho-pedagogical model. The implementation of educational processes is centered on students and learning communities.

7.1   The Organizational Model

The organizational structure of UNINETTUNO is akin to a "reticular" system: a central coordinating headquarters, technological poles and production centers distributed across national and international territories, all interconnected through telematics. The success and quality of online educational activities hinge on high-quality content, the lecturers' video communication model and the way the lecturer-tutor guides the students' learning processes on the Internet. All lecturers working for UNINETTUNO undergo mandatory training in the skills required by the teaching model that characterizes all distance teaching and learning processes. Lecturers who create video lessons receive specialized training in visual language and didactic methodologies for online teaching and for the creation of online teaching materials. Specific training is given to teacher-tutors who must guide their students' learning processes on the Internet. Teacher-tutors must learn how to foster collaborative learning by interacting with students through virtual classrooms, chats, forums, wikis, and AI systems like ChatGPT. They must know how to support student motivation to learn, encourage participation in interactive online activities, and monitor the learning performance of individual students and the entire classroom, to intervene in critical individual and collective moments. Thus, the new psycho-pedagogical model, implemented by UNINETTUNO International Telematic University, revolutionizes the traditional skills of university professors: from solitary lecturers to members of a learning team, from knowledge transmitters to advisors and guides in students' learning journeys. Professors must learn to conduct video lectures in telematic classrooms without physical student presence, to design and implement multimedia products, online tutorials, and materials to be published in educational cyberspace, and to guide students in self-directed learning using ever-evolving digital tools, methods, and technologies. A novel figure emerges: the telematic teacher-tutor. This mediator facilitates interactions between the actors in the teaching–learning process, proposing topics and presenting theses in forums, discussing them with learners in interactive virtual classrooms, setting achievable goals, and encouraging learners to participate in the active construction of new knowledge.

7.2 Technological Structure

The technological structure is the beating heart of UNINETTUNO's physical structures–buildings, classrooms, research labs, and secretariats–have been replaced by a robust technological infrastructure on the Internet. Virtual classrooms now replace traditional ones. Research labs span virtual laboratories and remote facilities of other universities or research centers worldwide. The UNINETTUNO model does not require much physical space: rather, it employs a powerful technological infrastructure and experienced staff to maintain the efficiency of the entire educational system.

As the UNINETTUNO University, established in 2015, we immediately utilized the internet and three-dimensional virtual environments to create the UNINETTUNO knowledge island on Second Life, where avatars of teachers and students collaboratively and constructively developed learning models together. Creating a technological platform, born from many research projects and based on Microsoft technology, has allowed us to design and implement a generative AI-based conversational system, founded on ChatGPT engines, which is currently working on all our content. This content includes almost one hundred thousand hours of video lessons and 30 million pages on subjects ranging from economics, law, psychology, neuroscience, cultural heritage, communication sciences, and various sectors of engineering. Currently, the company connected to Microsoft is creating the intelligent engine linked to the current platform. Next year, a research group from the University will be engaged in experimenting with how Generative AI will impact the processes and procedures of our distance teaching and learning models, once integrated into our platform characterized by a knowledge organization based on proven models applied to distance teaching processes.

7.3 Learning Environments: The Learning Cyberspace

The design and implementation of the educational cyberspace follows a systems approach grounded in established learning theories and the potential of evolving technologies. It centers on students and learning communities, fostering continuous psycho-pedagogical evolution.

The constructive and collaborative teaching and learning environments that have been achieved on the Internet, are essentially characterized by: (a) synchronous and asynchronous two-way and interactive communication; (b) active participation of students in knowledge construction.

UNINETTUNO's educational cyberspace design, students and faculty can engage in activities together with other students from different countries around the world with the goal of increasing knowledge and skills together. They cooperate to develop a collaborative learning model through which students, at various performance levels, work together toward a common goal.

Forums, chats, virtual classrooms on the Web and on UNINETTUNO's Island of Knowledge, where a virtual university with three-dimensional environments has been built, greatly amplify collaboration among students and between faculty and students from different parts of the world.

Technology is used to present a problem, stimulate study, connect topics in the real-world context, provide supports for group work, and enable learners to communicate their knowledge in the appropriate virtual spaces. Learning environments in the educational cyberspace have been designed to encourage collaborative construction of structured knowledge, focusing on individual and collective learning tasks. In cyberspace, students are continually challenged by teacher-tutors to become active knowledge builders and to identify the most appropriate paths and means to achieve their goals. Knowledge acquisition is dynamic rather than static, multimedia rather than linear, and systemic rather than systematic. The knowledge that each student constructs emerges as a summation of encounters and relationships with teacher-tutors within a virtual knowledge space.

The interaction between learning theories and technologies underlies the model that is used to place different types of multimedia and hypertext learning materials in cyberspace, including those related to the simulation of real cases. Virtual laboratories are one example, where students can manipulate and interact with virtual objects, formulating and testing hypotheses. In this learning activity, students, in addition to being guided by a human teacher-tutor, may also receive support from automated pedagogical agents. Students are continuously stimulated by the system to choose the optimal level of difficulty in their learning journey, linking it to prior knowledge of the subject.

Constant observation of the ways in which teaching and learning processes are carried out in the university's educational cyberspace has made it possible to verify the effectiveness of learning theories applied to specific technologies. University research consistently strives to refine conceptual frameworks with precision, thereby establishing a robust theoretical foundation that supports the ongoing accumulation of knowledge in the domain of technology applied to educational processes.

7.4 The Learning Environments of UNINETTUNO International Telematic University: The Teaching Cyberspace

The foundational idea of UNINETTUNO University's psycho-pedagogical and educational model, which is also its main feature, centers around the integration of the educational-pedagogical potential of different media. These media contribute to creating an open, flexible online learning environment by fostering new communication channels between students and teachers and enabling two-way, real-time interactions; and by empowering students to access, enrich, and share knowledge.

The new psycho-pedagogical model is characterized by the transition: (a) from teacher-centered to student-centered; (b) from knowledge transmission to knowledge construction; (c) from the integration of theory and practice; (d) from passive and competitive learning to active and collaborative learning.

The psycho-pedagogical model allows students to tailor their learning paths based on their own educational needs and competency levels. They play an active and constructive role in processing knowledge.

The systemically-designed educational cyberspace includes various learning environments, each implementing specific training methods to activate teaching–learning processes: (a) video lessons use a symbolic reconstructive learning model related to classical teaching methods; by linking to different types of teaching materials, it enables is multimedia and hypertextual study; (b) virtual labs offer the opportunity of reinforcing knowledge and skills through "learning by doing" practical application; (c) chat systems, forums and virtual classrooms in three-dimensional worlds, interactive classrooms based on online conversation, enable collaborative learning by sharing the stages of the educational process with other students from different linguistic, cultural, political, religious and social backgrounds.

Within each environment, students can seamlessly integrate different learning mode. The learning activity is structured in such a way as to facilitate the transfer of knowledge in different ways: (a) grom simple concepts to complex applications (video lectures and smart library); (b) from theory to practical application ("learning by doing" in virtual labs); (c) from guided exercises to web research (sitography and bibliography); (d) from individual study to interactive dialogue between teachers and students and among students (collaborative learning via synchronous and asynchronous communication and sharing tools).

This psycho-pedagogical model also emphasizes an ethical use of the web which emphasizes the value of teacher presence, countering the tendency of the Internet to overlook this aspect. University lecturers, in this context, becomes telematic lecturers-tutors who, with maieutic art, guide users-students in their learning journey not only in acquiring knowledge and skills, but also in becoming be the author of their own learning journeys. In summary, the teaching process embraces the complexity of the educational process. It shifts the focus from a dominant model-product approach to a model-process perspective.

7.5 Knowledge Organization in the Educational Cyberspace

The value and excellence of a university distance learning course is closely tied to several critical factors. These include the quality of the content, the interactive activities facilitated by teachers-tutors, and the technological and methodological architecture of the platform. The assurance of content quality rests on the expertise of the professors who are national and international authorities in their fields. These professors design and deliver video lectures for each subject. Additionally, they provide supplementary materials associated with these lectures, such as texts, handouts, essays, annotated bibliographies, selections of internet resources, virtual laboratories, and web-based exercises.

During course delivery, teachers-tutors conduct both qualitative and quantitative assessments to continuously monitor learning outcomes and allow for individualized support to help students succeed in their final exams.

Special emphasis is placed on collaborative learning. This occurs through two main channels: Interactive Classrooms and 3D Virtual Classrooms on the UNINETTUNO platform.

7.6 *Interactive Classrooms*

In UNINETTUNO's Interactive Classrooms, teachers and students engage with interactive tools, carry out practical exercises, intermediate assessment tests, and exam simulations. They debate and learn cooperatively and collaboratively and become active knowledge builders.

7.7 *3D Virtual Classrooms*

These three-dimensional environments manipulate variables like space and roles. Avatars representing real individuals interact with other avatars in this immersive setting, characterized by a strong sense of reality. Participants move beyond passive observation; they become actors in a new dynamic reality. Sensations and involvement are heightened as students and teachers-tutors meet virtually to develop immersive-collaborative teaching experiences.

Three-dimensional environments and other 3D immersive worlds allow users to transcend the confines of computer screens. Avatars of students and faculty engage in discussions, debates and conferences across universities worldwide. These immersive technologies connected to 3D virtual classrooms also enable experiments on the use of the metaverse in learning processes. Purpose-built spaces within these virtual realms–such as bars, dance halls, tennis courts and chessboards–facilitate socialization using avatars of faculty and students like in a real university. Other virtual spaces, including the Chancellor's Office and faculty buildings, enable remote work with avatars of university staff, teaching and administrative personnel. UNINETTUNO's educational cyberspace, which explores the use of generative artificial intelligence systems, aligns with data protection regulations, including the latest developments in European legislation such as the GDPR (General Data Protection Regulation).

8 Conclusion

The experience at UNINETTUNO International Telematic University highlights the potential for sharing curricula and creating new educational models collaboratively. These models emerge from universities across countries with diverse policies and cultures, responding to the shifts brought about by our globalized world. The interconnected intelligences of professors and students from both the northern and southern hemispheres contribute to educational content and new knowledge. Their collaboration transcends the imposition of cultural models, relying instead on intercultural and interlingual comparison and cooperation.

In this context, the Internet knows no borders, signifying continuity rather than conflict, and democratizing knowledge access to empower individuals. Today's knowledge networks hold the potential to create wealth and openly and democratically disseminate the wisdom of the world's foremost scientists and intellectuals. If, as we believe, quality branding will determine the competitive edge in global education markets, a distance university based on a network of the world's best traditional universities will undoubtedly win the challenge. The digital university can now address the demands of the new knowledge market by showcasing its quality branding, ensuring user satisfaction and transforming into an open system, capable of updating and integrating all available knowledge from the web, fostering global knowledge exchange. Moreover, it serves as the foundation for training generative AI systems, while granting user access to scientifically valid educational content online. Only through robust democratizing policies that grant universal access to knowledge can humanity forge a new path–one where individuals collaboratively uphold values of solidarity and respect for differences. These shared universal values, theoretically accepted by all, pave the way for justice and peace.

References

1. Garito MA (2001) The university for the new market of knowledge. World Futur: J Gen Evol 57(6):129–132
2. Lowyck J, Elen J (1993) Transitions in the theoretical foundation of instructional design. In: Designing environments for constructive learning. Heidelberg: Springer Berlin Heidelberg, Berlin, pp 213–229
3. Shuell TJ (1988) The role of the student in learning from instruction. Contemp Educ Psychol 13(3):276–295
4. Duffy TM, Lowyck J, Jonassen DH (eds) (1993) The design of constructivist learning environments—implications for instructional design and the use of technology. Springer, Heidelberg
5. Garito MA (2003) The Euro-Mediterranean distance university. In: "Public service review": European Union, public service communication agency Ltd: pp 129–132
6. Garito MA (2004) La perspective internationale. Une collaboration Europe-Pays de la Méditerranée. Rev Savoirs 5:97–103
7. Garito MA (2005) Towards the Euro-Mediterranean distance university. LlinE Lifelong Learn Eur Eur Adult Contin Educ J IX(1):47–49

8. Garito MA (2006) New technologies and distance teaching. University, Ministero dell'Istruzione, dell'Università e della Ricerca, Euro-Mediterranean Space of Higher Education and Research, Catania, pp 24–34
9. Garito MA (2007) Content sharing between NETTUNO and the Italian and Mediterranean universities. In: Van Petegem W (eds) European networking and learning for the future. The EuroPACE Approach, Annemie Boonen,Garant, Antwerpen-Apeldoom, pp 185–197
10. Garito MA (2008) Universities in dialogue in a world without distance. In: Education landscapes in the 21st century: cross-cultural challenges and multi-disciplinary perspectives. Cambridge Scholars Publishing, pp 35–368

Opening Up Teaching and Learning at Universities: OER, MOOCs and Microcredentials

Martin Ebner, Ernst Kreuzer, Sandra Schön, and Sarah Edelsbrunner

Abstract This paper presents the situation concerning open and digital learning and teaching opportunities with respect to three developments: Open Educational Resources (OER), Massive Open Online Courses (MOOCs), and the introduction of microcredentials. Firstly, the paper introduces these concepts. Then, before the background of the Austrian university landscape, the authors describe how these three measures are currently or will be implemented at Graz University of Technology (TU Graz). TU Graz is one of the first universities worldwide with an OER policy document and is as well host of the Austrian MOOC platform iMooX.at. Both provide the opportunity for new collaborations amongst universities and opportunities for learners. The introduction of microcredentials is just beginning at TU Graz and provides an opportunity for external learners and students to receive accredited confirmation of their qualifications from the university. The article not only discusses the opportunities and positive effects of these developments at TU Graz as part of the digitalization of teaching and lifelong learning activities, but also addresses the challenges in an increasingly globalized and competitive higher education landscape.

Keyword Higher education institution · Open educational resources · Massive open online courses · Microcredentials

---

M. Ebner · E. Kreuzer · S. Schön (✉) · S. Edelsbrunner
Graz University of Technology, Münzgrabenstraße 36, 5010 Graz, Austria
e-mail: sandra.schoen@tugraz.at

M. Ebner
e-mail: martin.ebner@tugraz.at

1 Facets of "Openness of Digitalization" in Higher Education

The increasing digitalization and most recently the phase of enforced distance learning at many universities worldwide has shown that the needs of learners and teachers concerning open and digital learning and teaching opportunities are evolving. For many, digitalization means the possibility of maintaining teaching at one's own university even though it is closed. This was also the case at the Austrian Graz University of Technology (TU Graz; Ebner [34]) Nevertheless, digitalization and the corresponding strategic decisions began several years earlier [5, 6]. In this article, we describe developments regarding the digitalization of teaching at TU Graz that are very similar in terms of one characteristic: All developments can be assigned to the principle of opening up academic teaching.

Opening up education is not always understood to mean the same thing - and in this article we also refer to different aspects. Opening up can refer to the opening of higher education institutions to individuals without college access. The Open Universities, for example, are in this tradition. Opening up can also refer to the possibilities of external persons using resources provided by universities, e.g., research laboratories and libraries. Especially in the context of Open Science, the term "open" refers primarily to the possibility of openly licensing data, research articles and educational resources. Prominent examples of open licenses are the Creative Commons licenses CC BY 4.0 or CC BY-SA 4.0. Open licenses allow re-use, modification, and re-publishing. Inamorato Dos Santos et al. [20] identified ten dimensions of open education in higher education, depending on where and how openness is possible and needed. This article presents the situation at TU Graz with respect to three developments concerning opening up education: Open Educational Resources (OER), Massive Open Online Courses (MOOCs), and the introduction of microcredentials.

OER are defined as educational resources where modification, revision, re-use, and publication are allowed, because they are available in the public domain or with open licenses [37]. Educational resources produced by universities are thereby made available to the public, such as self-organized learners, teachers in schools and beyond. Many different positive aspects and purposes of OER are known, among them inclusion, sustainability of investment, and less copyright infringements [3]. OER are also seen as a measure to support open educational practices [18].

MOOCs, short for Massive Open Online Courses, a term first coined in 2010, are freely accessible online courses that usually serve a very large number of registered learners at the same time [24]. Additionally, those MOOCs are hosted by MOOC-platforms like Udacity, edX, future learn or the Austrian pendant iMooX.at. These platforms are similar to typical Learning Management Systems, with special features as well as user interfaces allowing thousands of learners to interact in parallel. Each MOOC consists mainly of four components: (a) the main learning objects are (rather short) learning videos. Instructors present their topic by producing videos in different ways,(b) further learning content is presented by linking to other web content, uploaded documents, or interactive learning objects, (c) self-assessment is

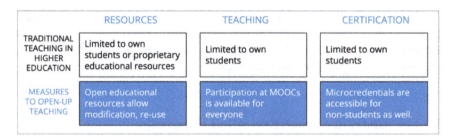

Fig. 1 Measures to open up teaching to the public in higher education versus traditional teaching in higher education

provided to assess learners' knowledge after watching the videos and interacting with the learning objects, and (d) a certificate is available if all self-assessments, tasks and homework has been successfully completed.

Finally, microcredentials as defined by the European Commission [15] "certify the learning outcomes of short-term learning experiences, for example a short course or training [and] offer a flexible, targeted way to help people develop the knowledge, skills, and competences they need for their personal and professional development." Typically, microcredentials in Europe are referring to the European Crediting Transfer System (ECTS) and certify a workload of 3–30 ECTS. As one ECTS credit equals about 25 working hours, a microcredential refers to 75–750 working hours. Microcredentials should be accepted by all (higher) education institutions. The main driving force in introducing microcredentials is the idea to make education more attractive to the labor market following the idea of professional continuing education in the sense of lifelong learning.

Figure 1 gives an overview of how these three developments differ from traditional teaching as measures to open up teaching to the public.

2 A Case Study on Opening Up Teaching at TU Graz

In this paper, we would like to present three measures in the context of the digitalization of teaching, which we suspect are not unique, but may represent a special feature in our combination at TU Graz. We would like to present these developments in the form of a case study, which we describe as those affected by them and actors ourselves. This is intended to provide a basis for third parties to compare our measures and developments with their decisions.

The guiding questions are:

1. What do the three selected measures currently mean for TU Graz and how are they dealt with?
2. In what way can the measures be described as "open"? How does this openness work and what does it change, if anything?

The article not only discusses the opportunities and positive effects of these developments at TU Graz as part of the digitalization of teaching [7] and lifelong learning activities, but also addresses the challenges in an increasingly globalized and competitive higher education landscape.

3 The Austrian Higher Education Landscape

Austria is a German-speaking European country with about 80 higher education institutions. Besides 22 public universities, Austria has 16 private universities. There are also 21 universities of applied sciences and 14 University Colleges of Teacher Education. Most students in Austria are enrolled at public universities which can be attended for comparatively low tuition fees; especially when compared internationally (Federal Ministry of Education, Science and Research [16]. As in other countries, higher education autonomy was introduced in Austria in 2002 [21], but in terms of implementation, Austria is in the international midfield, in other words, the universities are still clearly dependent on the Ministry of Education in financial terms, for example, through budgeting and guidelines [12].

4 Open Educational Resources at TU Graz

OER are and have been promoted, in demand, and increasingly published, not least because of Austrian copyright restrictions on the use of proprietary educational materials [29]. OER make university content available to a larger public. In Austria, OER has been mentioned in several national policies for higher education for several years [10], and is explicitly part of the "Open Science Policy Austria" [27]. Compared with the situation of OER policies in higher education institutions with neighboring Germany and Switzerland, it was stated that.

> (...) Austria has a significantly higher penetration of policies at institutional level compared to Germany and Switzerland. Now, we can only speculate about the reasons for this. One possible explanation could be that (according to the OER World Map) the number of national policy documents is also higher than in Germany and Switzerland, which could have supported the trend towards adopting institutional policies. Another explanation could be that the topic of OER was taken up earlier in Austria's higher education system than in Germany and Switzerland and therefore has been able to develop further than with its neighbours. [25, p. 141].

According to the latest analysis of the current performance agreements (2022–2024), all 22 Austrian public universities mention OER in their planned activities [11].

One of the stakeholders in the OER movement in Austria is TU Graz. As of 2010, there is evidence for a first strategic orientation of TU Graz towards open educational resources in the form of a publication entitled "Open Educational Resources as a Lifelong Learning Strategy using the example of TU Graz" [9]. Since the winter semester 2013/2014, there has been a dedicated continuing education course for OER, but due to lack of demand, it has only been held regularly since 2015. Since 2017, OER have also been anchored in the development plan as well as in the performance agreement. In the section on teaching and learning technologies of the development plan, this is described as follows.

> In recent years, TU Graz has been able to establish itself particularly in the area of Open Educational Resources (OER) and, in addition to intensive activity at the national level (co-author on the recommendation letter of the Forum Neue Medien in der Lehre association), can also point to collaboration in the Open Education Austria project (2016-2019). Furthermore, internal further education offers on OER and copyright should also be established. The offer of free educational materials should help fulfill the public educational mandate, make content easily accessible for future students or graduates and also actively support students and teachers on site by simplifying the handling of educational materials. (TU Graz, 2017, p. 60; translation by authors).

To elaborate further: In the performance agreement 2019–2021 of TU Graz, OER are mentioned in quite a few places, and similarly to the development plan, concrete activities are described, one of which was to point out an "OER Policy" planned for 2020 [35, p. 10, p. 59]. In the OER Policy [34], among other things, the organizational unit "Educational Technology" is listed as a contact for OER. For example, it was stipulated that the "Award for Excellent Teaching" will be linked to OER. To monitor and assess the impact of activities around OER, the Educational Technology team is aware of the need and possibilities of OER impact assessment, has carried out first research of existing approaches and developed measures for it [2, 4, 8, 31] (Fig. 2).

One important OER activity at TU Graz is the possibility to publish and archive OER in an OER repository: The development of this OER repository and a plug-in for teachers in the learning management system started in 2019. Now, after completing an OER further education, teachers are allowed to use the OER plug-in and their material is automatically published [23]. The OER repository is additionally connected (via API) with the Austrian-wide higher education OER referatory OERhub. at [19]. TU

Fig. 2 TU Graz activities, results, outcomes and expected key impact according to the TU Graz OER Policy

Graz, the "Forum Neue Medien in der Lehre Austria" (fnma) and partners are responsible for the development of an OER certification for higher education institutions and staff, which is a unique national development (see [30]).

The Educational Technology team serves as OER support and develops a variety of materials about OER for teachers and students, as well as other materials concerning technology-enhanced teaching, which are provided as OER as well. These materials are for example canvases for the creation of OER and the ReDesign of lectures [30]. One highlight is the TELucation collection, providing recommendations for technology-enhanced learning [2, 4, 8].

5 The National MOOC Platform iMooX Hosted by TU Graz

Massive Open Online Courses (short MOOCs) on the Austrian MOOC platform iMooX.at (hosted by TU Graz) allow all Austrian universities to offer online courses free of charge to their own students as well as to the public and lifelong learners [1]. This allows for new collaboration among universities and opportunities for learners and students to learn with MOOCs by other universities.

The platform was founded in 2013 by TU Graz and University of Graz, aiming to offer "education for all". Hence, from the very beginning, it was defined that each learning object—videos, documents, etc.—as well as the course must be licensed with a Creative Commons license. Therefore, the platform consists only of online courses for many learners on a higher education level as Open Educational Resources. Since 2013, more than 500 different courses have been offered in different fields of research and for different educational levels (bachelor, master, high school, adult education, and others) for learners of all ages.

In 2020, the MOOChub consortium was founded by TU Graz following the idea to bring together all MOOC platforms in the German-speaking area. As a follow up, a common standard for MOOC descriptions has been elaborated and coordinated between the different providers. As a result, the website moochub.org holds more than 600 online courses today, which are searchable via topics, providers, and interests. In 2021, iMooX.at also became a member of the European MOOC Consortium (EMC) and is currently one of the six European MOOC platforms committing to the EMC common microcredentials framework:

> The European MOOC Consortium is collaborating on a Common Microcredential Framework (CMF) to be used by these platforms on a voluntary basis, but which may under the right circumstances could be converted into a formal qualification or standard for use by a wider set of universities adhering to the framework, achieved in partnership with national and supranational agencies. [13]

The impact of the iMooX platform was investigated through a survey of MOOC creators, as part of the project "MooX—The MOOC Platform as a Service for All Austrian Universities" (2020–2023), which aims to expand iMooX.at as a national platform for higher education institutions. The study by Ebner et al. [5, 6] presents the findings and impacts of the project. The study collected data through an online survey with 17 course creators revealed that a significant majority affirmed that iMooX.at contributes to the dissemination of Open Educational Resources (OER) and has positive impacts on various groups. It is worth noting that staff from TU Graz have offered the most MOOCs on the iMooX platform so far, highlighting the university's active role in promoting and utilizing OER through this national initiative. This engagement not only underscores TU Graz's commitment to educational innovation but also its leadership in the broader adoption of OER across Austria.

6 Microcredentials Development at the TU Graz

Thirdly, microcredentials are presented, i.e. the possibility for external learners and students to receive an accredited confirmation of their qualifications from the university. It will be described how microcredentials are currently under development at the TU Graz.

The principle of lifelong learning is essential in today's world and must be spread multidirectionally and multidimensionally across all areas of life. For this reason, the organizational unit "Life Long Learning" (LLL) was established in 2005 at TU Graz as an interface between university (educational sector), economy and society [36].

Continuing technical education and training is becoming a constant, while at the same time the demands on continuing education are changing towards flexible, demand-oriented, and modularized learning settings. In the European Commission [14] "Recommendation on Effective Active Support to Employment", upskilling and reskilling opportunities aligned with labor market needs are seen as essential. Thus, new and/or realigned educational offers in the form of short training courses, certified by micro-credentials, could play an important role for companies as well as employees.

The work of the future will be more flexible, more mobile, and more digitally networked. This is a consequence of the current transformation processes toward digitalization affecting all areas of human life, especially that of continuing education. Against the backdrop of change in the world of work, it seems obvious that progressive specialization also requires adapted operational or organizational learning processes. This means that companies are increasingly required to design, manage, and integrate learning processes and learning environments into their organizations to create or be part of a life-long learning ecosystem to prepare employees for changing work fields and tasks. Companies need to become learning organizations, and employees also need to acquire lifelong knowledge. A sustainable change in the world of work also requires a new conception of academic continuing education or an adaptation in a holistic way, i.e., both on the institutional level and about transdisciplinary cooperation with the business community, as well as an acceleration of national and international cooperation [17]. A new kind of degrees is seen as a possible chance for universities all over the world: microcredentials [26].

The State University of New York [33] describes microcredentials as follows on their website: "At the most basic level, micro-credentials verify, validate, and attest specific skills and/or competencies have been achieved. They differ from traditional degrees and certificates in that they are generally offered in shorter or more flexible time spans and tend to be more narrowly focused. Micro-credentials can be offered online, in the classroom, or via a hybrid of both." In other words, microcredentials describe the awarding of certificates that comprise less than a study program but more than one course, i. e. roughly an effort of 5 to 30 ECTS. MOOC providers unfortunately use different names for these certificates, for example, edX speaks of "professional certificates", Udacity of "Nanodegrees", Coursera and others of "MasterTrack Certificates" [32]. These inconsistencies in designations and implementations of microcredentials, especially among MOOC providers, have been criticized [28] and there are numerous initiatives to ensure consistency. In Austria, for example, the Ministry of Education, Science and Research has made an initial proposal on a uniform understanding and descriptions of microcredentials that was sent to the universities.

Austrian higher education takes a positive view of the European developments on microcredentials, especially in the context of lifelong learning, which is becoming increasingly important, and see this as an opportunity for higher education institutions to qualitatively expand their educational offer, to develop new target groups and to expand cooperation with non-university partner institutions.

The following list names possible parameters of a microcredential in the sense of a short learning unit): title and short description of the micro-credential; references to the effort (in Austria, described with European Credit Transfer and Accumulation System, in short ECTS); target groups and participation requirements; performance assessment/examination modalities; description of learning outcomes, teaching/learning forms and formats (online, on site or blended); quality assurance; connection and credit options (integrated or standalone microcredential; continuing microcredential); and completion/credential.

The academic continuing education offering by TU Graz Life Long Learning (LLL) was reorganized and restructured with the continuing education campaign in the area of "Digitalization and Digital Transformation". Now learning offers and certificates are structured in "modules" that consist of learning units of five ECTS credits each. The learning effort to complete such a LLL module is calculated as about 150 h. As shown in Fig. 3, it combines blended learning activities such as a MOOC or online course, face-to-face training (which can also take the form of virtual presence) and a transfer assignment to prove application skills. If all components are demonstrated and met as sufficient, a microcredential will be awarded in the future [22].

In addition to a highly modularized program architecture based on 5 ECTS modules/microcertificates, which can be combined to form individual modules (mixed format: online phase/in-person phase and transfer phase) into a complete

Fig. 3 Components of a LLL module and microcredential

master's degree program ("up-scaling") or which also enable shorter target group-specific continuing education formats ("down-scaling"), the key features here are the selected three-phase and digitally supported teaching and learning arrangement. The flexibilization of continuing education is additionally supported by the increased use of methods and formats of technology-enhanced learning. These two new features enable a scalable and stackable continuing education portfolio (see Fig. 3), which better meets the requirements of contemporary academic continuing education and specifically promotes and supports life-long learning.

The objective of the further expansion of the continuing education offers of TU Graz LLL is to generate research-based continuing education content in a stackable manner in the sense of a modular system, to continuously expand it (also with modules/microcredentials/micro-certificates from partner institutions and universities) and thus to address specific target groups and their needs through a mix and match of micro- or also nano-certificates. MOOCs can play an important and leading role in designing flexible curricula and individual learning paths in continuing academic education, either as stand-alone offers or as part of a blended learning format.

As shown in Fig. 4, by combining individual microcredentials, a broad, flexible and topical offer can be developed in a structured and systematic way in the thematic continuing education clusters of TU Graz LLL, consisting of both shorter units (microcredentials), such as a module with a scope of 5 ECTS credits, and longer continuing education measures such as certificate courses (10/15/20/25 ECTS credits), up to multi-semester continuing education master programs (60/90/120 ECTS credits).

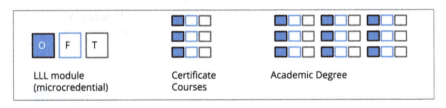

Fig. 4 Modular architecture of different LLL certificates at TU Graz (see [22])

7 Interrelationships of OER, MOOCs and Microcredentials at TU Graz and Their Effects

In this chapter, we explore the intricate connections between Open Educational Resources (OER), Massive Open Online Courses (MOOCs), and microcredentials at TU Graz. While these elements were discussed separately already in the first chapters, their interrelationships have significant implications for the digitalization of teaching and learning at the university: The following sections will detail how these components interact and the resulting impacts on business models, quality assurance, and TU Graz's position in a globalized higher education landscape.

7.1 Interrelationship of OER, MOOC and Microcredentials at TU Graz

Although the activities regarding OER, MOOCs and microcredentials were presented separately in this paper, interrelationsships became apparent. Figure 5 summarizes these interrelations that might be unique for TU Graz: MOOCs on the iMooX platform use OER, and microcredentials use MOOC participation or certificates as a component. Therefore, if people pay for LLL offers to get microcredentials, additional MOOCs might be required, which themselves require more OER.

7.2 Changes in Business Models Due to Openness

The introduction of microcredentials and MOOCs each pose challenges because teachers at TU Graz are expected to perform more advanced and novel activities. In fact, the essential peculiarity lies in openness: The fact that the individual materials may now also be used by others, and that MOOCs are free of charge per se, clearly changes the business model and sales potential for LLL offers. On the producer side, there are new development opportunities and possibilities to use existing materials but also to develop blended microcredentials very cost-effectively using AI generated learning videos. As far as the customers are concerned, it must be made clear and

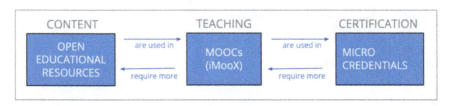

Fig. 5 Interrelationship of OER, MOOC and Microcredentials at TU Graz

communicated that, in the case of the LLL modules, TU Graz generates real qualitative added value for participants in LLL modules through the blended learning approach and the transfer orientation in the microcredentials, which justifies the participant fees.

In the described manner, two MOOCs have been pre-financed by microcredentials so far; the LLL courses were held 2 and 3 times respectively, and currently, a third MOOC is just opened for registration and part for another new microcredential course (June 2024). The costs of creating the MOOCs were thus already refinanced through the LLL courses.

These developments illustrate that the combination of financing MOOCs through the course fees of microcredentials represents a novel approach, showcasing TU Graz's innovative strategy in ensuring the sustainability and quality of lifelong learning offerings, combined with TU Graz's aim to ensure broad accessibility of their knowledge through MOOCs and the OER shared within.

7.3 Openness as an Opportunity to Show and Maintain Quality

At the same time, the use of OER and MOOCs also changes marketing: It is clearer for customers what the content and quality of the courses offered by TU Graz is in general. One can find out whether they like the content or the course management in the MOOCs, which provides a decent insight into the LLL module. OER and MOOCs are both free of charge and openly accessible, which means that more people can be reached directly and possibly also convinced of the quality of the offer. At the same time, this coupling of OER, MOOCs and LLL microcredentials also ensures that quality is strived for.

7.4 Development Before the Background of a Globalized and Competitive Higher Education Landscape

Finally, we believe that TU Graz is well positioned for an increasingly globalized higher education landscape with these developments around OER, MOOC and microcredentials, not least due to the increasing digitalization and good accessibility of online offers. OER and MOOCs potentially ensure a broad dissemination of content and awareness of TU Graz. Thus, OER and MOOCs are not only a component of microcredentials, but also the basis for the attractiveness and marketing of microcredentials, if OER and MOOCs by TU Graz are perceived as attractive. At the same time, we also see these offers as a flagship for future cooperation.

8 Outlook

Given the promising results from our initial implementations, we are optimistic about the continued development and integration of OER, MOOCs, and microcredentials at TU Graz. Since the launch of the iMooX platform in 2014 and the publication of our explicit OER strategy in 2020, we have amassed substantial experience in utilizing these resources to enhance our educational offerings. The introduction of microcredentials in autumn 2022 has further enriched our approach, demonstrating its potential through positive feedback and successful outcomes from the first rounds of implementation.

We believe that our strategy of financing specific MOOCs through the course fees of microcredentials is both innovative and sustainable, particularly for topics that are important for microcredentials and often aligned with the needs of companies for new and emerging fields. This targeted approach ensures the high quality and broad accessibility of our lifelong learning programs. While of course not all MOOCs can and will be funded in this manner, this model allows for the rapid development and financing of MOOCs that are essential for microcredentials. This is particularly beneficial given that alternative financing through third-party funding often involves longer lead times.

Looking forward, we remain open to adaptations and new ideas that can further leverage the synergy between OER and microcredentials. As the higher education landscape becomes increasingly globalized and competitive, TU Graz is well-positioned to continue leading in this space, using digitalization to expand access and improve the quality of our educational offerings. We hope that this text and our description of the approach will inspire others to adopt similar strategies, fostering a broader movement towards openness and innovation in education. These developments not only enhance our current programs but also serve as a flagship for future cooperation and advancement in higher education, ensuring the openness and sustainability of education.

References

1. Ebner M (2021) iMooX—a MOOC platform for all (universities). In: Proceedings of the 7th international conference on electrical, electronics and information engineering (ICEEIE), pp 1–5. https://doi.org/10.1109/ICEEIE52663.2021.9616685
2. Ebner M, Edelsbrunner S, Schön S (2022) Supporting learning and teaching with good design: report and lessons learned from learning experience design in higher education. In: Kang K (ed) E-service Digital Innovation. InTechOpen (preprint). https://www.intechopen.com/online-first/84343
3. Ebner M, Kopp M, Freisleben-Teutscher C, Gröblinger O, Rieck K, Schön S, Seitz P, Seissl M, Ofner S, Zimmermann C, Zwiauer C (2016) Recommendations for OER integration in Austrian higher education. Conference proceedings: the online, open and flexible higher education conference, EADTU 2016:34–44

4. Ebner M, Orr D, Schön S (2022) OER impact assessment: a framework for higher education institutions and beyond. Approaches to assess the impact of Open Educational Resources. Open Educ Stud 4(1):296–309. https://doi.org/10.1515/edu-2022-0018
5. Ebner M, Edelsbrunner S, Haas M, Hohla-Sejkora K, Leitner P, Lipp S, Mair B, Schön S, Steinkellner I, Stojcevic I, Zwiauer C (2023) Die Wirkung von MOOCs und iMooX.at aus Sicht von Kursersteller:innen. Zeitschrift für Hochschulentwicklung, vol 18 (Sonderheft Hochschullehre), pp 51–76. https://doi.org/10.3217/zfhe-SH-HL/04
6. Ebner M, Mair B, De Marinis C, Müller H, Nagler W, Schön S, Thurner S (2023) Digital transformation of teaching and perception at TU Graz from the students' perspective: developments from the last 17 years. In: Auer ME, Pachatz W, Rüütmann T (eds) Learning in the age of digital and green transition. ICL 2022. Lecture notes in networks and systems, vol 633. Springer, Cham, pp 366–377. https://doi.org/10.1007/978-3-031-26876-2_34
7. Ebner M, Schön S, Dennerlein S, Edelsbrunner S, Haas M, Nagler W (2021) Digitale Transformation der Lehre an Hochschulen—ein Werkstattbericht. In: Wilbers K, Hohenstein A (eds) Handbuch E-Learning. Expertenwissen aus Wissenschaft und Praxis—Strategien, Instrumente, Fallstudien, vol 94. Erg-Lfg. Dezember 2021, Beitrag, p 3.41
8. Ebner MS, Ebner S, Edelsbrunner S, Hohla K (2022) Potential impact of open educational resources and practices for good teaching at universities. The OER impact assessment at TU Graz. In: Auer ME, Pester A, May D (eds) Learning with technologies and technologies in learning. Experience, trends and challenges in higher education. Lecture notes in networks and systems, vol 45. Springer, Cham. https://doi.org/10.1007/978-3-031-04286-7_5
9. Ebner M, Stöckler-Penz C (2011) Open educational resources also lifelong-learning Strategie am Beispiel der TU Graz. In: Tomaschek N, Gronki E (eds) The lifelong learning University. Waxmann, Münster, pp 53–60
10. Edelsbrunner S, Ebner M, Schön S (2021) Strategien zu offenen Bildungsressourcen an österreichischen öffentlichen Universitäten. Eine Beschreibung von nationalen Strategien, Whitepapers und Projekten sowie eine Analyse der aktuellen Leistungsvereinbarungen. In: Wollersheim H, Karapanos M, Pengel N (eds) Bildung in der digitalen Transformation, Tagungsband der GMW 2021. Waxmann, Münster, pp 31–36
11. Edelsbrunner S, Ebner M, Schön S (2022) Strategien zu offenen Bildungsressourcen an österreichischen öffentlichen Universitäten. Eine Analyse der Leistungsvereinbarungen 2022–2024. In: Standl B (ed) Digitale Lehre nachhaltig gestalten, Medien in der Wissenschaft, Band, vol 80. Waxmann, Münster, pp. 209–214. https://doi.org/10.31244/9783830996330
12. Estermann T, Nokkala T, Steinel M (2011) University Autonomy in Europe II: the Scorecard. European University Association, Brüssel
13. European MOOC consortium (2019) EMC Common Microcredential Framework. https://emc.eadtu.eu/images/EMC_Common_Microcredential_Framework_.pdf
14. European Commission (2021) Commission recommendation (EU) 2021/402 Commission recommendation (EU) 2021/402 of 4 Mar 2021 on an effective active support to employment following the COVID-19 crisis (EASE). https://eur-lex.europa.eu/legal-content/EN/TXT/PDF/?uri=CELEX:32021H0402&from=EN. Accessed 26 Apr 2022
15. European Commission (2022) A European approach to micro-credentials. https://education.ec.europa.eu/education-levels/higher-education/micro-credentials
16. Federal Ministry of Education, Science and Research (2022). Homepage. https://www.bmbwf.gv.at
17. Fomunyam KG (2019) Education and the fourth industrial revolution: challenges and possibilities for engineering education. Int J Mech Eng Technol 271–284
18. Geser G (2007) OLCOS—Open eLearning content observatory services: open educational practices and resources. OLCOS Roadmap 2012. OLCOS, Salzburg. https://www.olcos.org/cms/upload/docs/olcos_roadmap.pdf
19. Gröblinger O, Ganguly R, Hackl C, Ebner M, Kopp M (2021) Dezentral bereitstellen—zentral finden. Zur Umsetzung hochschulübergreifender OER-Angebote. In: Gabellini C, Gallner S, Imboden F, Kuurstra M, Tremp P (eds) Lehrentwicklung by openness—open educational resources im hochschulkontext. Pädagogische Hochschule Luzern, Luzern, pp 39–44. https://doi.org/10.5281/zenodo.5004445

20. Inamorato dos Santos A, Punie Y, Castaño-Muñoz J (2016) Opening up education: a support framework for higher education institutions. JRC Sci Policy Rep EUR 27938 EN. https://doi.org/10.2791/293408
21. Janger J, Hölzl W, Hranyai K, Reinstaller A (2012) Hochschulen 2025: eine Entwicklungsvision. Wien, WIFO. http://www.wifo.ac.at/wwa/pubid/44698
22. Kreuzer E, Aschbacher H (2021) Universitäre Weiterbildung NEU denken. In: WINGbusiness Heft 04 2021, 6–9. https://issuu.com/beablond/docs/heft_04_2021_end
23. Ladurner C, Ortner C, Lach K, Ebner M, Haas M, Ebner M, Ganguly, R, Schön S (2020) The development and implementation of missing tools and procedures at the interface of a University's learning management system, its OER repository and the Austrian OER referatory. Int J Open Educ Res (IJOER), 3, 2. https://www.ijoer.org/the-development-and-implementation-of-missing-tools-and-procedures-at-the-interface-of-a-universitys-learning-management-system-its-oer-repository-and-the-austrian-oer-referatory/
24. McAuley A, Stewart B, Siemens G, Dave Cormier D (2010) Massive open online courses digital ways of knowing and learning, the MOOC model For Digital Practice. https://www.academia.edu/download/43171365/MOOC_Final.pdf
25. Neumann J, Schön S, Bedenlier S, Ebner M, Edelsbrunner S, Krüger N, Lüthi-Esposito G, Marin VI, Orr D, Peters LN, Reimer RT, Zawacki-Richter O (2022) Approaches to monitor and evaluate OER policies in higher education—tracing developments in Germany, Austria, and Switzerland. Asian J Dist Educ. http://asianjde.com/ojs/index.php/AsianJDE/article/view/619
26. Pelletier K et al (2021) 2021 EDUCAUSE Horizon Report, Teaching and Learning Edition. Boulder: Educause. https://library.educause.edu/resources/2021/4/2021-educause-horizon-report-teaching-and-learning-edition
27. Open Science Policy Austria—Österreichische Policy zu Open Science und der European Open Science Cloud, published, Feb 2022. https://www.bundeskanzleramt.gv.at/dam/jcr:93ebfe7e-9cfe-441b-8ed5-0dba32f38a81/7_9_mrv.pdf
28. Pickard L (2018) Analysis of 450 MOOC-Based microcredentials reveals many options but little consistency, MOOC report, class central, Jul 2018. https://www.class-central.com/report/moocs-microcredentials-analysis-2018/
29. Schön S, Ebner M (2020) Open educational resources in Austria. In: Huang R, Liu D, Tlili A, Gao Y, Koper R, Current R (eds) State of open educational resources in the "Belt and Road" countries. Lecture notes in educational technology. Springer, Singapore, pp 17–33
30. Schön S, Ebner M, Brandhofer G, Berger E, Gröblinger O, Jadin T, Kopp M, Steinbacher H (2021) OER-Zertifikate für Lehrende und Hochschulen. Kompetenzen und Aktivitäten sichtbar machen. In: Gabellini C, Gallner S, Imboden F, Kuurstra M, Tremp P (eds) Lehrentwicklung by openness—open educational resources im Hochschulkontext. Pädagogische Hochschule Luzern, Luzern pp 29–32. https://doi.org/10.5281/zenodo.5004445
31. Schön S, Braun C, Hohla K, Mütze A, Ebner M (2022) The ReDesign Canvas as a tool for the didactic-methodological redesign of courses and a case study. In: Bastiaens T (ed) Proceedings of EdMedia + Innovate learning. Association for the Advancement of Computing in Education (AACE), New York City. https://www.learntechlib.org/primary/p/221410/
32. Shah D (2019) Online degrees slowdown: a review of MOOC stats and trends in 2019, MOOCReport, 17 Dec 2019. https://www.classcentral.com/report/moocs-stats-and-trends-2019/
33. State University of New York (2021). Microcredentials at SUNY. https://system.suny.edu/academic-affairs/microcredentials/
34. TU Graz (2020). Richtlinie zu offenen Bildungsressourcen an der Technischen Universität Graz (OER-Policy), https://www.tugraz.at/fileadmin/user_upload/tugrazExternal/02bfe6da-df31-4c20-9e9f-819251ecfd4b/2020_2021/Stk_5/RL_OER_Policy_24112020.pdf
35. TU Graz and BMBWF—Technische Universität Graz und Bundesministerium für Bildung, Wissenschaft und Forschung (2018). Leistungsvereinbarung 2019–2021. Graz: TU Graz. https://www.tugraz.at/fileadmin/user_upload/tugrazInternal/TU_Graz/Universitaet/TU_Graz_kompakt/Leistungsvereinbarung_2019-2021.pdf

36. TU Graz (2022) Life long learning. https://www.tugraz.at/en/studying-and-teaching/degree-and-certificate-programmes/continuing-education/life-long-learning/
37. UNESCO (2019) Recommendation on Open Educational Resources (OER), 25 November 2019. http://portal.unesco.org/en/ev.php-URL_ID=49556&URL_DO=DO_TOPIC&URL_SECTION=201.html. Ebner M, Lorenz A, Lackner E, Kopp M, Kumar S, Schön S, Wittke A (2016) How OER enhance MOOCs—a perspective from German-speaking Europe. In: Jemni M, Kinshuk, Khribi MK (eds) Open education: from OERs to MOOCs. Springer. Lecture Notes in Educational Technology, pp 205–220

Computer-Based Methods for Adaptive Teaching and Learning

Ulrike Pado, Anselm Knebusch, and Konstanze Mehmedovski

Abstract In this chapter, we illuminate how the learning process in the initial phase of studies can be individualized through digital enrichment. To do this, we first summarize the results of a study on the concept of Computer Based Learning in Classroom (CBL). CBL, as defined in [5, 7], is a Blended Learning concept which addresses the heterogeneity of prior knowledge by allowing students to work independently in a highly structured digital learning environment under face-to-face guidance. Furthermore, we show how the enrichment of CBL with E-Assessments affects learning success. Finally, we delve into our current project of further individualizing the CBL concept. We offer the students adaptive tests that select test questions based on the students' ability level and investigate the usage and acceptance behavior of students regarding this technological innovation. In sum, we show that CBL is a flexible, digitally enriched didactic concept that allows universities to better serve the needs of their students.

1 Introduction

At the core of the digitally mature university are digital teaching methods. These allow teachers to adaptively and flexibly react to the learning needs of students and to changes in the learning environment (as for example through lockdown phases during the Covid pandemic). This is especially relevant for incoming students who may have heterogeneous previous knowledge due to different access paths to higher education.

U. Pado (✉) · A. Knebusch · K. Mehmedovski
HFT Stuttgart, Schellingstr. 24, 70174 Stuttgart, Germany
e-mail: Ulrike.Pado@hft-stuttgart.de

A. Knebusch
e-mail: Anselm.Knebusch@hft-stuttgart.de

K. Mehmedovski
e-mail: Konstanze.Mehmedovski@hft-stuttgart.de

Therefore, an effective didactic lecture concept should, on one hand, take into account the prior knowledge of individual students and, on the other hand, provide a framework for individual learning paces. Computer-based blended learning approaches especially lend themselves to the didactic design of such a learning environment, as this concept appears to particularly offer the possibility of individual guidance in the learning process [4, pp. 96–99].

In this chapter, we want to present the CBL approach developed at HFT Stuttgart, give insights into our study about the enrichment with E-Assessments, and discuss considerations on how the approach can be further developed through adaptive testing methods.

Adaptive testing is a well-established concept in psychometric testing [15]. Its goal is to individually select test questions (items) such that the test-taker's ability can be determined both more efficiently and accurately than with a standard test that contains the same number of items for all difficulty levels. For the CBL case, the possible reduction of test length (by up to 40%, [16]) is one attractive property of adaptive testing; its promise to avoid extended exposure of the test-taker to overly easy or hard items is another. We wish to present students with items that are roughly appropriate and not boring or de-motivatingly hard.

Additionally, we aim to examine student usage behavior for the adaptive tests and their acceptance of this methods in an initial test run.

We believe that a continuous exploration of didactic concepts like CBL supports the transformation of our universities towards digital maturity by flexibly serving our students' learning needs.

2 The Didactic Concept: Computer-Based Learning

The concept of computer-based learning in the classroom (CBL [7]) has been developed at HFT Stuttgart since 2016, to address the heterogeneity in the mathematical background knowledge of Engineering students at the beginning of their studies. However, it has also been successfully used in computer science courses and at more advanced stages of study, demonstrating the flexibility of CBL [5].

CBL further advances Blended Learning: With CBL, students work independently within face-to-face sessions while utilizing a highly structured learning environment. This may seem paradoxical at first glance (students could work on the digital content at home), but it is a very important part of the concept. On the one hand, learning in person ensures that students actually engage with the digital content (and not just plan to do so); on the other hand, they have a direct contact person who can help with difficulties, thereby preventing the student from abandoning the learning phase because of frustration. We aim to establish a solid foundation in Mathematics through active engagement with the content. Given the observed heterogeneity of students' prior knowledge, there is a need for a high degree of individualization.

Computer-Based Methods for Adaptive Teaching and Learning 299

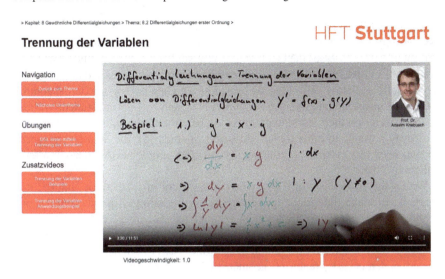

Fig. 1 Screenshot: Learning nugget in digital learning environment

This is realized in CBL by offering thematic resources in the form of digital *learning nuggets*, consisting of short instructional videos (approximately 10 minutes) and corresponding digital exercises (cf. the video screenshot in Fig. 1).

A learning segment (typically one to two lectures) begins with an (ideally practical) keynote lecture by the lecturer, introducing a problem statement. Students then transition into a self-learning phase, during which the lecturer is available for questions or discussion. At the conclusion of the learning unit, the problem is solved collectively in a plenary session. Thus, the lecturer transitions from being an information provider to a learning facilitator. CBL activates students and fosters a productive learning atmosphere by allowing student-lecturer interaction exactly when needed and at an individual level. This permits beginning learners to fill in any existing knowledge gaps as needed, while advanced learners can skip optional content and progress more quickly.

As part of the educational offerings in the introductory phase of studies, CBL provides a learning opportunity for students to pave the way for successful academic performance in the main study phase. This is because, in addition to acquiring subject-specific competencies, CBL enables and promotes students' self-regulated learning (e.g. Pintrich [14]). Based on the understanding that learning can be seen as a constructive process, it is assumed that both subject-specific and interdisciplinary competencies are acquired independently, and in a context of action [11]. We assume that in an appropriate didactic setting, self-regulated learning can also be learned independently and with reference to action.

To promote students' ability to study [12], students in CBL are consciously given freedoms within the self-learning phases. For example, students can decide for themselves whether they prefer to study alone, in pairs, or in groups, whether they acquire their knowledge using written or audiovisual media, or whether they work on all tasks or only on a predefined selection to reinforce their knowledge. The didactic concept of CBL addresses the fact that first-year students start their studies with different (subject-specific and interdisciplinary) learning experiences and thus also with varying self-regulatory skills [13], by providing a certain degree of external control, in terms of instructional guidance or support [11], and by offering students various feedback forms on their learning progress. As mentioned above, the CBL context is also easily portable to other topics and stages of study [5]; for example, advanced students need less external control and instead often appreciate the freedom offered to them by CBL.

Within the framework of CBL, where the variety of methods is limited, it is beneficial to make the videos interesting by incorporating different contexts and technologies. For instance, topics such as "number sequences" could be enriched with animations depicting bacterial multiplication, or "geometric series" could be enhanced with reflections on double-pane windows. Additionally, videos can be created as screencasts, featuring geometric animations. Alternatively, technology such as the lightboard can add visual impact Fig. 2.

Additionally, we would like to briefly mention that the materials created for CBL also serve as useful supplementary material for regular lectures. Since the content of Engineering Mathematics is largely similar across the various programs at HFT Stuttgart, we were able to supplement the materials where necessary. This has allowed us to develop a comprehensive digital teaching resource for the Mathematics 1 and 2 lectures of the Engineering programs through digital offerings. This was particularly

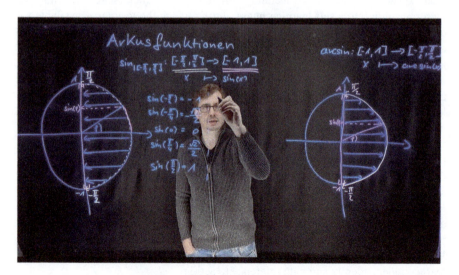

Fig. 2 Screenshot: Video generated using the lightboard

valuable during the "Covid semesters" (see [6]) and added to HFT Stuttgart's organizational resilience in the face of the pandemic. The teaching resource also ensures high quality assurance in regular semesters, for example for lectures conducted by adjunct lecturers.

In (CBL [7]), we examined the effects of CBL. For the study we divided a cohort of 103 students into two groups, with one group of 55 students undergoing a lecture in CBL format, while 48 students in the parallel group served as the control group. As a baseline, we used the orientation test of HFT Stuttgart, which is based on the *cosh* minimum requirements catalog created by a collaborative effort of schools and universities in the German state of Baden-Württemberg [1]. The students' learning progress was assessed in three paper-and-pencil tests during the semester.

We observed a significant deviation upwards in the CBL group (effect size for Test 1 and 2 was small, Test 3 was moderate). By the end-of-semester module examination (LN test) and following examinations, we could no longer detect a significant difference (see Fig. 3). We concluded that CBL promotes continuous learning throughout the semester, while the traditional format tends to encourage students towards "exam cramming". In order to evaluate the sustainability of the acquired competencies, the students participated in a re-test (repetition of the final exam of the previous semester)

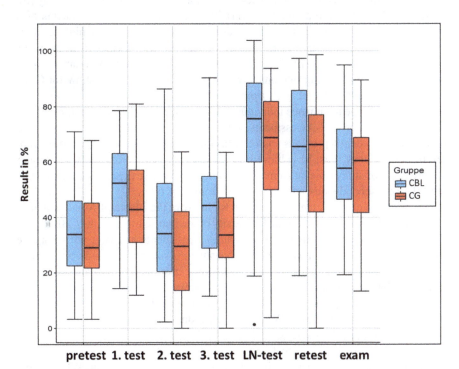

Fig. 3 Results of CBL-group (CBL) and control group (CG), see (CBL [7]), pretest: Evaluation before Semester, test 1–3: Paper pencil tests during semester, LN-test: Final exam of 1. semester (ungraded), retest: Test at beginning of 2. semester, exam: final exam of 2. semester (graded)

and we evaluated the results in the final exam of the second semester. But here we could not find a statistically significant difference in the results of the CBL and the control group. So, in this study, we could not prove that the continuous learning leads to more sustainable learning gains than "exam cramming" (in the 2nd semester there is no statistical evidence in favor of one of the groups). Anyway, we see softer indicators that the concept CBL leads to better results than in the control group (CBL [7]).

3 Goal

As outlined earlier, initial investigations into the effectiveness of the CBL concept in the introductory phase of Engineering Mathematics show positive results regarding students' subject competence, considering different performance levels of students [7]. Due to the high proportion of self-study, CBL also challenges and fosters students' self-responsibility and autonomy concerning their learning progress, addressing the overall student capability, which encompasses both subject-specific and interdisciplinary competencies necessary for a successful academic journey [3].

We now investigate the effects of enriching lectures with E-Assessments as a formative assessment, which we consider as a key element to establish high accountability in the CBL learning process. Enriching lectures with formative assessments has been well researched in [2]. There are indicators that formative assessments have positive effects on learning outcomes, which we now also wish to show for CBL. This investigation will be reported from Sect. 5 on.

Further, we hope to enhance individualization and learning outcomes through the integration of adaptive E-Assessments, where questions are algorithmically selected based on the student's previous answers. Instead of a predefined score, a minimum level of proficiency must be achieved. In doing so, we aim specifically for greater effectiveness of study time and a more motivating subjective experience for students in CBL. This second study will be reported from Sect. 6 on.

Our research question therefore is: Can we improve the adaptability and individualization of learning needed for the University of the Future by using the CBL lecture concept and enhancing it first with formative assessments and later with adaptive formative assessments in a way that benefits students subjectively and objectively?

We aim to determine whether these measures can positively influence students' learning behaviors and performance. In order to achieve insights into learning behaviors and student acceptance, we investigate, in addition to test results, how students utilize and perceive the tests. We also discuss data privacy concerns at the data collection and processing stages.

From our analyses, we seek to derive success factors and best practices for integrating (adaptive) tests and exercises into the CBL self-learning concept. What factors influence the effectiveness of these tests? How can they be most effectively integrated into the course curriculum to create individualized learning opportunities beyond the one-size-fits-all lecture and tutorial?

4 Data Privacy Considerations

Data privacy rules like the European General Data Protection Regulation (GDPR) protect data that refer to individuals identifiable through the data (e.g., through names or matriculation numbers). Data use for research has to abide by these regulations.

When collecting data in a teaching context to better understand student progress in class, no data privacy agreement with students will be necessary beyond any agreement they already made for using the university's learning management system (LMS). In this case, data collection and use are covered by the university's mission of teaching. However, in our studies, not all researchers were teachers in the participating cohort and provisions therefore had to be made to share data with the researchers in an appropriate way.

Out of consideration for possible privacy concerns of our students, we decided to collect the research data on a platform separate from the university's LMS. This allowed us to assign pseudonymized user names to the students, ensuring that the usage data available to the researchers contained no personally identifiable information. Usage data that was not originally pseudonymized was made available to the researchers in anonymized and aggregated form, so again no personally identifiable information (like students' names, university account names or matriculation numbers) was passed on.

Additionally, the students were provided with the option to grant explicit voluntary consent for the use of survey data collected under their real names and the integration of the collected usage data with their results from placement tests and end-of-semester examinations into individual learning paths, two types of data that are (potentially) traceable to an individual and therefore protected. This arrangement allowed the collection of this type of data to be conducted in compliance with the GDPR.

Students were offered the informed consent forms in a brief session where one of the researchers introduced the study, explained its goals and asked students for their help with data collection. Students were able to ask questions and were informed of the voluntary nature of participation and the option to opt out. We believe that these short information sessions helped ensure the observed high consent rate, as students understood the goals of the study and had met one of the researchers in person. We look into this effect more closely in Sect. 6.2.

5 E-Assessments in CBL

While the implementation of CBL has achieved initial successes, we concluded from our previous study (CBL [7]) that students were not yet engaging with the lecture content intensively enough on average. For example, in individual conversations, students stated that they wanted a higher level of commitment so that they could stay on track even in phases of lower motivation. This aligns with the tutors' opinion that the weekly exercise sheets were perceived more as optional offerings, and some students attended the exercise sessions without preparation, which in turn was demotivating for the better-prepared students. Also, a thorough review of the content can be a crucial factor for academic success in a self-learning concept like CBL, for instance (Ledermüller and Fallmann [9]).

This suggests the introduction of weekly formative assessments as a pre-examination requirement. We chose E-Assessments for the weekly tests due to the digitized nature of CBL and the availability of E-Assessments independently of time and place, which enables the students to also take the assessments as part of their study time at home.

The combination of 'moderate external guidance' and regular E-Assessments, has the goal of providing inexperienced students with sufficient orientation and support to achieve the desired learning goals more independently while also gaining experience with freedoms in the learning process and without jeopardizing learning success due to lacking self-regulatory skills. At the same time, the concept provides experienced learners with enough opportunities to learn in a self-directed or self-regulated manner and to achieve successful (subject-specific and interdisciplinary) competency acquisition.

5.1 Approach

Our investigation focuses on the participants of the lectures Mathematics 1 (and Mathematics 2) in the Bachelor's degree programs Civil Engineering and Industrial Engineering. The lectures each have a scope of 4 contact hours per week and are supplemented by a tutorial with a scope of 2 contact hours. The corresponding modules are weighted with 6 credit points (CP), while the tutorial, including associated exercises, contributes 1 CP and serves as a prerequisite for examination.

To ensure a sufficient number of participants, we include students from regular lectures in the study on E-Assessments and adaptivity, in addition to groups being taught with CBL.

We used the Learning Management System Moodle[1] to administer the E-Assessments. For the creation of the tasks, which aim to assess a comprehensive range of required competencies, the Moodle plugin Stack[2] is used. A database with approximately 600 randomized tasks has been gradually created. Figure 4 shows a sample question. Students successfully complete the test if they achieve 75% of the possible points. The tests can be repeated by students as many times as they wish within the allotted time, with a high variability of tasks intended to prevent the recognition of the test's solution through repeated attempts.

To address the research questions regarding the effectiveness of the implemented measures in the first study to E-Assessments, both quantitative and qualitative data

---

[1] www.moodle.org.

[2] https://moodle.org/plugins/qtype_stack.

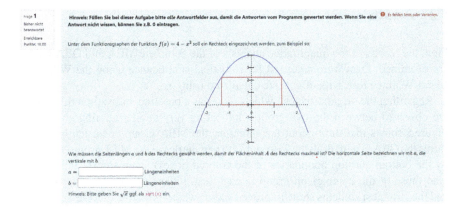

Fig. 4 Stack generated question in Moodle

are utilized. Due to the limited informativeness of data directly obtained from E-Assessments, stemming from permissible test interruptions, test retakes, and potential unauthorized aid with tests, data from in-person tests and surveys are used in the present study. Surveys were conducted among students in the foundational Mathematics lectures of the Civil Engineering and Industrial Engineering programs. Quantitative data were generated from the orientation test and three written tests (paper and pencil) conducted during the semester. The orientation test, conducted at the beginning of the semester, aims to assess the starting knowledge level of students. Participation in the three written tests was mandatory for all students in the examined courses, although a specific passing score was not required.

The analysis incorporates data from freshmen students across three winter semesters, beginning in the winter of 2017/2018 and ending with the last pre-Covid winter semester of 2019/2020. During the Covid pandemic, no data were collected due to the exceptional character of the teaching and learning situation.

The winter semester of 2017/2018, which precedes the introduction of E-Assessments, serves as the control group (without E-Assessment, OEA), comprising data from 103 students. The E-Assessment group (MEA) consists of data from the winter semesters of 2018/2019 (92 students) and 2019/2020 (110 students). Both the orientation test and the semester-long tests remained unchanged in content during the data collection period, ensuring the comparability of results. Each test measures the percentage of total points achieved.

5.2 Results

First, the results of the quantitative analysis of the complete MEA and OEA groups are examined. Data were analyzed for statistical significance using the Wilcoxon-Mann-Whitney test, with effect size calculated using η^2.

Regarding the orientation test, which serves as a baseline, only slight differences are observed between the two groups, indicating very similar starting conditions. Figure 5 shows that throughout the semester, the MEA group consistently outperforms the OEA group. Both the 25th and 75th percentiles of the MEA group (indicated by the bottom and top boundaries of the colored boxes) score significantly higher than those of the comparison group in each test, indicating that both the weakest and the strongest students show improved scores when formative E-Assessments are administered.

While the median (shown by the line in the colored boxes) is nearly identical between groups in the first test, in tests two and three, the median of the MEA group is notably higher than that of the OEA group. This is also reflected in the statistical data. A statistically significant difference is observed in test two ($p = 0.0099$), with a small effect size detected ($\eta^2 = 0.0225$). Similarly, in test three, the MEA group significantly outperforms the comparison group ($p = 3.248e - 06$), with a medium effect size observed ($\eta^2 = 0.0779$).

We turn to examining individual semesters. In Fig. 6, which shows the data separately for the three individual semester groups, differences within the MEA group are apparent, but the difference from the control group remains significant

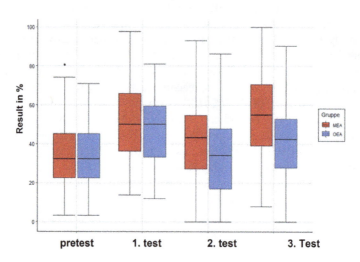

Fig. 5 Introducing E-Assessments: Cumulative results for the OEA (no E-Assessment) and MEA (with E-Assessment) groups

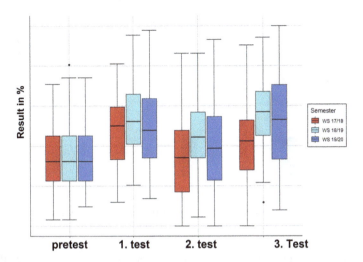

Fig. 6 Introducing E-Assessments: Results for individual semesters (2017/2018: OEA, no E-Assessment; 2018/2019 and 2019/2022, MEA: with E-Assessment)

in both cases. In comparison to the winter semester of 2017/2018 (without E-Assessments), the winter semester of 2018/2019 demonstrates significant improvements in both tests two ($p = 0.0030$) and three ($p = 6.688e - 07$), with a small effect size observed in test two ($\eta^2 = 0.0441$) and a large effect size in test three ($\eta^2 = 0.1444$). When comparing the winter semester of 2017/2018 (the OEA group) to the winter semester of 2019/2020, a significant upward deviation is again observed in test three ($p = 0.0009$), with small ($\eta^2 = 0.0150$) and medium ($\eta^2 = 0.0606$) effect sizes detected in tests two and three, respectively. Analyzing the two winter semesters comprising the MEA group, no significant difference or effect is observed in any test, although the range of results in the winter semester of 2019/2020 for the semester-long tests is slightly higher. Thus, the introduction of E-Assessments appears to be responsible for the improvements in the results of the two later winter semesters.

It seems unsurprising that mandatory E-Assessments lead to greater learning gains, but the results also clearly indicate, in our opinion, that the E-Assessments are only sporadically circumvented through computer algebra systems or similar means, even though the E-Assessments can be repeated and taken at home. Another important point is that stricter learning objectives do not overload the students. Only 15% of the students reported spending more than 5 h per week on their studies outside of lectures and tutorials, with 4–6 h approximately corresponding to the workload intended for the module per week.

This is important information, as stricter guidelines for learning progress can lead to student overload and there is also a risk of students neglecting other modules. Good communication with colleagues is crucial here. As part of the introduction of E-Assessments, a midterm exam was introduced in Mechanics 1 in coordination with

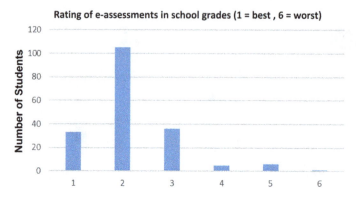

Fig. 7 Student ratings of the use of E-Assessments using (German) school grades

Mathematics 1 to avoid such effects, at least in the 'core subjects'. We see that the students' learning progress is good, and there is currently no overload of students. However, it is necessary to continuously monitor the development and stay in touch with colleagues teaching the same cohort.

Further, the students evaluated the extension of CBL with E-Assessments quite positively (Fig. 7), rating them predominantly at a grade of 2, which is the second best grade in the German system (five and six are failing grades). In our opinion, this indicates that the students wish to take responsibility for their own learning progress and see the E-Assessment as helpful for this goal.

In sum, we have been able to show a successful extension of the CBL didactic concepts through the first of our proposed measures, the addition of E-Assessments. The E-Assessments could be re-taken and taken at home, so their introduction does not reduce the flexibility of the CBL method. At the same time, students' learning outcomes profited demonstrably from the additional accountability introduced by the formative E-Assessments.

6 Adaptive Testing

We now go on to evaluating the second proposed measure, the addition of adaptivity to the formative E-Assessments. The adaptive testing approach is founded on Item Response Theory (IRT), which aims to place estimates of student ability and item difficulty on a common scale, so that each student's chance of success for a specific item can be predicted by the student's ability level in direct comparison with the item's difficulty [10]. If both levels are the same, a student has a 50% chance of succeeding at solving the item. Using known item difficulties, the student's ability is now estimated with growing accuracy from their past performance during the adaptive test and new items are chosen such that the estimate can be further improved. This approach to testing has several important implications for our use case.

The first implication is that item difficulties of course have to be known beforehand, so appropriate questions can be selected. Conventionally, item difficulties are established by norming, that is by collecting a large number of test-taker answers and deriving question difficulty from a test theoretical model [8].

Enriching the existing question bank with questions of varying difficulty levels and conducting empirical normalization was not feasible for us in advance. Therefore, in implementing adaptivity, we used two variants. Firstly, we actually selected different tasks perceived by the instructor as easier or more challenging. Secondly, for other tasks, we maintained the basic level of task difficulty but provided more or less assistance in the form of work instructions, hints, partial steps, and interim solutions, in line with "scaffolding." After the conclusion of our experiment, the second variant appears to us to be more suitable for many tasks. The difficulty level is more objectively assessable, likely leading to better results in a subsequent normalization process. Additionally, the workload for difficulty level creation is significantly reduced, which presents a crucial advantage given the resource-intensive nature of creating adaptive tests.

The second implication of using adaptivity is that the outcome of an adaptive test has to be interpreted differently from the outcome of a conventional test in which credit points are collected to reach a minimum passing threshold (e.g., 50% of all possible credits). For example, a student placed at ability level 3 is guaranteed to fail some items at this level, so the passing level has to be carefully defined.

Finally, from the point of view of the student, strategies acquired for conventional tests are often impossible or misleading. For example, it is not possible to revisit a hard question or to estimate one's ability from the amount of correct answers given.

6.1 Approach

The adaptive tests were realized using the Adaptive Quiz plugin[3] for Moodle. Item difficulties were induced on the basis of the concepts and strategies that students need to apply to solve each item, and the passing score was fixed above the score covering the content-based minimum requirements to ensure students had a higher than 50% chance to solve items presenting the minimum requirements.

We chose to extend the formative E-Assessments with adaptivity. Of course, it would also be possible to choose pre-test exercises adaptively in order to select appropriate training materials first. This is another promising avenue that we plan to explore in future work.

When taking the adaptive test using the Moodle plugin, there are two main differences to the conventional tests: (1) students cannot skip questions (since past performance is used to select the next question) and (2) they do not get feedback on which questions were correct and which were wrong.

---

[3] https://moodle.org/plugins/mod_adaptivequiz.

We offered the adaptive tests to a group of students in Mathematics, Civil Engineering and industrial Engineering over the course of an introductory Mathematics lecture that follows the CBL principles laid out above. Two out of the eleven E-Assessments in the course were offered in both adaptive and conventional format. Students were free to choose between the adaptive and traditional test format, and tests could be repeated within a certain time frame, as before.

Data was collected on three dimensions of observation: For insights into the **usage patterns** of the adaptive test options, we monitored the user trajectories over both tests within the context of our study. Among the recorded data there are access time and frequency, number of attempts, attempt completion status, and scores.

In order to more clearly understand the students' motivation for using the tests in the patterns we observe, we also conduct **qualitative investigations**, including two surveys and five in-person interviews, to gain insights into the student perspective on adaptive tests.

Finally, we investigate links of adaptive test usage to **academic performance** by connecting the usage of and performance in adaptive quizzes to the outcomes from (1) a placement test at the beginning of the semester and (2) the module exam at the end of the semester.

6.2 Results

We begin by presenting our observations on study participation and usage of adaptive and conventional tests, along with explanations for the observed patterns derived from our surveys and interviews. We also link up the results from adaptive and conventional tests with students' results from a placement test taken before starting the class. We organize our observations by the questions that are being considered, and roughly along the study progress, using qualitative and quantitative data as appropriate for each question in focus.

6.2.1 Student Motivation for Participation

Our study was conducted in a cohort of 140 students taught in four parallel classes that use the same materials and tests. Of these, 71 students agreed to linking up their pre-test and exam results with their test usage patterns (the data was processed by the resarchers in anonymized form). This translates to almost all of the students who were present on the days the project was introduced and agreement sheets were passed out. Conversely, we were not approached for additonal agreement sheets by students who had not been present on the day, although information about the study was available to them, as well.

This is an indication to us that students are open to trying new approaches and motivated to support research efforts into teaching methods if they have been personally approached and fully informed about the process. Acquiring informed student consent in our study was not just a step to protect students' GDPR rights, but also seems to have provided additional motivation for the students to participate in the study in the first place.

6.2.2 Student Motivation for Choosing Adaptivity

When the first adaptive test was offered, we observed three different user groups: 56 participants utilized only the adaptive test, 66 students only took the conventional test and 18 test takers used both versions.

This means that slightly more than half of the students chose the adaptive test offering. To better understand students' motivation for using the adaptive test specifically, we presented a survey after the first adaptive test option which was taken by 33 students, 21 of whom had chosen the adaptive version of the test. Students overwhelmingly explained that they used the adaptive test out of curiosity about the method (16 answers) or interest in innovative approaches to testing (16 answers; several answers could be selected). Seven students used the adaptive tests for extra practice and two used them in the hopes that they might be easier than the conventional test; one wanted to honor the initiative of the researchers (echoing our impressions from the GDPR data collection step).

Focusing on the 18 students who tried both the conventional and the adaptive tests, we observed several patterns: About half the students (eight) failed the conventional test at least once and then tried the adaptive version; no student who passed the conventional test at first try also tried the adaptive version. This indicates that some students hoped for an easier experience with the adaptive test (all these students at length succeeded, taking up to four attempts in total).

Conversely, six students passed the adaptive test at first try and then also took the conventional test (although they knew this setup). Possibly, they wanted to make sure that their passing score would be counted or they experimented with test difficulty. The adaptive test was not automatically easier for students: Four test-takers failed the adaptive test at first attempt and eventually switched to the conventional one (which they passed, again after up to four attempts in total). We discuss student behavior after success and failure in both test settings in more detail in Sect. 6.2.3 below.

Observed student behavior corroborates the impression that our student cohort is open to participation in the study and happy to experiment with the adaptive test format even though the familiar conventional test is always also available. This is both reassuring regarding our study, and also is a strong motivation in general for us as teachers to further develop our lecture concepts to include new technologies. Indeed, in our interviews, four out of the five interviewed students indicate that they agreed to participating in the study in general because they wanted to support the further development of teaching methods.

6.2.3 Test Taking Behavior

We now look at students' test-taking behavior, focusing on those who took several tests (conventional or adaptive). Students tried additional attempts predominantly if they did not succeed at the first try. Figure 8 shows adaptive test attempts and re-attempts only (recall that we know from Sect. 6.2.2 that students also switch modality for further attempts). All students who failed their first attempt tried again, and the vast majority remained in the adaptive modality (hardly any students stop the adaptive attempts after initial failure). Most students indeed succeeded at their second, or rarely third or more, attempt; those who appear to stop after their second attempt probably switch to the conventional mode instead.

However, we also see students who try to improve on a passing grade, going on to taking up to two more adaptive tests after passing at first try. The majority of these second attempts is never submitted for grading, though, and remains incomplete (grey bar in Fig. 8). This is consistent with requesting more practice instances or testing whether the adaptive test shows different questions each time.

Interestingly, we find this behavior only with the adaptive tests. Figure 9 shows repeated testing behavior within the the conventional tests. Here, again, the majority of test-takers tries again after a failed attempt (and the ones that appear to stop probably move to their first attempt in the adaptive paradigm). However, in the conventional mode, hardly any successful test-takers try additional attempts in the conventional mode, either stopping altogether or experimenting with the adaptive tests. Invalid tests were closed without answers after a maximum test duration.

This suggests that students do use the adaptive tests, as intended, to get more practice, while they do not use the conventional tests in this way.

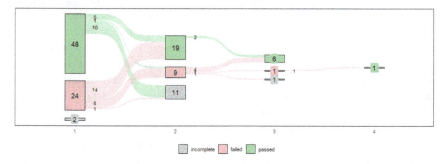

Fig. 8 Repeating the adaptive test after successful or failed attempts: Contoured bars represent successful (green) or failed (red) attempts (gray incomplete attempts were never submitted for grading). Data from the first test

Computer-Based Methods for Adaptive Teaching and Learning 313

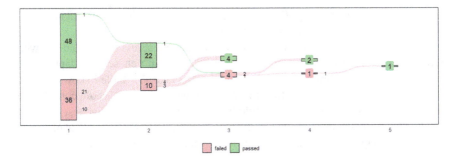

Fig. 9 Repeating the conventional test after successful or failed attempts: Contoured bars represent successful (green) or failed (red) attempts (invalid attempts timed out without answers were removed from data). Data from the first test

6.2.4 Placement Test Results and Test Choice

In order to further characterize the student groups who opt for the adaptive tests, we link up the test choice in the first adaptive trial to students' performance in the pre-class placement test.

We saw that roughly half the students in the class opted for adaptive testing (or tried both test types) in the first trial and that students repeated the adaptive test more often after initial success than the conventional test. It is possible to hypothesize that strong students (who do not have to worry about passing the formative test) might take the perceived risk of trying the new test type more often and that their ambition might show in the repeated successful adaptive tests. Alternatively, weaker students might flock to the adaptive test hoping to be presented with easier questions and avoid frustration. We therefore present aggregate data on students' placement test results, both for the whole cohort as well as focusing on adaptive test-takers (wherever we had permission to link these data).

We aggregated the placement scores for the whole cohort into three groups in Table 1: Group A is made up of students who scored at least 75% of the total achievable points for the placement exam, Group B has students who scored at least 50% (but below 75%), and Group C has students who scored below 50% of points. The test-takers are roughly equally distributed over the three ability groups, with a small

Table 1 All Students (N = 128) versus known Adaptive Test Takers (N = 71): Results of the placement test (aggregated by achieved percent of the total score)

	A	B	C
Achieved score %	≥75%	≥50%	< 50%
All students	31%	34%	35%
Adaptive test takers	30%	36%	34%

over-representation of Groups B and C (The placement test is informational in nature, so students of all result groups were able to take the class.).

We compare these overall numbers to the distribution across the 71 students who gave us permission to trace their learning paths and who chose the adaptive test format at the first opportunity ("known adaptive test takers"). Table 1 shows that the ability groups are distributed almost in the same way as for the whole cohort. The hypothesis that stronger students would be more open to the adaptive test does not seem to bear out—in this case, we would have expected a larger percentage of strong students (Group A) in our adaptive test-taker group than in the overall cohort.

6.2.5 Student Feedback: Hurdles

So far, we have discussed the first adaptive test trial. A second adaptive test option was offered three weeks later. Despite the large number of participants in the first trial, only 25 participants returned to the adaptive test for their first attempt, all others used the conventional test.

We can infer the reasons from our post-test survey after the first adaptive trial and two separate post-test surveys after the second trial, one for students who had taken the adaptive test and one for students who hadn't.

After the first adaptive trial, students' reasons to prefer the conventional test were mostly technical: Of the 12 participating students who did not take the first adaptive test, five gave as a reason that they had to log into the study LMS server with a special (pseudonymized) user name and password. Two more students said they had passed using the conventional test and did not want to do two tests; two students worried the test might be harder than the conventional one (or deceptively easy and thus not a good preparation for the final exam), and two said they wanted feedback on wrong answers. One did not want to experiment. Notably, of the 21 students in the survey who had used the adaptive test, another nine also commented on the lack of feedback on wrong answers.

After administrating the adaptive test for the second time, most of the students who chose the conventional test again gave as their reason that they did not receive feedback on wrong answers in the adaptive test, while they wanted to use this information to specifically target their own weaknesses. In the survey for users of the adaptive test, we had 10 responses from students. Their only suggested improvement was, again, to include feedback on wrong answers or a possible question review to learn from one's mistakes.

In sum, there were two hurdles for students to use the adaptive test: First of all, having to log onto a separate platform was inconvenient, and second, the adaptive test plugin only gives feedback on the ability level, but does not offer correct/incorrect feedback and a sample solution for incorrect answers. Predominantly, the lack of this latter feature seems to drive students away from the adaptive test, even though the continued users are very positive about the test format. We are preparing a follow-up study with post-test feedback to further investigate students' use of adaptive tests.

Computer-Based Methods for Adaptive Teaching and Learning 315

Even as it stands, this student feedback is very encouraging in that students clearly use the tests in CBL not just to fulfill the course requirements, but also to guide their own learning through feedback on wrong answers, as intended by the CBL method.

6.2.6 Student Feedback: Test Appropriateness and Usefulness

We now look at the post-test survey by the 10 students who took the adaptive tests both times they were offered. Figure 10 shows that these students were predominantly positive about the test (answering tend to agree or strongly agree): They felt motivated by the adaptive tests and thought that engaging with the tasks was helpful for them. This supports our hypothesis that choosing test tasks of appropriate difficulty should motivate students.

The task selection appears to have been appropriate as students felt that they could successfully apply all their knowledge and skills (and were therefore appropriately challenged). Despite the difference in test score interpretation between a percentage of the total points in the conventional tests and an ability level in the adaptive tests, all students but one reported that their test results aligned with their expectations.

Students were enthusiastic about the tests: 80% strongly agree that they would recommend the adaptive tests to their fellow students. 60% would always prefer an adaptive test option (and only 10% would always prefer a conventional test, Fig. 11).

In sum, those students who were undeterred by the technical hurdles strongly appreciate the adaptive test format and feel that it delivers appropriate questions. Future work will further explore this point once the technical issues are removed.

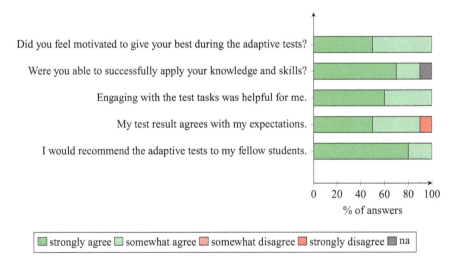

Fig. 10 Student feedback on adaptive tests. $N = 10$

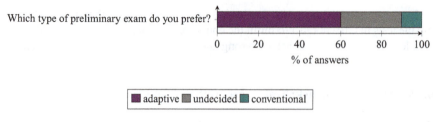

Fig. 11 Student feedback on adaptive tests, continued. $N = 10$

7 Outcomes, Conclusions and Recommendations

CBL is a proven teaching method that allows the University of the Future to flexibly adapt to a heterogeneous student body and a range of teaching settings (e.g., in-presence, hybrid or on-line) while putting students in the center of the learning process and enabling them to take their learning journey into their own hands.

CBL should be enriched with mandatory E-Assessments to encourage students to work continuously throughout the semester and achieve a higher learning outcome. We find that students wish to use the tests in CBL not just as course requirements, but also to guide their own learning through feedback on wrong answers.

However, required completion of E-Assessments provides time (but not place) constraints for the students, so that the freedom to determine their own learning progress is reduced. Furthermore, when using such methods, the impact of the workload on other modules must be considered. If the requirements are too time-consuming, students may neglect other modules. Good communication with colleagues is recommended here.

Making the assessments adaptive holds a lot of promise. Students are open to experimenting with adaptive E-Assessments and were satisfied with them. Using adaptivity also for exercise selection is an avenue for further work.

We find that students are open to new teaching methods that promise to help them reach their learning goals – we therefore strongly recommend that teachers should experiment with these measures and deliver them to students. Students, however, are easily discouraged from new methods if there are technical hurdles or the methods have perceived drawbacks compared to the established methods.

Finally, we find that addressing data privacy concerns does not stand in the way of teaching method development, but instead helps to motivate students to participate in our studies as our research goals become clear. We therefore recommend to approach study design from the start with data privacy concerns in mind, to be able to clearly communicate study goals and data needs to the participants.

Competing Interests The authors have no conflicts of interest to declare that are relevant to the content of this chapter.

Funding Ulrike Pado and Konstanze Mehmedovski acknowledge funding by Bundesministerium für Bildung und Forschung (BMBF), grant 16DHBKI072 (project KNIGHT).

Ethics Approval This study describes no experiments with humans. Students were observed during their regular completion of a semester-long course as part of their studies, with close attention to data privacy.

References

1. cosh Mindestanforderungskatalog Mathematik, Version 3.0 (2023). https://cosh-mathe.de/wp-content/uploads/2021/12/makV3.0.pdf
2. Black P, Wiliam D (1998) Assessment and classroom learning. Assess. Educ.: Princ. Policy Pract. 5(1):7–74. https://doi.org/10.1080/0969595980050102
3. Bosse E, Schultes K, Trautwein C (2013) Studierfähigkeit: Theoretischer Rahmen. Kolleg-Bote, Hamburg University, pp 83–101
4. Garrison DR, Kanuka H (2004) Blended learning—uncovering its transformative potential in higher education. Internet High. Educ. 2:95–105
5. Knebusch, A., Pado, U.: Working at your own pace: computer-based learning for computational linguistics. In: Proceedings of the 19th Conference on Natural Language Processing (KONVENS), Workshop on Teaching for NLP (2023)
6. Knebusch, A., Pado, U., Heintz-Cuscianna, B.: Online-Übungsphasen in den MINT-Vorlesungen an der HFT Stuttgart—Mehr Chancen als Hindernisse aus Studierendensicht. e-teaching.org (2020). https://www.e-teaching.org/etresources/pdf/erfahrungsbericht_2020_knebusch_pado_heintz-cuscianna_mint_hft-stuttgart.pdf
7. Knebusch A, Pfeiffer A, Wandler M (2019) Individualisiertes Lernen mit Computer begleitetem Lernen (Individualized learning with computer-assisted learning). Zeitschrift für Hochschulentwicklung 14:153–170
8. Lane, S., Raymond, M.R., Haladyna, T.M.: Handbook of Test Development, Routledge (2015)
9. Ledermüller, K., Fallmann, I.: Predicting learning success in online learning environments: self-regulated learning, prior knowledge and repetition. Zeitschrift für Hochschulentwicklung, pp. 79–99 (2017)
10. Lord FM (1980) Applications of Item Response Theory to Practical Testing Problems. Lawrence Erlbaum Associates, Hillsdale, NJ
11. Mandl, H., Gruber, H., Renkl, A.: Situiertes Lernen in multimedialen Lernumgebungen. Issing, Ludwig J. (Hrsg.): Information und Lernen mit Multimedia und Internet. Lehrbuch für Studium und Praxis. München: Verl. Internat. Psychoanalyse 138–148 (2002)
12. Merkt, M., Fredrich, H.: Studierfähigkeit - der Blick aus dem Magdeburger Schwesterprojekt: Studierfähigkeit in Weiterbildungsstudiengängen. van den Berk, K. Petersen, K. Schultes and K. Stolz (Hrsg.), Studierfähigkeit - theoretische Erkenntnisse, empirische Befunde und praktische Perspektiven. Universitätskolleg- Schriften, pp. 171–189 (2017)
13. Pfost, M., Neuenhaus, N., Kunter, P., Becker, S., Goppert, S., Werner, A.: Selbstständiges Lernen an der Hochschule - Diskussion eines computergestützen niedrigschwelligen Förderansatzes. Die Hochschullehre, pp. 83–101 (2020)
14. Pintrich, P.R.: The role of goal orientation in self-regulated learning. In: Boekaerts, M., Zeidner, P.M. (eds.), Handbook of Self-regulation (2000). https://doi.org/10.1016/B978-012109890-2/50043-3
15. Segall DO (2005) Computerized adaptive testing. In: Kempf-Leonard K (ed) Encyclopedia of Social Measurement. Elsevier, New York, pp 429–438
16. Veldkamp BP (2005) Optimal test construction. In: Kempf-Leonard K (ed) Encyclopedia of Social Measurement. Elsevier, New York, pp 933–941

Teaching Research Skills at the University. Does Digital Transformation Make a Difference?

María Isabel Pozzo

Abstract This paper aims to address the contributions brought up by digital transformation as regards teaching research skills at the university. In order to achieve this general goal, the following specific objectives are proposed: (1) to identify operational models of research skills based on a theoretical review, (2) to envision digital transformation in the field of teaching research skills at the university through a literature review, and (3) to analyze research skills teaching activities in light of digital transformation. These objectives are dealt with through secondary documents of different kinds: journal articles and Conference Proceedings, as well as a wider variety of sources, in search of research activities in different formats. After proposing two umbrella models for teaching research skills -here named "consecutive" and "progressive"-, a sequence of different tool for teaching research skills is presented in a chronological order, to envision digital transformation. Finally, different activities for teaching research skills are analyzed, concluding that digital transformation makes a difference as regards potentials for a more autonomous performance in doing research.

Keywords Research skills · University · Teaching practices · Digital transformation

1 Context and Goals

Since modernity, scientific research has become the best reputed mechanism for knowledge construction. As a result, numerous and diverse findings have changed the way we understand both nature and the social world. The relationship between

---

M. I. Pozzo (✉)
National Scientific & Technical Research Council, Rosario, Argentina
e-mail: pozzo@irice-conicet.gov.ar

National University of Rosario, Rosario, Argentina

National Technological University, Rosario, Argentina

science and technology has been a key factor in this process. Universities have played a major role in promoting research activities, both as the institutional framework and as the setting where researchers are raised. For doing so, research methodology courses are present in almost every university curriculum, no matter the knowledge area or educational level (undergraduate, graduate, and postgraduate). These courses are structured around established contents related to the research process. However, research training implies more than acquiring just declarative knowledge. Instead, it requires different skills that are useful for professional development even beyond academia.

On the other hand, digital transformation is changing our whole lives and, therefore, teaching and learning at the university. Both the concept of 'digital transformation' as well as 'advanced technologies', present in the title of this book, have different meanings along the technological advancements. Thus, they progressively comprise digitalization, internet, and artificial intelligence, just to mention the main milestones. In order to pose an ample definition, digital transformation is understood in this chapter as any technology-supported environment for university teaching. It is up to teaching staff to explore digital tools to expand the plethora of resources and provide the richest learning opportunities to their students.

In the described context, this work aims to address the contributions brought up by digital transformation as regards teaching research skills at the university. In order to achieve this general goal, the following specific objectives are proposed: (1) to identify operational models of research skills based on a theoretical review, (2) to envision digital transformation in the field of teaching research skills at the university through a literature review, (3) to analyze research skills teaching activities in light of digital transformation.

These objectives will be dealt with through secondary documents of different kinds. For specific objectives 1 and 2, scientific papers on the topic of the chapter will be reviewed: journal articles and Conference Proceedings. For specific objective 3, a wider variety of sources will be consulted in search of research activities that can shed light on the topics dealt with along the chapter.

As regards the formal organization of this chapter, each specific objective is displayed in a separate section:

Specific objective 1, in 'Theoretical framework'.
Specific objective 2, in 'Literature review'.
Specific objective 3, in 'Approach' and 'Observational outcomes'.

In a book section devoted to digital transformation, this chapter explores the contribution of advanced technologies to teaching research skills at the university.

2 Theoretical Framework: Identifying Operational Models of Research Skills

'Research skills' can be basically approached in two ways. On one hand, they can be considered as a diverse set of abilities related to the different tasks implied in each stage of a research project. There are several examples of this approach. An interesting one is a study by Stokking et al. [1], which explores the way upper secondary school teachers in the natural and social sciences assess the research skills of their students. Although this paper deals with assessment and not teaching, as it is the purpose of this chapter, this study provides the following list of teachers' goals, systematized from previous literature, which is useful for building students research assignments:

1. develop their knowledge about the concepts and content of the subject;
2. motivate them to work actively on the subject and facilitate exercise in independent learning;
3. develop their knowledge about the essentials of research, the language and argumentation and the research cycle: the commutation between theory, research. question, design, data and conclusions;
4. let them gain experience with the research steps or component skills;
5. introduce them to the tool box of the researcher;
6. introduce them to the 'ethos' of doing research;
7. develop the insight that knowledge is developed by people and is continuously being developed further.

According to the way each goal is presented, this list is proposed from the teacher's perspective and it aims to be translated into teaching activities. From the students' point of view, this paper maps the national Duch examination requirements concerning research skills for different subjects in secondary education, as they are blind to differences between knowledge areas. They are represented in ten consecutive steps:

1. identify and formulate a problem using subject-specific concepts;
2. formulate the research question(s), hypotheses and expectations (if any);
3. make and monitor the research plan: research design and time schedule;
4. gather and select information/data;
5. assess the value and utility of the data;
6. analyze the data;
7. draw conclusions;
8. evaluate the research;
9. develop and substantiate a personal point of view;
10. report (describe) and present (communicate) the research. (p. 99).

As Stokking et al. [1] state, this proposal has been a matter of criticism, as well. In that sense, Hodson [2] argued that science is holistic and can only be taught, learned, and assessed as such. Another drawback pointed out by the author is that the steps are not fully differentiated, and -unlike it is stated in the Dutch requirements-they are dependent on both the subject matter and the context in which the research is carried out. With other arguments, Kirschner [3] criticized the list for limiting research to practical execution and advocated the interdependency of the conceptual and procedural knowledge. Although these statements are appropriate, the list sound reasonable and indeed reflects the usual instrument for research/researcher assessment. In the scope of this work, this list -and similar ones- of consecutive steps will be called 'the consecutive model'.

Another way to reflect on research skills is through their progressive development, arranged in a system. In this sense, a team of Australian lecturers have proposed the "Researcher Skill Development Framework"-RSDF-[4], firstly web published in 2006. It consists in a double entry table: the horizontal axis presents the degree of students' autonomy when researching. Specifically, the increasing level of autonomy ascribes researching to the following progressive categories: prescribed (highly structured directions from educator), bounded (limited directions from educator), scaffolded, open-ended (the educator guides research initiated by students), and unbounded (students determined guidelines for researching are in accord with discipline or context). On the other side, in the vertical axis, actions performed by students and their behavioral characteristics are expressed. Each 'facet' (as named by the authors) and its correlated attitude occur when the students:

- initiate research and clarify what knowledge is required (curious)
- find and generate needed information using appropriate methodology (determined).
- determine the credibility of sources, information, and data and make own research processes visible (discerning).
- organize information & data to reveal patterns/themes, managing teams and processes (harmonizing).
- analyze information/data critically and synthesize new knowledge to produce coherent individual/team understandings (creative).
- discuss, listen, write, respond to feedback, and perform the processes, understandings, and applications of the research, heeding the needs of audiences (constructive) [4].

At the same time, each facet and level of autonomy is influenced by cross-sectional awareness of ethical, cultural, social, and team aspects. In the overall proposal, each component of this conceptual framework is carefully described to be a useful tool for educators from primary school to PhD level. It provides a systematization of the incremental development of the skills associated with researching, but also with problem solving, critical thinking and clinical reasoning.

The facets of the vertical axis may have some resemblance to the previous model, but they are truly formulated as skills, and not as activities. The degree of students' autonomy when researching expressed in the horizontal axis the most innovative feature of this proposal. Whereas the former model is consecutive in time, the latter is consecutive along professional growth, 'progressive' should be stated. Therefore, in the scope of this chapter, the RSD framework will be called 'the progressive model'.

A later theoretical instrument that can be included in this model is VITAE Researcher Development Framework, created by the Careers Research and Advisory Centre (CRAC) in 2010, a not-for-profit registered charity devoted to career development in the United Kingdom. It has a much more complex structure, represented in a circular diagram, encompassing four domains, twelve sub-domains, and sixty-three descriptors. The latter contains between three to five phases, which are the stages of development or levels of performance within each descriptor. Due to its structural complexity, no further analysis will be made here about it and the RSDF will be taken as the paradigm of the progressive model.

Among the conceptual models of research skills previously revised, the "Researcher Skill Development Framework"-RSDF-[4] shows the greatest reflection and the most sophisticated system. It really functions as a model rather than a mere list. In spite of the difficulty to fit every student of any subject, educational level, and cultural background in a standardized grid, the RSDF fosters awareness about every slight improvement in the long path of becoming a researcher. Although its categories may seem more related to learning than teaching, this matrix is a valuable resource, as stated by Gyuris [5], to design meaningful learning experiences for research skill training as well as tasks and assessments that are appropriate for expected levels of learners' development.

In the following section, research skills will be examined in combination with teaching in a context of digital transformation.

3 Literature Review: Teaching Research Skills in a Digital Environment

Teaching research at formal education in relation to digital transformation has been studied in different ways. A very ambitious project carried out in the United States in 2012 [6] presents the results of exploring middle and high school teachers' views of the ways today's digital environment is shaping the research and writing habits of their students. To do so, the research team implemented an online survey to more than 2,000 middle and high school teachers in selected communities. It was complemented with a series of online and offline focus groups with also middle and high teachers and some of their students. From this approach, the authors draw very interesting conclusions that, although focused on teenagers, shed lights also for university level nowadays. From their findings, they distinguish positive and negative effects of today's digital environment on students' research habits and skills according to the teachers who participated in this study. Among the positive impacts -which overpass the negative ones- they see that: the internet enables students to access a wider range of resources than would otherwise be available, as well as educational material in engaging multimedia formats, becoming more self-reliant researchers. As negative impacts, some teachers worry about increasing distractions and poor time management skills; potentially diminished critical thinking capacity; the ease with which today's students can borrow from the work of others; the difficulty judging the quality of online information; and students' overdependence on search engines to conduct research, discouraging the use of other resources such as online databases, the sites of respected organizations, printed books, or reference libraries. At the same time, a high number of teachers surveyed agree that internet search engines have conditioned students to expect to be able to find information quickly and easily, and -at the same time- the amount of information available online today is overwhelming to most students. Internet has also changed the very meaning of 'research'. In that sense, teachers and students alike report that for today's students, 'research' means 'Googling'. As a result, some teachers report that for their students, 'doing research' has shifted from a slow process of intellectual discovery to a fast-paced, short-term exercise aimed at locating just enough information to complete an assignment.

Moving on to another current digital tool, Williams [7] focuses on the advantages of using wikis to carry out project-based learning. The study was published in 2014, with classrooms characterized by digital devices such as laptops, phones and tablets becoming at those times as ubiquitous as the whiteboard. These digital devices allow students to review the lectures, take quizzes via a learning management system, and uploaded essays for instructors to assess and provide feedback for students' written work using the assessment tools available within the system. The author concluded a decade ago: "Slowly but surely much of the learning experience is increasing its digital footprint, with new teaching and learning systems and procedures being developed and tried out week by week." (p. 1).

Ten years later, that assumption has become true. These digital resources have not disappeared but are coexisting with new ones. Hence, the description about the use of wikis for project-based learning -one of the methods for teaching research skills [8]—is still useful nowadays. In this context, the author outlines five uses of wikis:

- information display and dissemination: all kinds of files can be uploaded/embedded onto a wiki page, and subsequently downloaded.
- collaboration: new pages can be created for students to collaborate.
- progress: lecturers can constantly monitor project progress simply by checking the page of each respective group of students.
- feedback: wikis provide the lecturer with various feedback methods that can assist students in completing the project more effectively.
- assessment: a lecturer can track updates made by each student to a single page and thus cam monitor who is actually doing the work for a more accurate assessments based on a student' contribution.

These features make wikis very useful for research skills development in university level, considering that writing a research project is a recursive process.

Other studies somehow refer to research training in a transformative context in different ways. For example, Omonovich et al. [9] refers to an innovative teaching approach for developing research skills in students. However, innovation is not associated with digital transformation but with extracurricular research activities built on the principles of free choice, reciprocity, and psychological comfort. Vaganova et al.

[10] do examine the role of technologies in teaching research, but their objective and data collected focus in students' motivation. Webster and Kenney [11] propose research activities as a complement for university training and not as the main goal, or what they call "research enhanced learning". However, this paper provides interesting web-based activities that will be retrieved later in this chapter.

In a study that directly relates research skills and technological skills, Castro Sandoval et al. [12] design an auto-perceived research skill teachers' survey, composed of the following comprehensive items: (1) information search, (2) technological mastery, (3) methodological mastery, (4) oral and written communication of research outcomes, and (5) ability to work in research teams. Within the technological dimension (2), the following domains were assessed: office tools (Excel, PowerPoint, Internet), bibliographic reference managers, qualitative data analysis software, quantitative data analysis software, computerized statistical packages for the analysis of information in general, and specialized databases for research. The study aims to design a plan for teachers' training from this pretest survey acknowledging the contribution of information and communication technology (ICT) within research skills. Although very insightful, this study takes technologies as a content and not as medium for teaching, as it is the aim of this chapter.

Specifically in the scope of blended learning, the study of Sáiz Manzanares et al. [13] explores the most efficient conditions of blended learning with Health Sciences students. It was found that the B-Learning environment in which the students obtained better learning results and a higher degree of satisfaction was the one that included the use of infographics and virtual laboratories based on self-regulated learning. The first ones are graphs that display information through words, phrases, images, videos, and interactive resources. Unlike the virtual laboratories, the infographics is considered a low-cost resource that facilitate dynamic conceptual understanding of the subject matter among students. This is a very interesting conclusion, although not tested specifically with research skills.

Continuing with a chronological revision, the last step up to nowadays in the ongoing digital transformation is situated at the end of 2022, with the public debut of Chat GPT (Chat Generative Pre-trained Transformer), a Generative Artificial Intelligent -GenAI- platform [14]. Its versatility to a broad range of uses made a great impact in different spheres of the public life. As regards the topic of this chapter -research skills- it mostly affected writing instruction, which relates to RSDF facet of organizing information and to Stoking et al. [1] tenth item: "report (describe) and present (communicate) the research" (p. 99).

Fontanelle-Tereshchuk [15] compares ChatGPT with editing tools such as "Grammarly", used as a typing assistant. Whereas both can be used to enhance our writing to support our research arguments, editing tools are less likely to completely write a paper on behalf of a student. This makes ChatGPT much more suitable to foster students' cheating and fall into plagiarism. This has been a major concern among educators, who are discussing policies and regulations to address the detrimental misuse of technology. Once again, as it was previously discussed about the negative impacts of internet [6], technological advancements bring up concern about their benefits and drawbacks. While these discussions are taking place in academic conferences, educational institutions, commercial and consultants' sites, some proposals are made for a beneficial and honest used of ChatGPT [16] applicable to teach research skills at the university, which will be part of the observational outcomes in this chapter.

To sum up, by the systematization of certain studies selected in a literature review, digital transformation can be envisioned in the field of teaching research skills at the university. As it was said in the introduction, the concept of advanced technologies undergoes fast changes along the time. For example, students' overdependence on search engines and online encyclopedias reported by Purcel et al. [6] in 2012 may be broaden by the challenges posed by artificial intelligent tools in 2024. In any case, the question is: if digital tools (such as internet, search engines, online encyclopedias, or AI tools like Chat GPT) are significantly impacting how students conduct research: what research assignments have been proposed for teaching research skills at the university? The following section will set an observational approach to answer this question.

4 Approach

As it was anticipated in the introduction, this work aims to address the contributions brought up by digital transformation as regards teaching research skills at the university. After fulfilling specific objectives 1 and 2, this section moves on to (3) *to analyze research skills teaching activities in light of digital transformation*. This objective will be addressed through observing a wide variety of sources in search of activities aimed to teach research skills at the university that can shed light on the main goal of the chapter. Some operational concepts are previously defined to carry out the analysis.

Research skills comprise declarative (conceptual) and procedural knowledge that involve both research methodology contents as well as research writing. Therefore, research skills are not restricted to research concepts. Although declarative knowledge is a fundamental component to teach research, a procedural dimension is essential in order to assure that their respective skills are in the way to be developed. Procedures are put into practice through action: a planned action for educational purposes is referred as 'activity'. A more complete activity, including planning and assessment, becomes a 'teaching practice'.

As stated by Omonovich et al. [9], the scope of the traditional lesson is limited so innovative actions are needed. In such a context, Aripin et al. [8] conduct a meta-analysis of academic papers published in national and international proceedings and journals to identify methods in the science learning of research skills. The systematization yields the following research skills methods for higher education: Inquiry-based learning, project-based learning, game-based learning, and experience study.

An activity has a more limited scope than a teaching method. In that sense, Vaganova et al. [10] adopt a definition of research activity as a special type of intellectual and creative activity implemented on the basis of search activity and the analysis of the results achieved, forecasting and reflection. Research activities perform several functions: educational, stimulating, and developmental. According to the form of research activity organization, these authors distinguish:

- individual (articles, term papers, graduation projects, final qualification works),
- group (competitions, olympiads, quizzes, trainings, discussions, games), and.
- mass research works (Conferences, symposia, and forums).

An overview of these categories indicates that individual activities proposed by these authors can also be accomplished in groups. In spite of the relative delimitation of these categories, the group activities they enumerate definitely add some innovative alternatives.

Whether in individual, group, or mass research works, getting acquainted with third-party production is a fundamental activity to become familiar with scientific outcomes. In that sense, Allison et al. [17] states that "one of the most important ways of learning about research is to read reports of research projects which have been completed and these are to be found mainly in libraries" (p. 1). This recommendation is then completed with indexes of thesis which -it should be added- nowadays are available online. It is indisputable that digitalization has been very beneficial to teaching research skills by allowing a greater access to academic bibliography.

Together with reading, writing is also essential in developing research skills. Dissertation/thesis writing has given place to several studies as regards the benefits of virtualization [18]. Indeed, digital transformation has expanded the horizons of teaching practices, bringing up additional educational resources. Considering that a dissertation/thesis is a research project, it can shed light to research skills development.

As previously anticipated, in the following section, some research skills teaching activities from different sources are presented: secondary documents (class reports published in academic journals), research handbooks, institutional websites, and academic books, selected by their potential to illustrate the digital transformation process within the limits of a book chapter. The exploration is carried out in two double-fold dimensions: (1) the relationship between activities and research skills, based on the conceptual frameworks previously described, and (2) the digital/analog nature of each teaching proposal, in order to address contributions brought up by digital transformation as regards teaching research skills at the university. The analysis is made in order to answer: *Does digital transformation make a difference in teaching research skills at the university?*

5 Observational Outcomes: Teaching Research Skills at the University, from Theory to Practice

5.1 Teaching Research Skills at the University Through Printed Handbook's Activities

The handbook titled "Research skills for students" [17] was first published in 1996 and was successively reprinted until 2016. Although some chapter previews are available online, this book is an example of the typical research printed handbook, that firstly appeared during an analog culture and 'resisted' in such a form during the digital transformation. As such type of educational resource, it contains the usual conceptual contents, plus activities at the end of each unit. A selection of them is transcribed below:

Activity 1.2

a. Write out three meaningful questions, one which involves a measure such as temperature, one which involves a relationship between two factors and one which involves possible causes of an event.
b. Ask two of your fellow students to read the questions and tell you what kind of answers they would expect. Did what they say agree with what you intended? (p. 5).

Activity 1.3

a. Write out three non-meaningful questions related to your own subject.
b. Ask two of your fellow students to read the questions and tell you what kind of answers they would expect. Did what they say agree with what you intended? (p. 6).

Activity 2.2

Look back at the examples of research questions which were given on pages 4–5. Sketch out the process by which you might follow steps a-e above (p. 11).

Activities 1.2 and 1.3 follow the same pattern: they consist in two-step activity (a and b), in which the first one is a written activity and the second, an oral one. The written activity (a) is not a writing assignment, but a reflection activity supported by the epistemic value of writing. The second part of the activity (b) consists in an oral peer-review, although expressed as "ask your fellow students". It is structured in three movements: students write—peers read and make comments—students compare what they meant and the feedback.

These activities rely on concepts previously presented in the book section, in a balance between declarative and procedural contents, as it was previously advocated.

Finally, in activities 1.2 and 1.3, there is an implicit aim to foster students' autonomy starting from the first stage: "prescribed researching", characterized by highly structured prompts. In this sense, activity 1.2a is more structured than 1.32.a:

it requires asking three questions, of a specific type each one, whereas 1.3.a. asks just for three questions of any type.

Activity 2.2 is also posed in two steps, although gathered in only one statement. The first part asks to recapitulate and reflect on research questions, not the ones provided by the students in activities 1.2 and 1.3, but ones provided by the book. This means there is no follow back activity on the questions elaborated by the readers. This would have provided an opportunity for self-assessment tending to the progressive horizontal axis of the RSDF. The research skills development in this activity seems to be expected just in its second part. Basically, it requests revising some book previous contents ("Steps a-e"), which refer to the general stages in the scientific method. In this book [17] they are enumerated as follows:

(a) A felt difficulty or the recognition of a problem; (b) location and definition of the difficulty or problem; (c) suggesting solutions to the problem. These normally take the form of hypothesis; (d) deductive reasoning out of consequences of suggested solutions or hypothesis; (e) testing the hypothesis by action (p. 11).

From the perspective of research skills development, these steps are another way to formulate the 'consecutive model' developed in the Theoretical framework, aligned with the procedural assignment "Sketch out the process…" (p. 11).

As it can be seen from these examples, the proposed activities are not technology-supported. There is no mention whether the writing activities should be written in a word processor file or on paper. The difference between the digital and the analog way is the ease for the editing process provided by the former, which is fundamental for the recursive nature of writing. At the same time, the peer review process benefits when it is done through a wiki, as described in Williams [7], but no wiki is mentioned in activity 1.2a nor 1.3a. The feedback provided by the student fellow is asked to be delivered orally probably due to the inconvenience of analog writing prevailing in the days the book was first published.

5.2 Teaching Research Skills at the University Through Blended Learning

In the journal article "Embedding research activities to enhance student learning", Webster and Kenney [11] report the implementation of class activities that rely on technology supports and mixed delivery modes to combine diverse theoretical perspectives and research methodologies. From the research skills point of view, the technology-supported environment described in this journal article allowed for online viewing of submitted research activities and provided students the opportunity to continually review, reflect and share their insights. The use of technology to create and share submission items was especially important for supporting face-to-face learning engagement and peer interaction in the authors' courses, given the large class sizes per subject per semester.

As all research tasks were submitted online to a shared web page, students took advantage of many different formats (such as audio, video, images, live sites) being used to construct knowledge in different shapes, such as collages, videos, audio, mind maps and online surveys. According to the authors, these possibilities inspire students' creativity, in the sense of fostering imagination towards different means of expressions. It is a different way to understand 'creativity' as posed in the RSDF, which states that students are creative when "they analyze information/data critically and synthesize new knowledge to produce coherent individual/team understandings." This is a more challenging assumption of creativity.

Another feature of work delivery is that students submitted their work when convenient within a weekly timeframe. Students were also encouraged to engage weekly in both informal and required peer-review tasks, describe the authors. From the research skills development perspective, asking for a regular performance both with deadline and without it avoids the negative impact of today's digital environment reported in the study by Purcel et al. [6], consisting in increasing distractions and poor time management skills.

Furthermore, open submissions for viewing by all students enables peer review. Students were encouraged to engage in both informal and required peer-review tasks as a learning method using web-based technology supports. By doing it, students could continually review submissions, reflect on their understanding and potentially gain meaning. In this way, they avoid the negative impact of digitalization reported in the study by Purcel et al. [6], consisting in potentially diminished critical thinking capacity. From the Researcher Skills Development Framework, peer review relates to the facet: "communicate & apply" to answer "how will we relate?" Students discuss, listen, write, respond to feedback and perform the processes, understandings and applications of the research.

A collaborative environment was fostered with voluntary posts made by students, which open for viewing by all students also facilitated a feeling of shared involvement. Finally, at the end of the semester, student engagement was further encouraged with a voted online award for the best research activity submitted. The technology-supported environment together with the embedded tasks provided different research methods to enhance learners' experiences within their university subject.

5.3 Teaching Research Skills at the University Through Asynchronous Activities Available in an Institutional Website

The great dissemination potential brought up by internet has made university teaching resources worldwide available. A good example in this sense is the Purdue University Online Writing Lab -OWL-, together with the Purdue University On-Campus Writing Lab. Besides the local community, "the Purdue OWL offers global support through online reference materials and services. ... It assists clients in their development as writers -no matter what their skill level-..." [19]. This excerpt from its website shows the awareness about different skills' levels, which coincides with the progressive development stated by the RSDF. Among its website tabs, there is one named "OWL exercises". The name 'exercise' clearly denotes the repetitive nature of the expected action, that differ from an activity as it was previously defined in this chapter. This site will be not further analyzed here because it is exclusively focused on writing (punctuation exercises, spelling exercises, sentence style, etc.).

An example of a research skills teaching proposal from a higher education institution allocated in its website is that by NorQuest College [20], in the city of Edmonton, Alberta, Canada. The College Library website offers "Library tutorials", with resources last updated in June 2023. The concept of "tutorial" has a closer meaning to 'procedural knowledge' than that of 'declarative knowledge'. In the same way, some Tutorials' topics are named with gerunds, thus denoting actions; for example: "Finding articles" and "Using databases".

Within the Library Tutorials menu bar there is a tab named "Research skills" whose denomination goes hand in hand with the topic of this chapter. Its contents address:

Research as conversation and communication,

Creating a research question,

Choosing relevant keywords,

Evaluating information,

Google better,

How to read an academic article,

Evidence informed research for nursing,

eBooks: Finding and Viewing,

Health Information Sources for Pathophysiology,

Grey Literature,

A Plan for Research: making sense of how and where to start.

Here, as well, there is a predominance of gerunds. However, when clicking on each item, the presented information tends more to concepts, rather than procedures. Almost all of them are audio-visual power point presentations, whereas two of them ("Research as conversation and communication" and "A Plan for research") are pdf files which include written texts, graphs, and minds maps. As regards the contents of each topic (named "Guide"), they contain some of the items of the consecutive model, but without a certain order. They do not follow the consecutive sequence of a research project, nor a progression of the researcher development. It is up to the reader to pick the needed item.

Back in the Library Tutorials menu bar, the tab "Research Skills: Practice Activities" -which is the most related to this chapter's goal- repeat almost the same topics as the "Research Skills" tab but with a different layout. First of all, the section is introduced in the following terms:

> "About the activities:
>
> These research activities are designed to help you learn important research skills. Read the associated PDF files for a skill overview and/or watch the recorded lectures. Then practice your research skills using the interactive activities."

Right by the side of the section presentation, there is one topic that -due to its localization and name- suggests being the first one in an implicit sequence: "Pre-Research: What type of Researcher are you?".

The other titles are: "Paraphrasing Skills", "Evaluating Sources", "Developing Research Questions", "Search Strategies: Keywords & Boolean Operators", and "Database Searching". No sequence is suggested here.

All titles are structured in the following items:

- Skill overview.
- Video (recorded lecture).
- Practice.

One of the topics adds a Card game.

Both the declarative knowledge (the concepts) as well as the activities are basically repeated in the different formats: in "Skill overview", as a power point presentation in a pdf file; "Practice", as an animated power point presentation: and "Video", as a recorded lecture in a YouTube video, much longer than the video "Practice".

The Skill overview of each topic always starts saying "Today we will learn about...", a phrase that resembles a synchronous class. The digital format enables accessing a class any time, from any place. Other than that, there is no exploitation of further digital possibilities.

The declarative contents (concepts) are usually at the beginning of the presentation, in order to provide the necessary knowledge to put into practice in the activities. It is worth highlighting that the activities are present along the 'class' and not just at the end, like it usually happens. For example, in the Guide for "Evaluating sources" there is an initial activity whose prompt is:

> "Activity: Evaluation Criteria.
>
> What do you look for in a Google source?".

The Guides devoted to writing skills (like "Paraphrasing"), propose an activity that was expected, which is paraphrasing a certain text that is provided in the guide. The activity is basically repeated in the different formats both in "Skill overview" as well as "Practice".

To sum up, these asynchronous research activities and resources allocated in a higher education institution website is another digital tool taken as a whole. It explicitly uses the term of 'research skills', which shows a more complex view of teaching how to do research. Although they are not defined, they implicitly show what is expected in each of them. As regards resources, they are mostly technology-supported self-made audiovisual materials. They do not include hyperlinks, so the recommendation posed by Allison et al. [17] about the importance of consulting research reports through online libraries and indexes of thesis is not exploited.

5.4 Teaching Research Skills with ChatGPT

As it was stated previously, ChatGPT has brought concern to educators for the potential plagiarism problem. In response to that concern, Dobrin [16] systematizes two kinds of contributions. On one hand, some recommendations to address the detrimental misuse of this technology in education; on the other hand, some teaching activities are suggested. In fact, they both go hand in hand, but still they can be distinguished from each other. According to the author [16], guidance for accounting GenAI in class critically and responsibly should be also provided.

(a) (a) The most important recommendations for instructors to avoid students' misuse of ChatGPT related to research skills posed by Dobrin [14] are:

1. Incorporate More On-Demand, In-Class Writing Assignments that Require Students to Engage Material with a Great Degree of Immediacy Than a Take-Home or Long-Term Assignment May Require
2. Require students to include materials that are only available in your classroom lessons, lectures or lab work, since such materials are less likely to be available in GenAI Large Language Models, if it is not repeated.
3. Require specific citations in written works. This will require that they locate the information they cite, an ability that by now is limited in GenAI.
4. Write assignments about current events, especially local ones.
5. Design assignments that focus on personal experiences with specific and narrow class objectives.
6. Focus on writing process rather than in products (pp. 18–19).

In a sense, these recommendations seem more concerned with replacing what ChatGPT can do than what students should do to learn. Still, they help educators prevent students' fraud. Besides, these recommendations have some intrinsic benefits. For example, in-class writing assignments (recommendation 1) will counteract monologic teacher-centered lectures. On the other hand, these recommendations will have an impact in the methodological approach: assignments focused on current events, especially local ones (recommendation 4) call for a more situated perspective of research, and personal experiences with specific and narrow class objectives (recommendation 5) brings out the subjectivity of the student/researcher, in line with the qualitative approach.

(b) Besides recommendations, the following suggested teaching activities from Dobrin ([16], p. 12) were selected:

Applied AI:

Choose a context and pick a topic that you find interesting. Prompt a GenAI program to produce a 500-word essay about that topic in a TED-talk style. Then, once you have the output, revise it and deliver it by making a video, again as a TED-talk. How well-suited is the GenAI essay to this purpose? Does the essay make interesting and insightful points? Is it repetitive? Is the language engaging?

For discussion:

1. Discuss as a Group How You Think You Might Use GenAI in Different Contexts (Academic, Professional, Civic, and Personal) and What You See as the Advantages and Drawbacks of Doing so.

   The activity on Applied Intelligence consists of different consecutive steps:

- The first one consists in choosing a topic and a context. The contexts enumerated by the author [16] refer to different spheres of human life. In the case of teaching research skills, that sphere is academia, and within it, the research activity. So, whereas choosing a context does not imply any real decision, because it is already decided, choosing a topic is always a challenge in any research endeavor. Besides being interesting for oneself, as the statement says, it has to be interesting also for the academic community, necessary, and innovative.
- Once the topic and context are chosen, the next step is the usual interaction between a person and this AI tool: to give a prompt, one that would be normally asked to the student. In this case, the roles are altered, and the student is the one who asks for the activity to be done, acting as an educator. According to the progressive level of autonomy proposed by the Researcher Skills Development Framework, this peculiar research activity could be categorized as "prescribed" (the lowest level of autonomy), in the sense that the prompt stipulates a certain length (500 words), a type of genre (an essay), and the style (a TED talk).
- The output is then revised, with no major detail except for the next step.
- The output should be delivered orally, in the same style as it was asked: a TED talk.
- The output is assessed as regards intrinsic quality (repetitive, interesting, insightful) and communicative purposes (suitability for the required style -a TED talk-, and interest for the audience).

As it can be observed from these steps, the first and last ones match with those of the consecutive model: to choose a research topic and, last, to communicate the research outcomes. Although in total these steps include introspective and procedural activities, the research skills corresponding to the central part of the consecutive model (designing a data collection approach, analyzing data, drawing conclusions) are missing.

As regards the activity for discussion, it is a group proposal to reflect on the way to use this AI tool, including advantages and drawbacks. In a way, it expands the last step of the previous activity (Applied IA) to a broader scope of contexts.

As expressed in both activities, a great emphasis is made on the context. In part, that is due to the fact that these activities are taken from the chapter devoted to context, but also, because adequacy is one of the greatest achievements of ChatGPT. This versatility could be exploited by asking the program to rewrite the output in different styles or audiences; for example, for amateurs and for specialists in the subject and then reflecting on the rhetoric features used in one case and the other. Although this extra activity would further the reflection on style, there would still be missing main research skills, because these activities reflect more on the instrument than in the skills themselves.

A possible way to enrich the Applied AI activity would be to ask the student to write the essay before asking the Program to do it, and then compare both productions. In this way, the students' activity would be more productive than just reflecting or performing what was produced by ChatGPT. Another way would be to rely on ChatGPT only for the communication stage of a research project as a training resource. In any way, as Dobrin [16] states, it is fundamental to make students aware about the need for a critical and responsible use of this tool. Developing research skills is important not only for the research outcomes but also as an educational mean for professional development. So, if the process matters as much as the product, the process cannot be delegated to a digital tool. At this point, it seems difficult to follow recommendation 6, which consists in focusing on writing process rather than in products.

6 Conclusions

As it was stated from the beginning of this chapter, this work aimed to address the contributions brought up by digital transformation as regards teaching research skills at the university. In order to achieve this general goal, the following specific objectives were proposed: (1) to identify operational models of research skills based on a theoretical review, (2) to envision digital transformation in the field of teaching research skills at the university through a literature review, (3) to analyze research skills teaching activities in light of digital transformation.

(1) In order to identify operational models of research skills, a theoretical review was carried out. Although the word 'model' did not appear, it was here introduced to gather the great variety of proposals based on two criteria: consecutive and progressive. Whereas the first one shows many proposals (some of which were presented in this chapter), the progressive is only represented by one complete and systematic proposal, the RSD framework.
(2) To envision digital transformation in the field of teaching research skills at the university, a reconstruction was done in an ad hoc way: a sequence of papers describing different tool for teaching research skills was presented following a chronological order of their publishing dates. This sequence starts with a printed research handbook first published in 1996 and finishes with digital books published end of 2023. Besides the publication date, the digital tool that was the focus in each commented work follows approximately the same chronological order. Presenting those tools through educational research projects, the digital transformation shows up in a concrete image. This review is not exhaustive, but rather takes some of the milestones of digital transformation. For example,

educational videos and podcasts also devoted, in one way or another, to the development of research skills, have not been discussed in this chapter.

(3) As regards the observational outcomes to analyze research skills teaching activities in light of digital transformation, the conclusions are:

(a) Research skills literature (journal articles and referenced books) as well as other educational sources (like educational websites) complements declarative with procedural knowledge, embedding activities and procedural contents to theoretical developments that are mutually necessary.
(b) Digital transformation has provided teaching practices with different and enriching alternatives to developing research skills. The distinction virtual / in-person education is not relevant when considering these varied resources, because they can be used in both.
(c) Digital transformation does not imply that one tool or activity replaces a former one. On the contrary, a coexistence of teaching possibilities is found along this process. For example, the peer review strategy in the embedding research activities to enhance student blended learning by Webster & Kenney [11] is also present in the printed handbook by Allison et al. [17].
(d) Digital transformation is not just a 'digital translation' of paper-based traditions. If exploited, it can expand enormously the teaching practices bringing up totally new possibilities for both staff and students to develop research skills. For example, peer-review activities (although with different terminology) were found even back in the printed research handbook. However, they benefit considerably when done through a wiki. This example shows that digital transformation does make a difference in teaching research skills at the university.

Since the digital transformation is still ongoing, more empirical studies and observations are needed. In particular, the last/current stage of the transformation that is taking place due to the rapid development of artificial intelligence is still posing numerous questions. Those concerns involve teaching research skill but also education in general and even all facets of life in society. However, it must be beard in mind that every technological change disrupts the current state of social practices. The chronological overview of the literature review has shown that, along with the positive impact of the Internet, educators have also reported negative effects. And yet, internet has not been discarded in education. Teaching practices have sought ways for internet to contribute to the teaching of investigative skills, despite its also existing adverse effects. A similar reflection is required with respect to artificial intelligence. In any circumstance, ethical considerations are needed to ensure the proper use of technologies.

While digital transformation continues, it is expected that the systematization done in this chapter will help practitioners to enrich teaching practices and course syllabus. However, as it can be concluded from Webster & Kenney [11] proposal, digital transformation only makes a positive difference depending on a reflecting assignment made by the lecturer or fostered by peer interaction. It is the educators' responsibility to explore and exploit the contributions brought up by digital transformation in any stage of it to teach research skills at the university under the most enriching educational settings.

References

1. Stokking K, Schaaf M, Jaspers J, Erkens G (2004) Teachers' assessment of students' research skills. Br Educ Res J (BERJ) (30)1:93–116. https://bera-journals.onlinelibrary.wiley.com/doi/; https://doi.org/10.1080/01411920310001629983
2. Hodson D (1992) Assessment of practical work. Sci Educ 1:115–144
3. Kirschner PA (1992) Epistemology, practical work and academic skills in science education. Sci Educ 1:273–299
4. Willison J, O'Regan K (2007) Commonly known, commonly not known, totally unknown: a framework for students becoming researchers. higher education research development 26(4):393–409. https://www.researchgate.net/publication/233444295_Commonly_known_commonly_not_known_totally_unknown_A_framework_for_students_becoming_researchers
5. Gyuris E (2018) Evaluating the effectiveness of postgraduate research skills training and its alignment with the research skill development framework. J Univ Teach Learn Pract (JUTLP) 15(4). https://doi.org/10.53761/1.15.4.5
6. Purcell K, Rainie L, Heaps A, Buchanan J, Friedrich L, Jacklin A, Chen C, Zickuhr K (2012) How teens do research in the digital world. a survey of advanced placement and national writing project teachers finds that teens' research habits are changing in the digital age. Pew internet project. Pew research center's internet American life project. http://pewinternet.org/Reports/2012/Student-Research
7. Williams S (2014) Using wikis to carry out project-based learning. J Interdiscip Res Educ (JIRE) 4(1):1–10
8. Aripin I, Hidayat T, Rustamanc N, Riandi R (2021) The effectiveness of science learning research skills: a meta-analysis study. Scientiae Educatia J Pendidik Sains 10(1):40–47. https://doi.org/10.24235/sc.educatia.v10i1.8486
9. Omonovich KD, Ilkhomugli SY (2023) Principles of development of research skills in students based on innovative approach science and innovation. Int Sci J 2:8. https://doi.org/10.5281/zenodo.8266084
10. Vaganova OI, Lapshova AV, Kutepov MM, Tatarnitseva SN, Vezetiu EV (2020) Technologies for organizing research activities of students at the university. Amazonia investiga 9(25):369–375
11. Webster CM, Kenney J (2011) Embedding research activities to enhance student learning. Int J Educ Manag 25(4):361–377. https://doi.org/10.1108/09513541111136649
12. Castro Sandoval JC, Silva Monsalve AM (2023) Fortalecimiento de las habilidades investigativas en docentes implementando un plan de formación apoyado en las tecnologías digitales. Páginas de Educación 16(2):20–38. https://doi.org/10.22235/pe.v16i2.3124
13. Sáiz Manzanares MC, García-Osorio CI, Díez-Pastor JF (2019) Differential efficacy of the resources used in B-learning environments. Psicothema 31(2):170–178. https://doi.org/10.7334/psicothema2018.330
14. Dobrin IS (2023) Talking about generative AI: a guide for educators. Broadview press
15. Fontanelle-Tereshchuk D (2024) Academic writing and ChatGPT: Students transitioning into college in the shadow of the COVID-19 pandemic. Discov Educ 3:6. https://doi.org/10.1007/s44217-023-00076-5
16. Dobrin IS (2023) AI and Writing. Broadview press
17. Allison B, Hilton A, O'Sullivan T, Owen A, Rothwell A (2016) Research skills for students. Routledge

18. Pozzo MI (2020) Virtual and face-to-face teaching practices for dissertation writing: current challenges and future perspectives. In Auer ME, May D (eds) Cross Reality and Data Science in Engineering. REV 2020 1016–1032. Springer Nature. https://doi.org/10.1007/978-3-030-52575-0_84
19. Purdue Online Writing Lab>OWL Exercises (2024). https://owl.purdue.edu/owl_exercises/index.html. Accessed 28th Mar 2024
20. NorQuest College Library Edmonton, Alberta, Canada (2023). https://libguides.norquest.ca/tutorials/research_activities. Accessed 28th Mar 2024

Gamification

Section Introduction

Carina Gabriela Lion

Current digital environments exhibit constant giddy pace and transformation. Debates around generative AI, adaptive learning, intelligent tutorials, and datafication raise questions about what young people we need to train in higher education, for what labor and professional world, and for what society. Confronted with these digital environments, why should we include as relevant a section on videogames in a book about the university of the future?

The section addresses, on the one hand, the value of videogames for strengthening the learning skills necessary in the coming years (autonomy, self-regulation, decision-making, hypothesizing, and experimentation, among others) and, on the other hand, gamification as an innovative didactic strategy for higher education. From a multidisciplinary perspective, it includes empirical evidence on the skills that videogames develop in students and their impact on the work world, gender perspective as a contemporary debate concerning videogames, and the development of serious videogames for teacher education, training, and professional development.

Video games have also become the most relevant options among leisure or entertainment activities. They are featured as resources in educational mediation at different levels and as an object of study for researchers in different fields who investigate their effects.

Research on video games [1, 2] has shown their value in strengthening certain cognitive skills such as critical thinking, hypothesizing, strategic decision making, problem solving, self-regulation, and autonomy and emotional skills such as self-confidence and empathy, and in the formation of critical subjects prepared for a changing work context that demands new skills. Video games also provide social interaction skills, critical reflection on stereotypes and the ability to solve complex problems from multiple and non-linear perspectives.

C. G. Lion (✉)
University of Buenos Aires, Buenos Aires, Argentina
e-mail: carinalion@gmail.com

We believe, therefore, the reading of the different articles in this section will open gateways to didactic creativity. It will incite reflection on the meaning of higher education regarding the skills of the future and will ask questions on current digital transformations and their impact on decision-making in institutional macro and micro policies. The articles in this section address, from different perspectives and research, video games as valuable cultural developments for higher education.

The chapter "Do war video games stimulate the critical thinking of university students towards war?" by Christian Rodríguez González, M. Esther Del Moral Pérez, Jonathan Castañeda Fernández, and M. Carmen Bellver Moreno examines the attitudes of university students towards war videogames, while examining aspects such as a cognitive dimension, personal attitude, ethical, argumentative, and expressive-communication dimensions. It makes a theoretical contribution to current debates and research related to war video games and critical reflection.

In "The role of gaming in HE: Strengthening distributed leadership and student commitment," Carina Lion and Verónica Perosi investigate the role of serious video games in HE, focusing on their potential to develop and distribute leadership and enhance student commitment. The research highlights the significance of technological mediation in learning and leadership within culturally and socially influenced contexts. The chapter contributes to the discussion of the use of serious games for the development of soft skills in young adults.

"Video games as an emerging consumer technology: Profiles, uses, and preferences of university students in Argentina" by María Gabriela Galli, María Cristina Kanobel, Diana Marín Suelves, and Donatella Donato discusses gender differences in playing video games with Argentine students. The study implements a quantitive approach to explain the underlying trends in student gaming habits. The findings point at video games as an emerging consumer technology for different age and gender groups.

In their contribution "Gamification within (higher) education," Christa Friedrich, Matthias Heinz, Josefin Müller, and Michelle Pippig present gamificaton as a learning-teaching adventure to support higher education. They emphasize the importance of gamification and discuss the benefits of incorporating gaming elements to expand teaching skills development in game-based learning. It considers the strengths and opportunities of an interactive learning environment and envisions didactically innovative applications of gamification elements in higher education.

References

1. Cunningham E, Green SE (2023). Cognitive Skills Acquired from Video Games. https://doi.org/10.1093/acrefore/9780190228613.013.1468
2. Koutsogiannis, D. and Adapmpa V. (2022) Video games and (language) education. Towards a critical post-videogaming perspective. https://doi.org/10.21248/l1esll.2022.22.2.366

Do War Video Games Stimulate the Critical Thinking of University Students Towards War?

Rodríguez González Christian🄳, Del Moral Pérez M. Esther🄳, Castañeda Fernández Jonathan🄳, and Bellver Moreno M. Carmen🄳

Abstract War video games have become popular digital entertainment platforms among young people. This research aims to determine the level of agreement among a sample of university students (N = 144) regarding the opportunities that these ludic artifacts offer to activate the intrinsic dimensions of critical thinking towards war. The methodology employed is quantitative, representing an empirical, non-experimental, descriptive, exploratory, and correlational study. A validated instrument ($\alpha = 0.887$) was utilized to gather their opinions as players. The results highlight that university students believe that these games contribute to stimulating the *cognitive dimension* of critical thinking towards war by providing information about relevant historical events. They also indicate that these video games stimulate the *personal-attitudinal dimension* to a lesser extent by portraying the cruelty of war. The *ethical dimension* is barely promoted, as war is considered a business and violence is glorified, while the *logical dimension* is enhanced by presenting war as necessary to achieve peace and making its causes visible. The *argumentative dimension* is also activated as these games reflect the ideological discourse inherent in wars. Finally, in the opinion of university students, these video games can encourage critical thinking towards war if they are considered more than just a game.

Keywords War · Video games · Critical thinking · University

R. G. Christian (✉) · D. M. P. M. Esther · C. F. Jonathan
Universidad de Oviedo, Oviedo, Spain
e-mail: rodriguez.christian@uniovi.es; christianrodriguez@facultadpadreosso.es

D. M. P. M. Esther
e-mail: emoral@uniovi.es

C. F. Jonathan
e-mail: castanedajonathan@uniovi.es

B. M. M. Carmen
Universidad de Valencia, Valencia, Spain
e-mail: m.carmen.bellver@uv.es

© The Author(s), under exclusive license to Springer Nature Switzerland AG 2025
E. Vendrell Vidal et al. (eds.), *Advanced Technologies and the University of the Future*,
Lecture Notes in Networks and Systems 1140,
https://doi.org/10.1007/978-3-031-71530-3_22

1 Introduction

Military video games constitute interactive narratives structured by violent episodes of current or historical wars, fictional combat, terrorist attacks, and territorial conquest threats among nations, among other elements, where the conflict between two opposing formations predominates [1]. These games replicate battles where the player's objective—often assuming the role of a soldier—is to defeat the enemy [2]. They prompt the execution of missions in a warlike scenario: eliminating enemy soldiers, bases, or vehicles [3], transporting allied equipment, defending allies, rescuing hostages, surviving war, etc. [4]. All of this occurs without questioning the ethical component, portraying a clean war devoid of brutality, where weapons and technology take the center of the stage, while victims are often rendered invisible [5].

In this context, it is worth questioning whether these video games can evoke a critical reflection about war or they merely serve as sophisticated forms of digital entertainment that anesthetize players about the real consequences of armed conflicts. This research aims to determine whether these video games promote critical thinking among users. Thus, based on Ennis' definition (1985) of *critical thinking* as a cognitive process that enables individuals to rationally explain certain facts or events, identifying their nature and ethical implications, the study seeks to analyze the opportunities that war video games provide for enhancing players' reflective capacity. Specifically, it investigates whether these games allow players to discern the nature of the war depicted and evaluate its consequences from an ethical perspective, as indicated by [6].

Some studies collect user opinions on the graphical aspects, gameplay, and social interaction of these video games [7], communication skills, strategical thinking, identity formation and leadership development acquired through gaming [8]. Imaz [9] identifies players' perceptions regarding the attitudes and values of characters in well-known war games. Another analysis express concern about the pro-violent behaviors exhibited by players [10]. Foster [11] analyzes the impact of represented violence on youth and its normalization. Tan [12] gathers players' regarding the portrayal of female characters.

Some video games assist in learning to think, solve problems and confront critical situations [13]. However, there is limited evidence directly related to *critical thinking about war*. Hence, this study examines players 'opinion on: how war is portrayed in the war video games they are familiar with and play, what socio-political issues they learn from them, what criticisms they perceive in these games, the extent to which they have contributed to changing their perception of war, etc.

2 Inherent Dimensions of Critical Thinking and War Games

These video games are considered violence artifacts, lacking sensitivity and empathy towards the victims. They immerse the players in fictional scenarios whose sole objective is to eliminate enemies [14]. While some research investigates the didactic possibilities to teach history [15], other authors, from an anthropological perspective, emphasize the dehumanizing and violent models they depict, rejecting their simplistic treatment of war and criticizing their lack of ethics [16]. Some authors criticize the transformation of the player into a killing machine, focused on thoughtless annihilation without empathy or feelings, and consider these games as platforms for indoctrination and militarization.

Other authors point out that these games can be tools that induce critical thinking about war and historical empathy [17] as they help to understand the nature of war and its ethical implications [18]. García-Moreno [2] notes that certain games from the *Metal Gear* series facilitate an understanding of the Cold War. Holdijk [15] asserts that games like *Valiant Hearts, Darkest Hour, Commandos: The Great War, Verdun*, and *Making History* can help to contrast historical facts. This disparity requires discriminating the true potential of these video games to activate critical thinking about war. Therefore, it starts with the dimensions that define it and its categories:

1. *Cognitive dimension*: This dimension is linked to the opportunity these video games provide to understand: (a) their treatment about war, including the simulation of real conflicts, historical adaptations, terrorist tactics, or science fiction scenarios; and (b) the approach they take towards war, whether it's critical, questioning the concept of war, reflective, inviting awareness and reflection, or playful.
2. *Personal-attitudinal dimension*: This dimension is related to the ability of these video games: (a) to promote empathy by recognizing others' feelings, stimulating solidarity, concern, and defense of the victims, or, conversely, inducing indifference [19], and (b) to evoke positive or negative emotions.
3. *Ethical Dimension:* This dimension is linked to the game's commitment: (a) to address ethical aspects [20], inviting players to reflect on their decisions without compromising enjoyment, allowing them to experience and react to ethical and moral problems or dilemmas arising from war (e.g., child soldiers, arms race, collateral damage, etc.),and (b) to criticize the representation of sexist stereotypes (male as soldiers and female as nurses), ethnic biases [21], or cultural prejudices.
4. *Logical dimension*: This dimension is associated with how these video games reflect the positions of opposing sides and depict the background and motivations of conflicts (economic and geopolitical interests, vengeance, enjoyment of destruction, etc.), which can either encourage players to interpret and understand current issues between nations or immerse them in mindless playful practices. Zagal [18] suggests analyzing the adopted perspective, the involved states, the scale and scope of the represented war, the relevance assigned to the war,

the identification of the military personnel involved, and the authenticity of the conflict.
5. *Argumentative dimension*: This dimension refers to the game's ability: (a) to help discern the ideological discourse and justification of the represented conflicts (whether they are depicted as necessary to maintain the status quo, inevitable for self-defense, or undesirable events); (b) the strategies used to situate the player within the events (verbalizing orders, contextualizing the history, using videos that place the player in the events, voiceovers, in-game documents, etc.); and (c) the utilization of characters to present different arguments regarding war (patriotic soldiers, surviving victims, civilians forced into war, nurses healing the wounded, indoctrinated child soldiers, etc.). The game determines the narrative and its progression, making the player feel part of the story and encouraging them to take sides in the conflict.
6. *Expressive-communicative dimension*: This dimension is related: (a) to the adopted audiovisual format of the game (photorealistic, cinematic, cartoonish, or comic aesthetics); (b) the proposed perspective (first-person, third-person, or a side view from an omniscient character); and (c) the allowed forms of expression for the player. It identifies the extent of freedom granted to the player to act (full freedom, limited to specific decisions, or limited to executing predetermined.

These dimensions serve to develop a systematic analysis of the contribution of war video games in stimulating critical thinking towards war. Thus, the aim is to gather players' opinions on the opportunities that these video games provide.

3 Methodology

The methodology adopted is quantitative, specifically an empirical non-experimental approach, as defined by [22]. It has a descriptive, exploratory based on the analysis of data obtained through a specifically designed and validated questionnaire. The objective of this investigation is to know the degree agreement of the university players about the potential of war video games to activate the critical thinking dimensions towards war.

3.1 Sample

The sampling method used is non-probabilistic. A total of 144 participants completed the online questionnaire, constituting a sample from the *Helanyah* Twitch gaming platform community, which comprises a total of 608 users. Among the participants, 78.5% are male, while 21.5% are female. In terms of age distribution, 16.0% are between 15–20 years old, 25.7% are between 21–25 years old, 29.2% are between 26–30 years old, and 29.2% are over 31 years old. Regarding education level, 53.5%

of the sample have completed Vocational Training (VT), 30.6% have a university degree or equivalent, and 16.0% have a Master or a Ph.D. In terms of gaming habits, 68.8% consider themselves regular gamers, 21.5% casual gamers, and only 9.7% identify as hardcore gamers. In terms of game preferences, 47.2% prefer FPS (First Person Shooter) games, such as the *Call of Duty* franchise; 20.8% choose Real-Time Strategy (RTS) games; 11.1% prefer TPS (Third Person Shooter) games; another 10.4% prefer Graphical Adventure Games (GAG), and an equal percentage prefer Turn-Based Strategy games (TBS).

3.2 Instrument

The Critical Thinking WVG Instrument measures the degree of agreement among players - using a Likert-type scale (1 = Strongly Disagree, 2 = Disagree, 3 = Agree, 4 = Strongly Agree) - regarding the opportunity offered by the war-themed video games they know and play to develop critical thinking about war. It consists of four items related to the classification variables of the subjects (gender, age, player profile, and preferred video games). Additionally, it includes 61 items associated with the six theoretical dimensions that define critical thinking (see Tables 1, 2, 3, 4, 5, 6, 7 and 8 included in the results section):

1. Cognitive Dimension (CD): 10 items - inferred from [19]—regarding the representation and treatment of war offered by the video games they know and play

Table 1 Descriptive Statistics: Opinion on the portrayal and approach of war offered by these video games

The war video games you know and play feature the war	Degree of agreement						
	1	2	3	4	5	X	SD
CD1. Simulating current conflicts	6.3	25.0	36.1	24.3	8.3	3.03	1.04
CD2. Adapting historical events	2.8	6.9	18.8	43.1	28.5	3.88	1.00
CD3. Identifying it with acts of terrorism	15.3	23.6	22.9	25.7	12.5	2.97	1.27
CD4. Portraying it as science fiction	4.9	10.4	25.0	34.0	25.7	3.65	1.12
CD5. Criticizing it	20.8	28.5	24.3	16.0	10.4	2.67	1.26
CD6. Inviting reflection on its consequences	22.2	21.5	18.1	21.5	16.7	2.89	1.41
CD7. Trivializing it	21.5	18.8	27.8	23.6	8.3	2.78	1.26
CD8. Highlighting socioeconomic inequalities between countries	20.8	18.8	16.7	22.9	20.8	3.04	1.45
CD9. As a technological challenge between countries	5.6	13.9	29.2	29.9	21.5	3.48	1.14
CD10. As the trigger for espionaje schemes, economic pressure, or political propaganda	7.6	9.7	27.8	31.9	22.9	3.53	1.17

Source Author's own elaboration

Table 2 Descriptive statistics: opinion on the emotions and feelings elicited by these video games

The war video games you know and play feature the war	Degree of agreement						
	1	2	3	4	5	X	SD
PAD1. Enable recognition of the feelings of the victims	17.4	29.9	16.0	21.5	15.3	2.88	1.35
PAD2. Promote solidarity with the victims	25.0	28.5	19.4	16.0	11.1	2.60	1.32
PAD3. Advocate for the defense of the victims	16.7	28.5	24.3	17.4	13.2	2.82	1.28
PAD4. Encourage hatred towards a culture/ethnicity or race	43.8	17.4	16.0	14.6	8.3	2.26	1.37
PAD5. Justify revenge between countries or peoples	29.9	22.9	18.8	18.8	9.7	2.56	1.35
PAD6. Show the cruelty of war	5.6	10.4	14.6	38.2	31.3	3.79	1.16
PAD7. Generate fear, distress, and helplessness in you	37.5	24.3	24.3	9.0	4.9	2.19	1.18
PAD8. Evoke sadness in response to certain events	13.9	18.8	25.7	22.9	18.8	3.14	1.31
PAD9. Leave you indifferent to war	34.7	18.8	20.1	17.4	9.0	2.47	1.36
PAD10. Evoke compassion and/or empathy for the victims	9.0	22.9	23.6	27.8	16.7	3.20	1.23

Source Author's own elaboration

   (simulation of real conflicts, historical adaptation, terrorist tactics, or science fiction), and the adopted approach (critical, reflective, or playful).
2. Personal-Attitudinal Dimension (PAD): 10 items—inferred from [19]—concerning the ability of these video games to promote empathy and solidarity with victims or, conversely, to encourage hatred, revenge, or indifference. It also includes items related to the feelings and emotions they evoke.
3. Ethical Dimension (ED): 10 items—adapted from [21, 23]—regarding underlying criticisms, reproduction of stereotypes, and the visibility of moral dilemmas associated with recreated wars.
4. Logical Dimension (LoD): 10 items—extracted from [24]—to discriminate the positions of each side, background, motivations of conflicts, and understanding of war.
5. Argumentative Dimension (AD): 11 items—inferred from [24]—concerning the justification of wars and the roles assigned to characters.
6. Expressive-Communicative Dimension (ECD): 10 items—inferred from [24]—about the audiovisual and technical aspects used by these video games to represent war (graphic format, gameplay perspective, and margins of freedom).

Table 3 Descriptive statistics: Opinion on the critiques, stereotypes, and moral dilemmas present in these video games

The war video games you know and play feature the war	Degree of agreement						
	1	2	3	4	5	X	SD
ED1. Criticize the presence of child soldiers in war	36.1	22.2	13.9	14.6	13.2	2.47	1.44
ED2. Portray the arms race as a business	6.9	13.2	20.8	33.3	25.7	3.58	1.20
ED3. Justify the use of violence to achieve a desired outcome	7.6	12.5	20.8	34.7	24.3	3.56	1.20
ED4. Bring visibility to the victims and collateral effects of war	16.0	23.6	22.9	23.6	13.9	2.96	1.29
ED5. Emphasize the suffering of the vulnerable	20.8	28.5	19.4	21.5	9.7	2.71	1.28
ED6. Depict female characters fighting in battle	11.8	20.8	23.6	25.7	18.1	3.17	1.28
ED7. Consider caring for the wounded as a feminine task	31.9	22.2	16.7	17.4	11.8	2.55	1.40
ED8. Represent characters of other races/ethnicities in a stereotypical manner	16.7	11.8	23.6	29.2	18.8	3.22	1.34
ED9. Denounce human rights violations	25.7	27.1	20.1	16.7	10.4	2.59	1.31
ED10. Offer solutions to minimize the consequences of war	41.0	29.9	15.3	9.0	4.9	2.07	1.17

Source Author's own elaboration

Table 4 Descriptive statistics: Opinion on the stance of these video games towards war

The war video games you know and play feature the war	Degree of agreement						
	1	2	3	4	5	X	SD
LoD 1. Depict the different positions of the opposing sides	6.9	6.9	27.1	31.9	27.1	3.65	1.15
LoD 2. Consider war essential to achieve peace	10.4	16.0	21.5	28.5	23.6	3.39	1.29
LoD 3. Defend war as a means of defense	4.9	10.4	20.1	36.8	27.8	3.72	1.12
LoD 4. Justify war as a business	13.9	9.7	27.8	27.8	20.8	3.32	1.29
LoD 5. Present war as a means of territorial control and dominance	2.8	4.2	15.3	41.0	36.8	4.05	0.97
LoD 6. Approve war as a form of revenge	7.6	9.0	21.5	31.9	29.9	3.67	1.21
LoD 7. Promote enjoyment through destruction	12.5	11.8	21.5	23.6	30.6	3.48	1.36
LoD 8. Depict the struggle for control of resources	1.4	7.6	17.4	35.4	38.2	4.01	1.00
LoD 9. Manifest superiority between countries	5.6	6.9	12.5	37.5	37.5	3.94	1.13
LoD 10. Encourage the pursuit of global hegemony	13.2	13.2	31.9	24.3	17.4	3.19	1.25

Source Author's own elaboration

Table 5 Descriptive statistics: Opinion on the justification, and argument of war in these video games

The war video games you know and play feature the war	Degree of agreement						
	1	2	3	4	5	X	SD
AD1. Consider war necessary to wield power	6.9	9.0	20.1	33.3	30.6	3.72	1.19
AD2. Present war as inevitable to maintain political relations and the international status quo	9.7	15.3	20.1	32.6	22.2	3.42	1.26
AD3. Identify war as something undesirable and illicit	14.6	35.4	26.4	16.0	7.6	2.67	1.14
AD4. Justify military interventions verbally	6.9	13.9	36.1	28.5	14.6	3.30	1.10
AD5. Contextualize history to justify war through videos, voiceovers, and in-game documents	3.5	9.7	22.9	39.6	24.3	3.72	1.05
AD6. Introduce history using videos, voiceovers, and/or in-game documents to justify war	2.8	9.7	25.0	36.8	25.7	3.73	1.04
AD7. Use characters with opposing views on war	9.7	15.3	22.9	27.8	24.3	3.42	1.28
AD8. Appeal to the patriotic sentiment of soldiers to defend their country	6.3	5.6	11.8	33.3	43.1	4.01	1.16
AD9. Tend to victimize the losing side	18.8	29.2	23.6	17.4	11.1	2.73	1.26
AD10. Glorify the winning side	2.8	6.9	11.8	31.9	46.5	4.13	1.05
AD11. Criticize the obligation of civilians to participate in war	19.4	25.7	28.5	17.4	9.0	2.71	1.22

Source Author's own elaboration

7. Perception of War Dimension (PWD): 6 items aimed at assessing how their perception of war has changed after playing these video games. Measured using a Likert-type scale (1 = Strongly Disagree, 2 = Agree, 3 = Strongly Agree).
8. Learning Dimension (LeD): 6 items that gather their opinions on what they have learned.

3.3 Procedure

The instrument was validated through an exploratory factor analysis, as the sample meets the minimum number of subjects (N = 144 > 50). The Bartlett's sphericity test yielded a significant result ($p = 0.000$), and the Kaiser-Meyer Olkin (KMO) measure of sampling adequacy showed a high value (KMO = 0.803). Following the approach of [25], the maximum likelihood method was chosen with an eigenvalue criterion > 1. It was found that with 14 items out of the 73 items included in the questionnaire, over 60% of the variance can be explained. The high goodness-of-fit of the data to the model is confirmed (chi-square = 2032.518, $p = 0.000$).

Regarding the communalities, all items have a high explanatory power of the variance, with only 5 out of the 73 items having a communality < 0.30. Therefore, this factor analysis allows us to affirm that the data obtained with the instrument,

Table 6 Descriptive statistics: Opinion about the audiovisual and technical resources of these video games

The war video games you know and play feature the war	Degree of agreement						
	1	2	3	4	5	X	SD
ECD1. Show photorealistic graphics	4.2	6.9	18.1	45.1	25.7	3.81	1.03
ECD2. Present war in a cinematic manner	2.8	1.4	18.8	41.0	36.1	4.06	0.93
ECD3. Have a cartoon or comic-like aesthetic	31.9	38.2	13.2	9.7	6.9	2.22	1.20
ECD4. Include realistic sound that enhances immersion	0.0	1.4	6.9	32.6	59.0	4.49	0.69
ECD5. Developed in the first-person perspective	7.6	5.6	19.4	31.3	36.1	3.83	1.20
ECD6. Developed in the third-person perspective	8.3	16.7	37.5	21.5	16.0	3.20	1.15
ECD7. Feature a side view to display all the details	10.4	18.1	29.9	21.5	20.1	3.23	1.26
ECD8. Allow the player to express themselves and act freely	17.4	29.9	21.5	18.1	13.2	2.80	1.29
ECD9. Empower the player to make decisions in real-time	8.3	21.5	23.6	29.2	17.4	3.26	1.22
ECD10. Control all the commands and actions of the player	14.6	25.0	24.3	22.9	13.2	2.95	1.26

Source Author's own elaboration

Table 7 Descriptive statistics: Opinion on aspects of these video games that have changed their perception of war

War video games have changed my perception of war to…	Degree of agreement				
	1	2	3	X	SD
PWD1. To become aware of the existence of different perspectives from each side	20.8	45.1	34.0	2.13	0.73
PWD2. To uncover the diverse interests that drive it	18.1	34.7	47.2	2.29	0.76
PWD3. To be conscious of its consequences on civilians	30.6	32.6	36.8	2.06	0.82
PWD4. To empathize with victims and refugees	36.1	29.9	34.0	1.98	0.84
PWD5. To acquire significant historical data	9.7	34.7	55.6	2.46	0.67
PWD6. To enable a more critical analysis	25.7	38.9	35.4	2.10	0.78

Source Author's own elaboration

Table 8 Descriptive statistics: Opinion on what these video games have taught them

War video games have taught me to...	Degree of agreement				
	1	2	3	X	SD
LeD1. To gain a better understanding of power struggles between nations	22.2	36.8	41.0	2.19	0.78
LeD2. To comprehend the arms trade as a source of income	22.2	31.3	46.5	2.24	0.80
LeD3. To be aware of territorial conflicts between countries	7.6	26.4	66.0	2.58	0.63
LeD4. To understand alliances between countries based on their historical backgrounds	12.5	36.1	51.4	2.39	0.70
LeD5. To foster greater empathy towards refugees and war victims	22.9	36.8	40.3	2.17	0.78
LeD6. To realize the consequences of acts of war	14.6	21.5	63.9	2.49	0.74

Source Author's own elaboration

structured into the eight previously defined theoretical dimensions, will be valid and reliable ($\alpha = 0.887$).

The data analysis is supported by descriptive analyses, followed by mean comparisons based on the classification variables, along with correlational analysis between the study variables. After confirming with the Kolmogorov–Smirnov test ($p < 0.001$) that the sample does not follow a normal distribution, non-parametric statistical tests were employed, as suggested by [26]. The Mann–Whitney U test was used for gender-related comparisons, and the Kruskal–Wallis test was used for age, educational level, player profile, and preferred type of war-themed video games. Differences between means are considered significant when $p < 0.05$. The SPSS-V27 software was used for the analysis.

4 Results

4.1 Cognitive Dimension (CD)

The players were asked about the portrayal of war in war-themed video games and the adopted approach (critical, reflexive, or playful) (Table 1).

The respondents agree on the portrayal of war in these video games as historical events and representations of science fiction. Regarding the approach, they indicate that these games perceive war as the trigger for espionage schemes, economic pressures, or political propaganda. The mean comparison based on the variable of educational level VT $\bar{x} = 2.65$; Bachelor's degree $\bar{x} = 2.66$; Master's or doctoral degree $\bar{x} = 3.48$) shows significant differences ($p = 0.022$), where higher-level students believe that these games trivialize war to a greater extent. Individuals who prefer shooter-type video games ($\bar{x} = 3.31$) identify war with acts of terrorism to a significantly higher degree ($p = 0.007$) compared to those who play other genres such as graphic adventures ($\bar{x} = 2.53$), RTS ($\bar{x} = 2.43$), and TBS ($\bar{x} = 2.60$). The variables of gender, age, and player profile do not yield significant differences.

4.2 Personal-Attitudinal Dimension (PAD)

The participants were asked about the emotions and feelings evoked by these war-themed representations and to what extent (Table 2).

In general, the respondents agree that these video games depict the cruelty of war, evoking sadness. While they do not explicitly promote hatred towards a culture or ethnicity, they do not allow for empathy towards the victims. There is polarization regarding the opportunity these games provide to recognize the feelings of the victims, promote solidarity, and defend them. The games do not generate fear or distress, but there are statistically significant differences based on gender (p = 0.038). Women are less indifferent to war ($\bar{x} = 2.03$) compared to men ($\bar{x} = 2.60$). On the other hand, individuals who prefer TBS games believe that these video games desensitize them more to war ($\bar{x} = 3.07$) (p = 0.037) compared to others ($\bar{x}$ GAG = 1.60 vs. $\bar{x}$ RTS = 2.30 vs. $\bar{x}$ FPS = 2.56 vs. $\bar{x}$ TPS = 2.69). Conversely, players of graphic adventures (p = 0.011) consider that this genre of video games elicits greater compassion and empathy for the victims ($\bar{x} = 3.80$) compared to the rest ($\bar{x}$ RTS = 3.07 vs. $\bar{x}$ TBS = 2.27 vs. $\bar{x}$ FPS = 3.34 vs. $\bar{x}$ TPS = 3.19). Age, level of education, and player profile do not show significant differences.

4.3 Ethical Dimension (ED)

Furthermore, it is interesting to ascertain to what extent players perceive the critiques expressed in these video games, the stereotypes and the underlying moral dilemmas (Table 3).

Video games, which are commonly played, portray the arms race as a business, justify the use of violence to achieve desired goals, depict stereotyped characters of different ethnicities, and feature female characters engaged in combat. Furthermore, unanimously, these games do not consider that they contribute to reducing the consequences of wars. However, despite women perceiving a higher degree of sexist bias in portraying caregiving as an exclusively female task (Women $\bar{x} = 2.97$ vs Men $\bar{x} = 2.45$), the mean contrast does not yield significant differences regarding the criticism, stereotypes, and moral dilemmas underlying these video games. On the other hand, in relation to their field of study (p = 0.001), undergraduate students particularly criticize the presence of child soldiers in war ($\bar{x} = 3.05$) compared to their peers (VT $\bar{x} = 2.36$ and Master/Ph.D. $\bar{x} = 1.70$). Meanwhile, Master/Ph.D. students (p = 0.036) perceive stereotypes in characters of other races/ethnicities to a greater extent ($\bar{x} = 3.87$) compared to students at other educational levels (VT $\bar{x} =$

3.03 and Undergraduate $\bar{x} = 3.20$). Gender, age, player profile, and preferred type of war game do not show significant differences.

4.4 Logical Dimension (LoD)

Similarly, their opinion on the contribution of these war video games to discriminate and understand the positions of each side, the background, the motivations of the conflicts, and the way war is perceived is collected (Table 4).

There is uniformity in the opinions of the respondents, as their responses converge in considering that these video games portray war as a formula for territorial dominance and the struggle for control of resources, emphasizing the superiority of some countries over others. War is also seen to a lesser extent as a means of defense, essential to achieve peace, showing the different positions of the opposing sides and justifying war as a means of revenge. On the contrary, players do not believe that these video games encourage the pursuit of global hegemony. Undergraduate students ($\bar{x} = 3.95$) significantly ($p = 0.044$) believe that these video games justify war as a means of defense, unlike their peers (VT $\bar{x} = 3.53$ and Master/Ph.D. $\bar{x} = 3.91$). They also believe that these games emphasize the struggle for control of resources (VT $\bar{x} = 3.87$; Undergraduate $\bar{x} = 4.34$; Master/Ph.D. $\bar{x} = 3.87$; $p = 0.022$) and reflect the superiority between countries (VT $\bar{x} = 3.82$; Undergraduate $\bar{x} = 4.25$; Master/Ph.D. $\bar{x} = 3.78$; $p = 0.043$). Gender, age, player profile, and preferred type of war game do not yield significant differences.

4.5 Argumentative Dimension (AD)

Similarly, it is relevant to know whether these games help discern the ideological discourse inherent in the represented war, its justification, and the arguments put forth to involve the characters (Table 5).

The data provides a homogeneous view, as these games appeal to the patriotic sentiment of soldiers to defend their country, glorify the winning side, and use videos, voiceovers, and/or in-game documents to contextualize war, considering it necessary and inevitable to wield power and maintain political hegemony. While there are no significant differences regarding gender, women criticize the fact that these video games do not portray war as something undesirable and illicit (Men $\bar{x} = 2.33$ vs. Women $\bar{x} = 2.74$). However, significant differences are observed based on the level of education ($p = 0.032$). Master/Ph.D. students consider these video games present war as a necessary means to wield power ($\bar{x} = 4.13$) compared to their peers (VT $\bar{x} = 3.48$ and Undergraduate $\bar{x} = 3.81$). On the other hand, university graduates indicate that the video games contextualize and justify war through videos and voiceovers to a greater extent ($\bar{x} = 4.05$) and significantly ($p = 0.029$) than their peers (VT $\bar{x}$

= 3.60 and Master/Ph.D. $\bar{x} = 3.48$). The rest of the variables do not significantly influence the results.

4.6 Expressive-Communicative Dimension (ECD)

The audiovisual format adopted by war video games, along with the gameplay perspective, expressive forms, and freedom they offer, are factors that influence the involvement of players and their emotional impact (Table 6).

Young people indicate that video games predominantly depict war in a cinematic manner, incorporating realistic sounds that enhance immersion. They promote gameplay in the first-person perspective and utilize photorealistic graphics, but they do not provide full freedom to act. It is worth noting the significant differences based on gender ($p = 0.045$); war video games played by women do not adhere to a cartoon or comic-like aesthetic ($\bar{x} = 1.83$) compared to those played by men ($\bar{x} = 2.33$). Similarly, undergraduate students ($\bar{x} = 3.57$) believe that video games promote interaction in the third-person perspective ($p = 0.031$) compared to their peers in VT and master's/doctoral programs ($\bar{x} = 3.04$ in both cases). Casual players ($p = 0.049$) state that they can express themselves and act more freely ($\bar{x} = 3.19$) compared to other players (Regular $\bar{x} = 2.77$ and Hardcore $\bar{x} = 2.14$). Similarly, respondents who prefer graphic adventures ($\bar{x} = 4.13$) significantly emphasize the realism of these games ($p = 0.002$) compared to others (RTS $\bar{x} = 3.20$; TBS $\bar{x} = 3.47$; FPS $\bar{x} = 3.82$; TPS $\bar{x} = 3.69$). Naturally, those who prefer FPS primarily play this type of video game (GAG $\bar{x} = 3.87$; RTS $\bar{x} = 2.90$; TBS $\bar{x} = 3.27$; FPS $\bar{x} = 4.49$; TPS $\bar{x} = 3.25$; $p = 0.000$). Likewise, the same applies to those who prefer third-person shooter games (TPS) (GAG $\bar{x} = 3.00$; RTS $\bar{x} = 3.47$; TBS $\bar{x} = 3.33$; FPS $\bar{x} = 2.88$; TPS $\bar{x} = 4.13$; $p = 0.001$). Young people who prefer FPS games indicate that the games they usually play do not feature a side view, preventing the display of all the details (GAG $\bar{x} = 3.53$; RTS $\bar{x} = 3.70$; TBS $\bar{x} = 3.33$; FPS $\bar{x} = 2.82$; TPS $\bar{x} = 3.69$; $p = 0.005$). The age variable does not yield significant differences.

4.7 Perception of War Dimension (PWD)

On the other hand, it was considered relevant to inquire the young participants about the changes that these video games have brought about in their perception of war (Table 7).

Players indicate that these video games have predominantly contributed to changing their perception of war by allowing them to acquire historical conflict data and uncover the interests that wars generate. It is worth noting that a significant group of both men and women, approximately 30%, do not believe that these video games activate their critical thinking, enable them to empathize with victims, or make them aware of the consequences on civilians. Regarding this matter, there are only significant differences based on the level of education, where undergraduate students affirm that these video games promote a more critical analysis of war compared to the others (VT x̄ = 2.06; Undergraduate x̄ = 2.32; Master's/Ph.D. x̄ = 1.78; p = 0.025).

4.8 Learning Dimension (LeD)

Furthermore, the university students were asked about their opinion regarding what they have learned from these video games (Table 8).

In general, young people indicate that these video games have helped them become aware of territorial conflicts between states, understand the consequences of wars, comprehend alliances between countries, grasp the business of the arms race, and gain a better understanding of power struggles. Significant differences are only observed when comparing means based on the level of education, with undergraduate students having higher means regarding the opportunity these games offer to understand the complex web of interests that motivate wars (LeD1: x̄ = 2.50, p = 0.004; LeD2: x̄ = 2.50, p = 0.030; LeD3: x̄ = 2.80, p = 0.025), especially compared to students in VT (LeD1: x̄ = 2.08; LeD2: x̄ = 2.09; LeD3: x̄ = 2.47).

5 Discussion and Conclusion

According to the respondents, war-themed video games enhance the *cognitive dimension* by discerning various perspectives through which war is presented, aligning with [27]. Additionally, concerning the *personal-attitudinal dimension*, players emphasize that the hyper-realism realism and cruelty represented in wars, foster compassion and empathy with the victims as noted by [28], contrary to the assertions of González-Vázquez and Igartua (2019). In terms of the *ethical dimension*, they also believe that it is strengthened by portraying the arms race as a business and denouncing sexist and ethnic stereotypes, in line with [29]. They also decry the presence of child soldiers as indicated by [30]. Regarding the *logical* and *argumentative dimension*, players consider that war is identified as means of territorial dominance, in line with [31]. Concerning the *expressive-communicative dimension* players highlight that these video games prioritize cinematic spectacle and special effects, as criticized by [32]. On the other hand, university students assert that these games have allowed them to acquire relevant information about historical conflicts, as indicated by [13]. Furthermore, players have delved into the economic, geopolitical and ideological interests underlying wars, aiding them in becoming more aware of their causes and consequences as concluded by [33].

Finally, mindful of the impact of these video games, teachers must take advantage using the potential of these video games to invite the students to reflect about the treatment and nature of wars. Thus the *cognitive dimension* can be enhanced through games addressing historical conflicts (*Call of Duty WWII*), comparing with those portraying war as a fictional or dystopian phenomenon (Metal Gear Solid), or those simulating terrorist attacks (*Call of Duty Modern Warfare 2*) or contemporary conflicts (*Spec Ops: The Line*). Educators may also discern the adopted approaches, whether they prioritize reflection on economic pressures between countries (*Alpha Protocol*), war propaganda (*Special Forces 2*), or the triggers of historical conflicts (*Valiant Hearts: The Great War*). It is crucial for educators to be familiar with these games to employ them for educational purposes, particularly in disciplines such as History, Economics, Politics or Sociology, to contemplate the factors determining wars. Additionally, it is suggested to contextualize military conflicts to help the youth to recognize the motivations that promotes wars and reflect about the moral dilemmas.

References

1. Jarvis L, Robinson N (2019) War, time, and military videogames: heterogeneities and critical potential. Crital Mily Stud 7(2):192–211. https://doi.org/10.1080/23337486.2019.157301
2. García-Moreno R (2017) La Historia a través de los videojuegos: La Guerra Fría en la saga Metal Gear (Trabajo Fin de Máster). Universidad de Valladolid. https://bit.ly/2W0jaKU
3. Bos D (2018) Answering the call of duty: everyday encounters with the popular geopolitics of military-themed videogames. Polit Geogr 63:54–64. https://doi.org/10.1016/j.polgeo.2018.01.001
4. Jørgensen K (2016) The positive discomfort of spec ops: the line. Game Stud 16(2). https://gamestudies.org/1602/articles/jorgensenkristine
5. Venegas A (2019) Memoria y representación de la guerra en el videojuego. Artículo en el blog Presura. https://bit.ly/3hWh5aG
6. Del Moral ME, Rodríguez C (2020) War video games: Edu-communicative platforms to develop critical thinking against war? J Comput Cult Herit 13(4):1–13. https://doi.org/10.1145/3404196
7. Robinson N (2019) Military videogames: more than a game. Rusi J 164(4):10–21. https://doi.org/10.1080/03071847.2019.1659607
8. Engerman JA (2016) Call of duty for adolescent boys: an ethnographic phenomenology of the experiences within a gaming culture. (Doctoral Thesis). The Pennsylvania State University. https://bit.ly/3ffHMp0
9. Imaz JI (2009) Videogames and education: a first empirical research in the Basque Country. In: Pivec M (ed), Proceedings of the 3rd European conference on games based learning. Academic Conferences Ltd, pp 195–201. https://www.academia.edu/25812608/Videogames_and_Educat ion_a_first_empirical_research_in_the_Basque_Country
10. Chang JH, Bushman BJ (2019) Effect of exposure to gun violence in video games on children's dangerous behavior with real guns: a randomized clinical trial. JAMA Netw Open 2(5):e194327. https://doi.org/10.1001/jamanetworkopen.2019.4327
11. Foster H (2016) How do Video Games Normalize Violence? A qualitative content analysis of popular video games (Doctoral Thesis). Northern Arizona University. https://bit.ly/2PenZf4
12. Tan V (2017) An examination of Singaporean video gamers perceptions of female video game characters. (Degree Thesis). University of Singapore. https://scholarbank.nus.edu.sg/handle/10635/134933
13. Metzger SA, Paxton RJ (2016) Gaming history: A framework for what video games teach about the past. Theory Res Soc Educ 44(4):532–564. https://doi.org/10.1080/00933104.2016.1208596

14. Cummins G (2010) Sex, violence and videogames. Commun Lawyer, 27(2):1–2. https://bit.ly/3maLk0m
15. Holdijk E (2016) Playing the Great war: getting historians involved in video games (doctoral dissertation). The University of Victoria, Canada. https://bit.ly/2tP1YMA
16. Dyer-Witheford N, De Peuter G (2009) Games of Empire: Global capitalism and video games. University of Minnesota Press
17. Boltz LO (2017) "Like hearing from them in the past": the cognitive-affective model of historical empathy in video game play. Int J Gaming Comput Mediat Simul (IJGCMS) 9(4):1–18. https://doi.org/10.4018/IJGCMS.2017100101
18. Zagal J (2017) War ethics: a framework for analyzing videogames. Proceedings of the 2017 digital games research association (DiGRA) International Conference, DIGRA. https://bit.ly/32jLMBFS
19. Engelhardt CR, Bartholow BD, Kerr GT, Bushman BJ (2011) This is your brain on violent video games: neural desensitization to violence predicts increased aggression following violent video game exposure. J Exp Soc Psychol 47(5):1033–1036. https://doi.org/10.1016/j.jesp.2011.03.027
20. Rollings A, Adams E (2003) Andrew rollings and ernest adams on game design. New Riders
21. Burgess MC, Dill KE, Stermer SP, Burgess SR, Brown BP (2011) Playing with prejudice: the prevalence and consequences of racial stereotypes in video games. Media Psychol 14(3):289–311. https://doi.org/10.1080/15213269.2011.596467
22. Cohen L, Manion L, Morrison K (2011) Research methods in education. Routledge
23. Sjoberg L, Via S (2010) Conclusion: the Interrelationships between Gender, War and Militarism. In: Enloe C (ed) Gender, war, and militarism: feminist perspective. Praeger, pp 231–237
24. Paul R, Elder L (2007) A guide for educators to critical thinking competency standards: standards, principles, performance indicators, and outcomes with a critical thinking master rubric. Rowman and Littlefield Publishers/The foundation for critical thinking
25. Lloret-Segura S, Ferreres-Traver A, Hernández-Baeza A, Tomás-Marco I (2014) El análisis factorial exploratorio de los ítems: una guía práctica, revisada y actualizada. Anales de Psicología 30(3):1151–1169. https://doi.org/10.6018/analesps.30.3.199361
26. Siegel S (1995) Estadística no paramétrica: aplicada a las ciencias de la conducta. Trillas
27. Burgess J, Jones C (2022) Exploring player understandings of historical accuracy and historical authenticity in video games. Games Cult 17(5):816–835. https://doi.org/10.1177/15554120211061853
28. Wulansari ODE, Pirker J, Kopf J, Guetl C (2020) Video games and their correlation to empathy. In: Auer M, Hortsch H, Sethakul P (eds) The Impact of the 4th industrial revolution on engineering education. ICL 2019. Advances in intelligent systems and computing, vol. 1134. Springer, pp 151–163. https://doi.org/10.1007/978-3-030-40274-7_16
29. Mou Y, Peng W (2009) Gender and racial stereotypes in popular video games. In: Ferdig R (ed), Handbook of research on effective electronic gaming in education. IGI Global, pp 922–937. https://doi.org/10.4018/978-1-59904-808-6.ch053
30. Mantello P (2017) Military shooter video games and the ontopolitics of derivative wars and arms culture. Am J Econ Sociol 76(2):483–521. https://doi.org/10.1111/ajes.12184
31. Salter MB (2011) The geographical imaginations of video games: diplomacy, civilization, America's Army and Grand Theft Auto IV. Geopolitics 16(2):359–388. https://doi.org/10.1080/14650045.2010.538875
32. González-Vázquez A (2019) La banalización de la guerra en los videojuegos bélicos. [Tesis doctoral]. Universidad de Salamanca. https://bit.ly/3N28RAp
33. Del Moral ME, Rodríguez C (2022) Oportunidades de los videojuegos bélicos para activar el pensamiento crítico: opiniones de los jugadores. Revista Colombiana de Educación 85:242–242. https://doi.org/10.17227/rce.num85-12561
34. Ennis RH (1985) A logical basis for measuring critical thinking skills. Educ LeadShip 43(2):44–48. https://jgregorymcverry.com/readings/ennis1985assessingcriticalthinking.pdf

The Role of Gaming in HE: Strengthening Distributed Leadership and Student Commitment

Carina Lion and Verónica Perosi

Abstract The cultural and social scenarios in which digital technologies, including videogames, play a part in learning-process changes mark the need to review formats and strategies currently applied in higher education (HE), as well as to plan changes that imagine and design a teaching method that is more in line with future challenges. Serious video games are used to educate, train, and coach, leveraging their ludic and motivating characteristics. The article presents the results of two serious video games in a population of young university students and young new faculty members, their influence in specific soft skills (leadership, commitment, self-regulation, autonomy). Data from both implementations were used to answer the following research questions: How can we increase creativity and leadership if the environment of the games itself can limit our possibilities? What types of learning should be promoted, which experiences are valuable, and what knowledge is relevant when technology plays a central role? How can we ensure continuous learning in this context? What tools can be provided to facilitate lifelong learning? The results show preparing students in soft skills, such as decision-making knowledge, work ethics, team work, and managing emotions is possible by integrating video games into their learning experiences. Including cultural trends currently occurring outside HE classrooms walls can enhance learning, motivation, and life-long learning. Conclusions formulate some theoretical considerations and powerful questions for the possible transformation of educational practices within HE.

Keywords Serious games · Leadership soft skills · AI-Data analysis

---

C. Lion (✉)
Facultad de Filosofía y Letras, Universidad de Buenos Aires, Puan 480 (1406), Buenos Aires, Argentina
e-mail: carinalion@gmail.com
URL: https://www.edutrama.com.ar

V. Perosi
Instituto Tecnológico de Buenos Aires, San Martín 202 (1004), Buenos Aires, Argentina

1 Introduction

1.1 The Problem

The breathtaking pace of technological development in recent years—machine learning, artificial intelligence (AI), internet of things (IoT), big data, 3 and 4D impressions, among others—offers a framework to rethink its relationship with HE, considering that changes in teaching are slow and not always visible. One of the main directions of modern didactics is a combination of traditional teaching methods and techniques with finding ways and means of activating the development of the students' pedagogical abilities. We are faced with increasingly digitized educational contexts in which digital data, code, and algorithms determine—to a greater or lesser extent—political agendas, commercial interests, business ambitions, a variety of scientific experiences and professional knowledge, and create an educational image realm based on efficiency, personalization, evidence-based learning, and continuous innovation [1].

In the past few years, many reports have studied trends observed as impacted by technology in educational settings, explicitly stating the learning possibilities offered by information and communication technologies in HE. These reports highlight the extended opportunities for continuous, active, personalized, and rhizomatic learning [2], game-based learning and the culture of doing [3, 4], story and event-based learning [5], and dynamic and incidental learning [6].

Regarding the worldwide impact of the pandemic, technologies have had greater visibility, particularly on remote education [7]. The UNESCO IESALC estimates show that the temporary closure affected approximately 23.4 million HE students and 1.4 million teachers in Latin America and the Caribbean, which represent more than 98% of the region's population of HE students and teachers (IESALC, 2020). According to a survey about e-learning in HE in Latin America developed by UNESCO before the COVID-19 pandemic, face-to-face education was still the predominant model in 65% of the universities, compared with 16% with a predominant hybrid model and 19% centered on e-learning.

Despite of the pandemic and its associated migration of classes to virtual environments, this shift has not necessarily brought about significant changes in university education, since the didactic model is still the classic one: explanation—application—verification [8]. We have observed that in online lessons this didactic format has been maintained without making significant changes.

Contemporary cultural scenarios impose the challenge of developing strategies to strengthen the development of soft skills with a critical perspective, in order to provide young university graduates with the tools the work world demands, which would favor their insertion in a constantly changing society [9]. According to [10], gaming methodology fosters students' independent and extracurricular work, leading to the deepening of their professional knowledge and the development of moral and volitional qualities. Additionally, it enhances their ability to anticipate, hypothesize, make strategic decisions, and collaborate as a team—capabilities highly valued today.

To date, there has been a shortage of research studying the relationships between play, decision making, and critical thinking. Some research shows that players who play strategy games score higher in open-minded thinking than other types of players [11]. Video games constitute a safe environment for practicing risky actions and solving real problems. This innovation in teaching and learning at university level promotes the development of soft and unique skills that are often overlooked in HE curricula. Additionally, they offer strategies for lifelong learning.

1.2 Background

Research on HE underscores the necessity of cultivating additional skills in students, stemming from innovative experiences. Gamification acknowledges the significance of games as drivers of experience and engagement for both students and young new faculty members. They strengthen self-esteem, facilitate meaningful learning, and generate communities of practice that extend beyond local boundaries, as they are global in nature. Research has shown that simulation games enhance critical thinking, problem solving abilities, creativity, and higher order learning [11–13].

With the rise of Web 2.0 technology, there has been a surge in opportunities for collaborative and participatory interaction with technological applications. This trend encourages improved access to information, facilitates the exchange of ideas and knowledge, and fosters content production [14, 15]. In HE, digital simulations, which engage students in interactive, authentic, and self-driven knowledge acquisition, are increasingly being adopted. Connolly and Stansfield [16] define game-based e-learning as a digital approach that delivers, supports, and enhances teaching, learning, assessment, and evaluation.

We understand that the preparation of future professionals is within the objectives of HE institutions. To achieve this goal, innovative teaching methods are often implemented, including games and simulations, which constitute the subject of this article. Research generally indicates that games and/or simulations have a positive impact on learning objectives. Yang [11] identifies three learning outcomes resulting from the integration of games into the learning process: cognitive, experiential, and affective outcomes.

In recent years, digital or web-based games have increasingly supported learning. In the context of online education, this research area attracts a significant amount of interest from the scientific and educational community; namely, from tutors, students, and game designers. With the growing expansion of technology, instructors and educational policy makers are interested in introducing innovative technological tools, such as video games, virtual worlds, and Massive Multi-Player Online Games (MMPOGs) [17].

Games and simulations show mixed effects across several sectors, such as student performance, engagement, and learning motivation. Different studies address the implementation of varied serious video games, [11, 18–20], and they reveal the potential of these technologies to engage and motivate beyond leisure activities [21].

Video games are a type of immersive experience that combines online and first-person involvement on the one hand, and the emotional impact of stories on the other—retelling the story, inhabiting it, and going to a greater depth level. Immersion is an absorbing state of mind that can be achieved with any medium and at any time. You can be as immersed in a book today as you were four hundred years ago and equally immersed in a series, a video game, or any type of experience in the metaverse. This reference brings us closer to an interesting question: Who am I? And how did I get here? In this way, as in many artistic, cultural, and technological expressions, it is important to connect with those who contemplate, perceive, enjoy, and discover the game, to connect with the person, their ways of looking, perceiving, feeling, and discovering the world.

In their meta-analyses, [14, 22] systematically reviewed articles to study the detailed effects of digital games on learning outcomes, concluding that games are important in supporting productive learning and highlighting the significant role of gaming design beyond its traditional medium. Prior to this review and along the same lines, in their meta-analysis, [23] reported positive outcomes in learning when using serious games in the educational process. Wouters and van Oostendorp [24], performing meta-analytic techniques, used comparisons as well to investigate whether serious games are more effective and motivating than conventional instructional methods. They found higher effectiveness in terms of learning and retention, but less motivation compared to traditional instructional methods. Indeed, serious games tend to be more effective when regarded as a supplement to other instructional methods and involve students in groups and multiple training sessions.

Lin [25] developed a multi-dimensional approach to categorize games and offered a review of 129 papers on computer games and serious games, explicitly targeting cognitive, behavioral, affective, and motivational impacts, as well as engagement. The most frequent outcomes were knowledge acquisition and content understanding, alongside affective and motivational outcomes. Lion [26] in their meta-analysis of the cognitive domain, examine how design elements in simulation-based settings affect self-efficacy and transfer of learning. Their findings conclude that gathering feedback post-training rather than during the process results in higher estimates of self-efficacy and learning transfer.

Researchers have also looked at games and simulations from a theoretical perspective. For instance, [27] examined the theoretical background and models employed in the study of games and simulations. They focused primarily on the theories of cognitivism, constructivism, inactivism, and the socio-cultural perspective. This literature review indicates an increasing recognition of the effectiveness of digital games in promoting scientific knowledge and concept learning. However, there is comparatively less emphasis on their role in facilitating problem-solving skills or exploring outcomes from the perspective of scientific processes, affect, engagement, and socio-contextual learning. This view is echoed by other researchers, such as [28], who systematically reviewed and demonstrated the effectiveness of simulation games on satisfaction, knowledge, attitudes, skills, and learning outcomes within nurse practitioner programs.

Research on serious video games has shown advances in perceptual, superior thinking, and metacognitive skills [13, 29, 30], in self-appraisal [31], in problem resolution, [32, 33], in social (teamwork) skills and collaborative construction [34–37], in leadership [37, 38], in self-confidence and motivation [39], which show their value as tools to strengthen learning in HE. Beyond these investigations on video games, it must be considered that, currently, most technological developments have AI engines. Real-time interactivity and fast responses from AI applications can generate timely feedback in video games as well. Today´s mindtools (devices, apps, platforms, and video games) can create dilemmas for education that we should analyze.

1.3 Research Questions

This article aims to describe two serious video games developed to strengthen young leadership and explore what type of skills they generate to answer some of those questions:

RQ1. How can we increase creativity and leadership if the environment of the games itself can limit our possibilities?
RQ2. What types of learning should be promoted, which experiences are valuable, and what knowledge is relevant when technology plays a central role?
RQ3. How can we ensure continuous learning in this context?
RQ4. What tools can be provided to facilitate lifelong learning?

1.4 Purpose

The study showcases relevant ludic experiences through two serious video games (Emerging Leaders & Bold Leaders) that have been implemented with young university students and new faculty and have shown initiative, self-esteem, critical thinking, distributed leadership, and decision making [22]. These soft skills contribute to the improvement of student performance and could define the transformations that the close future university may need. The questions asked for each of the videogames focused on distributed leadership, soft skills (e.g. decision making, critical thinking), gameplay, and their relationship with deep learning.

2 Methodology

2.1 *The Video Game Emerging Leaders*

The video game Emerging Leaders [40] proposes a neighborhood with different missions (taking care of a park, improving accessibility to public buildings, raising funds for the municipality, etc.). Each mission presents different levels of complexity. Several followers and funds are required (which are acquired as the player progresses) for resolution. It is a computer game. As the missions are accomplished, community recognition is acquired, and it is scored up on the leadership scale, which is composed of several factors: recruitment of followers, funds, and resolution of missions, which account for commitment to the community. The leadership scale enables the player to engage in increasingly complex and socially relevant functions. The players themselves can create alternative solutions for each of the missions. The decisions and the search for these creative and effective solutions are vital for leadership growth, which is the goal of the videogame, along with the recognition of the community in social media networks.

The players are not alone but are part of a network. Followers can be characters in the game (neighbors, teachers, doctors, pedestrians who walk all the time on the screen and who can be potential followers—they are always real players who are part of a network—but also, other activists who support the player in their missions. In this way, and as network games are designed, when members identify themselves as such and share common interests with the rest, communities are strengthened.

Belonging to a community does not necessarily imply that all its members form a compact group. The video game community provides the social structure in which connections and peer support are developed. They are communities that learn from this virtual space and consolidate links that strengthen their members and the production of knowledge among them [41].

The game proposes the following mechanics: The player must choose a situation to solve (e.g. missions) and may participate in the construction and protection of their community, contributing to improving key aspects of the problems that will be raised and selected. According to their participation, each player will gain experience in the form of points of expertise in each area of knowledge they use, and they will have higher social visibility in media and networks until they become a benchmark. The missions increase their scope of influence zone from the neighborhood to the country and cover different problems: school, club, semi-urban space, parks, waterfront, etc. The missions have multiple options that favor the decision-making processes, as in a decision tree.

The game has 4 levels:

Level 1: Definition of zone to operate, problem diagnosis based on indicators.
Level 2: Search for creative solutions, interaction with colleagues and experts in social networks.

Level 3: When the problem is solved, it goes to the level of positioning and expertise to be consulted by other players. The queries add more points.
Level 4: A similar problem of greater complexity is solved again, or other problems are solved to gain versatility in the fields of knowledge.

This proposal is associated with alternative reality modalities, which aim to improve quality of life by directly engaging the video player as a co-creator of the game narrative. Rather than simulating reality, these modalities integrate real-life problems into the immediate reality of video game players. In addition, as players gain confidence in their decisions, they can propose alternative solutions. If they have followers, these can be added from the programming to the mission tree, and they may then add possible solutions. Therefore, there is a continuous source of missions and solutions.

The missions are growing in the scope of the influence zone, from the neighborhood to the country and travel through different problems: school, club, semi-urban space, parks, waterfront, etc. The missions have multiple options that as a tree favor the decision-making processes.

Finally, the game proposes emerging events. Every x minutes the game triggers events that must be resolved in a limited time, which differ from the missions that the player normally resolves. These events are simultaneous for all players and may present a difference with normal missions: instead of the player choosing a solution, the player votes for a solution and when the event time ends, the most voted solution is implemented. Players who chose the most voted solution will get a popularity bonus, consequently growing their popularity level, based on values related to the resolution of real community problems and the interweaving of a network that constructively resolves together (Fig. 1).

2.2 Data Collection for Emerging Leaders

The video game was played by four groups of students from two different university majors during their first year. We selected a young audience at the University of Buenos Aires, who were in the first year of their majors to identify whether certain soft skills (strategic decision making, strengthening self-esteem, and creativity) could be developed through the use of this video game. Subsequently, we aim to investigate whether these skills are sustained throughout their academic trajectory in the following years. The research tracking these students is currently under way.

In both cases, the video game was introduced during a one-hour session, with each student having access to their own computer for gameplay. During this time, non-participant observation of the players was conducted, recording their gameplay actions, time allocation, decision-making processes, and innovative solutions created. Additionally, an interpretation of video game analytics derived from the game's backend data was performed. For each, a postgame survey was administered.

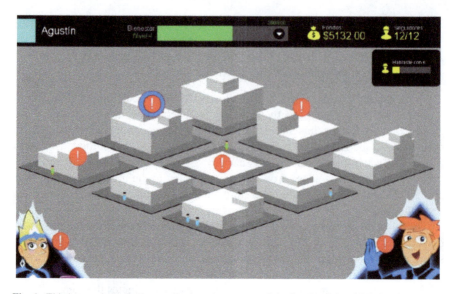

Fig. 1 This image shows both central game characters and the first level, in which the player must decide on one zone and one type of problem to resolve. The upper margin displays the leadership the player is achieving based on decisions made

One hundred students from two different majors (Law and Economics) were willing to participate the research team will follow up with these students in the upcoming years.

We opted for the following procedures:

- Introducing the video game in an hour session (one computer was available per student, and they were playing for an hour).
- Non-participant observation of video players (while playing, it was recorded how they played, what they spent time on, which decisions they made, what new solutions they created).
- Interpretation of video game analytics (number of cases resolved; number of proposed solutions, selected cases, number of followers achieved, amount of score on each of the items that the videogame proposes: health, education, recreation and environment; scope of leadership achieved in the neighborhood). These data come from the game's own backend.

2.3 The Video Game Bold Leaders

The video game Bold Leaders [42, 43] aims to strengthen leadership in organizational contexts (among whom are teachers, coordinators, area directors, department directors, deans, etc.) in an innovative and disruptive way from the perspective of distributed leadership. The first two levels can be played on a mobile phone or on the computer, and the third level is played on the computer.

Distributed leadership from a professional development viewpoint implies empowering the voice of decision-makers in an organization. Distributed leadership is more than a form of management. The aim is to ensure that activities that boost innovation are shared and distributed within and among organizations. This involves having enough individuals and teams within organizations who carry out leadership tasks and responsibilities in a coordinated manner. To make this possible, it is necessary to facilitate participation through dialogue, create opportunities for members to weigh up each other's reasons and to critically analyze and review their own. The goal is to give way to honest conversations, so that narratives of all the experiences can be retrieved so as to favor environments for reflection, agency, shared work, and enrichment. Distributed leadership implies a joint vision, attentive listening, accepting and delegating functions to other leaders, which are projective ways of thinking, among other features that have been considered for professional practice.

From this position, the video game Bold Leaders proposes three levels:

1. Guardian: Based on photos that are uploaded and the ranking criteria of each one, that allows the users to playfully show the image of their organization. Each user chooses a photo of their organization that represents it, and they also choose the criteria (efficiency or quality), to weigh in on their own and other users' photos. This level is dynamic and helps to recognize how the members themselves perceive their reference institution in real time (Fig. 2).
2. Selfie: At this level, the goal is to answer trivia questions or perhaps, based on the answers to the trivia questions, leadership profiles intersect. These profiles are not linear, so it is complex to plan the response hoping to obtain a specific type of leadership. The user can leave the trivia quiz, and the next time they connect, they will see the last unanswered question. The questions are random, and the number of questions is varied and multidimensional. Twenty questions must be answered to determine the type of management leadership. The purpose is to define a user's leadership profile according to the answers given in each case. The questions are closed, and the user must select the answer they deem appropriate. After completing the trivia, each answer option will assign a score to each of the possible profiles. Once the game is completed, users will no longer have access to view the questions, answers, or their score. However, they will be able to see the profile that resulted from their gaming experience. The game also provides the characteristics of the type of leadership that emerges from user responses.
3. Hacker (Labyrinth): The objective of this level is to overcome a maze and take objects that will later be necessary for the design of a collaborative activity. Its main objective is the survey of five themes that were defined in the video game as priorities for the approach of teaching leadership in the university:

Fig. 2 Guardian, Selfie, and Hacker levels in Bold Leaders video game

- Adversity management.
- Effective communication
- Emotional intelligence
- Conflict resolution
- Delegation of tasks and monitoring

Some are timed and others present diverse complexities. The design of each of the labyrinths is linked to its specific theme. Each labyrinth moves its avatar based on questions that users must answer, providing insights into each player's decision-making process and knowledge about the subject matter. In some mazes, the players cannot leave without the help of the team, which serves to measure the degree of collaboration between the players.

2.4 Data Collection for Bold Leaders

As this is a video game aimed at empowerment and leadership in teaching and institutional management, it was decided to use it with forty young new faculty members and ten department directors from the Economics and Law majors at the University of Buenos Aires. In this game, soft skills are associated to management and decision making, including emotional intelligence aspects (in contrast to the previous game). What this game emphasizes are the abilities of task management, delegating, teamwork, and conflict resolution during the task.

The dynamics were:

- To introduce the videogame in an hour session (one computer was arranged per student and they were playing one hour).
- Non-participant observation of video players (while playing, their actions, time, and decision were recorded).
- Analysis of video game analytics (leadership trends resulting from trivia; number of objects collected in mazes; resolution time of mazes; number of requests for help in mazes).

These data come from the backend of the game itself. In both cases, a postgame survey was administered. Data systematization is shown in the next point.

3 Results

3.1 Emergent Leaders Results

The following table presents the data collected with the procedures explained above for the categories studied (Table 1).

3.2 Bold Leaders Results

From the Bold Leaders game, after the trivia (selfie that detects leadership styles) we have the data shown in the following table, which shows the difference between young new faculty members and heads of department as regards leadership style and soft skills developed as a result of videogaming.

Table 1 Categories studied and data obtained

Category	Data
Gender	Women 60%; men 40%
Age group	19–20 years old
Creation of new solutions	85%
Observed delays	choice of missions (more than 2 min), in the decision about solutions (more than 2 min)
Number of cases resolved at level 1	12 out of 20 (no gender differences)
Number of cases resolved at level 2	6 out of 20 (no gender differences)
Followers collected	Average of 70
Gender differences	Women have higher scores in education and environment. Men have higher scores in health and environment

In order to win at the labyrinths, users were asked to solve problems, play on the net and share the findings of clues that were being obtained. By doing so, they would be able to generate soft skills in the leadership of the university chairs.

What emerges from this video game is that the Department heads, especially, are considering distributed leadership. This is noted in the following issues:

- They delegated the search for objects to leave the maze to other players; but they also searched for the objects themselves; They tracked the found and missing objects to exit the maze. In the backend of the game, they appear as those who traveled the most in the labyrinth, and even after having solved it, they returned to the labyrinth to check that everything was solved.
- In the post-test survey, they marked the items of leadership strengthening and problem resolution as the most relevant with respect to the game and their own experiences of playing it (Table 2).

3.3 Post-Test Survey Results

The surveys reported some positive effects of the use of video games in young new faculty members and department directors. Among them (Table 3):

On the other hand, video games were built to allow for experimentation. Also, they allow us to see the consequences of the players' actions in an almost immediate way in simulated environments that favor decision-making. This granted the ability to play with ethical standpoints, to hypothesize with work strategies, to make mistakes, to fail, and to start over; They are laboratories for experimentation and lifelong learning. They allow a multidimensional work in which the real, the projected, and the virtual actions converge. Surveys in this case yielded more than 70% in students, as well as in teachers and department directors:

Table 2 Differences between young new faculty members and heads of department

Young new faculty members (up to age 25)	Heads of department
35% of young new faculty members have imitative leadership, whose work is a demanding model that asks students to copy; they aspire to a high standard because the search for excellence is permanent	60%: orientational leadership; 23% formative leadership; 9% democratic leadership; 7% affiliate leadership
25% of young new faculty members have a formative leadership, whose main characteristics are to "get" the students out of their comfort zone and lead them to learning and continuous improvement; to detect strengths and weaknesses and provide timely feedback for the improvement in their knowledge construction processes	Professors and deans said they had achieved the following from the labyrinths: · Adversity management · Effective communication · Emotional intelligence · Conflict resolution
13% of young new faculty members have an affiliate leadership, whose main characteristics are to prioritize communication; they have a flair for sharing ideas and resolving conflicts; they prioritize people to the detriment of organizational tasks and objectives	
9% of young new faculty members have an orientational-style leadership that seeks to inspire a vision and guide students to participate and engage with said vision	
8% of young new faculty members have a democratic-style leadership, whose main characteristic is to offer room for decisions to be taken in a participatory and consensual way, with little centrality in the teacher's decision	

Table 3 Positive effects related to RQ1 and RQ3

Positive effects	%
Creativity (contribution of new solutions to problems)	70
Improvement in strengthening self-perception in relation to management and leadership	75
Increase in coordination, integration and interpretation of visual information with the motor reaction (This is interesting in the face of current multitasking pressures)	85
Confidence in decision making (in both games)	95

- Strengthening/Reinforcement of self-perception about leadership ability
- Strengthening experimentation and the search for solutions to problems.
- Time optimization for decision-making
- Motivation and engagement

4 Discussion

A vital element in achieving learning goals is the relationship between motivational processing and the outcome processing (satisfaction). We have seen this as a result of the implementation of both games for students as well as teachers.

Research interest in the incorporation of games in HE education is constantly developing [44, 45]. The pedagogical shift, from lecture-centered to student-centered environments and the increasing use of games as innovative learning technologies, calls for a transformation in HE. In this respect, games and simulations are expected to play a significant role in the learning process.

Furthermore, reality-based scenarios and action-oriented game activities promote fruitful interactions and meaningful feedback. These lead to collaborative construction of knowledge.

The high cost of designing games and simulations is still a significant challenge. The development of each of these video games involved seven months of work and a high cost in programming and design.

However, we believe that games are already, to a certain extent, integrated into educational systems to achieve a variety of learning outcomes [3, 25]. Furthermore, it is one of the changes considered to be introduced in HE to develop and strengthen soft skills, which are presently necessary in the articulation between the academic and the professional realm (problem solving, leadership, creativity, critical thinking).

We believe that four keys are central when designing video games:

- stories with content. To paraphrase Steven Spielberg, we must put the person in a scenario where, no matter where they look, they are surrounded by a three-dimensional experience. That's the future.
- the spectator, the participant, the student is the one who takes some role in the story.

Creators are beginning to think about active, diverse and sophisticated, but at the same time subtle, ways to do it. It is about telling a great story, including an audience that wants to place themselves in it and inhabit that world through their imagination, which means leaving the parameters of the environments and beginning to interact through other means.

- the power of video games designed to enrich, expand, and recreate the sensory experience.

What does it mean to educate in the world and in the other world [46] in the real and digital world [47] at the same time? Improving knowledge experiences involves not only embracing the mental revolution of our time but also understanding the scope and opportunities of a reality that is not only physical (as if nothing else existed). Reconstructing in a critical and creative way to become part of that reality a complex and mutant reality. Not doing so means educating through forms of the past. It involves educating for the past.

- the intersections between the field of performing arts and performance, the field of cognition to think about the device.

Opening our minds to disruptive thoughts from other fields of knowledge, demanding the development of complex and critical transversal competencies (specialized skills in a specific area + interdisciplinary skills), and assuming the risk of thinking up actions that expand the current limits in which the learning process is conceived.

We have acknowledged that in the face of datafication and algorithmization, which produce performance results through data interpretation, one challenge lies in playful environments that invite to think outside the box that encourage flexibility in decision making, autonomy, and critical thinking, as well as the critical ability to empathize in the years to come. We believe that the intelligent systems which are growing in adaptive and contextualized learning environments can work in detection and recommendations but give little opportunity for creative and critical thinking. These video games are not yet related to any specific discipline or content. (This will have to be investigated in future research). They are generic and defined in terms of skills. However, their value lies in their original authorship, allowing their design and functionalities to be manipulated according to the hypotheses we wish to test.

From our perspective and experience as educators, we believe teaching means exploring, reviewing, discussing, questioning, or merely, inventing something different both on the level of knowledge and on that which can be transformed into reality from it. Each proposal of knowledge deployed must be thought to transcend the world and the other world, i.e., physical and digital reality.

Here comes the cultural and gnoseological force of educational video games: in first person, sensory environment, putting all our senses and bodies into play in an integral way, with open routes and possibilities, where we know where to start but not where to end, with imaginative narrative engineering and with playful emotion—an epic knowledge experience.

Something related to revising university purposes seems to prioritize academic content but disregards preparing students in soft skills, which are necessary today for professional development: decision making knowledge, work ethics, teamwork, managing emotions, and so on. This calls for a revision in time allotment in which, together with subject-matter contents, skills such as self-efficacy, creativity, leadership, and self-regulation are included. Gaming is associated with early childhood. However, videogames are played by youngsters at the university as well. Incorporating cultural trends occurring outside university classroom walls can enrich learning, improve motivation, and boost the desire to continue learning for life.

5 Conclusions

Both existing literature and our own surveys indicate that the introduction of serious video games is both valid and relevant in current research, particularly for developing necessary soft skills in university education in the years ahead. We refer to serious video games that are linked to the acquisition of knowledge and required skills in the context of today's society, in articulation with the demands of the business and organizational world.

There are games whose primary purpose is not entertainment, enjoyment or fun. Serious games, educational gaming, as well as virtual worlds developed for educational purposes reveal the potential of these technologies to engage and motivate beyond leisure activities [21, 48].

Video games can provide cognitive, behavioral, and attitudinal results that are fundamental, interactivity and feedback, motivation and competition, playfulness and problem-based learning, collaborative learning, progression and repetition, as well as realism and immersion.

Internet in HE urges students to build digital and collaborative skills for the twenty-first century through gaming. Also, the emergence of a participatory culture in education spurs researchers to get involved with digital games.

Engagement and motivation are major factors in enhancing HE learning objectives [14, 25, 49–51]. Motivation is considered a central factor in most of the reviewed studies.

Nowadays, videogames embody a highly productive moneymaking machine. Additionally, youths merge into them emotions and skills strengthening not usually valued in academia: self-confidence, decision making, hypothesizing and anticipation strategies, as well as the epic triumph achieved through the feeling of accomplishment when winning the game. Why not realize what is given into what is needed?

References

1. Williamson B (2018) Big data en Educación. El futuro digital del aprendizaje, la política y la práctica. Morata, Madrid
2. Sharples M, McAndrew P, Weller M, Ferguson R, Fitzgerald E (2012) Innovating pedagogy 2012: open University innovation report 1. The Open University, Milton Keynes
3. Cunningham E, Green SE (2023). Cognitive skills acquired from video games. https://doi.org/10.1093/acrefore/9780190228613.013.1468
4. Koutsogiannis D, Adapmpa V (2022) Video games and (language) education. Towards a critical post-videogaming perspective. https://doi.org/10.21248/l1esll.2022.22.2.366
5. Sharples M (2013) Mobile learning: research, practice and challenges. Distance Educ China 3(5):5–11
6. Sharples M (2015) Seamless learning despite context. In: Wong LH, Milrad M, Specht M (eds) Seamless learning in the age of mobile connectivity. Springer, Singapore, pp 41–55. https://www.researchgate.net/publication/283689863_Seamless_Learning_Despite_Context#fullTextFileContent
7. Lion C, Cukierman U, Scardigli M (2022) The emergence of the emergency in higher education in Argentina. IJET 17(11). https://doi.org/10.3991/ijet.v17i11.31113
8. Litwin E (1997) Las configuraciones didácticas. Paidós, Buenos Aires
9. Kalimullin AM, Vlasova VK, Sakhieva RG (2016) Teachers' training in the magistrate: structural content and organizational modernization in the context of a federal university. Int J Environ Sci Educ 11(3):207–215

10. Ezrokh YS (2014) Gaming method for the stimulation of the motivation and success of activities of students-economists of junior classes. Educ Sci 7:87–102
11. Yang J (2012) A digital game-based learning system energy for energy education: an energy conservation pet. Turk Online J Educ Technol 11(2):27–37
12. Oblinger D (2006) Simulations, Games, and Learning. http://www.cameron.edu/~lindas/DianaOblingerSimsGamesLearning.pdf
13. Sattar S, Khan S, Youssaf R (2022). Impact of playing video games on cognitive functioning and learning styles. https://doi.org/10.30537/sjcms.v5i2.885
14. Lie M, Stephen A, Supit L, Achmad S, Sutyo R (2022). Using strategy video games to improve problem solving and communication skills: a systematic literature review. https://doi.org/10.1109/ICORIS56080.2022.10031539
15. McLoughlin C, Lee MJW (2008) The three P's of pedagogy for the networked society: personalization, participation, and productivity. Int J Teach Learn High Educ 20(1):10–27
16. Connolly T, Stansfield M (2006) Using games-based eLearning technologies in overcoming difficulties in teaching information systems. J Inf Technol Educ 5(1):459–476
17. Buckless FA, Krawczyk K, Showalter DS (2014) Using virtual worlds to simulate real-world audit procedures. Issues Account Educ 29(3):389–417
18. Justo R, Ramos R, Llandelar S, Sanares R, Rodelas N (2022). Game-based learning for student engagement: a paradigm shift in blended learning education. https://doi.org/10.1063/5.0109625
19. Van Roessel L, Van Mastrigt-Ide J (2011) Collaboration and team composition in applied game creation processes. In: DiGRA '11, proceedings of the 2011 DiGRA international conference, think design play, 1–14
20. Willoughby T (2008) A short-term longitudinal study of internet and computer game use by adolescent boys and girls: Prevalence, frequency of use, and psychosocial predictors. Dev Psychol 44(1):195–204
21. Anderson EF, McLoughlin L, Liarokapis F, Peters C, Petridis P, Freitas SD (2009) Serious games in cultural heritage. In Ashley M, Liarokapis F (eds), VAST 2009: 10th international symposium on virtual reality, archaeology and cultural heritage 22–25 Sept 2009. Eurographics Association, St. Julians, Malta, pp 29–48
22. Clark DB, Tanner-Smith EE, Killingsworth SS (2015) Digital games, design, and learning a systematic review and meta-analysis. Rev Educ Res 86(1):79–122
23. Backlund P, Hendrix M (2013) Educational games-are they worth the effort? A literature survey of the effectiveness of serious games. In Games and virtual worlds for serious applications (VS-GAMES), 1–8
24. Wouters P, van Oostendorp H (2013) A meta-analytic review of the role of instructional support in game-based learning. Comput Educ 60(1):412–425
25. Connolly TM, Boyle EA, MacArthur E, Hainey T, Boyle JM (2012) A systematic literature review of the empirical evidence on computer games and serious games. Comput Educ 59(2):661–686
26. Gegenfurtner A, Quesada-Pallarès C, Knogler M (2014) Digital simulation-based training: A meta-analysis. Br J Edu Technol 45(6):1097–1114
27. Li MC, Tsai CC (2013) Game-based learning in science education: a review of relevant research. J Sci Educ Technol 22(6):877–898
28. Warren JN, Luctkar-Flude M, Godfrey C, Lukewich J (2016) A systematic review of the effectiveness of simulation-based education on satisfaction and learning outcomes in nurse practitioner programs. Nurse Educ Today 46:99–108
29. Helle L, Nivala M, Kronqvist P, Gegenfurtner A, Björk P, Säljö R (2011) Traditional microscopy instruction versus process-oriented virtual microscopy instruction: A naturalistic experiment with control group. Diagn Pathol 6(1):1
30. Siewiorek A, Gegenfurtner A, Lainema T, Saarinen E, Lehtinen E (2013) The effects of computer-simulation game training on participants' opinions on leadership styles. Br J Edu Technol 44(6):1012–1035

31. Arias Aranda D, Haro Domiguez C, Romerosa Martinez MM (2010) An innovative approach to the learning process in management: the use of simulators in higher education 353:333–334
32. Hou HT (2015) Integrating cluster and sequential analysis to explore learners' flow and behavioral patterns in a simulation game with situated-learning context for science courses: a video-based process exploration. Comput Hum Behav 48:424–435
33. Hou HT, Li MC (2014) Evaluating multiple aspects of a digital educational problem-solving-based adventure game. Comput Hum Behav 30:29–38
34. Lin YL (2016) Differences among different DGBLs learners. Int J Bus Manag 11(1):181–188
35. Stanley D, Latimer K (2011) 'The Ward': a simulation game for nursing students. Nurse Educ Pract 11(1):20–25
36. Tiwari SR, Nafees L, Krishnan O (2014) Simulation as a pedagogical tool: measurement of impact on perceived effective learning. Int J Manag Educ 12(3):260–270
37. Wang C (2016) Using multimedia tools and high-fidelity simulations to improve medical students' resuscitation performance: an observational study. BMJ Open 6(9):e012195
38. Siewiorek A, Saarinen E, Lainema T, Lehtinen E (2012) Learning leadership skills in a simulated business environment. Comput Educ 58:121–135
39. Chang YC, Peng HY, Chao HC (2010) Examining the effects of learning motivation and of course design in an instructional simulation game. Interact Learn Environ 18(4):319–339
40. Lion C, Perosi V (2016) Los videojuegos serios: puentes de creatividad y expansión educativa. Revista Anales de la Educación Común, Publicación de la Dirección General de Cultura y Educación de la Provincia de Buenos Aires, Sección Artículos en torno al tema: Los videojuegos y la creatividad. Año 2 http://revistaanales.abc.gov.ar/
41. Wenger E (2001) Comunidades de Práctica Aprendizaje significado e Identidad. Paidós, Buenos Aires
42. Lion C, Perosi V (2019) Didácticas lúdicas con videojuegos educativos. Escenarios y horizontes alternativos para enseñar y aprender. Novedades Educativas, Buenos Aires
43. Lion C, Perosi V (2018) Los videojuegos serios como escenarios para la construcción de experiencias. Educadores del Mundo, Revista Telecolaborativa Internacional, Mayo 2018:4–8
44. Girard C, Ecalle J, Magnan A (2013) Serious games as new educational tools: How effective are they? A meta-analysis of recent studies. J Comput Assist Learn 29(3):207–219
45. Subhash S, Cudney E (2018). Gamified learning in higher education: a systematic review of the literature. https://doi.org/10.1016/j.chb.2018.05.028
46. Baricco A (2019) The Game. Anagrama, Buenos Aires
47. Rose F (2011) The art of immersion: how the digital generation is remaking hollywood, madison avenue, and the way we tell stories. W W Norton & Company, New York
48. Zhang X, Chan Y, Hu L, Wang U (2022) The metaverse in education: definition, framework, features, potential applications, challenges, and future research topics. https://doi.org/10.3389/fpsyg.2022.1016300
49. Erhel S, Jamet E (2013) Digital game-based learning: impact of instructions and feedback on motivation and learning effectiveness. Comput Educ 67:156–167
50. Ke F, Xie K, Xie Y (2015) Game-based learning engagement: a theory-and data-driven exploration. Br J Edu Technol. https://doi.org/10.1111/bjet.12314
51. Nadolny L, Halabi A (2015) Student participation and achievement in a large lecture course with game-based learning. Simul Gaming 47(1):51–72. 1046878115620388
52. IESALC, (2020) Instituto Internacional para la Educacion Superior en America Latina y el Caribe: COVID-19 and higher education: today and tomorrow; Impact analysis, policy responses and recommendations. unesdoc.unesco.org/ark:/48223/pf0000375693
53. URL online-engineering.org/dl/iJET/iJET_vol17_no11_2022.pdf
54. Liu CC, Cheng YB, Huang CW (2011) The effect of simulation games on the learning of computational problem solving. Comput Educ 57(3):1907–1918

Video Games as an Emerging Consumer Technology: Profiles, Uses, and Preferences of University Students in Argentina

María Gabriela Galli, María Cristina Kanobel, Diana Marín Suelves, and Donatella Donato

Abstract The use of video games is linked to current social changes and to inherent characteristics such as their interactive nature. They are a lens through which we can observe new dynamics and models. The video game industry is central to the entertainment and technology sector and is growing year after year in terms of audiences and profits. Recent studies from the technological, social and business sectors have attempted to analyse the effectiveness of the use of video games as tools for education, information, training, participation and interaction. A total of 286 students from two educational institutions in Argentina participated in this study, which utilized a quantitative approach with a descriptive correlational scope and a non-experimental cross-sectional design. Data collection occurred in May 2022 through non-probabilistic convenience sampling. Sociodemographic information and data on video game usage were gathered using a survey technique administered to students from both public institutions located in the Metropolitan Area of Buenos Aires. The results show that men have a greater preference for video games, confirming the relevance of such cultural products for them. There are also significant gender differences in age, hours of use, game modes, platform, and video game genre. We therefore conclude that video games are an important emerging consumer technology for different age and gender groups, and that there are gender differences and similarities in the use of video games as an entertainment activity that deserve to be explored in order to exploit the potential of these resources.

M. G. Galli
Universidad de Tres de Febrero, Maipú 71, C1084ABA Cuidad Autónoma de Buenos Aires, Argentina

M. C. Kanobel (✉)
Universidad Tecnológica Nacional, R. Franco, V. Domínico, 5050, B1874ABY Buenos Aires, Argentina
e-mail: mkanobel@fra.utn.edu.ar

D. M. Suelves · D. Donato
Universitat de València, Blasco Ibáñez 30, 46010 Valencia, España

© The Author(s), under exclusive license to Springer Nature Switzerland AG 2025
E. Vendrell Vidal et al. (eds.), *Advanced Technologies and the University of the Future*,
Lecture Notes in Networks and Systems 1140,
https://doi.org/10.1007/978-3-031-71530-3_24

Keywords Video games · Technology · Students · Higher education · Gender

1 Introduction

Playing is part of everyone's development and fulfils different functions at different stages of development and throughout life. Playing is a voluntary activity that takes place within specific spatial and temporal boundaries, with precise rules that are freely accepted by the players [1]. People play from the moment they are born, and play mediates and shapes intersubjective spaces in relation to their environment, helping them to appropriate and transform the complex reality around them by interacting with different objects and subjects. In this way, play goes beyond mere entertainment and is linked to the development of a person's fundamental abilities and skills: social, communicative, emotional, cognitive and motor [2].

Games are part of culture, and since time immemorial people of all ages have played games for different reasons: entertainment, competition, testing their skills, or social cohesion, among others. In recent decades, video games have become an emerging market that is constantly evolving, immersed in a digital society. In 2022, the video game industry will generate $184.4 billion worldwide, with 3.2 billion players [3]. It is estimated to be the largest entertainment medium of the twenty-first century [4]. Argentina is one of the countries with the highest participation in the video game industry, with an annual turnover higher than that of cinema and music combined. The Undersecretariat of Knowledge Economy promotes citizen participation in various government calls for video game development projects. Specifically, in 2022, [5] showed that the industry had a turnover of more than 72 million dollars. Its main markets were the United States and Canada, and companies employed around 1,500 people, 23% of whom were women. The most common types of games in the country are casual/social games (37%), adventure games (34%), action games (29%), educational games (27%) or puzzle games (26%). They are targeting increasingly wider audiences, with 89% of companies developing products for 19–35-year-olds, 60% for teenagers aged 14–18, 50% for adults aged 36–60, 32% for children aged 7–13, 19% for people over 60, 14% for children aged 4–6, and 6% for children under 4. These data show that the national video game industry has been gradually targeting older users, rather than focusing exclusively on children and adolescents. On the other hand, a survey [6] which covers the population over 13 years of age living in urban areas in Argentina with more than 30,000 inhabitants, showed that most of a third of the sample participating in the study played video games, most of them several times a week (90%). Women played less than men. Those who played the most were teenagers between 13 and 17 years old (82%), and young people between 18 and 29 years old (52%). The survey also found that those who played the most belonged to the highest socio-economic levels.

The study also revealed that 26% played alone, 20% played with friends—either face-to-face or online—and 10% played with family members. It was observed that the most popular game genres were action, adventure, sports, and puzzle games (e.g., Candy Crush and Tetris). In this sense, video games are a cultural industry phenomenon associated with mass consumer appeal and have become cultural objects in today's society.

Video games of different genres and subgenres [7] have become the most relevant objects among leisure or entertainment activities. They are part of the culture of children, adolescents and even young people and adults. In addition, they are now used as inputs in the treatment of diseases and rehabilitation [8]. They are also incorporated as resources in educational mediation at different levels and as an object of study for researchers in different fields, who investigate their effects, both positive and negative, on users [9]. It is worth mentioning that video games, as an emerging artistic hypergenre, have an impact on cognitive, emotional and kinesthetic development and intervene in the construction of subjectivity in the gamer generation [10]. Through games, users can exercise their digital skills, solve problems individually or collaboratively, evaluate processes, explore and manipulate a virtual environment without real-world consequences, develop creativity, engage in compelling narratives and decision-making, strengthen communication skills, and even generate their own language, allowing them to create an identity within their community [2, 10]. In turn, they "arouse enough motivation in children and adolescents to make them feel connected to their inner dynamics" [11], can improve cognition [12–16] and affect the affective and emotional domains [11, 17, 18]. They are also mediators of the learning environment [2, 19–22]: working with video games in collaborative environments can improve students' learning and motivation [23]. In addition to the potential benefits of video games, several studies have highlighted possible negative effects, such as the influence of violent content on user behaviour [24–26], hyperactivity and inattention [27], the relationship between game type and behaviour [26, 28, 29] and the relationship between academic performance and excessive video game use [30–33]. It is worth noting that the World Health Organisation (WHO) has included gaming disorder in the 11th revision of the International Classification of Diseases (ICD). However, studies on the subject show that gaming disorder affects only a small proportion of people and is manifested when the player prioritises gaming over other daily activities, when they have no control over their gaming behaviour, and when their behaviour causes physical or psychological harm [34].

We must also mention that these technological and cultural products have important implications for social interaction and coexistence. They reproduce ideas, values and gender stereotypes associated with narratives and aesthetics: strong, robust, brave and dominant men versus weak, passive and submissive women. Therefore, some games tend to be categorised as "video games for boys" and others as "video games for girls", thus categorising play practices in childhood and adolescence [35]. Similarly, from a gender perspective, some studies show that men have better visual and spatial skills than women [36], play more frequently and for longer hours [6, 37–40], and prefer sports games, racing games and shooters. Conversely, women prefer less competitive and more social, affective, and educational games, arcade games, and casual games [37, 40, 41], and are less prone to addiction than men [37, 42–44]. In contrast, other studies show that women can be just as experienced as men [40].

Differences can also be seen in terms of content: male characters predominate, with little presence of female protagonists and many hypersexualised depictions [40, 45]. Furthermore, female characters are overrepresented in casual games compared to other formats [46]. Many women have also experienced harassment during their gaming sessions [45, 47]. To alleviate some of these problems, the My Game, My Name campaign against sexism in video games and e-sports returned in 2022. It raises awareness of the harassment and humiliation women experience when playing games, and the fact that they must hide behind male nicknames to avoid being attacked and to level the playing field.

1.1 Objective

Video games are one of the emerging consumer technologies among people of different ages and genders. Therefore, the aim of this paper is to describe some characteristics of higher education (HE) students in Argentina and to analyse possible gender differences and similarities in the use of video games as a leisure activity.

2 Method

The focus of the present study is quantitative, with a descriptive correlational scope. We try to specify the profiles and characteristics of the students to be analysed.

This research project is non-experimental and cross-sectional, as the data were collected at a single point in time (May 2022) and no deliberate intervention was made in the environment.

We used non-probability sampling [48], also known as convenience sampling. The inclusion criterion for participants was that they had to be studying a degree or undergraduate studies at a public institution in the Greater Buenos Aires Area (AMBA, in its Spanish initialism), an urban area that includes the city of Buenos Aires and 40 municipalities in the province of Buenos Aires.

2.1 Participants

The research focused on a population comprised of 303 students. To determine the necessary sample size, a 5% margin of error and a 95% confidence level were established. This translates to a required sample size of approximately 170 students. Subsequently, the actual sample in this study is composed of 286 HE students from two public institutions in Argentina, one for undergraduate studies and the other for university degrees: Mathematics and ICT in Education. This augments the representativeness of the sample concerning the target population. The number of women (n = 84; 29.37%) and men (n = 202; 70.62%) was significantly different, with a predominance of men. The age ranged from 17 to 49 years (M = 26.83 years, SD = 7.99 years and CV = 29.80%) and 59% lived in the Autonomous City of Buenos Aires, while the rest lived in the Province of Buenos Aires.

2.2 Instruments

The technique used to collect information on socio-demographic data and video game use was an ad hoc survey using Microsoft Forms®.

The questionnaire consisted of 17 questions: 14 closed and 3 open questions. It asked for information on qualitative variables (gender, area of residence, level of education, current degree, use of social networks, use of video games, use of video games at weekends, game mode, type of connection, video game genre, name of video game, platform, use of strategies) and quantitative variables (age, frequency of video game use, daily hours of video game use, self-assessment of gaming competence). The closed questions were answered with yes or no, and the multiple-choice questions, with a single score on a rating scale. For the variable frequency of video game use, the following scale was used: once a month, once a week, between 2 and 3 times a week, and every day or almost every day. For the self-assessment of gaming competence, we decided to rate it on a scale from 1 to 6 (1 being the lowest and 6 being the highest). Open-ended questions were used to collect data on the name of the most frequently played video games, hours of daily use and age.

2.3 Procedure

This work consisted of four stages. The first consisted of the development and validation of the instrument by experts (3 men and 3 women) with experience in research and professional links with emerging technologies in education or video games. Based on the expert review, some of the survey indicators were revised. A second evaluation showed a level of agreement of 85%. Reliability was ensured by limiting the distribution of the survey to a single researcher who forwarded it to the students' email accounts.

The second stage was the collection of information. Firstly, the teaching staff of the programs were asked for permission to conduct the study. The researcher explained the aims of the study and showed them the questionnaire so that they could give their consent. Then, during a one-hour class, the questionnaire was distributed, and the researcher explained to the students the purpose of the study and how to complete the questionnaire. Participation in the study was voluntary and anonymous, and each respondent agreed to the use of the information provided for academic purposes, thus covering the ethical aspects of privacy and confidentiality.

The questionnaire was then refined and tabulated. Finally, we analysed the data and drew conclusions.

2.4 Data Analysis

We used visualisation methods and univariate, bivariate and multivariate statistical techniques for data processing and analysis.

We first carried out a descriptive analysis of the data based on the variables mentioned above. We used Mann–Whitney and Pearson's chi-square tests to investigate possible relationships between variables. We also used the Multiple Correspondence Analysis (MCA) technique to combine the categories of variables according to the contribution of each participant. The procedure consisted of analysing the eigenvalues and variances in the list of data. By analysing their projections on each axis, the explained part of the variance and the values that contributed most to the definition of the dimensions, we constructed a bar chart. In addition, we constructed a biplot with the most representative values of the planes, formed by the contributions of the first axes. We grouped the variable categories into clusters using the levels of the gender variable. In this way, the association between the categories of the variables age, career, number of hours of use, game mode, video game genre, platform and type of connection allowed us to build profiles of groups of students with similar characteristics in their responses.

Data processing was carried out using spreadsheets and R software (version 4.0.2), considering a significance level to determine whether there were significant differences equal to or greater than 5%.

3 Results

The participants in the study came from two public institutions in Argentina. Of them, 78% were pursuing the first year of their undergraduate studies (HTCS: Higher Technician in Computer Science). The rest were doing the second year in a university degree (CC_BDES: Complementary Curriculum Cycle for the Bachelor's Degree in Education Sciences; CC_BDEM: Complementary Curriculum Cycle for the Bachelor's Degree in Education Management) and they already hold a HE degree. There was a greater presence of women in education-related studies (100% in CC_BDES and 64% in CC_BDEM), while men predominated in computer science (82%). There were also significant gender and age differences (Pearson chi-square test, p-value < 001).

As can be seen in Fig. 1, the age distribution of the participating students was largely dependent on career and gender. In CC_BDES all participants were women over 30, in CC_BDEM men were younger than women, whereas in HTCS both groups were more homogeneous.

Fig. 1 Age distribution of participating students by field of study and gender

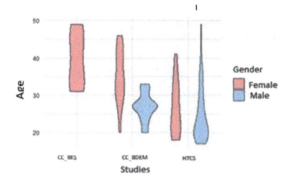

Fig. 2 Distribution of students who played video games by gender

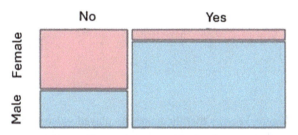

We found highly significant differences between students who used social networks and those who did not (Pearson chi-square test, p-value < 001), with similar behaviour in both genders.

The results of the analysis show that 63.63% of the participants in the study currently play video games in their leisure time, and only 11% of them are women. Figure 2 shows significant differences between those who play video games and those who do not, with men showing a greater preference for video games than women (Pearson chi-square test, p-value <001).

3.1 Profile of Students Who Play Video Games

Eighty-nine per cent of the sample of participants who played video games were in undergraduate studies related to the field of computer science, while the rest were in degrees related to education (CC_BDEM and CC_BDES). In both fields of knowledge, we observed that men played more than women in their free time.

The age of the study participants who played video games (n = 182) ranged from 17 to 49 years old (M = 24.32 years, SD = 6.50 years, CV = 26.75%). More specifically, 64% were between 17 and 24 years old, 22% were between 25 and 32 years old, and 14% were over 32 years old. There were also significant differences in the relationship between gender and age (Pearson chi-square test, p-value < 05).

Fig. 3 Age distribution of participants who played video games by gender

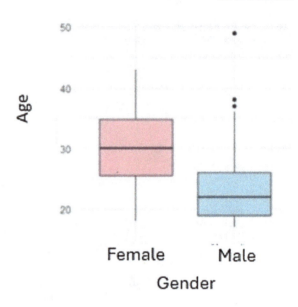

Figure 3 shows that the men who played video games the most were the youngest, while there were no age differences among women.

When respondents were asked about the frequency of video game use in their leisure time, 77% said they played at least twice a week (37% between 2 and 3 times and 40% every day or almost every day), 14% played once a week, and 9% played once a month.

Data related to frequency of use showed gender differences, although no statistical significance was found between frequency of use and gender (Pearson chi-square test, p-value = 0.1851). The frequency of daily or almost daily use was higher among men than among women: 43% compared with 10%. In addition, 60% of women and 79% of men played at least twice a week. On the other hand, the percentage of women who played only once a month was higher: 20% compared with 7% of men. The same was true for those who did it once a week: 20% compared to 14%.

When analysing the frequency of use by age range, Table 1 shows that women aged 17–24 played most frequently (two to seven days a week), while the rest did not play more than three times a week. In the case of men, on the other hand, a certain frequency of playing was maintained regardless of the age of the user. From these data we can see that frequency decreases with age.

In terms of how participants rated the average time they spent playing video games, the average was 2.46 h per game session, but the score was highly variable (CV = 65.55). When analysing this variable in relation to the age range of the group, significant differences were observed (Pearson chi-square test, p-value < 001). In this respect, Table 2 describes some specificities of the analysis by gender. It is noteworthy that most women spent no more than 3 h per session. On the other hand, there are groups of men in all age groups who played for 4 h or more; this was only observed for women in the 17–24 age group.

Table 1 Distribution of frequency of students' use of video games by gender and age

Gender	Age range	Frequency of use			
		At least once a month (%)	Once a week (%)	Between 2 and 3 times per week (%)	Every day or almost every day (%)
Women	17–24	0	0	67	33
	25–32	0	33	67	0
	33–49	50	25	25	0
Men	17–24	7	9	36	47
	25–32	6	18	41	35
	33–49	11	33	22	33

Table 2 Distribution of number of hours per day students play video games by gender and age

Gender	Age range	Number of hours per play session		
		1 h or less (%)	Between 2 and 3 h (%)	4 h or more (%)
Women	17–24	33	33	33
	25–32	67	33	0
	33–49	100	0	0
Men	17–24	18	58	24
	25–32	41	53	6
	33–49	22	56	22

In terms of time spent playing video games during the week, 47% of participants said they played more hours at weekends than on weekdays, 23% spent the same amount of time and a smaller percentage spent less time playing at weekends (20%). Notably, 80% of women and 54% of men reported spending more time on this activity at weekends (Fig. 4).

Another aspect that characterises the profile of students who play video games is related to the way they play. For the two main game modes (single player and multiplayer), the percentage of women playing alone was higher: 90%, compared to 46% of men. On the other hand, 54% of men played with other people (multiplayer), compared with only 10% of women. The data show a significant difference in game mode according to gender (Pearson chi-square test, p-value <01). Preferences in game mode also vary by age and gender: while younger and adult women preferred to play alone, no such differences by age group were found for men. In addition, 73% of the participating students played online, and the proportion was much lower among women than among men: 9% compared to 91%.

The results show that the most commonly used physical platforms or devices for playing games were desktop computers (52%), smartphones (25%) and, to a lesser extent, game consoles (15%) and netbooks/notebooks (8%). The percentage of women playing on their smartphones was three times higher than that of men (60%

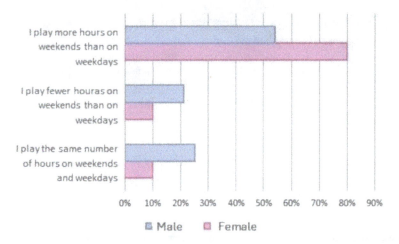

Fig. 4 Distribution of weekly time spent on video games by gender

compared to 21%), in contrast to desktop computers, which were used by 56% of men compared to 20% of women. The data show a significant difference by gender (Pearson chi-square test, p-value < 001).

In terms of age, desktop computers were the most common platform among the youngest participants, while the higher the age, the more participants played on their smartphones. Differences in device type by age were significant (Pearson chi-square test, p-value <01).

When analysing gender preference for video game genre, as Fig. 5 shows, men preferred shooters (29%), role-playing games (RPGs) (24%), sports games (16%) and casual games (8%). Significant differences were found between gender and video game genre preference (Pearson's chi-square test, p-value < 001).

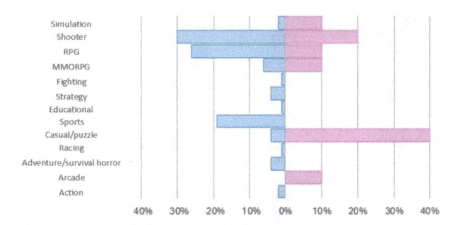

Fig. 5 Preferred video game genres by gender

Fig. 6 Video games preferred by participating students

Most women (40%) preferred casual games, while men preferred shooters (30%) and RPGs (26%). It is worth noting that 20% of women also said they liked shooters (20%), RPGs (10%) and MMORPGs (10%).

Students were also asked which game they played most often. Figure 6 shows the video games reported, with the strategy game League of Legends (15%), that is a Multiplayer Online Battle Arena (MOBA), the sports game FIFA (10%) and the first-person shooter Valorant (8%) standing out.

People of both genders reported playing the massively multiplayer online role-playing game (MMORPG) World of Warcraft (10%). The top three games preferred by men were FIFA (10%), League of Legends (17%) and Valorant (9%). Women preferred casual games such as Candy Crush (27%) or Bubble Shooter Genies (10%) and rol games such as Baldur's Gate (10%). Significant differences were found between gender and video game genre preference (Pearson's chi-square test, p-value <001).

We also investigated the players' self-perception of competence, which received scores between 3 and 6 (M = 4.35, SD = 0.98 and CV = 22.56%). The Mann–Whitney-Wilcoxon test (p-value = 0.1079) showed no significant differences between gender and self-perceived competence. However, the self-perceived competence of men was higher than that of women (Mv = 4.41, Mm = 3.90).

On the other hand, students were consulted about game strategies, including movement planning, strategic decision making, anticipation of opponent's actions, resource management, time management, and coordination with peers. 95% of respondents reported using strategies when playing (66% always, 29% sometimes). Table 3 shows gender differences in the use of strategies: 69% of women and only 40% of men said they always used strategies when playing video games.

Table 3 Distribution of students by gender according to the frequency with which they used strategies

Uses strategies	Gender	
	Women (%)	Men (%)
Always	69	40
Sometimes	26	50
Never	5	10

Table 4 Relationship of each variable to each dimension (eta2)

	Dimensions (eta2)		
	Dim.1	Dim.2	Dim.3
Gender	0.392	0.005	0.076
Age	0.296	0.234	0.307
Degree	0.414	0.027	0.122
Play_hr_day	0.374	0.133	0.131
Game_mode	0.288	0.279	0.066
Connection_mode	0.348	0.043	0.127
VG_genre	0.786	0.689	0.557
Device	0.408	0.280	0.248

3.2 Multidimensional Synthesis

In order to build a profile of the sample of HE students participating in the study who play video games in their leisure time, we carried out a multiple correspondence analysis (MCA) to identify the relationship between the variables gender, age, degree, number of hours of use, game mode, video game genre, platform, and type of connection.

Table 4 shows the eta2 values, which measure the relationship between each variable and each dimension.

In this sense, the first dimension (Dim.1) is most strongly related to preferred video game genre (0.786), followed by a group of variables such as device, degree, gender, average number of hours of gaming per day, and connection mode. Dimensions two and three (Dim.2, Dim.3) are related to preferred video game genre too (0.689, 0.557) but less than first dimension.

MCA reveals distinct dimensions reflecting the interplay of various factors within the realm of gaming behavior and preferences. Dimension 1 appears to encapsulate a multifaceted construct, suggesting a complex interrelationship among these variables in shaping gaming behavior. Dimension 2 and Dimension 3 highlight the preferences on video game genre, indicating how these factors might intersect with gaming engagement patterns, emphasizing their collective impact on gaming tendencies. Together, these dimensions offer valuable insights into the intricate dynamics underlying gamers' behaviors and preferences, providing a comprehensive framework for understanding the multidimensional nature of gaming-related phenomena.

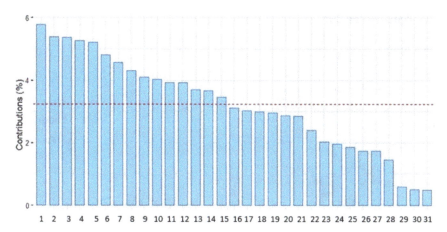

Fig. 7 Contribution of variable categories. References: 1: 33–49; 2: 4 h or more; 3: Action; 4:Net/notebook; 5: Smartphone; 6: Education; 7: does_not_play_online; 8: 1 h or less; 9:Woman; 10: Casual/puzzle; 11: Educational; 12: 25–32; 13: Strategy; 14: Sports; 15: Shooters; 16: MMORPG; 17: plays_with_others; 18: 2–3 h; 19: plays_alone; 20: Computer; 21: Console; 22: Adventure/survival horror: 23: Arcade; 24: 17–24; 25: Racing; 26: Simulation; 27: plays_online; 28. RPG; 29: ITC; 30: Man; 31: Fighting

Figure 7 shows the contributions of each category (from highest to lowest) of the three dimensions. The dashed line corresponds to the value that the contribution of each category would have at 100% inertia of the corresponding subspace.

The data on the number of variables (and categories) that accumulate the highest percentage of information can be seen in Fig. 8, where 74% of the inertia is accounted for.

In this structure of relationships, we could distinguish three groups of students (marked in the diagram with two colours, according to gender, to facilitate their identification).

Group 1 (blue area, excluding the part that overlaps with the pink area): this group includes male computer science students aged between 25 and 32, who play online games with other people, preferring shooters and sports games.

Group 2 (pink area, excluding the part that overlaps with the blue area): this group includes women in education, aged 33–49, who play 1 h or less per day on their smartphone or netbook/notebook, preferring simulation, casual and racing games.

Group 3 (overlapping pink and blue areas): this group includes men and women doing IT studies, aged 25–32, who play 2–3 h a day alone and offline on a computer, preferring role-playing games.

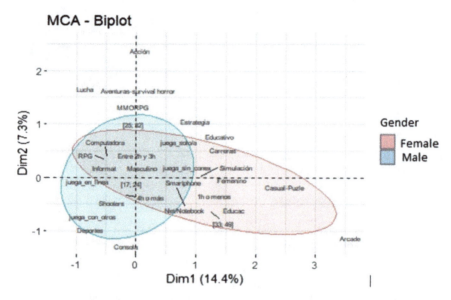

Fig. 8 Multidimensional representation of variables

4 Discussion

The main objective of this study was to analyse the relationship between variables such as gender, age, studies, hours played, game mode, video game genre, platform and type of connection in a sample of Argentine students who play video games in their free time.

First of all, it should be noted that the participants of the study were studying in two public HE institutions in Argentina. Most of them were related to IT studies, and the rest were related to education. In this respect, the distribution of students by gender is very different. The results of a higher or lower representation of women in the fields of education and IT are in line with the fact that women in Argentina tend to study social sciences rather than applied sciences [49] and with the results of studies carried out in other contexts, such as Spain [50].

Secondly, looking at the data, we can confirm that there are gender and age differences in this sample of university students in terms of consumption habits and preferences regarding the genre of video games. Neither group is unfamiliar with the medium, and more than half of the students interviewed play video games in their free time, although there are significant gender differences between those who play and those who do not (men have a greater preference for these cultural products). The data obtained on the use of video games is higher than in the study conducted [6] and is consistent with the gender bias, confirming the relevance of this cultural product for men [6, 51]. This result contrasts with other international studies [52, 53], which confirm that in recent years there has been a trend towards gender parity in use, a fact that is not yet evident in Argentina.

Thirdly, regarding the use of video games and the analysis of the profile of the sample of students who play video games, there are significant gender differences in terms of age, hours of use, game mode, platform and video game genre. The participants who played the most video games were those in the youngest age group (17–24 years). Gaming decreases with age for women, while it remains the same for men; these findings are consistent with previous studies [6, 52, 53].

Almost half of the participants indicated that they played more hours on weekends than on weekdays, which is consistent with other work [54]. Most of them played in their free time at least two days a week, and the average play time was 2.46 h per play session. The frequency of use differed between genders, with men spending almost four times as much time on this activity as women, every day or almost every day. In addition, women mostly played video games as an individual exercise, in contrast to men who mostly played online in multiplayer game modes, as seen in previous studies [6, 55].

In terms of the use of specific devices or platforms, the group interviewed expressed a strong tendency to use desktop computers for their gaming sessions and, to a lesser extent, smartphones, game consoles and netbooks/notebooks. The percentage of women using smartphones was much higher than that of men, who preferred desktop computers. These results are in line with other paper [54]. In addition, the use of computers is also more frequent among younger people, and the use of mobile devices increases with age.

In general, the results obtained in the group of men participating in this study are in line with previous studies, such as another carried out [56] in the Spanish context. Like the results of the present study, these researchers found that men spend more leisure time playing video games, play at an older age and are ahead of women in the number of hours spent playing video games on the PC, both during the week and at weekends.

The preferred video game genres were shooters, role-playing games, sports games and casual games, in line with [6], although there were differences by gender. While men's main choices were shooters and RPGs, women preferred casual games [57]. and played shooters, RPGs and MMORPGs to a lesser extent. This may be related to the content of these video games, their different treatment of gender roles, and the existence of games "for boys" and games "for girls" [58].

Finally, most subjects claimed to use strategies when playing games, especially women. In turn, men perceive themselves as more competent than women when playing games (or in what other studies have called video game culture [59]. This is consistent with the so-called idealisation of self-competence observed in other domains, such as digital literacy [60].

5 Conclusions

Play is an inherently human activity. "Everyone, regardless of age, likes to play" [50]. Play is universal and has multiple functions. People of all ages, backgrounds and eras have voluntarily participated in games for purely recreational purposes. As well as being fun, games motivate people to act, to take risks, to persevere and to set and achieve increasingly challenging goals. These aspects play a central role in the development process of every human being [61]. Through play, people develop ethical skills, such as respect for rules and other participants; relational skills, such as socialisation, teamwork and sharing; and cognitive skills, such as creativity, attention, memory, logic and reasoning [1, 62].

Today, video games are one of the most important social, cultural and technological expressions of contemporary life. It is the world's number one entertainment industry, both in terms of budget and time spent. It is the medium and language of today's society. Through their interactive and participatory experiences, video games contribute to the creation of a new cultural imagination. Their influence on contemporary life is evident in the economic, social, cultural and artistic spheres. They favour the progressive disappearance of the real and the advent of the era of the simulacrum and derealisation [63].

This work has made it possible to identify the profile of video game players based on the analysis of a sample of university students in Argentina. The data obtained allows us to understand the habits and preferences in the choice and use of video games. This makes it possible to better understand a reality that affects our leisure time and our relationships and interactions with other people and with games.

This reality should not be ignored; it will be essential for the education of young people and for the training of those who accompany them as teachers or family members, since playing video games has proved to be independent of age.

This paper provides an analysis of video game usage among HE students in Argentina. By examining sociodemographic variables and usage patterns through a quantitative approach, the study uncovers critical insights into the intersection of gender, age, and current degree with gaming habits. These findings highlight the prevalent cultural and social role of video games among students. This research offers valuable data that can contribute to the existing literature on contemporary digital culture, to outline educational strategies aimed at integrating video games into pedagogical practices, to improve student participation and learning outcomes in HE. In this way, we agree about the idea that video games as cultural significance, communicative power, motivational potential, skill development benefits, and experiential learning opportunities, are valuable tools that should be integrated into educational practices [64].

As far as the limitations of this study are concerned, they are mainly related to the sample selection: the size of the sample is not large enough and it was purposively selected. Therefore, as lines of future or prospective research, we intend to include other higher education institutions to try to overcome this limitation. Furthermore, it would also be interesting to go beyond national borders and carry out similar studies with samples from other countries to make a comparative study, find differences or similarities and extrapolate the results.

Acknowledgements This paper was developed within the framework of the Universitat de Valencia and Universidad Nacional de Tres de Febrero (Argentina) agreement Code RRII2023-24201. We appreciate the collaboration of Manuel Gil Fernández who carried out the translation of the text, that was funded by the Department of Education and School Management, University of Valencia.

References

1. Huizinga J (2007) Homo Ludens, Alianza, España. (Original work published 1938).
2. Esnaola Horacek G, Galli M (2016) Juegos. Juguetes y Videojuegos. Para Juanito 9:17–23
3. Newzoo The Games Market in 2022: The Year in Numbers, https://newzoo.com/insights/articles/the-games-market-in-2022-the-year-in-numbers. Accessed 21 Dec 2022
4. Wong L (2021) América Latina Juega. Historia del videojuego latinoamericano. Heroes de papel
5. Revale HJ, Minotti J, Cerutti O (2022) Resumen anual 2022. Observatorio de la Industria Argentina de Desarrollo de Videojuegos, Universidad Nacional de Rafaela
6. Sistema de Información Cultural de la Argentina Encuesta Nacional de consumos culturales 2013/2023, https://www.sinca.gob.ar/VerDocumento.aspx?IdCategoria=10. Accessed 2023/3/28.
7. López-Gómez S (2018) Análise descritiva e interpretativa do deseño e contido dos videoxogos elaborados en Galicia (Tesis doctoral). Universidad de Santiago de Compostela, España. https://minerva.usc.es/xmlui/handle/10347/16695
8. Vidal MI, Castro MM, Caamaño T, Marín D (2021) Describiendo experiencias con videojuegos en contextos diversos. In: Rodríguez J, López S (coords) Los videojuegos en la escuela, la universidad y los contextos sociocomunitarios, pp 193–202. Octaedro, Barcelona
9. López Gómez S, Rial Boubeta A, Marín Suelves D, Rodríguez Rodríguez J (2022) Videojuegos, salud, convivencia y adicción. ¿Qué dice la evidencia científica?. Psychol Soc Educ 14(1):45–54
10. Esnaola Horacek G (2006) Claves culturales en la construcción del conocimiento: ¿qué enseñan los videojuegos? Alfagrama, España
11. González González C, Blanco Izquierdo F (2008) Emociones con videojuegos: incrementando la motivación para el aprendizaje. Revista Electrónica Teoría de la Educación. Educación y Cultura en la Sociedad de la Información, vil 9, pp 69–92
12. Brilliant T, Nouchi R, Kawashima R (2019) Does video gaming have impacts on the brain: evidence from a systematic review. Brain Sciences 9(10)
13. del Moral M, Guzmán A (2016) Jugar en red social: ¿Adicción digital versus comunicación e interacción en CityVille? Cuadernos de Información (38)
14. López Ortiz IA, Mosquera Angulo H (2016) Percepción de los Estudiantes sobre la Utilización de Videojuegos en Cursos de la Universidad Nacional Abierta y a Distancia -UNAD. Publicaciones E Investigación 10:163–175
15. Nouchi R, Taki Y, Takeuchi H, Hashizume H, Nozawa T, Kambara T, Sekiguchi A, Miyauchi CM, Kotozaki Y, Nouchi H et al (2013) Brain training game boosts executive functions, working memory and processing speed in the young adults: a randomized controlled trial. PLoS One 8(2)
16. Shams T, Foussias G, Zawadzki J, Marshe V, Siddiqui I, Müller D, Wong A (2015) The effects of video games on cognition and brain structure: potential implications for neuropsychiatric disorders. Curr Psychiatry Rep 17(9)
17. Revuelta I, Guerra J (2012) ¿Qué aprendo con videojuegos? Una perspectiva de metaaprendizaje del videojugador. Revista de Educación a Distancia (33)

18. Revuelta F, Pedrera M (2018) Bases neuroeducativas y socioemocionales para trabajar con videojuegos en contextos de aprendizaje. Edmetic 7(2):5–9
19. del Moral M (2014) Videogames: Opportunities for learning. J New Approaches Educ Res 3(1)
20. Gee J (2004) Lo que nos enseñan los videojuegos sobre el aprendizaje y el alfabetismo. Aljibe, España
21. Gee J (2007) Good video games + good learning. Peter Lang, Suiza
22. Prensky M (2014) No me molestes mamá, ¡estoy aprendiendo. SM, España
23. Chen C, Law V (2016) Scaffolding individual and collaborative game-based learning in learning performance and intrinsic motivation. Comput Hum Behav 55:1201–1212
24. Anderson CA, Shibuya A, Ihori N, Swing EL, Bushman BJ, Sakamoto A et al (2010) Violent video game effects on aggression, empathy, and prosocial behavior in Eastern and Western countries: a meta-analytic review. Psychol Bull 136:151–173
25. Hassan Y, Bégue L, Bushman B (2013) The more you play, the more aggressive you become: a long-term experimental study of cumulative violent video game effects on hostile expectations and aggressive behavior. J Exp Soc Psychol 49(2). Social Psychology
26. Peralta L, Torres M (2020) Adicción a videojuegos en relación con la conducta antisocial y delictiva en adolescentes de un colegio estatal de Lima. Revista de Investigación y Casos en Salud 5(3):118–130
27. Gentile DA, Swing EL, Lim CG, Khoo A (2012) Video game playing, attention problems, and impulsiveness: evidence of bidirectional causality. Psychol Pop Media Cult 1:62–70
28. Kim D, Nam JK, Keum C (2022) Adolescent internet gaming addiction and personality characteristics by game genre. PloS One 17(2)
29. Lobel A, Engels R, Stone L, Burk W, Granic I (2017) Video gaming and children's psychosocial wellbeing: a longitudinal study. J Youth Adolesc 46:884–897
30. Chacón Cuberos R, Zurita Ortega F, Martínez Martínez A, Castro Sánchez M, Espejo Garcés T, Pinel Martínez C (2017) Relación entre factores académicos y consumo de videojuegos en universitarios. Un modelo de regresión. Píxel-Bit. Revista de Medios y Educación (50), pp 109–121
31. López O, Pinargote O, Cedeño E, Zambrano W, Mendoza M (2021) El uso de videojuegos en el confinamiento y su incidencia psicológica y social en las personas. Revista Ibérica de Sistemas e Tecnologias de Informação 43(3):633–646
32. Sanchez M, Mari M, Benito A, Rodriguez F, Castellano F, Almodovar I (2021) Personality traits and psychopathology in adolescents with videogame addiction. Revista Adicciones 2(3):1629–1638
33. Sharif I, Wills T, Sargent J (2010) Effect of visual media use on school performance: a prospective study. J Adolesc Health 46(1):52–61
34. Organización Mundial de la Salud. Addictive behaviours: Gaming disorder, https://www.who.int/news-room/questions-and-answers/item/addictive-behaviours-gaming-disorder#:~:text=Gaming%20disorder%20is%20defined%20. Accessed 31 Oct 2022
35. Benítez Larghi S, Duek C, Moguillansky M (2017) Niños, nuevas tecnologías y género: hacia la definición de una agenda de investigación. Revista Fonseca (14):167–179
36. Griffiths MD (1991) Amusement machine playing in childhood and adolescence: a comparative analysis of video games and fruit machines. J Adolesc 14(1):53–73
37. Chóliz M, Marco C (2010) Pattern of use and dependence on video games in infancy and adolescence. Anales de Psicología 27(2):418–426
38. Esposito M, Serra N, Guillari A, Simeone S, Sarracino F, Continisio GI, Rea T (2020) An investigation into video game addiction in pre-adolescents and adolescents: a cross-sectional study. Revista Medicina 56(5):200–221
39. Rodríguez M, Padilla F (2021) El uso de videojuegos en adolescentes. Un problema de Salud Pública. Revista Enfermería Globa 20(2):557–591
40. Shen C, Ratan R, Cai D, Leavitt A (2016) Do Men Advance Faster than Women? Debunking the Gender Performance Gap in Two Massively Multiplayer Online Games. J Comput-Mediat Commun 21(4):312–329

41. Salas E, Merino C, Chóliz M, Marco C (2017) Análisis psicométrico del test de dependencia de videojuegos (TDV) en población peruana. Revista Universitas Psychologica 16(4):1–13
42. Mentzoni RA, Brunborg GS, Molde H, Myrseth H, Mar Skouverøe KJ, Hetland J, Pallesen S (2011) Problematic video game use: estimated prevalence and associations with mental and physical health. Cyberpsychol Behav Soc Netw 10(14):591–596
43. Schou C, Billieux J, Griffiths MD, Kuss DJ, Demetrovics Z, Mazzoni E, Pallesen S (2016) The relationship between addictive use of social media and video games and symptoms of psychiatric disorders: a large-scale cross-sectional study. Psychol Addict Behav 30(2):252–262
44. Sánchez-Domínguez J, Telumbre Terrero J, Castillo L (2021) health and addictions/salud y drogas. Descripción del uso y dependencia a videojuegos en adolescentes escolarizados de Ciudad del Carmen, Campeche. Health and Addictions/Salud Y Drogas 21(1)
45. Fox J, Tang W (2017) Women's experiences with general and sexual harassment in online video games: rumination, organizational respon-siveness, withdrawal, and coping strategies. New Media Soc 19(8):1290–1307
46. Wohn D (2011) Gender and race representation in casual games. Sex Roles: A J Res 65(3):198–207
47. Chess S, Shaw A (2014) A conspiracy of fishes, or, how we learned to stop worrying about #GamerGate and embrace hegemonic masculinity. J Broadcast & Electron Medi 59(1):208–222
48. Patton M (2002) Qualitative research and evaluation methods. Sage Publications, Thousand Oaks
49. Secretaría de Políticas Universitarias: Mujeres en el Sistema Universitario Argentino 2020–2021. Secretaría de Políticas Universitarias (2022)
50. Martín A et al (1995) Actividades lúdicas. El juego, alternativa de ocio para jóvenes. Popular, Madrid
51. Guerra-Santana M, Rodríguez-Pulido M, Artiles-Rodríguez J (2021) Use of social networks by university students from a personal and educational sphere. Aula Abierta 50(1):497–504
52. Asociación Española de Videojuegos: La industria del videojuego en España en 2021. Anuario. Asociación Española de Videojuegos (2022)
53. Entertainment Software Association. The 2022 Essential Facts About the Video Game Industry (2022)
54. Ricoy Lorenzo M, Ameneiros Estévez A (2016) Preferencias, dedicación y problemáticas generadas por los videojuegos. Revista complutense de educación 27(3):1291–1308
55. Hinojal H, Pirro A (2020) Adolescentes y los videojuegos. Realidades, percepciones y posibilidades. In: Jiménez Alcázar JF, Rodríguez GF, Massa SM (Coords), Historia, videojuegos y educación: nuevas aportaciones, pp 31–46. Universidad de Murcia. Servicio de Publicaciones, Murcia.
56. Iglesias-Caride G, Domínguez-Alonso J, González-Rodríguez R (2022) Influencia del género y la edad en el uso de los videojuegos en la población adolescente. Psychol Soc Educ 14(2):11–19
57. Carvalho AA, AraÃējo I, Zagalo N (2014) A framework for gamified activities based on mobile games played by Portuguese University Students. International Association for the Development of the Information Society
58. Padilla-Walker LM, Nelson LJ, Carroll JS, Jensen AC (2009) More than just a game: video game and internet use during emerging adulthood. J Youth Adolesc 39:103–113
59. Oceja J, González-Fernández N (2020) University students and video games: perceptions, use, and preferences according to gender. Educ Policy Anal Arch 28:66
60. Marín D, Gabarda V, Ramón-Llin JA (2022) Análisis de la competencia digital en el futuro profesorado a través de un diseño mixto. Revista de Educación a Distancia (RED) 22(70):1–30. https://doi.org/10.6018/red.523071
61. Rivoltella PC (2011) Filosofia del videogioco: capovolgendo McLuhan. Vita e Pensiero 2011(6):110–115
62. Juul J (2011) Half-real: video games between real rules and fictional worlds. MIT Press, Cambridge
63. Baudrillard J, del Solar JJ (2000) Pantalla total. Anagrama, Barcelona, p 239

64. Scolari C (2018) Alfabetismo transmedia en la nueva ecología de los medios. Universitat Pompeu Fabra, Libro blanco

Gami|cation: Gamification Within (Higher) Education

L. Christa Friedrich, Matthias Heinz, Josefin Müller, and Michelle Pippig

Abstract The serious game Gami|cation—the learning-teaching adventure was developed to support (higher education) teachers in designing teaching–learning arrangements that promote motivation by enriching educational processes with game elements. The different types of using gamification, the promotion of intrinsic and extrinsic motivation and the creation of an interactive learning environment are also considered. The paper focuses on didactically innovative applications of gamification elements in higher education. Furthermore, the scientific foundation, the development process and the developed application scenarios of gamification are described. Previous experiences from workshops are reflected upon to gain ideas for further development.

Keywords Gamification · Educational game · Learning experience · Higher education didactics · Game pedagogy

---

L. C. Friedrich · M. Pippig (✉)
Technische Universität Dresden, CIDS CODIP, 01062 Dresden, Germany
e-mail: michelle.pippig@tu-dresden.de

L. C. Friedrich
e-mail: linda_christa.friedrich@tu-dresden.de

M. Heinz
Technische Universität Dresden, Zentrum für Interdisziplinäres Lernen und Lehren, 01062 Dresden, Germany
e-mail: matthias.heinz@tu-dresden.de

J. Müller
Fachhochschule Dresden, Güntzstraße 1, 01069 Dresden, Germany
e-mail: j.mueller@fh-dresden.eu

© The Author(s), under exclusive license to Springer Nature Switzerland AG 2025
E. Vendrell Vidal et al. (eds.), *Advanced Technologies and the University of the Future*, Lecture Notes in Networks and Systems 1140,
https://doi.org/10.1007/978-3-031-71530-3_25

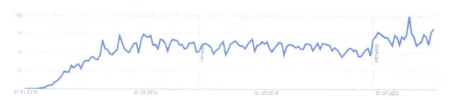

Fig. 1 Google trend analysis on gamification 2010 to February 2024

1 Introduction

In recent years, the integration of gamification into higher education has become increasingly important. Gamification has not only established itself in the higher education area. Interest in the topic has even intensified in the past years, as illustrated by the Google trend analysis on gamification in Fig. 1. Since 2010, gamification has seen a consistently high level of search interest with an upward trend.

This paper contributes by presenting an in-depth analysis of Gamilcation. Gamilcation—The Learning-Teaching Adventure is an innovative approach to the gamified design of teaching–learning arrangements in higher education. This paper describes the current evaluation of gamification, the scientific foundation of Gamilcation and the detailed presentation of concrete application scenarios. The focus ranges from the support that gamification offers teachers in integrating game elements into their teaching–learning arrangements in different educational contexts. Practical experiences are reflected upon, adaptation possibilities are shown and insights into the lessons learned from this learning-teaching adventure are provided. The aim is to demonstrate the depth and complexity of the synthesis between gamification and education.

2 Gamification in Higher Education

Current research findings on gamification in higher education indicate that game elements can improve student learning experience and outcomes. A meta-analysis by Khaldi et al. [8] shows that gamification in higher education can positively impact students' learning behaviour, performance and satisfaction if the users' individual needs and personality profiles are also considered. Game elements can have a varying, motivating effect. Gamification can be implemented in different ways in university teaching [5]. Gamification of teaching–learning arrangements can support learning processes, increase students' motivation and performance and boost their engagement [14]. Gamification and personalisation of content are particularly important here, so machine learning techniques are used to adapt the selection of game elements to the gamified content [8]. If implemented, gamified teaching–learning arrangements can positively affect cognitive, motivational and behavioural learning outcomes [12]. Bai and colleagues even go so far to declare that gamified teaching–learning arrangements deliver better learning outcomes than teaching–learning arrangements without gamification [1]. The fact that gamification has arrived in the higher education is also illustrated by the establishment of gamification professorships and degree programmes. Nevertheless, it is important to be aware of the potential risks of gamification and to consider these when applying it. "Using game elements can be associated with a variety of risks and side effects, e.g. manipulation, demotivation, encouragement to show behaviour only when rewarded, replacement of intrinsic motivation by striving for extrinsic rewards, game optimization instead of concentration on the task and increased disclosure of information" [7].

Gamification in higher education refers to integrating game elements and principles into teaching and learning activities to increase student engagement, motivation and learning success. Among other things, this innovative method is used in pedagogy to increase the intrinsic motivation of users through game mechanics, such as scoring systems and progress indicators, and to achieve learning objectives. A central aspect of gamification in higher education is creating an interactive learning environment that maintains students' attention and encourages them to actively participate in the learning process [4]. The use of gamification in teaching offers a variety of benefits to improve learning. The integration of game elements intensifies the feedback in the interaction with students. This enables teachers to track students' learning progress better and provide individualised feedback. With challenges, competitions and rewards, students are motivated to deepen their knowledge, develop critical thinking skills and improve problem-solving abilities [9]. In addition, using playful elements encourages students to become more involved in the learning process and improves their performance [6]. Another advantage is the simplification of teaching and learning tasks. By using gamification, complex teaching content can be presented in a playful and entertaining way, which makes it easier to understand and absorb the information. Finally, gamification helps to increase student satisfaction. The interactive and motivating learning environments created by gamification lead to a positive learning experience and greater retention of the learning material [13]. To enable learners to benefit from these positive effects, it is necessary to sensitise teachers to gamification and demonstrate the broad horizon of possible game elements.

3 The Development Process of Gamilcation

Gamilcation is based on the results of the project LOS (Learning Experiences in OPAL with Game Elements), which was linked to the Center for Open Digital Innovation and Participation (CODIP) and the former Media Centre. The project aimed to enhance the learning experience in digital learning, also with the learning management system (LMS) OPAL (online platform for academic teaching and learning) by using game strategies and elements to expand the skills development of teachers in this regard to network and sensitise interested stakeholders and to develop the technical infrastructure strategically [10]. This should increase the quality of digital teaching and learning scenarios in the long term, support the *shift from teaching to learning* and promote student motivation and participation in digital learning. To achieve the objectives mentioned above, the project LOS developed further training and transfer programmes for integrating and using game elements and strategies within the Saxon learning platform and made them accessible to all stakeholders at Saxon universities. These consist of interlinked online and face-to-face elements and ensure that the topics of learning experience and gamification are established in the state's higher education landscape. The central element of the transfer and further education concept was an open knowledge and qualification base with free teaching and learning materials within the LMS OPAL. Tutorials, best practices, didactic design patterns, course templates and a community were created through LOS. In the design pattern development context, the game elements of the Marczewski taxonomy [11] were also taken up, selected and rewritten for the higher education area and in the OPAL context. The game elements were compiled in close cooperation with experts from the Saxon institutions of higher education for game-based teaching and learning formats, as well as the e-learning supporters and with the involvement of the target group (lecturers at Saxon institutions of higher education). This preliminary work was taken up and further developed by the Digital Learning and Gaming Cultures (DLGC) cluster at the CODIP.

Due to the increasing challenge of inspiring and sustainably motivating young people to demand learning content, the DLGC cluster, consisting of media design experts and specialists in the field of media education and didactics, focused on the synthesis of gamification and education. This intensive dialogue resulted in the development of the Serious Game Gamilcation—The Learning-Teaching Adventure, which conveys the use and characteristics of gamification elements in a low-threshold way. Serious games provide a better quality of learning as well as a variety of positive results compared to conventional learning-teaching arrangements [2]. "One established method of teaching is the use of Serious Games, as it has various positive effects in terms of motivation and engagement" [3]. The central objective of Gamilcation is to enrich educational processes by integrating game elements, which is a first step for additional learning motivation. This considers different types of use and aims to promote intrinsic and extrinsic motivation and create an interactive learning environment.

The development of the serious game was based on the periodic table for gamification elements developed by Marczewski [11]. Concerning educational contextualisation, the interdisciplinary DLGC cluster identified a total of 46 elements as particularly enriching and applicable in the education area. These selected elements are recorded on playing cards and act as an invitation for users to explore the world of gamification.

The decision to arrange the playing cards into a playing field in the form of a hexagonal grid was based on several considerations. The approach of a periodic table, as chosen by Marczewski, was abandoned because no fixed arrangement of the elements could be recognised. In contrast, each element in a classical periodic table has a fixed position, both horizontally and vertically. For this reason, a more suitable layout had to be found. Figure 2 illustrates the development process for the game design. The initial consideration was to retain the linear approach (with rectangular

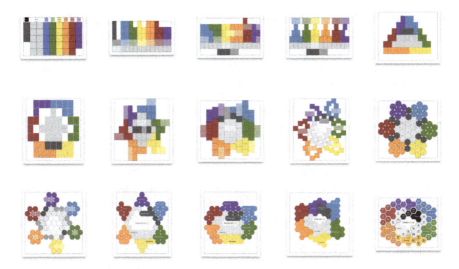

Fig. 2 Development process of the game design of Gamilcation (own illustration)

cards), which was based on Marczewski's periodic table. An attempt was made to visualise this linearity in the form of a table or pyramid. In the process, the layout developed more and more towards a star and circle shape, which finally ended with an oval game board (with a hexagonal grid) and hexagonal-shaped playing cards. The hexagonal shape can also be found in Marczewski's player types.

Another critical reason the hexagon was chosen as the basic shape was the number of player types—there are six. This ensures an even distribution without any tendencies towards order. In addition, the hexagon can save unnecessary space when arranging the playing cards on the game board, which is particularly advantageous when using the analogue serious game. The arrangement of playing cards on a board as a hexagonal grid is similar to the game board of *The Settlers of Catan*, a design feature familiar to many games. This game design is often chosen because of the following essential aspects:

- **Aesthetics and visual appeal**: The regular arrangement of hexagons can create a visually appealing pattern that encourages players to engage more deeply with the game and have a positive play experience.
- **Clear structure and orientation**: The hexagonal layout provides a clear and easily recognisable structure for the playing field, which can help players orientate themselves and plan their strategies. This clear structure can help reduce confusion and ensure a smooth gaming experience.
- **Variability and flexibility**: The hexagonal layout allows a flexible design of the playing field and offers space for various game elements and strategies. Players can develop and try out different tactics, resulting in a rich and varied gaming experience.

- **Cognitive challenge and learning effects**: The complex structure of the game board can provide cognitive challenges that encourage players to think strategically, solve problems and make decisions. This can lead to improved cognitive engagement and learning effects as players develop their skills in areas such as planning, decision-making and problem-solving.

The developed serious game Gamilcation consists of eight different areas, which are made up of six user types and two general gamification categories. These are based on the Marczewski taxonomy [11], which was mentioned before. The graphical representation in Fig. 3 in the form of a circle illustrates how the game design has developed in its arrangement of user types and reflects the ongoing development. At the top of the cycle is the moment of upheaval, symbolised by the **Disruptor**, which questions established structures and seeks innovative paths. The **Free Spirit** likes to start on uncharted territory to discover and explore new things. **Philanthropists** identify with their actions and offer support and help. **Socialisers** are motivated by interacting and connecting with other people. The **Achiever** challenges themselves to achieve personal goals, while the **Player** pushes their limits and masters' new challenges. The two general categories, **General** and **Rewards,** do not require their own side on the hexagonal playing cards, as they can be applied to all user types.

The interlocking of the user types in a circle or hexagon is also intended to illustrate those players or learners never belong to just one user type [11]. They are more strongly characterised by one or two specific types, but there is also a mixture of the characteristics of the other user types. The colours were assigned to the different user types using the colour wheel (Fig. 3). This colour wheel, which consists of the primary and complementary colours, served as a guide for the selection of colours to ensure a harmonious and visually appealing design. No specific derivation of the

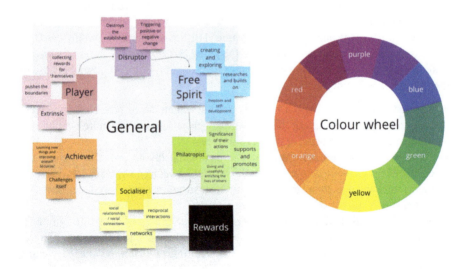

Fig. 3 Considerations on the arrangement with colour reference of the user types (own illustration)

colour meanings was deliberately made for the assignment to the user types. Rather, it was important to use different colours to clearly distinguish the user types on the one hand and achieve a visually appealing presentation on the other. Users should thus be able to identify the different types easily and, at the same time, be visually stimulated, which enriches the overall gaming experience.

Overall, the arrangement of the hexagon playing cards into a hexagonal playing field can have a positive impact on players and learners by providing aesthetic appeal, clear structure, variability and flexibility as well as cognitive challenges. These features should contribute to a valuable gaming experience while supporting mental skills and learning processes.

This resulted in a game board (Fig. 7) and 46 game element cards (exemplary Fig. 8), which can be used in different application scenarios (see Chap. 4). The eight areas are shown on the game board, including areas for the corresponding game element cards. The game element cards are placed on this—depending on the scenario, the cards are already on the game board (Fig. 4) or the players have to assign the game cards to the areas on the game board. The playing cards consist of a front side with a description of the element and a back side with examples of applications. These can also be separated from each other for the applications of a special scenario (scenario 4) and put together by the players.

4 Application Scenarios of Gamilcation

Gamilcation is based on the project LOS product described above. The special feature of Gamilcation is its versatility and adaptability, which make it possible to use the serious game in both German and English as well as analogue and digital. The serious game can be used as a playful introduction to the various user types (Free Spirit, Philanthropist, Socialiser, Achiever, Player and Disruptor) or for creative support in the development of individual teaching–learning arrangements. The interactive playing field not only introduces users to the world of gamification, but also actively challenges them. The flexibility makes it possible to target different competencies and use the scenarios according to requirements. To illustrate this, three developed scenarios are presented below as examples.

Scenario 1—Overview

The fully equipped Gamilcation landscape (playing field) is used for scenario 1 (Fig. 4). All game elements are placed in their corresponding position in the Gamilcation landscape.

At a first glance, the elements show their element abbreviation and name as well as their affiliation to a user type or a general category (General, Rewards, Free Spirit, Philanthropist, Socialiser, Achiever, Player and Disruptor). The information on the description and the learning objectives can be viewed by activating the game elements.

Fig. 4 Gami|cation landscape for analogue and digital use in scenario 1 (own illustration)

In the digital version, this is done by clicking on the element, while in the analogue version it is done by turning the game card (Fig. 5). This scenario provides a comprehensive overview of the available elements and their significance in the context of gamification and thus sensitises the user to a wide range of gamification possibilities.

Fig. 5 Cover sheet, front and back of the gamification element Easter Egg (own illustration)

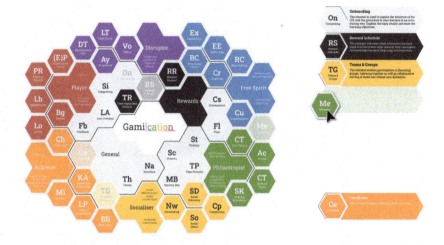

Fig. 6 Digital utilisation of scenario 2 (own illustration)

Scenario 2—Designing Your Own Course with Gamification Elements

The setup and interaction mechanisms for viewing the game element information are identical to scenario 1. In addition, users in this scenario have a **planning area** at their disposal for designing their own course. The selected gamification elements are arranged by removing, dragging and dropping them into the planning area (Fig. 6). This scenario gives users the creative opportunity to design a course by specifically selecting gamification elements and integrating them into their planning.

Scenario 3—Assign Utilisation Types

No game elements are visible on the Gamilcation landscape in this scenario. Only the areas of user types and general categories (General, Rewards, Free Spirit, Philanthropist, Socialiser, Achiever, Player and Disruptor) are shown on the playing field (Fig. 7).

Next to the playing field, the game elements are arranged individually and unsorted with their element abbreviations and names, but without any assignment to an user type (Fig. 8).

The task of the players—whether alone or in a group—is to assign the game elements on the gamification landscape to the correct user types and general categories. Interaction occurs in both the analogue and digital versions by moving and placing the elements. In the digital version, the implementation of the drag-and-drop function significantly increases the interactivity of the serious game. It is evaluated using a points system, with each correct allocation being rewarded with a point. The person or group with the most points wins. Figures 9 and 10 below show examples of the game introduction and the legend for the user types and categories, distributed over seven analogue playing cards.

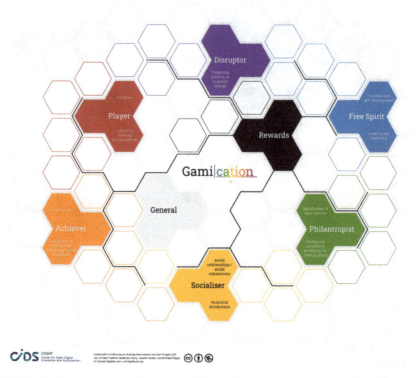

Fig. 7 Gami|cation landscape for scenario 3 (own illustration)

Fig. 8 Front and back of the gamification element Easter Egg (own illustration)

Gamilcation: Gamification Within (Higher) Education

Fig. 9 Game introduction distributed over four analogue playing cards (own illustration)

Fig. 10 Legend of the user types and general categories distributed over three analogue playing cards (own illustration)

These scenarios offer teachers flexible and versatile application options in various educational contexts. They promote the understanding of game elements and enable the targeted development of teaching–learning arrangements. Gamilcation is thus intended to offer a playful method of imparting knowledge, promoting teamwork skills and creating an entertaining learning environment in the education area through the simultaneous integration of game elements, particularly the **pedagogical double-decker.** A detailed description of the experience of Gamilcation is described in Chap. 5. Even the use of a few easily integrated game elements in non-game-based teaching–learning arrangements can make a significant difference. Gamilcation sensitises teachers to this innovative perspective to make teaching–learning more effective and enriching.

Gamilcation is constantly being further developed based on continuous practical experience gained through its use in workshops. The scenarios are adapted and expanded in line with the needs of the target group. This flexible approach makes it possible to make quick and needs-based adjustments. For example, it is possible to deliberately leave the back of the playing cards blank—without specifying application examples—so that users can develop their own use cases for their specialist area and/or their teaching. This makes it possible to create a constantly growing repertoire of different practical applications that can be made available to higher education teachers.

Another application is currently work in progress, whereby the correct allocation of the description and learning objectives of the game elements is to take place. In this special game variant (scenario 4), the front (description) and back (learning objectives and application examples) of the game elements are separated. The users must find the matching front and back sides, put them together and then check them. In the analogue version, the check is carried out using a free field on the backside. If the two sides are stapled together correctly, the abbreviation of the element is merged in this field on the back. If the pages do not match, only dots and dashes can be recognised without any connection. This playful activity promotes precise knowledge of the game elements and allows participants to check and validate their own performance. The haptic experience of interacting with the playing cards reinforces the learning experience and provides an interactive way to consolidate the acquired knowledge.

5 Experience and Further Development

The first teaching and learning strategies have already been trialled in practice as part of previous projects, such as LOS. The starting point for this was the annual *Community in New Media conference* in Dresden, Germany in 2020. As part of this face-to-face event, a workshop was held with participants from the higher education area, who were introduced to gamification and Marczewski's player types. The participants were able to compete against each other in a kind of matching game. The groups first had to assign the corresponding characteristics to the player types (Round 1), based on scenario 4. Afterwards (Round 2), the various game elements were assigned to the respective player types, as described in scenario 3 (Chap. 4). The implementation of this workshop met with positive feedback from the participants.

Building on the experiences of the first workshop, the interactive and playful character was utilised and expanded in other formats. Marczewski's revised taxonomy for the higher education context and its use in the LMS OPAL was trialled with teachers from the Saxon higher education area in several further education courses offered by the LOS project. After the end of the LOS project, the concept was revisited and adapted by members of the DLGC cluster. After a one-year revision of the concept, the developed learning-teaching adventure Gamilcation was finally integrated into various contexts and tested for its acceptance and usability in higher education.

An initial trial of the revised concept took place as part of the Teacher Training Bootcamp, which was organised in the context of the EU project *Game4Change: Playful Learning in Next Generation Entrepreneurship*. During this online event, participants had the opportunity to test their knowledge of the gamification elements in relation to the user types in small groups. The implementation of the serious game led to a significant entertainment factor for all participants. However, the implementation of Gamilcation is not only intended to provide the teachers with enjoyment of the games themselves, but also to help them engage more intensively with the various gamification elements and develop an understanding of how they can integrate these elements into their own teaching. To make this possible, teachers are not only introduced to the method of gamification in the sense of the **pedagogical double-decker**, but the workshop itself is gamified, which is also perceived as a positive effect by the users. Understanding the different types of use enables teachers to adapt their teaching methods to the different learning styles, preferences and abilities of students and to support them accordingly.

After further revisions to the graphics and content, the second testing phase of Gamilcation took place in an online workshop with teachers ($n = 11$) from various disciplines at *Dresden University of Technology, Leipzig University and Zittau/ Görlitz University of Applied Sciences*. The learning and teaching adventure was used on a digital whiteboard (miro). The respective game elements were to be assigned to the user types (scenario 3). Regarding the evaluation of the serious game, individual participants expressed their wish for a better introduction to the task and more information on the individual user types so that the scenarios could be assigned more easily. Overall, all participants perceived the workshop method as positive. The majority of teachers (88 per cent) also stated that the workshop had inspired them to try out new things or use gamification in their teaching activities.

Previous experience from digital and analogue workshops has shown that Gamilcation is extensive and should be made available in a condensed form, especially for analogue events. Possible starting points here would be the following.

- Assignment of only one card per category (eight cards in total) to avoid excessive demands and generate quick successes for users
- Reduce (e.g. focus only on user types or only on general and rewards) or increase the number of cards depending on the participant's level of knowledge
- Consideration of different levels, such as beginner to expert mode
- Provision of initial starting points for participants in the form of pre-placed cards on the playing field to avoid excessive demands, counteract demotivation and still enable an effective expansion of the field of vision
- Implementation of strategies for both analogue and digital applications

Fig. 11 Gami|cation in use at a German face-to-face workshop (own illustration)

- Need for a needs-orientated approach to complexity (it does not always have to be all elements)

The experience gained from the practical applications (Fig. 11) shows how the design of the scenarios can be optimised for different target groups.

Participants' reflections from the field of educational science highlighted critical aspects of competition, particularly regarding the assessment of the *best* or *fastest* performance and the presentation of rewards. The importance of integrating competition and rewards in a way that promotes the collaborative nature of gamification was particularly emphasised. The participants would like to see a greater emphasis on cooperation aspects in the context of gamification. The focus should be on how the results of individual groups can lead to an overall result and how individual progress and development can be supported within the gamified framework. These critical voices and development proposals reinforce the importance of a balanced and collaborative approach when integrating gamification into teaching–learning arrangements.

6 Conclusion and Outlook

This paper illustrates how the integration of gamification into teaching–learning arrangements in higher education can succeed in an innovative way through the specially developed serious game Gamilcation—The Learning-Teaching Adventure. Current research results and the development process of Gamilcation emphasise the profound relevance of this gamified method.

The application scenarios illustrate the versatility of Gamilcation, which ranges from imparting basic knowledge to the creative design of individual courses. The flexible applicability in different educational contexts highlights Gamilcation as an adaptive tool that supports teachers in using gamification elements in a targeted manner to (additionally) motivate their students, encourage (more) interaction (and thus promote retention) and specifically support the desired behavioural development.

The practical experiences and further developments reflect an ongoing process in which Gamilcation is adapted based on workshop experiences. This iterative development enables optimal alignment with the needs of the target group and emphasises the lively development of this learning-teaching adventure.

In the context of the ongoing development of Gamilcation, the focus is on the planned implementation of the game for digital use in *Unity*. This step includes realising the previous scenarios in 2D models and providing a version in English and German. In addition, the aim is to ensure that the game is accessible and usable for both individual players and groups. Through the implementation in *Unity*, teachers and students can enjoy an interactive and entertaining learning experience that helps to increase motivation, engagement and learning success and can also be easily adapted to different learning needs.

The implementation of Gamilcation in practice-orientated workshops and similar applications has led to a significant gain in knowledge, enabling a diverse broadening of horizons. On the way to a comprehensive integration of Gamilcation in higher education, however, further testing is required in order to fully utilise the potential of this innovative teaching and learning method and to be able to adapt it to different specialist target groups.

Overall, this paper makes it clear that gamification is not only a tool for imparting knowledge, but also contributes to the promotion of teamwork, intrinsic motivation and a positive learning experience. However, the path to a comprehensive integration of gamification in higher education requires a continuous sensitisation of teachers as well as a needs-based adaptation of gamified teaching–learning arrangements. Gamilcation will continue to contribute to this.

References

1. Bai S, Hew KF, Huang B (2020) Does gamification improve student learning outcome? Evidence from a meta-analysis and synthesis of qualitative data in educational contexts. Educ Res Rev 30(100, 322)
2. Bakhuys Roozeboom M, Visschedijk G, Oprins E (2017) The effectiveness of three serious games measuring generic learning Features. Br J Educ Technol 48(1):83–100
3. Brandl LC, Schrader A (2024) Serious games in higher education in the transforming process to education 4.0—systematized review. Educ Sci 14:281. https://doi.org/10.3390/educsci14030281
4. Deterding S, Dixon D, Khaled R, Nacke L (2011) From game design elements to gamefulness: defining "gamification". In: Lugmayr A (ed), Proceedings of the 15th international Academic Mindtrek conference: envisioning future media environments. ACM, New York, USA, pp 9–15
5. Fischer H, Heinz M, Schlenker L, Münster S, Follert F, Köhler T (2017) Die Gamifizierung der Hochschullehre — Potenziale und Herausforderungen. In: Strahringer S, Leyh C (Eds.) Gamification und Serious Games. Grundlagen, Vorgehen und Anwendungen. Edition HMD. Springer Vieweg, Wiesbaden, Germany, pp 113–125
6. Hamari J, Koivisto J, Sarsa H (2014) Does gamification work?—A literature review of empirical studies on gamification. In: 2014 47th Hawaii international conference on system sciences, pp 3025–3034. IEEE
7. Heinz M (2021) Gamification's Dark Side Horizon. In GATE: VET (ed). Let them play: a journey to game-based learning. Unitedprint.com Vertriebsgesellschaft mbH, Academy for Vocational Education and Training, Dresden, pp 24–25
8. Khaldi A, Bouzidi R, Nader F (2023) Gamification of e-learning in higher education: a systematic literature review. Smart Learn Environ 10:10
9. Kapp KM (2012) The gamification of learning and instruction: game-based methods and strategies for training and education. Wiley
10. Lehmann C, Fischer H, Heinz M, Mueller J (2020) PonG: Parcours on gamification—how to get educators gamification-ready. In: Proceedings of the 12th international conference on computer supported education, 2–4 May 2020—vol 1. Prague, GonCPL, pp 687–693. https://doi.org/10.5220/0009804406870693
11. Marczewski A (2015) User types. Gamification, game thinking and motivational design. CreateSpace Independent Publishing Platform, In Even Ninja Monkeys Like to Play
12. Sailer M, Homner L (2020) The gamification of learning: a meta-analysis. Educ Psychol Rev 32(1):77–112
13. Sosniuk O, Ostapenko I (2016) Gamification as an element of active learning in higher education. In: DisCo 2016: towards open education and information society. papers of 11th international conference. Centre of Higher Education Studies, Prague, pp 72–77
14. Tolks D, Sailer M (2021) Gamification as a didactic tool in higher education. In: Hochschulforum Digitalisierung (ed), Shaping digitalisation in studies and teaching together. Innovative formats, strategies and networks. Springer, Wiesbaden, Germany, pp 515—532

Emerging Trends in Higher Education

Section Introduction

Dominik May

Amid a significant shift in educational paradigms, it is crucial to acknowledge and navigate the multifaceted landscape that defines this transformative era. Higher education is undergoing a profound evolution, driven by the pervasive influence of globalization and digitization. These forces are reshaping various dimensions of human existence, compelling educational institutions to adapt and innovate. As technology advances rapidly and global market dynamics evolve, educational frameworks must remain responsive and flexible to meet the demands of a changing world.

One notable trend in this evolving landscape is the increasing emphasis on shorter-term learning engagements, such as micro-credentials and continual education, catering to diverse age groups. This shift reflects a growing recognition of the need for lifelong learning and skill development in an era where knowledge and competencies must be continuously updated. Additionally, the contemporary student populace, characterized by digitally native learners accustomed to constant online connectivity, presents unique challenges and opportunities for educators. As these students bring new expectations and learning behaviors, educational institutions must evolve to effectively engage and support them. Moreover, the advent of Artificial Intelligence (AI) is poised to transform educational practices across various tiers. While AI offers unprecedented opportunities for personalized learning and administrative efficiencies, it also poses significant challenges that must be addressed. Navigating the implications of AI in education requires careful consideration of ethical, practical, and pedagogical dimensions.

This section encompasses a comprehensive spectrum spanning research endeavors, application development, first-hand accounts, and detailed descriptions of educational tools. Through these efforts, the goal of this section is to bridge the divide

---

D. May (✉)
University of Wuppertal, Wuppertal, Germany
e-mail: dmay@uni-wuppertal.de

between "pure" scientific inquiry and the day-to-day activities of educators. By integrating theoretical research with practical applications, this section aims to foster an adaptive, inclusive, and forward-thinking educational environment. This synthesis of knowledge and practice will empower educators to harness emerging technologies effectively, ensuring that the transformative potential of these advancements is fully realized in the classroom.

The study "EFL Learning in the Digital Era: Navigating Language and Culture in Jordanian Universities" by Rababah explores the interplay between language, technology, and culture in EFL education at Jordanian colleges. It investigates the impact of digital communication technologies, artificial intelligence, and the internet on language usage, comprehension, and cultural preservation. Utilizing qualitative methods, including semi-structured interviews with 15 Jordanian university EFL teachers, the research highlights the significant transformation in language training facilitated by interactive and multimedia resources. The study offers practical guidelines and techniques for educators, governments, and institutions to effectively integrate technology into English language instruction. Recommendations for future research include involving a broader range of EFL instructors and students, developing comprehensive teacher training programs, advocating for clear online ethics and data security policies, and promoting digital inclusion and equal access to technology for all students. By implementing these strategies, Jordanian educational institutions can enhance language learning and foster cross-cultural understanding.

Nursing practice often involves critical situations, such as receiving patients with severe prognoses in the emergency operating room, resuscitating cardio-respiratory arrest, and managing post-partum hemorrhage. In these contexts, inter-professional interactions significantly impact care outcomes, and deficiencies in non-technical skills can be fatal. Salih's study "Impact of simulation on the development of non-technical skills" aims to assess the impact of simulation on developing non-technical skills in student nurses at ISPITS-FES Morocco. A prospective, quasi-experimental monocentric study was conducted with 52 student nurse anesthetists from two different semesters. Fifteen days before the experiment, pre-observations were conducted using the TEAM grid. The study began with three pre-tests on emergency reception, cardiopulmonary arrest, and post-partum hemorrhage, followed by theory classes and a formative simulation session with three scenarios. Post-simulation, a satisfaction questionnaire and three post-tests of theoretical knowledge were administered. Two months later, a post-observation of student behavior was conducted. The results showed that simulation significantly improved student satisfaction and behavior, although knowledge retention in the post-tests was not significant. The development of non-technical skills is crucial, highlighting the importance of simulation in health sciences training programs.

In "Identifying AI Generated Code with Parallel KNN Weight Outlier Detection", Karnalim discusses the increasing concern about plagiarism in programming. The concern is exacerbated by generative AI, which can easily provide solutions for assignments intended to be completed independently. Many existing AI detectors focus solely on general text. This study introduces an AI detector for programming based on KNN weight outlier detection. Despite some students continuing to use AI for assignments where it is prohibited, AI-generated code differs noticeably from undergraduate work. The detector utilizes six intermediate representations: text strings, token strings, generalized token strings, expanded token strings, linearized syntax trees, and linearized parse trees. Experiments conducted on three datasets using MAP and processing time as performance metrics demonstrate that the detector is both effective and efficient. Optimal performance is achieved when using token strings as the intermediate representation, with n-grams set to 110, and three clusters.

Enterprise resource planning (ERP) systems have become crucial in managing organizational operations, including those in the education sector. Among the most widely used ERP systems in education is the SAP system. The chapter "The Use of SAP in Education: A Review of Current Practices and Future Directions" by Amin et al. reviews the current practices and future directions of SAP in education, exploring its applications in student information systems, financial management, and human resource management. While SAP offers significant advantages to the education sector, its successful implementation requires meticulous planning, adequate training, and consideration of the unique needs of educational institutions. Additionally, addressing privacy and security concerns is essential to protect student and institutional data. This paper discusses the challenges and benefits of using SAP in education and provides recommendations for future research and practice.

Vargas-Mendoza explores in "Detection of the Creativity Potential of Engineering Students" creativity as a crucial training requirement and professional skill for engineers, influenced by various factors including environment, approach, model, teacher, and institution. In engineering, creativity is typically linked to problem-solving and product generation. This study evaluated the creative potential of university students using engineering-specific tools and methodologies to generate innovative ideas. Students were trained in TRIZ, USIT, and SCAMPER techniques to enhance their creativity in problem-solving and projects in mechanics, mechatronics, and nanotechnology. The initial and final creative potential were assessed using Carter's creative profile and the Creatrix assessment, respectively, and the creativity enhancement hypothesis was tested with parametric Z and t-tests. Results showed that most students had intermediate creative potential at the course's start and improved to high creative potential by the final evaluation. Notable improvements were observed in certain groups segmented by program, semester, gender, and academic average. These findings suggest that the deliberate use of these methodologies in class significantly enhances the creative competency of engineering students, particularly younger students pursuing scientific careers.

In "The next step in Challenge-based Learning: Multiple challenges in a Single Course in Engineering: A New Model in Experiential Education", Membrillo-Hernández et al. explore Challenge-Cased Learning (CBL) as a widely utilized teaching technique in experiential engineering education, known for exposing students to real-life problems and developing both disciplinary and transversal skills. Typically, CBL experiences are designed around a specific topic that aligns with the course syllabus, presenting a challenge for the entire class. Often, a training partner collaborates to generate the challenge. Traditionally, a single challenge is solved by the entire class, generating multiple potential solutions through team efforts. This report examines a CBL experience where different training partners were involved simultaneously, proposing multiple challenges for the same class. The authors demonstrate that competencies can be effectively developed using various challenges concurrently within the same group, supported by several training partners. The evaluation utilized a strict competency development rubric. The findings indicate that this expanded CBL model, which incorporates multiple simultaneous challenges, enriches discussion and significantly strengthens the designated competencies.

EFL Learning in the Digital Era: Navigating Language and Culture in Jordanian Universities

Luqman M. Rababah

Abstract This study examines the relationship between language, technology, and culture in EFL education in Jordanian colleges. It examines how digital communication technologies, artificial intelligence, and the internet affect language usage, understanding, and cultural preservation across cultures. The study uses qualitative research methods, including semi-structured interviews with 15 Jordanian university EFL teachers, to accomplish these goals. The research revealed that the integration of interactive and multimedia re-sources in language training has resulted in a significant transformation facilitated by technology. This study emphasizes the research's practicality. It provides guidelines and techniques for educators, governments, and institutions to use technology in English language training. Future research should involve a diverse range of EFL instructors and students, develop comprehensive teacher training programs, advocate for clear online ethics, data security, and inclusivity rules, and promote digital inclusion and equal technological access for all students. Jordanian educational institutions can enhance language learning and cross-cultural understanding.

Keywords Language technology · Culture interaction · EFL Instruction · Jordanian universities · Educational technology · Cultural preservation

1 Introduction

The relationship between language, technology, and culture is dynamic and transformational in an era of ubiquitous technology and worldwide communication. This phenomena changes how humans communicate, learn, and engage across borders. EFL teachers at Jordanian institutions are at the vanguard of a digital revolution that surpasses conventional language teaching. Jordanian higher education is marked by its diversified culture and the need of English for academic and professional success.

L. M. Rababah (✉)
Jadara University, Irbid 733, 21110, Jordan
e-mail: Rababah80@gmail.com

This provides a unique chance to examine how the internet domain has affected language usage, cultural norms, and EFL classroom pedagogy. This interdisciplinary research examines how technology affects English teaching and learning, focusing on EFL teachers at selected Jordanian institutions. As technology is incorporated into language training, issues emerge about its effects on cross-cultural communication, language competency, and cultural identity. We use semi-structured interviews with 15 EFL teachers from Yarmouk, Jordan, Jadara, The Hashemite, and Irbid National universities to understand this dynamic. We investigate how the digital world has enhanced cross-cultural communication, including linguistic and cultural norms convergence and divergence. We also examine the ethics of technology-mediated intercultural contacts at Jordanian universities and their effects on culture preservation. We want to inform educators, policymakers, and academics by studying EFL teachers' experiences in this changing environment. The changing relationship between language, technology, and culture affects language instruction, Jordan's socio-cultural fabric, and its place in the globalized world [1, 2].

Research questions:

1. How does the integration of technology into English as a Foreign Language (EFL) instruction in selected Jordanian universities influence cross-cultural communication, language proficiency, and cultural preservation among EFL instructors and students?
2. What are the ethical dimensions of technology-mediated intercultural interactions within the Jordanian university context, and how do they impact the preservation of Jordanian culture within EFL classrooms?

1.1 Problem Statement

In today's digital environment, language, technology, and culture interact more, affecting language education and communication. In Jordan, where English fluency is valued academically and professionally, technology in EFL training has received attention. The complex effects of this integration are unexplored. EFL professors are leading the digital transformation as Jordanian colleges educate students for global involvement. Technology in language schools brings potential and difficulties that require careful consideration. Understanding how technology integration affects language education, cross-cultural communication, language competency, and Jordanian culture preservation is the issue. EFL classroom linguistic and cultural norms may change with digitalization. Jordan's cultural preservation and identity are crucial, thus technology's influence on education must be examined. The ethical aspects of technology-mediated intercultural contacts in Jordanian universities must also be examined to provide responsible and culturally sensitive language instruction. This research examines EFL teachers' experiences and viewpoints at Yarmouk, Jordan, Jadara, The Hashemite, and Irbid National universities to address these essential challenges. I want to illuminate the complex and changing relationship between language, technology, and culture in Jordanian EFL teaching. Educators, politicians, and scholars in Jordan and abroad must comprehend the problems and potential of this language-technology-culture trifecta. We want to improve language teaching, cultural identity, and technology usage in Jordanian university EFL classes by addressing this issue [3–5].

1.2 Significance of the Study

This research is important for Jordanian universities and the worldwide field of language instruction, technological integration, and cultural preservation. Define the significance:

1. Educational Impact: EFL teachers, language educators, and educational institutions at Jordanian universities such Yarmouk, Jordan, Jadara, The Hashemite, and Irbid National universities would benefit from this research. Understanding how technology affects language teaching and cultural preservation may improve English language education by informing pedagogy and curriculum.
2. Cultural Preservation: In a globalized society, cultural identity is crucial. Technology and cultural preservation in Jordan are examined in this research. It illuminates obstacles and possibilities to responsibly integrate technology while protecting Jordanian culture and identity.
3. Technology-mediated intercultural contacts' ethical implications are important. This work advances ethical and culturally aware communication in a globalized environment. It provides educators and policymakers with ethical standards for a more inclusive and respectful education.
4. Research Contribution: The research adds to language, technology, and culture literature. It provides a sophisticated knowledge of these processes in Jordanian higher education, useful for future study.
5. Worldwide Relevance: This study has worldwide implications. In an age where technology impacts language instruction globally, the study's results may help educators and researchers in many languages and cultures.
6. Policy implications: teaching policymakers might use the study's results to encourage technological integration in language teaching while protecting cultural identity and encouraging ethical communication.
7. The study's findings may help EFL teachers improve professionally. Understanding technology's effects allows educators to tailor their lessons to students' requirements.

2 Literature Review

Modern educational research and practice focus on language, technology, and culture. This literature review examines major topics and results connected to this dynamic interaction, laying the groundwork for the study's inquiry of Jordanian universities' EFL training.

Integrating Language and Technology

Technology-based language training has revolutionized language learning worldwide. Online resources, language learning apps, and digital platforms have transformed language teaching [6]. Technology-enhanced language learning improves student engagement, access to genuine resources, and self-directed learning [7].

Cultural factors in language learning

Culture is fundamental to language development and communication [8]. Culture is inextricably linked to language, which affects cultural norms, values, and identities. For effective communication, language learners must negotiate the target language's culture [9]. Thus, EFL training commonly includes cultural elements to improve intercultural competency [10].

Cultural Preservation in Language Education

Language education preserves and revitalizes indigenous languages and cultures in multicultural and multilingual situations [11]. To preserve cultural identity and legacy, language educators include cultural components into language training [12]. Jordan's rich cultural past makes this aspect of language teaching especially essential.

Language and Technology Ethics

The internet age has raised language education ethics. Technology-mediated communication increases privacy, security, and online responsibility concerns [13]. The ethics of internet intercultural exchanges are crucial [14]. The ethical use of technology to overcome cultural barriers and respect cultural diversity is debated.

2.1 Related Studies

Numerous research have examined novel ways to improve language learners' international competency and cultural awareness. These studies use different methods and technology to examine language learning and cultural immersion. This review analyzes these works and highlights their contributions to the field while providing smooth transitions across research endeavors.

Perez and Gadelha [15] examines how digital storytelling might improve language learning intercultural competency. Digital storytelling immerses students in other cultures via multimedia and tales. This immersive method increases cultural knowledge and sensitivity, deepening cross-cultural communication comprehension. Moving to mobile-assisted language learning, [16] examine how mobile technology affects EFL learners' cultural competency. Their study shows that mobile devices improve learners' cultural understanding and navigation. Mobile platforms provide learners the freedom and resources to authentically interact with other cultures. Using WhatsApp in a Chinese language course, [17] demonstrate how ordinary digital media may improve language and cultural acquisition. Their research shows that popular messaging apps may aid cultural integration. This method connects language learning to real-world culture by using familiar technologies.

Online interactions boost cultural competency in the digital age. Ruiz de Zarobe and Zalbidea [18] use case studies to examine how virtual dialogues and interactions improve cultural competency. Their study emphasizes the need of real-time communication in comprehending other cultures. Lin and Lan [19] examine how social media improves intercultural competency. They demonstrate how social media helps learners legitimately connect with other cultures. This immersive digital world helps students expand their cultural perspectives. Wang and Sun [20] conducted a case study to see whether online platforms improve intercultural ability. Virtual environments help language learners develop cultural awareness and sensitivity, according to their study. Interactive online platforms allow students to discover and connect with various cultural ideas.

Lamy and Hampel [21] studied how technology-mediated communication affects intercultural competency using mixed methodologies. Their extensive study shows how digital technologies and communication strategies alter language learners' cultural understanding. Their research illuminates the complexity of cultural immersion via technology by evaluating quantitative and qualitative data. Ke and Shen [22] explore language instruction using VR. Virtual reality may generate immersive cultural encounters, helping people comprehend other cultures, according to their case study. This method deepens cultural understanding by immersing students in virtual surroundings.

Blake [23] emphasizes utilizing technology to educate culture in language classes. The research shows how digital instruments may teach culture. Technology in the curriculum may provide pupils new cultural perspectives and expand their language learning. In a case study on social media for intercultural language acquisition, [24] emphasize the importance of online interactions in cultural awareness and international language learning. Social media lets students interact with global communities and improve their cross-cultural communication abilities.

In conclusion, these research show how technology may alter language teaching by promoting international competency and cultural awareness. Each study adds to our knowledge of how digital tools and methods may help language learners connect with varied cultural situations.

2.2 Research Gap

One notable gap in the extant literature on language education, technology, and culture is the absence of attention on the unique setting of English as a Foreign Language (EFL) teaching in Jordanian institutions. While there is a growing corpus of literature that covers the larger landscape of technology-enhanced language learning and its cultural consequences, there is a significant scarcity of research that dive into the details of technology integration in EFL classrooms in Jordan. This gap is especially pertinent since Jordanian institutions such as Yarmouk, Jordan, Jadara, The Hashemite, and Irbid National universities have their distinct cultural, linguistic, and educational variations that need specialized consideration [35]. Knowledge how technology is deployed and the problems and possibilities it brings in this environment is vital for a thorough knowledge of language teaching in Jordan. Moreover, this study deficit makes it tough to detect how cultural preservation and ethical issues are formed and maintained in Jordanian EFL classes, which are vital parts of language instruction in a culturally rich society like Jordan.

Furthermore, the present literature lacks a full analysis of the influence of technology on the preservation of Jordanian cultural identity within the area of language teaching. While numerous studies stress the significance of cultural identity preservation in language instruction, there is minimal research that analyzes how technology influences this element, notably in EFL programs in Jordan [37]. Given Jordan's rich cultural past and the role of cultural identity in language teaching, bridging this gap is crucial for appreciating the intricate interaction between language, technology, and culture in Jordanian higher education. Additionally, the ethical implications of technology-mediated intercultural interactions in language training remain comparatively unexplored in this context [36]. The ethical elements of technology use in Jordanian EFL classrooms deserve greater attention, as they have become more relevant in language instruction, particularly in a globalized culture where cultural sensitivity and ethical concerns are crucial.

Therefore, there is an urgent need for context-specific research that looks into technology integration into EFL education in Jordanian college [25, 26]. Such study should concentrate on cultural preservation, ethical challenges, and technological breakthroughs in the area, especially from 2018 to 2021, to give a more up-to-date knowledge of the language-technology-culture trifecta within this unique educational context. This study will play a vital role in bridging the present research gap and contribute to a more nuanced knowledge of language instruction in Jordanian institutions.

3 Methodology

Research Design: The study used a mixed-methods approach to examine the complex relationship between language, technology, and culture in modern society, focusing on Jordanian university EFL instructors. Yarmouk, Jordan, Jadara, Hashemite, and Irbid national universities. The target demographic was EFL teachers at selected Jordanian institutions. Given their cultural settings, deliberate selection was used to guarantee varied representation from these colleges. Semi-structured interviews were conducted with 15 EFL teachers. These lecturers from the institutions above represented a variety of experiences and opinions. The research included semi-structured interviews, questionnaires, and linguistic analysis to improve validity and reliability. Triangulation was used to validate data from many sources. A pre-test and structured interview approach ensured the interview questions' dependability and consistency. 15 EFL teachers were interviewed semi-structured. These interviews examined how digital communication technologies, AI, and the internet affect language use, understanding, and preservation across cultures. In semi-structured interviews, qualitative data was examined using thematic content analysis. To show how technology affects language instruction and culture, themes and patterns were established. The synthesis of these interviews examined how the digital environment aids EFL teachers and students' cross-cultural communication.

4 Results and Discussion

4.1 Results for Research Question 1

How does the integration of technology into English as a Foreign Language (EFL) instruction in selected Jordanian universities influence cross-cultural communication, language proficiency, and cultural preservation among EFL instructors and students?

The semi-structured interviews with 15 EFL teachers from various Jordanian institutions showed some critical findings about how technology affects language usage and cultural norms in English language instruction:

4.1.1 Instructors' Perspectives on Technology Integration

Jordanian university EFL professors said online learning platforms and digital tools are essential to language education. Many teachers use digital technologies.

1. Participant A3: "Online platforms have made English lessons more engaging and accessible."

2. Participant A7: "Digital resources are essential. Students respond well to multimedia materials. They add dynamism to language instruction."
3. Participant A10: "I've noticed a significant shift towards digital tools in recent years. They allow us to tailor language lessons to individual student needs, making learning more personalized."
4. Participant A12: "Technology has a huge impact on language pedagogy. Our students are more tech-savvy than ever, and we must use this to improve their language skills."
5. Participant A13 "The integration of technology has made cultural exploration more accessible. Students can engage with English language and culture in ways that were not possible before."

4.1.2 Technology Has Made Language Learning Easier with Interactive and Multimedia Materials

Instructors said that students now have a variety of digital tools to connect with English in different ways.

1. Participant A2: "Interactive lessons, videos, and online language games make language learning fun and engaging."
2. Participant A5: "Multimedia materials have revolutionized language instruction. It's amazing how technology has expanded students' listening, speaking, and comprehension practice."
3. Participant A8: "With the wide variety of digital materials, we can accommodate different learning styles. Technology lets us give students a customized learning experience."
4. Participant A11: "I've seen firsthand how technology can bridge the gap between classroom learning and real-world language use. Students can now explore authentic English language content online, making their learning journey more immersive."
5. Participant A14: "Technology has made language learning more dynamic. We can incorporate multimedia content that reflects current trends and real-life communication, which is crucial for our students' language development."

4.1.3 Technology Has Changed EFL Classroom Culture

According to interviewees, digital communication technologies allow students and teachers to engage across cultures, improving cultural understanding.

1. Participant A1: "Online discussions and collaborations with students from different backgrounds give our students a broader worldview and a deeper appreciation for cultural differences."
2. Participant A4: "Digital communication tools have enabled cross-cultural dialogues. It's amazing to see students share ideas and insights about their own and other cultures, enriching the learning environment."

3. Participant A6: "As instructors, we've seen a shift in cultural awareness. Technology has allowed students to interact with English speakers from around the world, breaking down cultural barriers and promoting inclusive learning."
4. Participant A9: "Technology has transformed our classrooms into vibrant cultural hubs. Through virtual exchanges and collaborative projects, students improve their language skills and become global citizens, embracing diverse cultural perspectives."
5. Participant A13: "Technology encourages students to explore and share their cultural backgrounds with peers and instructors, fostering unity in our language learning community."

4.1.4 An Ethical Perspective

Technology-mediated intercultural exchanges have ethical implications. In Jordanian universities, instructors realized the necessity to cover ethical concerns including online conduct and cultural sensitivity.

1. Participant A2: "Intercultural interactions online present ethical challenges that we, as instructors, must address. You must teach students proper online behavior and instill a deep respect for cultural differences."
2. Participant A5: "Technology opens doors to global connections, but it also demands responsibility. Instructors teach students online communication ethics and foster a respectful online learning community."
3. Participant A8: "As technology blurs geographical boundaries, we must emphasize ethical conduct in EFL classrooms and ensure a safe and respectful online environment."
4. Participant A11: "The digital realm exposes students to diverse perspectives, but it also requires a keen sense of ethics. We must equip our learners with the tools to navigate this complex terrain."
5. Participant A14: "We must balance cross-cultural understanding and respectful, inclusive online dialogues."

4.1.5 Cultural Variety Versus Homogenization

The effects of technology on culture were multifaceted. Technology allowed students to connect with other cultural settings via digital information, yet Western cultural influences through digital resources might homogenize cultures.

1. Participant A3: "Technology can celebrate cultural diversity by providing a window into different cultures, but we must also consider Western cultural dominance in digital resources."
2. Participant A6: "Cultural diversity is a treasure, and technology can help us preserve and celebrate it. However, Western norms can inadvertently homogenize our cultural landscape."

3. Participant A9: "The digital world lets us explore cultural diversity like never before, but we must be vigilant to prevent cultural homogenization and make all voices heard."
4. Participant A12: "Cultural diversity is a strength, and technology helps us share and appreciate it. However, cultural homogenization is a risk, especially in language learning."
5. Participant A15: "In our interconnected world, technology promotes cultural diversity, but we must be vigilant against local culture erosion. It's a delicate balance."

4.2 Discussing the First Research Question

Following interviews with 15 EFL professors at Jordanian institutions, this research illuminates the many ways technology affects English as a foreign language instruction and classroom culture. The author's careful examination and interpretation of the data yielded useful insights about Jordan's complicated relationship between technology, language, and culture.

4.2.1 Technological Integration

EFL teachers' recognition that technology is essential to language teaching shows how digital technologies have changed education. This study shows that instructors see technology as a crucial part of language training, as the author states. The author recognizes technology's importance in language instruction.

4.2.2 Language Learning

Interactive and multimedia tools have made language learning easier, highlighting the value of digital materials in engaging pupils and broadening language acquisition. The author's explanation of this data shows an awareness for how technology improves teaching and student involvement.

4.2.3 Cultural Influence

The fact that technology has changed cultural norms in the EFL classroom shows how digital communication tools affect cross-cultural relationships. Intercultural competency in language instruction is stressed by the author's emphasis on comprehending various cultural views.

4.2.4 An Ethical Perspective

In a digitally linked society, educators have ethical duties, as the author acknowledges ethical implications of technology-mediated intercultural exchanges. This conclusion supports the author's ethical considerations regarding language training technology.

4.2.5 Cultural Variety Versus Homogenization

The author's sophisticated examination of technology's influence on cultural variety vs homogeneity shows his knowledge of the complicated dynamics. The author balances cultural variety enrichment and Western cultural domination in this important problem. The author's presentation of the study's findings recounts the results and shows a profound comprehension of their consequences. The Jordanian EFL classroom's complex investigation of technology, language, and culture sparks a vibrant dialogue that encourages more language education research.

4.2.6 These Findings' Relevance to the Theoretical Framework

The Cultural-Historical Activity Theory (CHAT), Technological Pedagogical Content Knowledge (TPACK), and Cultural Dimensions of Learning Framework all influence the study's conclusions. These frameworks provide a thorough analysis of the cultural, historical, and technical factors at play in Jordanian universities' technology-mediated EFL training.

CHAT: CHAT is a core theory for understanding cultural and historical dimensions of learning and human action. CHAT helps the research examine how cultural and historical influences have affected Jordanian university EFL teachers' practices and activities. It contextualizes technological integration in society and shows how culture, technology, and language learning interact. CHAT highlights how Jordan's culture and technology have affected EFL instruction. It illustrates how cultural attitudes about technology in education have evolved and how they impact teachers' pedagogy.

The TPACK Framework addresses technology, pedagogy, and subject knowledge. Jordanian university EFL instructors are assessed on their technical, pedagogical, and linguistic subject skills to provide culturally appropriate teaching using TPACK. TPACK may be used to study the difficult balance between technical expertise and pedagogical skills required for technology-mediated language teaching in a culturally diverse context. Digital technology integration by instructors for cultural preservation and language acquisition is examined.

Based on Hofstede and Hall, the Cultural Factors of Learning Framework stresses cultural influences on learning and communication. This method examines how collectivism vs. individuality and high vs. low context communication styles effect technology integration and Jordanian culture preservation in EFL teaching. This paradigm lets researchers study how culture affects teachers' technology use and cultural preservation in the classroom. The extensive study of cultural factors that impact technology-enhanced language learning improves our understanding of culture, technology, and language education. Multiple theoretical frameworks offer the study a complete picture of Jordanian EFL classrooms' complex relationship between technology, language, and culture. You may study how historical, cultural, and pedagogical factors impact technology-mediated language acquisition.

4.3 Results of Research Question 2

What are the ethical dimensions of technology-mediated intercultural interactions within the Jordanian university context, and how do they impact the preservation of Jordanian culture within EFL classrooms?

The research found numerous ethical issues connected to technology-mediated intercultural encounters at Jordanian universities and their impact on Jordanian culture in EFL classrooms:

4.3.1 Ethics of Technology-Mediated Intercultural Interactions

Online Behavior and Cultural Respect: EFL teachers understood the need of teaching students' online etiquette and cultural respect. They stressed that online interactions must promote respect, empathy, and cultural awareness since they lack face-to-face indications.

Participant A5 said, "We need to teach our students not just the language but also how to behave ethically online. Respect for different cultures is a must in our interconnected world."

Participant A3: "In the digital world, manners matter more than ever. We're teaching English and cross-cultural communication."

Participant A7: "Online interactions can bridge gaps or create barriers. We foster cultural respect in our students."

Participant A9: "I tell my students, 'Your words online represent not just you but also your country and culture. Let's make it positive.'".

Participant A11: "Teaching cultural sensitivity is part of our duty. It's about making the online world a space where diverse cultures coexist harmoniously."

Participant A14: "Jordan has a rich heritage, and we want our students to embrace it online. It's about embracing our culture and others'."

4.3.2 Online Students' Privacy and Data Security Were a Major Concern

Teachers stressed the ethical importance of protecting students' personal data and securing communication channels.

Participant A8 said, "Privacy is a big concern. We must ensure that students' information is safeguarded. It's an ethical responsibility:"

Participant A12 said, "We need to be aware of the digital divide. It's our ethical duty to ensure that all students can participate, regardless of their access to technology."

Participant A13 "In this era of technology, preserving our pupils' privacy is important. We must build a secure digital environment."

Participant A6: "Online learning requires trust. We must reassure students that their personal data is safe."

Participant A10: "Privacy breaches can have serious consequences. It's our ethical obligation to protect student data."

Participant A13: "Technology should empower, not compromise. We must protect our students' privacy online."

Participant A15: "Our students' trust is our most valuable asset. We're committed to data security ethics."

4.3.3 Digital Divide Awareness

Not all students have equal access to technology, therefore ethics applied. Instructors pledged to provide inclusive activities for students without digital tools.

Participant A4: "We take the digital divide into account and promote inclusive learning."

Participant A7: "The digital divide is a real challenge, but we're determined to bridge it. No student should be left behind."

Participant A9: "Inclusion is about equity. We're working to ensure that every student, regardless of resources, can fully participate."

Participant A11: "Technology can amplify inequalities. Our ethical duty is to minimize them in online education."

Participant A14 "Our commitment to inclusivity extends beyond the classroom. It's about creating equal opportunities for all."

4.3.4 Impact on Jordanian Culture Preservation

These ethical considerations affected Jordanian culture in EFL classrooms:
Teaching students to appreciate cultural diversity online was seen as maintaining Jordanian culture. Promote Jordanian values and traditions while cultivating cultural awareness, instructors stressed.

"By teaching respect for other cultures, we also instill pride in our own. It's a way to preserve our Jordanian identity," said A3.

Participant A1: "Respecting other cultures reflects our cultural values and preserves Jordanian identity."

Participant A4: "As Jordanians, we want our students to be proud of their culture. Teaching cultural respect online helps preserve it."

The participant A13 said, "Preserving our culture isn't about isolation; it's about sharing and understanding. Online respect fosters cultural pride."

Participant A11: "Respecting others preserves our culture's beauty."

Participant A14: "Online respect preserves our Jordanian heritage, which gives us pride."

4.3.5 Inclusive Approaches

The digital divide awareness linked to attempts to keep technology-mediated language learning inclusive. Inclusion maintained cultural variety in Jordanian classrooms, instructors said.

Participant A2: "Inclusion isn't just about technology; it's about valuing and preserving every Jordanian cultural perspective."

Participant A6: "Inclusion enriches our cultural mosaic and preserves our diverse heritage."

Participant A9: "Inclusion isn't just an ethical issue; it preserves Jordanian cultures in our classrooms."

Participant A13: "Inclusion protects the richness of our Jordanian culture from being overshadowed."

Participant A15: "Inclusion protects our cultural diversity and shows our dedication to Jordanian identity."

According to participant A10, "Inclusion is not just an ethical concern; it's also about preserving our cultural diversity. We want all voices to be heard, regardless of technology access."

4.3.6 Digital Cultural Exchange

Online ethics and cross-cultural respect facilitated meaningful interactions. Instructors felt that such exchanges allowed students to share Jordanian culture and learn about others, conserving and enhancing their heritage.

Participant A15 said, "Our students become cultural ambassadors online. They share Jordanian traditions and learn about the world. It's a beautiful way to preserve our culture."

Participant A3: "Online cultural exchanges preserve culture modernly. Our students represent Jordanian culture."

Participant A5: "The digital realm is a canvas for preserving and sharing our cultural heritage. It's a dynamic way to ensure our traditions endure."

Participant A8: "Our students share and learn Jordanian culture online, enriching our heritage."

A12: "Digital cultural exchanges are bridges to the world. We're not just preserving; we're actively engaging with our culture."

Participant A10: "Online cultural exchange enriches our cultural tapestry by evolving and growing with each interaction."

4.4 Discussing the Second Research Question

Research Question 2's findings on the ethical aspects of technology-mediated intercultural interactions and their impact on Jordanian culture in EFL classrooms support the selected theoretical frameworks and help explain the complex dynamics at play: Cultural-Historical Activity Theory (CHAT): Research Question 2 emphasizes cultural and historical aspects of human activity, supporting CHAT. They emphasize online ethics and cultural diversity. CHAT helps us place these ethical issues in Jordanian EFL instruction's cultural and historical context. CHAT helps us understand how technology-mediated intercultural exchanges affect cultural preservation by acknowledging technology's influence on cultural norms and ethics.

Technological Pedagogical Content Knowledge (TPACK) Framework: Research Question 2 supports the TPACK framework, which emphasizes technology in education. Ethical considerations including online conduct and privacy are crucial to appropriate technology usage in language training. TPACK helps us understand how EFL teachers use technical, pedagogical, and linguistic content knowledge to handle ethical challenges. It stresses the ethical implications of technology-mediated language training for cultural preservation.

Cultural Dimensions of Learning Framework: The results support the framework, especially in cross-cultural respect and inclusion. This paradigm highlights cultural elements that affect learning and communication, and the study's findings show how ethical online conduct and cultural diversity concerns mirror these characteristics.

This approach helps us comprehend language acquisition and preservation's cultural components by identifying the ethical aspects of technology-mediated intercultural encounters.

Under the theoretical frameworks of CHAT, TPACK, and the Cultural Dimensions of Learning Framework, Research Question 2's findings give a complete assessment of technology-mediated intercultural exchanges in EFL classrooms' ethical implications. They stress the necessity of ethics in technological integration, cultural preservation, and the complicated relationship between technology, language, and culture in Jordanian universities. These frameworks help us understand and apply the study's results to language instruction and cultural preservation.

4.4.1 Link to Previous Studies

1. CHAT-based research like Vygotsky and Leontiev's stress the cultural and historical aspects of human action, which matches the findings. Like similar research, the results acknowledge how technology impacts Jordanian EFL culture and language learning methods [27].

2. Alignment with TPACK Framework: The research emphasizes ethical issues in technology integration in education. This supports prior study that highlights educators' need to understand and handle technological ethics in the classroom [28].

3. The findings emphasize cross-cultural respect and inclusion in technology-mediated language learning, aligning with the Cultural Dimensions of Learning Framework. Previous research have shown that cultural aspects affect learning and communication [29].

4.4.2 Novel Insights and Extensions

1. Ethical Considerations: Previous studies have touched on the topic of ethics in technology-mediated education, but this study explores the Jordanian EFL setting. It examines how instructors see and handle online conduct, privacy, and the digital divide in their culture [30, 31].
2. Cultural Preservation: The study clearly examines how these ethical elements affect Jordanian cultural preservation in EFL classes, surpassing prior research. It shows how ethical online conduct promotes cultural respect, inclusion, and preservation [32, 33].
3. Contextualization at Jordanian institutions: The research's particular emphasis on Jordanian institutions illuminates their distinct difficulties and prospects. The Jordanian cultural and pedagogical context informs this research on technology and ethics in education [34].

In conclusion, Research Question 2 shows that ethical and cultural factors in technology-mediated education are important, but it goes further by providing a nuanced understanding of these factors in the Jordanian EFL context. Focusing on ethics and cultural preservation in this setting is new to the discipline.

5 Conclusion and Recommendations

This study investigates the effects of technology on the process of teaching and learning English in educational institutions in Jordan. This study employs the Cultural-Historical Activity Theory (CHAT), the Technological Pedagogical Content Knowledge (TPACK) Framework, and the Cultural Dimensions of Learning Framework to analyze two primary subjects: the impact of technology on language education and its ethical ramifications for the preservation of Jordanian culture.

The research revealed that the integration of interactive and multimedia resources in language training has resulted in a significant transformation facilitated by technology. This transformation encompasses several aspects, including the promotion of cultural awareness and the facilitation of intercultural exchanges. However, it also brought attention to the ethical concerns around internet use, privacy, and the need to address the digital divide. In order to provide equal opportunities for technology-mediated learning, it is imperative for educators to prioritize the instruction of online ethics, cultural diversity, and the safeguarding of student privacy and data.

The report also emphasized the need of fostering cultural reverence and substantive digital engagements as means of safeguarding Jordanian culture. Educators assume a pivotal role in fostering cultural sensitivity and inclusivity within the realm of online education. However, the study has some shortcomings. These limitations include a limited sample size, a focus on Jordanian colleges, and instructor self-report data.

To address these constraints, future research should include a wider spectrum of EFL instructors and students. Developing comprehensive teacher training programs that provide educators with the knowledge and abilities to ethically and culturally use technology in classroom is also important. Advocate for clear online ethics, data security, and inclusivity rules at educational institutions. In conclusion, digital inclusion and equal technological access for all students must be actively promoted. Additionally, English as a Foreign Language (EFL) instructors and students from various cultures should collaborate to build a greater awareness for global viewpoints and cultural heritage. In conclusion, Jordanian educational institutions may improve language learning and cross-cultural understanding by integrating technology into EFL teaching.

References

1. Alshare F, Alkhawaldeh AM, Eneizan BM (2019) Social media website's impact on moral and social behavior of the students of university. Int J Acad Res Bus Soc Sci 9(3):169–182
2. Banikalef AA (2019) The impact of culture and gender on the production of online speech acts among Jordanian facebook users. Int J Arab-Engl Stud (IJAES) 19(2):395–410
3. Banihani ANM (2021) The effect of using blended learning in teaching English on direct and deferred achievement of primary school students. Int Bus Res 14(4):1–50
4. Hatamleh IHM (2024) Exploring the multifaceted influences of social media motivation on online relationship commitment among young adults in Jordan: an empirical study. Hum Behav Emerg Technol
5. Hatamleh IHM, Safori AO, Ahmad AK, Al-Etoum NMI (2023) Exploring the interplay of cultural restraint: the relationship between social media motivation and subjective happiness. Soc Sci 12(4):228
6. Chen N-S, Fang Y-H (2014) Effects of different video lecture types on sustained attention, emotion, cognitive load, and learning performance. Comput Educ 79:197–209
7. Hockly N (2016) Mobile learning. ELT J 70(2):166–175
8. C. Kramsch (1998) Language and culture. Oxford University Press
9. Byram M (1997) Teaching and assessing intercultural communicative competence. Multiling Matters
10. Alptekin C (2002) Towards intercultural communicative competence in ELT. ELT J 56(1):57–64
11. McCarty TL (2011) Language planning and policy in Native American communities. Wiley-Blackwell
12. Kumaravadivelu B (2003) Beyond methods: macrostrategies for language teaching. Yale University Press
13. Leask M, Younie S (2013) National education technology standards for teachers: From TPACK to SAMR. J Digit Learn Teach Educ 29(4):127–134
14. Kramsch C (2014) Second language acquisition, applied linguistics, and the teaching of foreign languages. Mod Lang J 98(3):645–655
15. Perez B, Gadelha LM (2020) Digital storytelling as a tool for intercultural competence development in language learning. J Multiling Multicult Dev
16. Al-Azawei A, Parslow P, Lundqvist KO (2019) The impact of mobile-assisted language learning on EFL learners' cultural competence. Comput Educ 138:104–120
17. Kurek M, Zhang J (2018) Digital media in language education: a case study on whatsapp in a chinese language course. Comput-Assist Lang Learn 31(5–6):457–475
18. Ruiz de Zarobe Y, Zalbidea JG (2019) Promoting cultural competence in language learning through online exchanges: a case study. ReCALL
19. Lin CH, Lan YJ (2018) The role of social media in promoting intercultural competence in language learning. Educ Technol Soc 21(3):78–90
20. Wang C, Sun X (2020) Enhancing intercultural competence in online language learning: a case study. J Interact Online Learn
21. Lamy MN, Hampel R (2018) Technology-mediated communication and intercultural competence development: a mixed-methods study. Lang Learn Technol 27(4):497–509
22. Ke IC, Shen YC (2019) The role of virtual reality in language learning and cultural understanding: a case study. Interact Learn Environ 27(4):497–509
23. Blake RJ (2020) Cultura and beyond: teaching culture through technology in the language classroom. Foreign Lang Ann 53(2):254–268
24. Chen MH, Tzeng JY (2018) Using social media for intercultural engagement in language learning: a case study. Lang Cult Curric 31(3):257–272
25. Yunus K, Hmaidan MAA (2021) The influence of idioms acquisition on enhancing English students' fluency. Int J Educ, Psychol Couns 6(40):124–133
26. Yunus K, Hmaidan M (2021) The strategies used by lecturers in teaching translation of idiomatic expressions. Int J Educ Psychol Couns 6(40):134–144

27. Vygotsky LS, Cole M (1978) Mind in society: development of higher psychological processes. Harvard University Press
28. Koehler MJ, Mishra P (2008) Introducing TPCK, in handbook of technological pedagogical content knowledge (TPCK) for educators, vol 1. AACTE Committee on Innovation and Technology, pp 3–29
29. Hall ET (1976) Beyond Cult 3(1):56–89
30. Mishra P, M. J. Koehler (2006) Technological pedagogical content knowledge. Framework Teach Knowl 5(3):123–140
31. Selwyn N (2010) Looking beyond Learning: notes towards. Crit Study Educ Technol 7(4):321–335
32. Kim Y (2008) Communication and cross-cultural adaptation: an integrative theory, vol 2(3), pp 45–67
33. Byram M (1997) Teaching and assessing intercultural communicative competence, vol 4(2), pp 89–107
34. Bates AW (2015) Teach Digit Age 8(2):112–130
35. Hatamleh IM (2021) The association of social media motivation, trust, culture restraint and relationship benefit towards online relationship commitment. USM, Malasia.
36. Masadeh R, Almajali D, Majali T, Hanandeh A, Al-Radaideh A (2022) Evaluating e-learning systems success in the new normal. Int J Data Netw Sci 6(4):1033–1042
37. AL-Sous N, Almajali D, Al-Radaideh A, Dahalin Z, Dwas D (2023) Integrated e-learning for knowledge management and its impact on innovation performance among Jordanian manufacturing sector companies. Int J Data Netw Sci 7(1):495–504

Luqman M. Rababah is an associate professor of Applied Linguistics at Jadara University in Jordan. He teaches sociolinguistics, pragmatics, and second language acquisition. His research interests include pragmatics, semantics, discourse analysis, and sociolinguistics. Email: luqman@jadara.edu.jo

Impact of Simulation on the Development of Non-technical Skills

Fatima Zahra Salih, Rajae Lamsyah, Abdelkrim Shimi, and Yasser Arkha

Abstract Nursing practice is confronted with critical situations such as: the reception of patients with a serious vital prognosis in the emergency operating room or the resuscitation of cardio-respiratory arrest, as well as the management of post-partum haemorrhage. Under these conditions, inter-professional interactions affect care. And when the quality of these so-called non-technical skills is deficient, the damage can lead to death. The aim of this study is to assess the impact of simulation on the development of non-technical skills in student nurses at ISPITS-FES Morocco. A prospective, quasi-experimental monocentric study involving 52 student nurse anesthetists from 2 different semesters. 15 days prior to the experiment, the population was pre-observed at the training site, using the TEAM grid. The study began with the distribution of 3 pre-tests on the following topics: Reception in the emergency operating room, cardiopulmonary arrest and post-partum hemorrhage. They then attended 3 theory classes. They then took part in a formative simulation session comprising 3 scenarios. Afterwards, a questionnaire was administered to assess satisfaction, followed by 3 post-tests of theoretical knowledge. And 2 months later, a post-observation of student behavior. The results showed a significant impact of the simulation on the degree of satisfaction, as follows thus improving behavior. However, knowledge retention was not significant in the post-test. The development

---

F. Z. Salih (✉) · Y. Arkha
Mohammed V University in Rabat, Rabat, Morocco
e-mail: fatimazahra_salih@um5.ac.ma

Faculty of Medicine, Pharmacy and Dentistry in Rabat, Rabat, Morocco

Ibn Sina University Hospital Center in Rabat, Rabat, Morocco

Y. Arkha
e-mail: y.arkha@chis.ma

R. Lamsyah · A. Shimi
Sidi Mohamed Ben Abdellah University in Fez, Fez, Morocco

Faculty of Medicine, Pharmacy and Dentistry in Fez, Fez, Morocco

R. Lamsyah
Higher Institute of Nursing and Health Techniques in Fez, Fez, Morocco

© The Author(s), under exclusive license to Springer Nature Switzerland AG 2025
E. Vendrell Vidal et al. (eds.), *Advanced Technologies and the University of the Future*,
Lecture Notes in Networks and Systems 1140,
https://doi.org/10.1007/978-3-031-71530-3_28

of non-technical skills is crucial, hence the interest of simulation in health sciences training programs.

Keywords Simulation · Skills · Behavior · Nursing students

1 Introduction

Nursing education represents a complex journey, involving both the acquisition of in-depth academic knowledge and the development of essential clinical skills [1–3].

However, for novice nurses, the challenge of reconciling theory and practice often manifests itself as a significant gap, creating tension between the skills demonstrated in the clinical setting and the theoretical knowledge acquired during the course of training. This disparity, known as the theory–practice gap, has major implications for patient safety, job satisfaction, and reducing healthcare errors [4–8].

Concern about the theory–practice gap is not limited to the clinical setting; it also has its roots in the educational environments where nurses are trained. Faculty play a crucial role in closing this gap by identifying contributing factors and integrating strategies to overcome it. Nursing educators have the responsibility to socialize students to clinical practice while introducing theoretical concepts, emphasizing the importance of linking theory to the reality of practice [7–10].

In this context, simulation emerges as a promising strategy to reduce the theory–practice gap. Although simulation-based training is widely recognized for improving the quality of healthcare, its specific impact on nurses' non-technical skills (NTS) is an area that requires further exploration. This study aims to evaluate the impact of simulation on the development of non-technical skills among nurses in intensive care anesthesia, with particular emphasis on satisfaction, knowledge, behavior, and communication and work skills in team.

Through a single-center quasi-experimental evaluative approach, this study targets anesthesia and intensive care nurses at the Higher Institute of Nursing and Health Technology Professions (ISPITS) of Fez. The hypotheses formulated examine the correlation between simulation as a teaching technique and the development of soft skills, while the research question specifically explores the impact of simulation on soft skills among these students.

In the following sections, we will address the research sub-questions, detail the study methodology, present the results, and discuss their relevance in light of current knowledge in the field.

The objective of our study The study aims to evaluate the impact of simulation on Kirkpatrick's three levels (satisfaction, knowledge, behavior) in the field of nursing education, focusing particularly on non-technical skills and crisis management. The four chapters of the article provide a bibliographic synthesis on the impact of clinical simulation, detail the methodology of the study, present the results obtained, and discuss these results in light of current knowledge. In conclusion, recommendations are made for the effective integration of simulation into nursing education to improve the acquisition of non-technical skills and CRM.top of form.

2 Materials and Methods

2.1 Type of Study

The study adopts a single-center quasi-experimental cohort approach.

This is a prospective longitudinal study aimed at evaluating the impact of a simulation-based educational intervention using simulated patient (SP).

2.2 Location of the Study

The large amphitheater of ISPITS in Fez was chosen as the main location to conduct the study.

2.3 Population and Sampling

The target population is made up of all ISPITS Fez students enrolled in the 2nd and 3rd year of the Bachelor of Nursing cycle, option Nurse in Anesthesia-Resuscitation (IAR), during the 2021–2022 academic year. The sample includes 52 students (31 in IAR-S4 and 21 in IAR-S6), meeting the inclusion criteria, having accepted and signed the consent form (Table 1).

2.4 Study Design

The study includes a group of 52 students in the 2nd and 3rd year of the Bachelor's degree at ISPITS in Fez.

Table 1 Size of the study population according to study option and sex

Study option	Effective	Man	Women
IAR-S4	31	14	17
IAR-S6	21	06	15

2.5 Measuring Instruments

Three measurement instruments were used to collect empirical data:

1. **A satisfaction questionnaire**: evaluating the evolution of the degree of satisfaction and the feeling of self-efficacy in three distinct sections.
2. **Cognitive tests**: covering knowledge relating to the reception of the patient in a BOU, the resuscitation of an ACR and the management of PPH.
3. **A grid**: Emergency Team Assessment (TEAM): Used to assess participants' non-technical skills and resource management.

3 Organization and Conduct of the Study

3.1 Information for Students

An audience information session preceded the study, where students were informed about the purpose and signed a consent form.

3.2 Process

An observational pre-test was conducted two weeks before the experiment to assess students' non-technical and CRM skills [11]. Participants then completed three cognitive pre-tests before receiving a theoretical lesson on the specific topics [12].

3.3 Building Scenarios

The clinical simulation scenarios were designed following the recommendations of SOFRASIM and MOROCCO SIM to mobilize the technical and non-technical skills of learners.

3.4 Choice of Simulation Type

Standardized patient (SP) simulation was favored to train non-technical skills and strengthen caregiver-patient relationships [11–14].

3.5 Progress of the Simulation Session

The session took place in the large ISPITS amphitheater, comprising three stages: briefing, simulation, and debriefing, promoting a climate of trust.

3.6 Student Assessment

In the assessment of students in this study, Kirkpatrick's three-level model was used [15, 16], as the fourth level was impractical due to methodological and temporal constraints. The three levels assessed are as follows:

1. **Reaction**: Assessment of the degree of satisfaction and feeling of self-efficacy using a 14-item questionnaire, covering aspects such as confidence, expectations, organizational satisfaction, and educational interest.
2. **Learnings**: Evaluation of the retention of theoretical knowledge through three tests relating to the content of the theoretical course, with closed questions and some multiple choice.
3. **Behavior**: Assessment of non-technical skills, including crisis resource management (CRM), using an emergency team assessment measurement grid (TEAM) validated internationally since 2015 [11, 17, 18].

The evaluation aimed to measure the impact of the simulation on the development of students' non-technical skills, with a focus on cognitive, social skills, and personal resources necessary for effective and safe operations in crisis situations (Fig. 1).

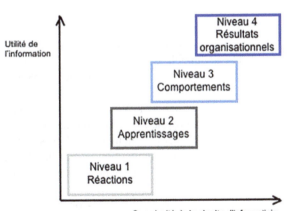

Fig. 1 Evaluation model (1967)

4 Statistical Analysis

The statistical processing of the data in this study was carried out with the SPSS software version 23. The evolution of the scores over time (between the pre-test and the post-test) was compared, using the Student's t test to quantitative variables and the Pearson chi-square test for qualitative variables. Descriptive statistics, such as mean, standard deviation, and frequency, were used to summarize the population data. Significant results were defined at an uncertainty level of 5% ($p < 0.05$).

5 Result

This chapter reveals the results of the present research, focusing on several aspects such as the degree of satisfaction, the feeling of self-efficacy, the retention of theoretical knowledge, individual and collective technical performance, as well as non-technical skills, particularly teamwork, in the context of welcoming a patient through to the management of PPH. It also explores the development of appropriate behaviors in the face of critical situations observed among nursing anesthetist students in the 2nd (S4) and 3rd (S6) year of the License cycle at ISPITS in Fez. The study focuses on the comparison of the evolution between the pre-test and the post-test, as well as on the differences between the two groups (S4 and S6).Top of form.

5.1 Satisfaction with the Simulation Session

Analysis of the responses in our study reveals that most students (74.51%) had specific expectations before the simulation session. On the other hand, 43.14% admitted to having felt fears, while 58.86% were without apprehension. Concerning the reception, more than half of the students (58.82%) rated it as excellent, while 41.18% rated it as average.

The duration of the simulation session was considered appropriate by a large majority (94.23%) of the students interviewed. Confidence during debriefing was reported by the majority, although 3.92% of students did not express confidence. The reality of the clinical situation during the debriefing was noted by 90.2% of the students, and more than 94% believed that the debriefing placed them in a real clinical situation.

Overall, most students reported growth in their clinical reasoning. Although 33.33% did not wish to participate in another simulation session, a significant majority (66.7%) would welcome the idea of participating again. These results suggest a positive reception of the simulation session by the majority of students, highlighting its positive impact on their learning and professional development (Table 2).

Table 2 Descriptive results of satisfaction on the simulation session

Satisfaction with the simulation session		Number	Percentage (%)
Special expectations	Yes	39	74.51
	No	13	25.49
Fears	Yes	22	43.14
	No	30	56.86
Welcome by trainers	Poor reception	0	0.00
	Home Medium	21	41.18
	Excellent welcome	30	58.82
The duration of the simulation sessions	Appropriate	49	94.23
	Not appropriate	3	5.77
Confidence during the Debriefing	Yes	49	96.08
	No	2	3.92
The reality of the clinical Situation during debriefing	Yes	46	90.20
	No	5	9.80
The simulation completes The practical training	Yes	48	94.12
	No	2	5.88
The progression of clinical Reasoning	Yes	49	96.08
	No	2	3.92
Being an actor in a new Simulation session	Yes	34	66.67
	No	17	33.33

5.2 Analysis of Proportions Reception Phase at the BOU, Before and After the Simulation

Concerning the reception phase, Students in semesters 4 and 6 benefit from the welcome simulation, with an increase in positive post-test responses. However, the low value of the Student's test does not make it possible to confirm a significant difference between the pre-test and post-test phases at the 5% threshold.

Despite perceived improvements among students in semesters 4 and 6 after the reception simulation, the statistical analysis fails to significantly demonstrate the difference between the pre-test and post-test phases, due to the low value of the Student test (Table 3).

The intra-group analysis results show a slight increase in the proportion of respondents who chose "yes" after the reception simulation for students enrolled in S6 compared to those enrolled in S4, with respective proportions of 0.90 and 0.87. This difference, although modest, suggests a possible trend where Semester 6 students benefit slightly more from the simulation than their Semester 4 counterparts (Table 4).

Table 3 Comparison of the proportions of the reception phase at the BOU, before and after the simulation

Statistical group					
ID		Average	Standard deviation	Average standard deviation	T.Student
Home S4	Pretest	0.8484	0.31017	0.09809	0.36
	Post test	0.9	0.31623	0.1	
Home S6	Pretest	0.89	0.31429	0.09939	0.071
	Post test	0.9	0.31623	0.1	

Table 4 Comparison of proportions between groups

	Proportion	Standard deviation	Average standard deviation
HomeS4	0.87	0.31	0.07
HomeS6	0.90	0.31	0.07

5.3 Analysis of ACR Phase Proportions Before and After Simulation

Comparative analysis of the proportions before and after the ACR simulation reveals an interesting trend. The results show an increase in the proportion of participants who acquired knowledge after the simulation for both groups, but this increase is more marked among the students of semester 6. Indeed, the average of people who developed knowledge after the simulation is higher among students in semester 6 than among those in semester 4.

This suggests that the ACR simulation appears to have a more significant impact on learning among Semester 6 students (Table 5).

The data presented in the inter-group proportion comparison table highlights subtle differences between the groups of students enrolled in ACR for semesters 4 and 6. Although the average proportion of students in semester 6 (0.9450) is slightly higher than that of semester 4 students (0.9355), these differences remain relatively modest. The average standard deviation for the two groups also remains quite close, indicating consistency in the dispersion of the data within the groups. These results

Table 5 Comparison of ACR phase proportions before and after simulation

ID		Proportion	Standard deviation	Average standard deviation	T Student
ACR S4	Pretest	0.91	0.19	0.06	0.77
	Post test	0.96	0.08	0.03	
ACR S6	Pretest	0.91	0.16	0.05	1.59
	Post test	0.99	0.03	0.01	

Table 6 Comparison of inter-group proportions

	Proportion	Standard deviation	Average standard deviation
ACR S6	0.9450	0.11687	0.02613
ACR S4	0.9355	0.14690	0.03285

suggest some homogeneity in how the two groups responded to the ACR simulation, with differences in performance that were not significantly marked (Table 6).

5.4 Analysis of Phase Proportions of Postpartum Hemorrhage PPH Before and After Simulation

Following our statistical analysis, it appears that the proportion of individuals who confirmed having acquired knowledge after the simulation phase is higher than that observed before said simulation. For semester 4 students, the proportion after the simulation was 0.89, compared to 0.68 before the simulation. Similarly, among semester 6 students, the proportion after the simulation reached 0.91, compared to that of 0.82 before the simulation (Table 7).

The analysis of the proportion of respondents regarding the improvement of knowledge during the management of hemorrhage before and after the simulation reveals that it is the students in S6 who improved their knowledge more compared to their peers in semester 4 (Table 8).

Comparing the proportions before and after the simulation for hemorrhage management reveals interesting results. For S4 students, the proportion increased significantly, from 0.6895 (pretest) to 0.8952 (posttest). The standard deviation also showed a reduction, indicating greater consistency in responses after the simulation.

Table 7 Comparison of the phase proportions of the HPP before and after the simulation

ID		Proportion	Standard deviation	Average standard deviation	T Student
HemorrhageS4	Pretest	0.6895	0.31367	0.1109	1.45
	Post test	0.8952	0.24612	0.08702	
HemorrhageS6	Pretest	0.825	0.20354	0.07196	0.932
	Post test	0.9125	0.17061	0.06032	

Table 8 Comparison of inter-group proportions

	Proportion	Standard deviation	Average standard deviation
HemorrhageS4	0.7923	0.29234	0.07308
HemorrhageS6	0.8688	0.18697	0.04674

The mean standard deviation difference was 0.1109, and the Student's t test gave a value of 1.45, suggesting a statistically significant improvement.

In contrast, for S6 students, although the proportion increased from 0.825 (pretest) to 0.9125 (posttest), the standard deviation showed a less marked decrease. The average standard deviation decreased from 0.07196 to 0.06032. However, the Student's t test indicated a value of 0.932, suggesting that the difference is not statistically significant.

In conclusion, simulation seems to have a positive impact on improving knowledge related to hemorrhage management, particularly for S4 students, where the results are statistically significant. This highlights the potential effectiveness of simulation as an educational tool in the context of PPH management.

5.5 Analysis According to Behavioral Change

5.5.1 According to the Different Categories

The use of the Chi-square test made it possible to evaluate the improvement in the participants' behaviors before and after the simulation session. The results confirmed a significant transformation in several aspects.

In the pretest phase, most participants experienced difficulty following the team leader's orders. However, after the simulation, a large part acquired skills to effectively complete the ordered tasks, thus confirming the improvement in expectations with a probability associated with the chi-square test of less than 5%.

Analysis of the overall outlook revealed a significant improvement, with 40.4% of post-test participants opting for the "always" option. The probability associated with the Chi-square test shows a statistical link between the test phases and this improvement.

Communication within the group improved significantly, as indicated by 55% of respondents saying the improvement became steady after the session. The probability associated with the Chi-square test confirms the significant impact of the session on communication.

The session also had a significant impact on teamwork, going from 59.6% saying it was not a priority before the session to 65% post-test confirming the opposite. The probability associated with the Chi-square test shows a statistical link with this improvement.

The team's ability to act with composure was significantly improved from 61.5% to 88.5% after the session, as confirmed by the probability associated with the chi-square test.

Regarding the adaptability of teams to changes in situations, the percentage of respondents saying they can always adapt increased from 26.9% before the test to 61.5% post-test, with a probability associated with the test of Chi-square confirming the significant impact of training.

Table 9 Cross-analysis according to the different categories of behavior top of form

		Never	Rarely	About half the time	Often	Always	Chi square
Expectations	Pretest	0.00%	0.00%	77.00%	23.00%	10.00%	0.014
	Post test	0.00%	0.00%	0.00%	59.60%	40.40%	
Global outlook	Pretest	0.00%	0.00%	59.60%	40.40%		0.000
	Post test	0.00%	0.00%	0.00%	59.60%	40.40%	
Communication	Pretest	13.50%	11.50%	34.60%	40.40%		0.000
	Post test		5.00%	5.00%	35.00%	55.00%	
Team work	Pretest	0.00%	0.00%	59.60%	40.40%		0.000
	Post test	0.00%	0.00%	10.00%	25.00%	65.00%	
Cold blood	Pretest	0.00%	31.50%	34.60%	23.80%	10%	0.046
	Post test	0.00%	0.00%	10.00%	31.50%	58.50%	
Team morale	Pretest	0.00%	9.60%	15.00%	30.40%	45.00%	0.058
	Post test	0.00%		6.00%	25.20%	68.80%	
Adaptation	Pretest	26.90%	32.70%	0.00%	20.40%	20.00%	0.037
	Post test			1.00%	38.50%	51.50%	
Control	Pretest	0.00%	36.50%	23.10%	40.40%	0.00%	0.000
	Post test	0.00%	0.00%	0.00%	44.20%	55.80%	
Anticipation	Pretest	38.50%	0.00%	21.20%	40.40%	0.00%	0.000
	Post test	0.00%	0.00%	0.00%	0.00%	100.00%	
Identifying priorities	Pretest	7.70%	13.50%	38.50%	30.40%	10.00%	0.001
	Post test	0.00%	0.00%	0.00%	53.80%	46.20%	
Monitoring standards	Pretest	21.20%	13.50%	25.00%	40.40%	0.00%	0.000
	Post test	0.00%	0.00%	0.00%	26.90%	73.10%	

The training also had a significant effect on the teams' controlling behavior, anticipation of possible actions, identification of priorities and monitoring of standards, as indicated by the results of the chi-square test, thus strengthening the capacity overall teams to effectively manage critical situations (Table 9).

5.5.2 According to the Overall Score of the Behavior Categories

The overall score on non-technical team performance was higher among students in the post-test phase. This result reflects, in some way, the effectiveness of the training in developing the non-technical performance of the team.

The overall score confirms our conclusions, training through simulation significantly and positively modified the behavior of candidates (Fig. 2).

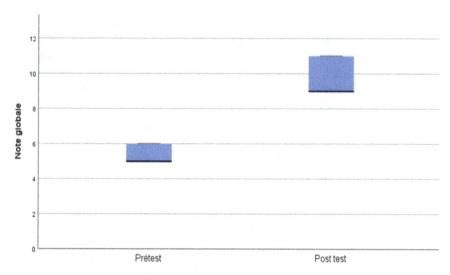

Fig. 2 Overall rating of behavior categories

5.5.3 Themes Proposed for Future Simulation Sessions

Concerning the themes proposed for the next simulation sessions, a clear trend emerged: 71% of students expressed an interest in simulation sessions focusing on CNT in critical situations. In contrast, half of the responses suggested sessions on topics such as preoperative ACR, multiple trauma patients, epilepsy and shock states. Finally, 35% of respondents expressed an interest in sessions covering drowning, electrification and burn scenarios (Fig. 3).

6 Discussion

6.1 The Simulation Effect in Administration (Reception, Medication)

Aliner and colleagues [19] demonstrated the effectiveness of scenario-based simulation training on nursing students' clinical skills and competencies. The sample of 99 students in the UK, divided into control and experimental groups, underwent patient simulation training using the Laerdal SimMan. Results showed a statistically significant difference in mean scores from pretest to posttest, with the experimental group demonstrating higher overall scores.

Bremner et al. [20] examined the use of the human patient simulator as a teaching strategy with novice nursing students. From a sample of 41 students, 95% rated the simulator session as good to excellent, with 68% recommending it as a mandatory

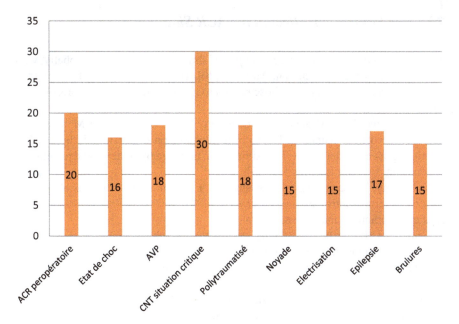

Fig. 3 Distribution according to the themes proposed for future simulation sessions

component of their training program. More than 60% indicated that the simulation experience increased their confidence in physical assessment skills.

A 2021 study by Wang et al. [21] found that the average score for overall communication skills and hospitality increased from 27.25 to 35.94 after a simulation session.

A meta-analysis of 19 articles [22–39] showed that patient simulation was used to facilitate learning and provide feedback to nurses. Studies have focused on specific skills such as mental health communication, communication skills with hearing-impaired patients, end-of-life care, and assessment of communication skills related to various scenarios.

Literature reviews confirm the study results, highlighting improved skills among nursing students in various areas, including reception, communication, and patient satisfaction.

In conclusion, medication administration and greeting are not only psychomotor skills, but also require critical thinking and clinical decision-making. Patient simulation offers a realistic approach in a safe environment. Effective communication is crucial for patient safety, and simulation improves nursing students' skills on multiple levels, contributing to better quality of care.

6.2 The Simulation Effect in the ACR Situation

In healthcare education, the increasing use of simulation is observed globally, with significant investments, such as in Australia [40]. These initiatives aim to respond to clinical challenges, particularly those related to resuscitation after cardiorespiratory arrest [40].

Studies have demonstrated the effectiveness of the Self-learning Methodology in a Simulated Environment (MAES) in the acquisition of nursing skills [41, 42]. Debriefing, an essential phase of the simulation, is valued by students and recognized as beneficial for their learning [43–45].

Debriefing, in particular, allows students to learn from their mistakes, thus promoting a constructive learning process [46–48]. Clinical simulation also addresses gaps in the practical and theoretical training of nursing students [49].

It provides a bridge between theory and practice, thereby reducing the gap between classroom learning and clinical placements [50–52]. The simulated scenarios, enriched by the input of the judges, make it possible to assess the technical and non-technical skills of the students [53].

Recent studies, such as those carried out in Brazil [54] and Spain [55], confirm the benefits of clinical simulation in terms of acquiring practical knowledge and nursing skills.

Clinical simulation effectively prepares students for the clinical challenges encountered in their future careers, ensuring a smooth transition between theory and practice [52, 54]. Thus, it contributes to training competent nurses, ready to meet the challenges of their profession [52].

6.3 The Simulation Effect in the Situation of Postpartum Hemorrhage

Obstetric simulation is an effective approach to developing management and communication skills, thereby improving patient outcomes in uncommon clinical scenarios. A study conducted by Signe Egenberg in Tanzania (2017) [56] showed significant improvements in the clinical management of postpartum hemorrhage (PPH) after training in multi-professional teams during obstetric emergencies. A significant change was observed in the pre- and post-test scores, demonstrating the effectiveness of the simulation.

Another study by Lathrop et al. [57] evaluated the impact of a simulation on the management of PPH in four midwifery students. Self-assessments showed improved skills after simulation, and students recommended adding simulation sessions to course content. The simulation sessions allowed students to put their theoretical knowledge into practice.

Simulations also provide students with the opportunity to think critically about the management of obstetric complications, thereby aligning practice with theory. This approach is supported by the American College of Nurse Midwives [58]. Additionally, simulations play a crucial role in maintaining the skills of practicing midwives, preparing them to handle rare but potentially serious scenarios.

Finally, the study carried out in North Africa on the effect of simulation in the management of PPH gave positive results, highlighting the importance of this approach to increase knowledge and control of hemorrhage situations. postpartum among a large number of nurses, thus contributing to the improvement of obstetric care. Recommendations from the Joint Commission on Accreditation of Healthcare Organizations (JCAHO) emphasizing the need for simulation training for obstetric emergencies are also highlighted in this study [56–59].

6.4 The Effect of Simulation on TEAM (Leadership, Teamwork, Communication, Stress Management, Etc.)

Simulation and team-based learning are crucial educational strategies in nursing. Several recent studies have examined the effects of programs integrating simulation and team-based learning on different dimensions of nursing skills.

A 2020 Korean study evaluated a nursing simulation program with team-based learning, showing significant post-test improvements in nursing students' knowledge, team performance, and teamwork [60]. A 2021 Canadian study identified correlations between communication, teamwork, and leadership after simulation sessions [61].

In the United States, a 2017 study highlighted a more positive attitude toward nurse-physician collaboration after simulation events [62]. Another 2020 US study used an Escape Game experiment to teach teamwork, showing statistically significant improvements in teamwork and leadership post-simulation [63]. A 2021 New Zealand study found statistically significant improvements in nursing leadership after simulation training [10]. For pediatric critical care nurses in the United States, a 2020 study showed post-simulation improvements in knowledge, confidence, and clinical teamwork performance scores [64].

These studies conclude that simulation is an effective teaching technique for developing interprofessional soft skills such as leadership, teamwork, and stress management. Effective communication, teamwork, and leadership are also identified as essential skills by the Quality and Safety Education for Nurses (QSEN) initiative [65]. To prepare future nurses for safe, quality practice environments, programs incorporating these skills must be integrated into pre- and post-graduation nursing programs [66, 67].

6.5 Strengths and Limitations of the Study

Strong points: This research work addresses three major themes in the field of simulation in nursing science, thus marking a significant step forward for nursing research in Morocco. These include the reception of the patient in the BOU, the resuscitation of an ACR and the management of PPH, areas little explored among nurses in the literature, especially in Morocco. This study represents the first in Morocco to achieve KirkPatrick Level 3 in nursing studies, using varied data collection methods and focusing on participant satisfaction. The quality supervision and contributions of experts in intensive care anesthesia and medical simulation were also decisive for the success of this research.

Boundaries: This study offers important conclusions despite its limitations, including the short duration and small sample size, as well as the single-center nature of the research. Human, temporal and geographical constraints influenced the choice of this approach. The unavailability of the simulation center led to opting for standardized simulations rather than hybrids, despite efforts to overcome this limitation. The conduct of activities in a single location and the presence of evaluators could have an impact on the results, thus highlighting the need to take into account the Hawthorne effect.

7 Conclusion

In conclusion, Critical situations in healthcare services require a team approach to ensure the safety of the patient and healthcare staff.

Nursing simulation, recommended by QSEN and SFAR, provides a safe environment to develop crucial skills such as safety, communication and teamwork.

Our study evaluates this method in a context where standards and technical developments differ. The results show its practical and psychological effectiveness, although its theoretical impact is limited.

The new SFAR recommendations emphasize the importance of debriefing after critical events. Clinical research must continue to improve the quality of care and patient safety. Optimizing teamwork is a major challenge, requiring specific training and the use of validated tools such as ANTS, NOTECHS and Team STEPPS.

The credibility of the results and the recommendations reinforce the potential of this technique in education, prompting future studies to assess its long-term impact in the field of health.

References

1. Scully NJ (2011) The theory-practice gap and skill acquisition: an issue for nursing education. Collegian 18(2):93–98. https://doi.org/10.1016/j.colegn.2010.04.002
2. Ajani K, et Moez S (2011) Gap between knowledge and practice in nursing. Procedia—Soc Behav Sci 15:3927–3931. https://doi.org/10.1016/j.sbspro.2011.04.396
3. Hussein MH, et Osuji J (2017) Bridging the theory-practice dichotomy in nursing: the role of nurse educators. J Nurs Educ Pract 7(3):20–25
4. Gardiner I, et Sheen J (2016) Graduate nurse experiences of support: a review. Nurse Educ Today 40:7–12, mai 2016, https://doi.org/10.1016/j.nedt.2016.01.016
5. Salifu DA, Gross J, Salifu MA, et Ninnoni JP (2019) Experiences and perceptions of the theory-practice gap in nursing in a resource-constrained setting: a qualitative description study. Nurs Open 6(1):72–83, janv 2019. https://doi.org/10.1002/nop2.188
6. Esmaeili M, Cheraghi MA, Salsali M, et Ghiyasvandian S (2014) Nursing students' expectations regarding effective clinical education: a qualitative study. Int J Nurs Pract 20(5):460–467. https://doi.org/10.1111/ijn.12159
7. Saifan A, AbuRuz ME, et Masa'deh R (2015) Theory practice gaps in nursing education: A qualitative perspective. J Soc Sci/Sosyal Bilimler Dergisi 11(1):2015, Consulté le: 17 avril 2024. [En ligne]. Disponible sur. https://www.researchgate.net/profile/Mohannad-Aburuz-2/publication/278811628_Theory_Practice_Gaps_in_Nursing_Education_A_Qualitative_Perspective/links/55866dcc08aef58c039ef82d/Theory-Practice-Gaps-in-Nursing-Education-A-Qualitative-Perspective.pdf
8. Montayre J (2015) Critical thinking in nursing education : addressing the theory-practice gap. Nurs Rev 25–25
9. Hegland PA, Aarlie H, Strømme H, et G Jamtvedt (2017) Simulation-based training for nurses: systematic review and meta-analysis. Nurse Educ Today 54:6–20
10. Armstrong P, Peckler B, Pilkinton-Ching J, McQuade D, et Rogan A (2021) Effect of simulation training on nurse leadership in a shared leadership model for cardiopulmonary resuscitation in the emergency department. Emerg Med Australasia 33(2):255–261, avr. 2021, https://doi.org/10.1111/1742-6723.13605
11. Medical consultation simulations and the question of the actors—simulated or standardized patients.—Abstract—Europe PMC. Consulté le: 17 avril 2024. [En ligne]. Disponible sur: https://europepmc.org/article/med/29879336
12. Standardized Patient Program, University of Fribourg/Faculty of Sciences, 5.
13. Granry JC, Moll MC (2012) state of the art (national and international) in terms of simulation practices in the field of health in the context of continuing professional development (CPD) and the prevention of risks associated with care. Has mission report of 10 Jan 2012
14. Gilibert D, Gillet I (2010) Review of models in training evaluation: individual and social conceptual approaches. Psychol Pract 16(3):217–38. S1269176309000261
15. Adamson KA, Kardong-Edgren S, et Willhaus J (2013) An updated review of published simulation evaluation instruments. Clin Simul Nurs 9(9):e393–e400 (2013)
16. Flin R, et O'Connor P (2017) Safety at the sharp end: a guide to non-technical skills. CRC Press. Consulté le: 17 avril 2024. [En ligne]. Disponible sur. https://www.taylorfrancis.com/books/mono/; https://doi.org/10.1201/9781315607467/safety-sharp-end-paul-connor-rhona-flin
17. Teamwork and patient safety in dynamic domains of healthcare: a review of the literature—MANSER—2009—Acta Anaesthesiologica Scandinavica—Wiley Online Library. Consulté le: 17 avril 2024. [En ligne]. Disponible sur. https://onlinelibrary.wiley.com/doi/full/; https://doi.org/10.1111/j.1399-6576.2008.01717.x
18. Cronenwett LR, Bootman JL, Wolcott, et Aspden P (2007) Preventing medication errors. National Academies Press. Consulté le: 17 avril 2024. [En ligne]. Disponible sur: https://books.google.fr/books?hl=fr&lr=&id=fsqaAgAAQBAJ&oi=fnd&pg=PT17&dq=Bootman+JL,+Cronenwett+LR,+Bates+DW,+et+al.+Preventing+medication+errors.+Washington,+DC:+National+Academies+Press%3B+2006.&ots=0IHQS0yoOl&sig=2L2Yv3TBW2w8kXeE6bDfR0oUjPE

19. Alinier G, Hunt WB, et Gordon R (2004) Determining the value of simulation in nurse education: study design and initial results. Nurse Educ Pract 4(3):200–207
20. Bremner MN, Aduddell K, Bennett DN, et VanGeest JB (2006) The use of human patient simulators: best practices with novice nursing students. Nurse Educator 31(4), 170–174 (2006)
21. Using Simulation Teaching to Improve the Communication Skills of Clinicians—ProQuest. Consulté le: 17 avril 2024. [En ligne]. Disponible sur: https://www.proquest.com/openview/0c33f4f4a096baa5d10c8d951abc1cd0/1?pq-origsite=gscholar&cbl=866377
22. Becker KL, Rose LE, Berg JB, Park H, et Shatzer JH (2006) The teaching effectiveness of standardized patients. J Nurs Educ 45(4), Consulté le: 17 avril 2024. [En ligne]. Disponible sur: https://www.researchgate.net/profile/John-Shatzer/publication/7151904_The_Teaching_Effectiveness_of_Standardized_Patients/links/56aa383408ae2df82166d0fd/The-Teaching-Effectiveness-of-Standardized-Patients.pdf
23. Evaluating the use of standardized patients in undergraduate psychiatric nursing experiences—ScienceDirect. Consulté le: 17 avril 2024. [En ligne]. Disponible sur: https://www.sciencedirect.com/science/article/abs/pii/S187613990900499X
24. Communication and human patient simulation in psychiatric nursing: issues in mental health nursing, vol 30, No 8. Consulté le: 17 avril 2024. [En ligne]. Disponible sur: https://www.tandfonline.com/doi/abs/; https://doi.org/10.1080/01612840802601366
25. The influence of teaching method on performance of suicide assessment in baccalaureate nursing students—Rebecca Luebbert, Ann Popkess (2015). Consulté le: 17 avril 2024. [En ligne]. Disponible sur: https://journals.sagepub.com/doi/abs/; https://doi.org/10.1177/1078390315580096
26. Mental Health Clinical Simulation: Therapeutic Communication—ScienceDirect. Consulté le: 17 avril 2024. [En ligne]. Disponible sur: https://www.sciencedirect.com/science/article/abs/pii/S1876139916000256
27. Webster D (2013) Promoting therapeutic communication and patient-centered care using standardized patients. J Nurs Educ 52(11):645–648. https://doi.org/10.3928/01484834-20131014-06
28. Adib-Hajbaghery M, et Rezaei-Shahsavarloo Z (2015) Nursing students' knowledge of and performance in communicating with patients with hearing impairment. Jpn J Nurs Sci 12(2):135–144, avr. 2015. https://doi.org/10.1111/jjns.12057
29. Use of simulated patient method to teach communication with deaf patients in the emergency department—ScienceDirect. Consulté le: 17 avril 2024. [En ligne]. Disponible sur: https://www.sciencedirect.com/science/article/abs/pii/S1876139916300147
30. Dang BK, Palicte JS, Valdez A, et O'Leary-Kelley C (2018) Assessing simulation, virtual reality, and television modalities in clinical training. Clin Simul Nurs 19:30–37
31. Bloomfield JG, O'Neill B, et Gillett K (2015) Enhancing student communication during end-of-life care: a pilot study, Palliative & supportive care, vol 13, n 6, pp 1651–1661
32. Innovative approach to teaching communication skills to nursing students | J Nurs Educ. Consulté le: 17 avril 2024. [En ligne]. Disponible sur: https://journals.healio.com/doi/abs/; https://doi.org/10.3928/01484834-20090918-06
33. Glatard T et al (2012) A virtual imaging platform for multi-modality medical image simulation. IEEE Trans Med Imaging 32(1):110–118
34. Canivet J, Bonnefoy J, Daniel C, Legrand A, Coasne B, et Farrusseng D (2014) Structure–property relationships of water adsorption in metal–organic frameworks. New J Chem 38(7):3102–3111
35. Langewitz W, Heydrich L, Nübling M, Szirt L, Weber H, et Grossman P (2010) Swiss Cancer League communication skills training programme for oncology nurses: an evaluation. J Adv Nurs 66(10):2266–2277. https://doi.org/10.1111/j.1365-2648.2010.05386.x
36. Brown RF et al (2009) Identifying and responding to depression in adult cancer patients: evaluating the efficacy of a pilot communication skills training program for oncology nurses. Cancer Nurs 32(3):E1–E7

37. Schlegel C, Woermann U, Shaha M, Rethans J-J, et Van Der Vleuten C (2012) Effects of communication training on real practice performance: a role-play module versus a standardized patient module. J Nurs Educ 51(1):16–22, janv. 2012. https://doi.org/10.3928/01484834-20111116-02
38. Paans W, Müller-Staub M, et Niewneg R (2013) The influence of the use of diagnostic resources on nurses' communication with simulated patients during admission interviews. Int J of Nurs Knowl 24(2):101–107, juin 2013. https://doi.org/10.1111/j.2047-3095.2013.01240.x
39. Hsu L-L, Chang W-H, et Hsieh S-I (2015) The effects of scenario-based simulation course training on nurses' communication competence and self-efficacy: a randomized controlled trial. J Profess Nurs 31(1):37–49
40. Health Workforce Australia. Health Workforce Australia (2012)
41. Díaz JL, Leal C, García JA, Hernández E, Adánez MG, et Sáez A (2016) Self-learning methodology in simulated environments (MAES\copyright): elements and characteristics. Clin Simul Nurs 12 (7):268–274
42. Díaz Agea JL, Megías Nicolás JA, García Méndez M, de Adánez Martínez G, et Leal Costa C (2019) Improving simulation performance through Self-Learning Methodology in Simulated Environments (MAES©). Nurse Educ Today 76:62–67, mai 2019. https://doi.org/10.1016/j.nedt.2019.01.020
43. Dufrene C, et Young A (2014) Successful debriefing—best methods to achieve positive learning outcomes: a literature review. Nurse Educ Today 34(3):372–376
44. Levett-Jones T, et Lapkin S (2014) A systematic review of the effectiveness of simulation debriefing in health professional education. Nurse Educ Today 34(6):e58–e63
45. Neill MA, et Wotton K (2011) High-fidelity simulation debriefing in nursing education: a literature review. Clin Simul Nur 7(5):e161–e168
46. Shin S, Park J-H, et Kim J-H (2015) Effectiveness of patient simulation in nursing education: meta-analysis. Nurse Educ Today 35(1):176–182
47. Ricketts B (2011) The role of simulation for learning within pre-registration nursing education—a literature review. Nurse Educ Today 31(7):650–654
48. Norman J (2012) Systematic review of the literature on simulation in nursing education. Assoc Black Nur Faculty Found J (ABNFF) 23(2), Consulté le: 17 avril 2024. [Enligne]. Disponible sur: https://search.ebscohost.com/login.aspx?direct=true&profile=ehost&scope=site&authtype=crawler&jrnl=10467041&AN=75292183&h=jkgYS83TYlWpYQn4jGFjeKSNqij1QGR1Ywu6xNQgTDRgrVI10jNzgqi3h2zXJTN1g%2BvXlIFGGMDT9HXCwb5H5g%3D%3D&crl=c
49. King A, Holder MG, et Ahmed RA (2013) Errors as allies: error management training in health professions education. BMJ Qual Safety 22(6):516–519
50. Hope A, Garside J, et Prescott S (2011) Rethinking theory and practice: pre-registration student nurses experiences of simulation teaching and learning in the acquisition of clinical skills in preparation for practice. Nurse Educ Today 31(7):711–715
51. Lisko SA, et O'dell V (2010) Integration of theory and practice: Experiential learning theory and nursing education. Nur Educ Perspect 31(2):106-108
52. SciELO–Brasil—Impact of high-fidelity simulation in pediatric nursing teaching: an experimental study impact of high-fidelity simulation in pediatric nursing teaching: an experimental study. Consulté le: 17 avril 2024. [En ligne]. Disponible sur: https://www.scielo.br/j/tce/a/K84tN8vMMMKZGPCgkYgPVjD/
53. Vézina K, Roch G, Paradis V, Dallaire D, et Simon M (2018) Simulation interdisciplinaire en réanimation cardiorespiratoire: innover pour sauver des vies en milieu hospitalier. Revue Francophone Internationale de Recherche Infirmière, 4(3), e167–e178. https://doi.org/10.1016/j.refiri.2018.06.002
54. The design and validation of a clinical simulation scenario in the management of a cardiac arrest during hemodialysis session. Consulté le: 17 avril 2024. [En ligne]. Disponible sur: https://www.scirp.org/journal/paperinformation?paperid=117787

55. Arrogante O, González-Romero GM, Carrión-García L, et Polo A (2021) Reversible causes of cardiac arrest: nursing competency acquisition and clinical simulation satisfaction in undergraduate nursing students. Int. Emerg Nur 54:100938, janv. 2021. https://doi.org/10.1016/j.ienj.2020.100938
56. Egenberg S, Karlsen B, Massay D, Kimaro H, et Bru LE (2017) "No patient should die of PPH just for the lack of training!" Experiences from multi-professional simulation training on postpartum hemorrhage in northern Tanzania: a qualitative study. BMC Med Educ 17(1):119, déc. 2017. https://doi.org/10.1186/s12909-017-0957-5
57. Lathrop A, Winningham B, et VandeVusse L (2007) Simulation-Based learning for midwives: background and pilot implementation. J Midwifery Women's Health 52(5):492–498. https://doi.org/10.1016/j.jmwh.2007.03.018
58. Core competencies for basic midwifery education. The American College of Nursing-Midwives
59. Sentinel event alert #30: preventing infant death and injury during delivery. Joint commission on accreditation of healthcare organizations
60. Roh YS, Kim SS, Park S, et Ahn J-W (2020) Effects of a simulation with team-based learning on knowledge, team performance, and teamwork for nursing students. CIN: Comput Inf Nur 38(7):367, juill. 2020. https://doi.org/10.1097/CIN.0000000000000628
61. Kleib M, Jackman D, et Duarte-Wisnesky U (2021) Interprofessional simulation to promote teamwork and communication between nursing and respiratory therapy students: a mixed-method research study. Nurse Educ Today 99:104816, avr. 2021. https://doi.org/10.1016/j.nedt.2021.104816
62. Krueger L, Ernstmeyer K, et Kirking E (2017) Impact of interprofessional simulation on nursing students' attitudes toward teamwork and collaboration. J Nurs Educ 56(6):321–327, juin 2017. https://doi.org/10.3928/01484834-20170518-02
63. Valdes B, Mckay M, et Sanko JS (2021) The impact of an escape room simulation to improve nursing teamwork, leadership and communication skills: a pilot project. Simul Gam 52(1):54–61, févr 2021. https://doi.org/10.1177/1046878120972738
64. Karageorge N, Muckler VC, Toper M, et Hueckel R (2020) Using simulation with deliberate practice to improve pediatric ICU nurses' knowledge, clinical teamwork, and confidence. J Pediatr Nur 54:58–62. https://doi.org/10.1016/j.pedn.2020.05.020
65. Quality and Safety Education for Nurses (QSEN) (2003) QSEN skills. qsen.org/competencies/
66. Nieva VF et, Sorra J (2003) Safety culture assessment: a tool for improving patient safety in healthcare organizations. BMJ Qual Saf 12, n suppl 2, II17–II23. https://doi.org/10.1136/qhc.12.suppl_2.ii17
67. Ballantyne JE (2008) Cultural competency: highlighting the work of the American Association of Colleges of Nursing-California Endowment Advisory Group. J Prof Nurs 24(3):133–134. https://doi.org/10.1016/j.profnurs.2008.04.002

Identifying AI Generated Code with Parallel KNN Weight Outlier Detection

Oscar Karnalim

Abstract Plagiarism is an emerging issue in programming, and it becomes more difficult to identify due to generative AI; it can easily provide solutions on assignments that are supposed to be completed without any help. Many AI detectors are only focused on general text. We present an AI detector for programming based on KNN weight outlier detection. On assignments disallowing AI help, only few students still insist to use AI. Further, AI generated code is different to that of undergraduates. The detector employs six intermediate representations: text strings, token strings, generalized token strings, expanded token strings, linearized syntax trees, and linearized parse trees. According to our experiment on three data sets with MAP and processing time as the performance metrics, our detector is satisfactorily effective and efficient. The most optimal performance occurs when the intermediate representation is the token strings, the n for n-grams is 110, and the number of clusters is 3.

Keywords Clustering · AI detection · Programming education · Academic integrity

1 Introduction

Plagiarism and collusion are common in programming [33]. Both are about copying and reusing one's code without appropriate credit [11]. The only difference is about the awareness of the owner of the copied work: plagiarism occurs when the owner is unaware about the act. In some literature, collusion is considered as part of plagiarism [16].

There are three common ways to deal with programming plagiarism [1]. Instructors can reduce pressure to plagiarize by breaking down large assignments to smaller assignments [2], or allocating bonus marks on early submissions [37]. Instructors

O. Karnalim (✉)
Faculty of Smart Technology and Engineering, Maranatha Christian University, Surya Sumantri Street No. 65, Bandung 40164, Indonesia
e-mail: oscar.karnalim@it.maranatha.edu

© The Author(s), under exclusive license to Springer Nature Switzerland AG 2025
E. Vendrell Vidal et al. (eds.), *Advanced Technologies and the University of the Future*,
Lecture Notes in Networks and Systems 1140,
https://doi.org/10.1007/978-3-031-71530-3_29

can reshape student misconception of plagiarism by informing them about academic integrity [34]. Sometimes, the delivery is aided by automated systems [20, 40].

The third approach and perhaps the most common one to deal with plagiarism is to reduce the opportunity. Instructors can confirm the authenticity of student submissions by having interviews [14] or asking for oral presentation [15]. Further, they can update the assignments for each offering to prevent cross-semester plagiarism (i.e., copying from a senior's work) [32], or they can allow students to use their own case studies, preventing in-class plagiarism (i.e., copying from a colleague's work) [10].

Imposing penalty and identification mechanism are another way to reduce the opportunity of plagiarism [19]. Students who are involved in plagiarism need to be penalized for a deterrent effect. The penalty varies from informal warning to zero marks for the course. Since accusing plagiarism should be supported with strong evidence [17], plagiarism needs to be identified. Instructors often employ a code similarity detector to report excessive similarity, and then manually justify whether the similarity is evident enough for intentional plagiarism. A number of code similarity detectors developed for that purpose [4]. Some of them (e.g., JPlag [28] and MOSS [29]) are stable and they are commonly used in academia.

Generative AI [7] like Github Copilot [24] and ChatGPT [22] introduces a new form of plagiarism [39]. It can be misused on individual assignments where such help is not permitted. Students can just copy and paste code generated by AI to solve several tasks. They will not learn how to write code from scratch. The misuse is considered as plagiarism since the generated code is based on other people's work included in the training data. Nevertheless, the misuse cannot be detected with conventional code similarity detectors that pairwise compare student submissions; AI generated submissions will be unique and have low similarity.

There are a number of automated detectors for AI generated work [26]. However, many of them are focused on text, not code. GPTZero[1] employs a predicting mechanism and sentence distribution. GLTR [13] employs word-based features including probability, rank, and entropy. AI Text Classifier[2] employs 34 existing models. GPT-2 Output Detector [36] employs RoBERTa English corpus data.

Code authorship detectors [18] might be helpful to detect AI generated work if the involved students rarely use AI help. Student independent submissions will be different to those generated by AI. The latter tends to have more complex syntax and to be less readable than the former [39]. Several examples of code authorship detectors are [3, 5, 42]. Nevertheless, the code authorship detectors expect previous submissions to be available, which might not be practical.

In response to the aforementioned gap, we present an automated detector for AI generated code. It is based on outlier detection in clustering [35], assuming that on assignments that disallow AI help, only few students will try to breach the rules and still use AI. Further, the AI generated code will be different to student code [39]. The detector converts student programs to one of six intermediate representations,

---

[1] https://gptzero.me/.

[2] https://beta.openai.com/ai-text-classifier.

generates n-gram vectors [9], and lists the outliers with the parallel implementation of KNN weight outlier detection [31].

Our study has two research questions:

- RQ1: Which setting of KNN weight outlier detection does entail the highest effectiveness and efficiency for detecting AI generated code?
- RQ2: Which intermediate representation does entail the highest effectiveness and efficiency for detecting AI generated code?

The two research questions were addressed with three data sets of introductory programming under two metrics: Mean Average Precision (MAP) [9] for effectiveness and processing time for efficiency.

AI can increase work productivity in industries [43] and we are not against that. However, in academia, some assignments expect students to write their solutions from scratch without any helps. Introductory programming for example, often employs many small assignments with straightforward solutions [2]. The aim is to let students experiencing writing a working program and learning how to use program syntax. Learning programming is similar to learning math at which exercises are needed. The expected skills might also be useful for using AI help later in the workplace. AI generated code is not always correct and complete [39]. The code needs to be modified and adapted.

It is worth noting that our presented detector might also be useful for detecting contract cheating [23] where illegal help from third parties are considered. If the third parties are professional programmers or they are unaware with the course, the submissions might be different to those written by students.

2 Method

2.1 The AI Detector

Our AI detector works in five steps. At first, all submissions are tokenized with ANTLR [27] and converted to an intermediate representation at which all comments and white space are excluded. If a submission has more than one code files, the content will be concatenated. There are six available intermediate representations, inspired from those commonly used in plagiarism detection [19]: text strings, token strings, generalized token strings, expanded token strings, linearized syntax trees, and linearized parse trees. All graph-related representations are excluded as they do not work on incomplete and/or incompilable submissions.

Text strings are sequences of alphanumeric tokens, which are keywords (e.g., 'if' and 'while') and identifier names (e.g., variable and function names). String constants are excluded as they are expected to be the same for all submissions; the constants are often used to form the output. Text strings are partly inspired from [41].

Token strings are sequences of code tokens. They include not only keywords and identifiers, but also constants and symbols. This representation can sufficiently show the program content and how it works. Generalized token strings are similar to the token strings except that easily disguised tokens are generalized to their own types. These tokens include identifier names and constants. Generalized token strings are quite common for plagiarism detection [19]. Expanded token strings are a combination of token strings and their generalized version. It is expected to provide richer information than using only one representation alone. Token strings are inspired from [38] while the generalized ones are inspired from [28]. Expanded token strings are exclusively introduced in this study.

Linearized syntax trees capture more structural information from the program since they include the parsing rules [8]. Linearization enables the trees to be treated as token strings. Directly comparing syntax trees is relatively slow [12]. In our case, we employ preorder linearization and it is inspired from [21].

Linearized parse trees are similar to the linearized syntax trees except that the former considers all parsing rules including intermediate rules for compilation [8]. They usually have larger trees than the latter. It is expected to provide comprehensive information and the representation is inspired from [25].

Second, n-grams are formed for each submission at which one n-gram consists of n subsequent tokens. To illustrate this, a token string with 'if', '(', 'x', '>', '3', ')', ':', 'print', '(', '"a"', and ')' will be converted to five 7-g:

- 'if', '(', 'x', '>', '3', ')', and ':'
- '(', 'x', '>', '3', ')', ':', and 'print'
- 'x', '>', '3', ')', ':', 'print', and '('
- '>', '3', ')', ':', 'print', '(', and '"a"'
- '3', ')', ':', 'print', '(', and '"a"', and ')'

Third, based on the n-grams, an index is formed. Index refers a dictionary with key-value tuples, at which the keys are the n-grams and the values are the occurrence frequencies [9]. The representation is efficient as retrieving a particular value based on a key is constant-time.

Fourth, distinct n-grams from all submission indexes are listed. Each index is then converted to a vector: the features are the distinct n-grams and the values are the occurrence frequencies for given submission index.

Fifth, outliers are identified based on a parallel implementation of KNN weight outlier detection [31] with ELKI, a data mining library [30]. The algorithm is derived from KNN algorithm, which groups similar instances to k clusters. According to literature [6], the algorithm is relatively effective and the effectiveness is quite stable across data sets. It is weighted so that k does not heavily affect the performance. However, k still needs to be set.

2.2 Addressing The Research Questions

Our study has two research questions. RQ1 is about the most optimal setting of KNN weight outlier detection for identifying AI generated code. It was addressed by searching the most optimal n for n-grams and the most optimal number of clusters (k, though it should not have large impact due to our weighting mechanism). We employed token strings as the default intermediate representation.

Two performance metrics were employed: MAP for effectiveness and processing time for efficiency. MAP is the average of precision at top-X [9] where X refers to the rank position of each AI generated submission. Precision at top-X is the proportion of AI generated submissions retrieved in top-X. For instance, if there are two AI generated submissions from a total of four submissions and they are ranked the first and the third, the MAP will be 83.3%. It is resulted from $(1/1 + 2/3)/2 = (100\% + 66.6\%)/2 = 166.6\%/2$. Higher MAP refers to higher effectiveness.

Processing time is the amount of time required to process given data set. It was measured in nanoseconds but reported in seconds for readability. Shorter processing time refers to higher efficiency.

Three data sets were considered. They were from a Python introductory programming course for undergraduate students. Exam data set contained two assignments: mid and final exams. Both assignments had 17 submissions each, half of them (nine for the mid and eight for the final) were independent submissions while the rest were intentionally written with the help of AI for the purpose of analysis. Each assignment needed to be completed in a physical laboratory in two hours. No discussions among colleagues were allowed. Internet access was provided only to students assisted by AI for accessing ChatGPT.

Weekly 1 data set consisted of 17 weekly assignments. Each assignment was issued to a class with either 26 or 28 students, at which one to three of them employed ChatGPT. Unlike the exam data set, the weekly 1 data set allowed discussions among colleagues and internet access. Further, AI was only used for partial help. The data set had a total of 1136 independent submissions and 35 AI generated submissions.

Weekly 2 data set had 52 weekly assignments with comparable nature to those of the weekly 1 data set. However, no AI generated submissions identified in the weekly 1 data set were reconsidered. Further, each assignment had six additional AI generated submissions that were intentionally created for analysis. For these submissions, AI was used for writing the whole solutions, not only some parts of it. The data set had a total of 1136 independent submissions and 312 AI generated submissions.

For the most optimal n for n-grams, we started with the first ten positive integer ($n = 1$ to $n = 10$). However, since the pattern did not show the 'peak' (reaching the highest MAP and then getting reduced), we introduced larger n: ten first multiples of five (5, 10, 15, 20, ... 45, 50). Still, the pattern did not show the 'peak'. Finally, we set twenty multiples of ten, from 10 to 200. The 'peak' occurred and we reported these results. Unigram ($n = 1$) was exclusively included in the analysis to show the impact of not using n-grams (e.g., bag-of-words approach).

For the most optimal number of clusters (k in the KNN weight outlier detection), we employed the first ten positive integers. We did not test larger k like in n-grams since the resulted MAPs are the same and the processing time showed no interesting patterns. Further, larger k might force few clusters to only have one submission since each assignment only had up to 34 submissions.

RQ2 is about the most optimal intermediate representation for detecting AI generated code. The six intermediate representations introduced in our detector were considered: text strings, token strings, generalized token strings, expanded token strings, linearized syntax trees, and linearized parse trees. They were compared one another with two performance metrics (MAP and processing time) on the three data sets (exam, weekly 1, and weekly 2). All scenarios employed the most optimal setting obtained from RQ1 findings.

3 Results and Discussion

3.1 Optimal Setting for Identifying AI Generated Code

Among the three data sets, Fig. 1 shows that the highest MAP was generally obtained on the exam data set. It is expected as the assignments were more restricted: no access to the internet and no discussions among colleagues. The lowest MAP on the contrary, was generally obtained on the weekly 1 data set. The assignments allowed both access to the internet and discussions among colleagues. Further, AI assisted students did not use AI for writing the whole submissions.

Employing n-grams positively affects effectiveness; having n set larger than one increased MAP. However, increasing n larger than 10 did not show any substantial changes. No trend patterns were observed. The most suitable value for n was 110; it resulted in the highest total MAP (164%): 73% on the exam data set, 44% on the weekly 1 data set, and 47% on the weekly 2 data set. The optimal n was reasonably

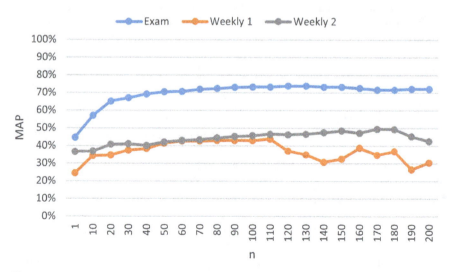

Fig. 1 MAP of various n values in n-grams on the three data sets

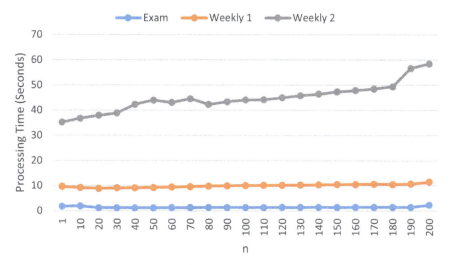

Fig. 2 Processing time of various n values in n-grams on the three data sets

high. AI generated code was only identifiable if it had a bunch of uncommon statements for undergraduates. One or two program statements might be insufficient for evidence.

In terms of processing time, dealing with the exam data set was the fastest as it only had two assignments (as shown in Fig. 2). The slowest processing time was obtained on the weekly 2 data set as it had 52 assignments.

Larger n tended to result in longer processing time. More distinct features were considered as long n-grams are more difficult to match with others. However, the trend was only shown on the weekly 2 data set, which had the largest number of assignments. Other data sets were quite small and the processing time was heavily affected by software and operating system dependencies.

Changing the number of clusters (k) in the KNN weight outlier detection did not affect effectiveness. The resulting MAPs are the same: 73% for the exam data set, 44% for the weekly 1 data set, and 47% for the weekly 2 data set. It was only different on $k = 2$ where its MAP on the weekly 2 data set was slightly reduced by 1%.

Figure 3 shows that changing the number of clusters also had no substantial effect on the processing time. Although larger k meant more clusters to check for each submission's membership, the checking process was relatively fast. Changes in the processing time might be more affected by software and operating system dependencies.

Based on the experiment, it was clear that the most optimal n for n-grams was 110 and the most optimal number of clusters (k in the KNN weight outlier detection) was 3. The number of clusters can be anything from 3 to 10 but we believe having fewer clusters are preferred for faster checking of the clusters' membership.

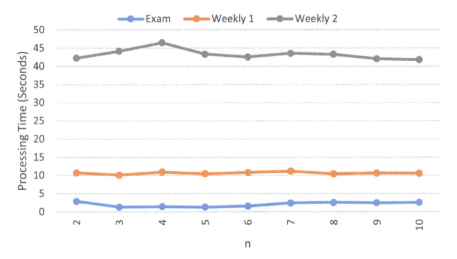

Fig. 3 Processing time of various numbers of clusters (*k*) on the three data sets

3.2 Optimal Intermediate Representation for Identifying AI Generated Code

In this evaluation, *n* for *n*-grams was set to 110 and number of clusters (*k* in the KNN weight outlier detection) was set to 3. Figure 4 depicts that text strings resulted in the lowest MAP. It was expected as program semantic was not only reflected by identifiers and keywords. Some of them were symbols, such as '>' and ':'.

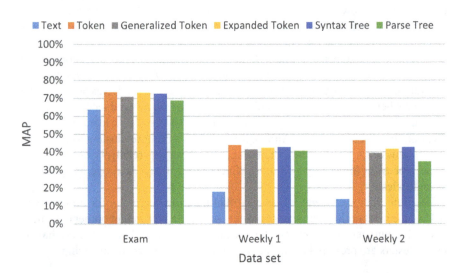

Fig. 4 MAP of various intermediate representations on the three data sets

The most effective intermediate representation was the token strings with a total of 164% MAP: 73% on the exam data set, 44% on the weekly 1 data set, and 43% on the weekly 2 data set. Generalizing the tokens or expanding the tokens with their generalized version slightly reduced the effectiveness by 12 and 7%. AI generated code might have uncommon identifier names and such hints might be removed when the tokens were generalized. Employing linearized syntax and parse trees did not positively affect the effectiveness either. The former reduced MAP by 6% while the latter reduced MAP by 20%. AI generated code might result in similar parsing rules in some parts of syntax or parse trees (e.g., program entry, function calls, and statements). Linearized parse trees experienced larger reduction as it included intermediate rules for compilation.

Figure 5 shows that employing either text or token strings was the fastest since their preprocessing was only about simple tokenization. The processing time became longer if the tokens were generalized (additional translation preprocessing) or expanded with the generalized version (more features to consider). Employing linearized syntax trees took longer than employing the first four representations as syntax trees needed to be generated and traversed. The same applied to employing linearized parse trees, and it even took far longer. Parse trees included many intermediate rules for compilation in the indexes.

Among the six intermediate representations, token strings were the most effective and one of the fastest. Text strings were the least effective while linearized parse trees were the slowest.

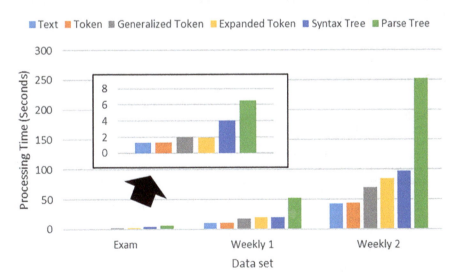

Fig. 5 Processing time of various intermediate representations on the three data sets

4 Conclusion and Future Work

This paper presents an automated detector for AI generated code with KNN weight outlier detection and six intermediate representations: text strings, token strings, generalized token strings, expanded token strings, linearized syntax trees, and linearized parse trees. According to our evaluation on the three data sets, the detector performs best when n for n-gram is 110 and the number of clusters is 3. Among the six intermediate representations, token strings are the most effective with a total of 164% MAP from the three data sets. They are also quite efficient as it only take 44 s to process 54 assignments with 1448 submissions. The performance of our detector is considered satisfactory.

Our study has a number of limitations, which might be part of the future work. First, the effectiveness of our detector is not really high. Further improvements are needed before it can be used in academia. Second, our detector only works on assignments at which the use of AI is prohibited. Other approaches might be needed to recognize AI generated code on assignments that allow the use with proper citation. Third, our study was conducted in an introductory course from a particular institution. Reconducting the study on other courses from other institutions with larger data sets might strengthen the findings. Fourth, though the performance metrics are quite common, we acknowledge that other metrics might provide richer perspective of the findings.

Acknowledgements The research presented in this paper was supported by the Research Institute and Community Service (LPPM) at Maranatha Christian University Indonesia.

References

1. Albluwi I (2019) Plagiarism in programming assessments: a systematic review. ACM Trans Comput Educ 20(1):6:1–6:28
2. Allen JM, Vahid F, Downey K, Edgcomb AD (2018) Weekly programs in a cs1 class: experiences with auto-graded many-small programs (MSP). In: ASEE annual conference and exposition
3. Bhattathiripad PV (2012) Software piracy forensics: a proposal for incorporating dead codes and other programming blunders as important evidence in AFC test. In: 2012 IEEE 36th annual computer software and applications conference workshops, pp 206–212. IEEE
4. Blanchard J, Hott JR, Berry V, Carroll R, Edmison B, Glassey R, Karnalim O, Plancher B, Russell S (2022) Stop reinventing the wheel! promoting community software in computing education. In: Working group reports on innovation and technology in computer science education, pp 261—292
5. Burrows S, Uitdenbogerd AL, Turpin A (2009) Application of information retrieval techniques for source code authorship attribution. In: Proceedings of the database systems for advanced applications: 14th international conference, DASFAA 2009, Brisbane, Australia, 21–23 Apr 2009, vol 14, pp 699–713. Springer (2009)
6. Campos GO, Zimek A, Sander J, Campello RJ, Micenková B, Schubert E, Assent I, Houle ME (2016) On the evaluation of unsupervised outlier detection: measures, datasets, and an empirical study. Data Min Knowl Discov 30:891–927

7. Carlini N, Tramer F, Wallace E, Jagielski M, Herbert-Voss A, Lee K, Roberts A, Brown TB, Song D, Erlingsson U et al (2021) Extracting training data from large language models. In: USENIX security symposium, vol 6
8. Cooper KD, Torczon L (2012) Engineering a compiler, 2nd ed (2012)
9. Croft WB, Metzler D, Strohman T (2010) Search engines: information retrieval in practice
10. Fowler M, Zilles C (2021) Superficial code-guise: investigating the impact of surface feature changes on students' programming question scores. In: 52nd ACM technical symposium on computer science education, pp 3–9
11. Fraser R (2014) Collaboration, collusion and plagiarism in computer science coursework. Inform Educ 13(2):179–195
12. Fu D, Xu Y, Yu H, Yang B (2017) WASTK: a weighted abstract syntax tree kernel method for source code plagiarism detection. Sci Program 2017:1–8
13. Gehrmann S, Strobelt H, Rush AM (2019) GLTR: statistical detection and visualization of generated text
14. Grunwald D, Boese E, Hoenigman R, Sayler A, Stafford J (2015) Personalized attention @ scale: talk isn't cheap, but it's effective. In: 46th ACM technical symposium on computer science education, pp. 610–615
15. Halak B, El-Hajjar M (2016) Plagiarism detection and prevention techniques in engineering education. In: 11th European workshop on microelectronics education, pp 1–3
16. Joy M, Cosma G, Yau JYK, Sinclair J (2011) Source code plagiarism-a student perspective. IEEE Trans Educ 54(1):125–132
17. Joy M, Luck M (1999) Plagiarism in programming assignments. IEEE Trans Educ 42(2):129–133
18. Kalgutkar V, Kaur R, Gonzalez H, Stakhanova N, Matyukhina A (2019) Code authorship attribution: methods and challenges. ACM Comput Surv 52(1)
19. Karnalim, O., Simon, Chivers, W.: Similarity detection techniques for academic source code plagiarism and collusion: a review. In: International Conference on Engineering, Technology and Education (2019)
20. Karnalim O, Simon Chivers W, Panca BS (2022) Educating students about programming plagiarism and collusion via formative feedback. ACM Trans Comput Educ 22(3):31:1–31:31
21. Kikuchi H, Goto T, Wakatsuki M, Nishino T (2014) A source code plagiarism detecting method using alignment with abstract syntax tree elements. In: 15th IEEE/ACIS international conference on software engineering, artificial intelligence, networking and parallel/distributed computing, Las Vegas, pp 1–6 (2014)
22. Kocoń J, Cichecki I, Kaszyca O, Kochanek M, Szydło D, Baran J, Bielaniewicz J, Gruza M, Janz A, Kanclerz K et al (2023) Chatgpt: jack of all trades, master of none. Inf Fusion 101861
23. Lancaster T (2018) Academic integrity for computer science instructors. In: Higher education computer science, Cham, pp 59–71
24. Nguyen N, Nadi S (2022) An empirical evaluation of github copilot's code suggestions. In: 19th international conference on mining software repositories, pp 1–5
25. Nichols L, Dewey K, Emre M, Chen S, Hardekopf B (2019) Syntax-based improvements to plagiarism detectors and their evaluations. In: 24th conference on innovation and technology in computer science education, pp 555–561
26. Orenstrakh MS, Karnalim O, Suarez CA, Liut M (2023) Detecting llm-generated text in computing education: a comparative study for chatgpt cases
27. Parr T (2013) The definitive ANTLR 4 reference
28. Prechelt L, Malpohl G, Philippsen M (2002) Finding plagiarisms among a set of programs with JPlag. J Univers Comput Sci 8(11):1016–1038
29. Schleimer S, Wilkerson DS, Aiken A (2003) Winnowing: local algorithms for document fingerprinting. In: International conference on management of data, pp 76–85
30. Schubert E (2022) Automatic indexing for similarity search in ELKI. In: Skopal T, Falchi F, Lokoc J, Sapino ML, Bartolini I, Patella M (eds) Proceedings of the similarity search and applications—15th international conference, SISAP 2022, Bologna, Italy, Oct 2022. Lecture notes in computer science, vol 13590, pp 205–213. Springer (2022)

31. Schubert E, Zimek A, Kriegel HP (2014) Local outlier detection reconsidered: a generalized view on locality with applications to spatial, video, and network outlier detection. Data Min Knowl Discov 28:190–237
32. Simon: designing programming assignments to reduce the likelihood of cheating. In: 19th Australasian computing education conference, pp 42–47 (2017)
33. Simon CB, Sheard J, Carbone A, Johnson C (2013) Academic integrity: differences between computing assessments and essays. In: 13th Koli calling international conference on computing education research, pp 23–32
34. Simon SJ, Morgan M, Petersen A, Settle A, Sinclair J (2018) Informing students about academic integrity in programming. In: 20th Australasian computing education conference, pp 113–122
35. Singh K, Upadhyaya S (2012) Outlier detection: applications and techniques. Int J Comput Sci Issues (IJCSI) 9(1):307
36. Solaiman I, Brundage M, Clark J, Askell A, Herbert-Voss A, Wu J, Radford A, Krueger G, Kim JW, Kreps S, McCain M, Newhouse A, Blazakis J, McGuffie K, Wang J (2019) Release strategies and the social impacts of language models
37. Spacco J, Fossati D, Stamper J, Rivers K (2013) Towards improving programming habits to create better computer science course outcomes. In: 18th ACM conference on innovation and technology in computer science education, pp 243–248
38. Sulistiani L, Karnalim O (2019) ES-Plag: efficient and sensitive source code plagiarism detection tool for academic environment. Comput Appl Eng Educ 27(1):166–182
39. Toba H, Karnalim O, Johan MC, Tada T, Djajalaksana YM, Vivaldy T (2023) Inappropriate benefits and identification of chatgpt misuse in programming tests: a controlled experiment
40. Tsang HH, Hanbidge AS, Tin T (2018) Experiential learning through inter-university collaboration research project in academic integrity. In: 23rd Western Canadian conference on computing education
41. Ullah F, Wang J, Farhan M, Jabbar S, Wu Z, Khalid S (2018) Plagiarism detection in students' programming assignments based on semantics: multimedia e-learning based smart assessment methodology. Multimedia Tools Appl
42. Ullah F, Wang J, Jabbar S, Al-Turjman F, Alazab M (2019) Source code authorship attribution using hybrid approach of program dependence graph and deep learning model. IEEE Access 7:141987–141999
43. Yang CH (2022) How artificial intelligence technology affects productivity and employment: firm-level evidence from Taiwan. Res Policy 51(6):104536

The Use of SAP in Education: A Review of Current Practices and Future Directions

Md. Tanvir Amin, Md. Tahsin Amin, Fahmida Haque Mim, and Jarin Sobah Peu

Abstract Enterprise resource planning (ERP) systems have become increasingly important in managing organizational operations, including those in education. One of the most widely used ERP systems in education is the SAP system. This paper presents a review of the current practices and future directions of SAP in education. The paper examines the various ways in which SAP has been used in education, including student information systems, financial management, and human resource management. SAP can bring significant advantages to the education sector, successful implementation requires careful planning, adequate training, and consideration of the unique needs of educational institutions. Additionally, privacy and security concerns must be addressed to ensure the protection of student and institutional data. The paper also discusses the challenges and benefits of using SAP in education and provides recommendations for future research and practice.

Keywords SAP · Education ERP · Administrative and Academic Management · Student Information Systems

---

Md. T. Amin (✉)
Faculty, Department of ICT, Rabindra Maitree University, Kushtia, Bangladesh
e-mail: aminmd.tanvir@gmail.com

Md. T. Amin
Department of CSE, Patuakhali Science and Technology University, Patuakhali, Bangladesh

F. H. Mim
Faculty, Department of CSE, Rabindra Maitree University, Kushtia, Bangladesh

J. S. Peu
Department of ICT, Bangladesh University of Professionals, Dhaka, Bangladesh

© The Author(s), under exclusive license to Springer Nature Switzerland AG 2025
E. Vendrell Vidal et al. (eds.), *Advanced Technologies and the University of the Future*, Lecture Notes in Networks and Systems 1140,
https://doi.org/10.1007/978-3-031-71530-3_30

1 Introduction

ERP systems, such as SAP (Systems, Applications, and Products), have gained prominence in various industries for their ability to streamline and integrate business processes. In recent years, there has been a growing interest in utilizing SAP in the field of education [1]. This research paper aims to provide a comprehensive review of the current practices and applications of SAP in education, highlighting its benefits, challenges, and potential future directions. The SAP system is a widely used ERP system that has been adopted by various organizations, including those in education. SAP provides a range of modules that support different business processes, including financial management, human resource management, procurement, and supply chain management [2]. In education, SAP has been used to support various processes, including student information systems, financial management, and human resource management [3].

2 Background

SAP is a leading ERP software suite designed to integrate and streamline various business processes within an organization. Originally developed in the early 1970s by German engineers at SAP AG, the software has evolved into a comprehensive solution used by businesses worldwide [3]. The integration of technology in education has become essential for enhancing efficiency, productivity, and overall educational outcomes. SAP, a leading ERP system, offers a robust platform for managing various administrative and academic functions within educational institutions [4]. This paper explores the current landscape of SAP implementation in education and the potential it holds for transforming educational processes.

2.1 Evolution of ERP in Education

The evolution of ERP systems in educational institutions has undergone significant changes over the years, driven by the need for streamlined processes, improved efficiency, and better management of resources. The adoption of SAP in the education sector is part of this evolution. Here's a general timeline of the key developments [4].

Early Management Information Systems

In the early days, educational institutions primarily relied on manual and paper-based systems for managing administrative tasks such as student enrollment, financial accounting, and human resources. Basic Management Information Systems (MIS) started to emerge, incorporating early computer technologies to automate certain processes [5].

Integration of Modules

As technology advanced, educational institutions began integrating different modules to manage various aspects of their operations. These modules included student information systems, finance, human resources, and more. This integration aimed to improve data accuracy, reduce redundancy, and enhance overall efficiency in administrative processes [5].

Introduction of ERP System

In the late 1990s and early 2000s, ERP systems gained popularity in various industries, including education. ERP solutions offered a comprehensive suite of integrated applications that covered a wide range of functions, from finance and HR to student services. These systems allowed for real-time data access, improved collaboration, and better decision-making [6].

Customization of ERP Sector

ERP vendors recognized the unique needs of educational institutions and started customizing their solutions to address these specific requirements. Features such as student enrollment, grading, scheduling, and academic reporting became integral components of ERP systems tailored for the education sector [6].

Adaptation of SAP in Education

SAP, a leading ERP vendor, gained prominence in the education sector due to its robust and scalable solutions. Education institutions began adopting SAP ERP to streamline their operations, improve data accuracy, and enhance overall organizational efficiency. SAP's solutions for education cover a range of modules, including Student Lifecycle Management, Finance, HR, and Analytics [6].

Cloud Based ERP Solutions

With the advancement of cloud computing, educational institutions started moving towards cloud-based ERP solutions. This shift offered benefits such as flexibility, scalability, and reduced infrastructure costs. SAP introduced cloud-based solutions, including SAP S/4HANA Cloud, which catered to the evolving needs of educational institutions [7].

Focus on Analytics and Business Intelligence

Recent trends involve a stronger emphasis on analytics and business intelligence within ERP systems for educational institutions. This enables institutions to make data-driven decisions, enhance student outcomes, and optimize resource allocation [7].

2.2 Overview of SAP in Education

The software has evolved into a comprehensive solution used by businesses worldwide. SAP provides a suite of integrated education modules applications that cover a wide range of functions within an organization. These applications are organized into modules, each addressing specific sectional processes. Some of the key SAP modules include [8].

New Higher Education Technology

Proactive institutions are transforming, using smart education technology to deliver the new skills, tools, and job experiences students will need—and the personalized, interactive tools they expect [8].

Intelligent Finance and Accountability

Using next-generation finance solutions, higher educational institutions can make the most of finite resources—and meet demands for transparency around affordability, value, and learning outcomes [8].

Smart Campus

To keep pace with digitally native students, and to stay sustainable, institutions are evolving into smart campuses transforming everything from entrance admissions and student experiences to parking [8].

Always on Student Engagement

Colleges and universities create student information systems resembling video gaming to gauge student interests and sentiments using the data to tailor meaningful recruitment messaging and outreach [8].

Changing Faculty and Staff Paradigm

Teaching talent from the business world is rebalancing tenured university faculty models. Contingent workforces support learning areas most in need and departments are integrating to form agile hubs [9].

Accelerate Research Process

Advancements in AI, analytics, and computing power help research institutions to automate functions, integrate digital processes, streamline projects and finance and accelerate vital discoveries [9] (Fig. 1).

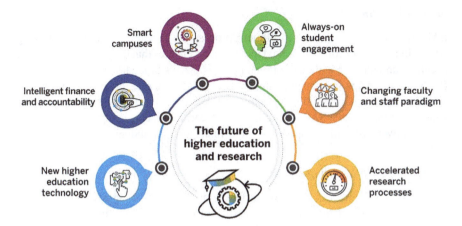

Fig. 1 SAP in Education [2]

3 Current Practices of SAP in Education

3.1 Administrative Process

SAP is a widely used enterprise resource planning (ERP) software suite that plays a crucial role in streamlining administrative tasks across various industries, including education. In the context of education, SAP Analytics Cloud and SAP Success factor solutions can be applied to streamline administrative tasks, by integrating analytics and planning functionality with SAP applications, this solution can provide direct insights into administrative, HR, Payroll, and financial data [10].

Finance

SAP helps educational institutions manage their finances more efficiently by providing tools for budgeting, accounting, and financial reporting. It helps in tracking expenses, managing revenue streams, and ensuring compliance with financial regulations. SAP enables educational institutions to streamline procurement processes, automate purchase orders, and manage vendor relationships [10].

Human Resources

AP facilitates the management of employee data, including personal information, job roles, and employment history. This helps in automating HR processes and ensures accurate and up-to-date records. SAP's payroll module automates payroll processing, including calculations, deductions, and compliance with tax regulations. SAP supports talent acquisition, performance management, and employee development. It helps educational institutions attract, retain, and develop skilled personnel, contributing to overall organizational success [10].

Student Enrollment

SAP streamlines the student enrollment process by providing tools for managing admissions, registration, and student records. This ensures that the process is efficient, accurate, and aligned with institutional policies. SAP's Student Information System (SIS) module helps in managing student data, academic records, and grading [11]. It enables educational institutions to track student progress, manage transcripts, and generate reports. Some SAP solutions support learning management systems, aiding in the delivery of online courses, content management, and student collaboration [12].

Integrating and Reporting

SAP allows for integration across different modules, ensuring seamless flow of information between finance, HR, and student enrollment systems. This integration reduces manual data entry, minimizes errors, and enhances overall efficiency. The reporting capabilities of SAP enable educational institutions to generate custom reports for various stakeholders. This helps in data-driven decision-making and compliance reporting [12].

Mobile Accessibility

Many SAP solutions offer mobile accessibility, allowing administrators, faculty, and staff to access relevant information and perform tasks remotely. This contributes to increased flexibility and responsiveness [12]. (Fig. 2).

3.2 Academic Management

The integration of SAP in academic management can streamline various processes, including course planning, scheduling, and grading. SAP S/4HANA Cloud offers a

Fig. 2 SAP Module for Education [3]

comprehensive suite of ERP solutions that can be customized to meet the specific needs of educational institutions [13]. Modern ERP system with embedded analytics, AI, and machine learning supports efficient and agile processes to maximize resource value and help you deliver measurable cross-campus and online campus results. Here's how SAP can be integrated into academic management [14].

Student Information System (SIS)

SAP can automate the student enrollment and registration process. It can manage student data, track academic history, and ensure accurate information for course planning. SAP SIS maintains detailed student records, including personal details, academic achievements, and attendance. Instructors and administrators can track student progress through integrated dashboards and reports [14]. Implementing analytics modules could provide insights into student performance trends, helping educators identify areas for improvement. It extended to include parents, SAP could enable communication about student progress, events, and other relevant information [15].

Course Planning

SAP enables academic institutions to define and manage academic programs, courses, and curriculum. It facilitates the planning of courses and ensures alignment with academic goals. The ERP system helps in allocating resources such as classrooms, faculty, and materials based on the course requirements. SAP can automate the process of creating class schedules based on factors like faculty availability, room availability, and student preferences. Any changes to the schedule can be reflected in real-time, and notifications can be sent to relevant stakeholders, reducing confusion [15].

Grading

SAP can automate the grading process, reducing the manual effort required by instructors. SAP can provide a centralized gradebook where instructors can enter grades for assignments, exams, and other assessments. It can calculate grades based on predefined criteria and ensure consistency in grading practices. The system can generate reports and transcripts, making it easier for students to access their grades and for institutions to comply with reporting requirements [16]. SAP can provide a centralized gradebook where instructors can enter grades for assignments, exams, and other assessments. SAP can assist in generating official transcripts and academic records for students, ensuring accuracy and consistency. SAP's reporting tools enable educational institutions to analyze data related to student performance, resource utilization, and overall academic management. Key performance indicators (KPIs) can be defined and monitored to assess the efficiency and effectiveness of academic processes [17].

Integration with LMS System

SAP can provide a centralized portal for students, faculty, and administrators, facilitating communication and collaboration. It can integrate with LMS platforms to

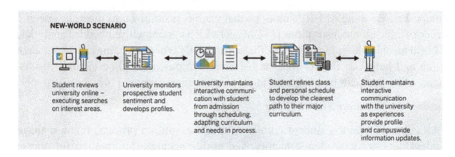

Fig. 3 Academic management system module of SAP [17]

ensure a seamless flow of information between academic management and online learning systems providing a centralized platform for communication between students, faculty, and administrators. Integration with an LMS could enhance the educational experience by providing a platform for course content, assignments, and assessments. Integration with communication tools could streamline announcements, discussions, and notifications for students, parents, and faculty. Automated notifications can be set up to inform students and faculty about important dates, changes in schedules, or academic updates. Ensuring the security of student data is crucial. SAP, if adopted, would need robust security measures to protect sensitive information. The system must comply with relevant educational regulations and data protection laws [17] (Fig. 3).

3.3 Case Studies

These examples showcase the diverse ways in which educational institutions globally have leveraged SAP to enhance their operations, streamline processes, and make data-driven decisions. However, in Bangladesh AIUB, Asian University for Woman, Bangladesh University of Textiles, University of Dhaka, University of Liberal Arts are adopting SAP base on December 2023 SAP university alliances report [18]. Here are some case studies given below for some specific university:

Case-1: The University of Kentucky implemented SAP to streamline its financial and administrative processes. Improved efficiency in financial management, reduced manual workload, enhanced reporting capabilities, and better decision-making support [18].

Case-2: The University of Pretoria adopted SAP ERP to integrate its various systems and improve overall operational efficiency. Enhanced communication and collaboration across departments, streamlined administrative processes, and better data-driven decision-making [18].

***Case*-3**: The University of Sydney implemented SAP to modernize its student information systems, human resources, and finance operations. Increased transparency, improved data accuracy, enhanced reporting capabilities, and a more user-friendly interface for students and staff [18].

***Case*-4**: The University of Duisburg-Essen implemented SAP ERP to integrate its administrative processes and improve resource management. Streamlined workflows, reduced duplication of efforts, improved financial reporting, and better alignment with industry best practices [18].

***Case*-5**: King Saud University adopted SAP to enhance its financial management and human resources processes. Improved budgeting and forecasting, enhanced payroll and HR processes, increased transparency, and better compliance with regulatory requirements [18].

***Case*-6**: The University of São Paulo implemented SAP to improve its financial management and human resources functions. Increased efficiency in processing financial transactions, better tracking of expenses, and improved overall financial control [18].

***Case*-7**: UC Berkeley implemented SAP to modernize and integrate its financial, procurement, and HR systems. Increased efficiency in financial operations, reducing manual processes. Improved data accuracy and reporting for budgeting and forecasting. Enhance compliance with regulatory requirements [18].

4 Benefits of SAP in Education

4.1 Improved Efficiency

SAP helps organizations streamline their business processes by providing a comprehensive suite of integrated applications. These applications cover various functions such as finance, human resources, supply chain, manufacturing, and more. By automating and integrating these processes, SAP eliminates redundancies, reduces delays, and ensures smoother operations [9].

SAP allows for the automation of routine and repetitive tasks across different departments. This automation not only saves time but also minimizes the likelihood of errors associated with manual data entry and processing. For example, SAP can automate invoice processing, purchase order creation, and inventory management [11].

SAP provides a centralized platform for data management, ensuring that information is stored in a consistent and accessible manner. This centralized approach eliminates the need for manual data entry into multiple systems, reducing the risk of errors and ensuring data integrity. This, in turn, improves decision-making processes and overall efficiency [12].

By digitizing and automating various business processes, SAP contributes to the reduction of paperwork. For instance, electronic document management systems within SAP enable organizations to move towards paperless workflows. This not only saves physical storage space but also facilitates quicker document retrieval and sharing [12].

SAP provides powerful reporting and analytics tools that enable organizations to generate real-time insights into their operations. This allows decision-makers to access accurate and up-to-date information, facilitating faster and more informed decision-making. This real-time visibility enhances overall operational efficiency by identifying areas that require improvement or optimization [13].

SAP's integrated platform enhances collaboration among different departments and teams. This integration ensures that information flows seamlessly across the organization, reducing the need for manual communication and coordination. Improved collaboration leads to faster decision-making and more agile responses to changing business conditions [14].

SAP includes features for regulatory compliance and risk management. By automating compliance checks and incorporating risk management functionalities, organizations can ensure that they adhere to industry regulations and standards. This reduces the risk of legal issues and associated paperwork [14].

40% of higher education institutes all over the world now adopt and use this SAP solution. Most of the institutes are renowned worldwide and in best university based on QS based and the times institute world rankings.

4.2 Data Integration and Accuracy

Integrating data from different departments allows organizations to create a comprehensive and holistic view of their operations. This integrated perspective enables a deeper understanding of how different departments and functions interconnect, fostering better strategic decision-making [11].

Timely and accurate information is essential for effective decision-making. Data integration ensures that decision-makers have access to up-to-date and relevant data from various sources, helping them make informed choices that align with the organization's goals and objectives [11].

Data integration facilitates collaboration among different departments by breaking down silos and promoting a culture of information sharing. When departments can access and utilize each other's data seamlessly, it fosters better teamwork and communication, leading to more cohesive and collaborative efforts [13].

Integrated data systems streamline business processes by reducing manual data entry, duplication, and errors. Automation and synchronization of data across departments lead to more efficient workflows, saving time and resources that can be redirected towards more value-added activities [14].

A unified view of customer data is critical for delivering a personalized and seamless customer experience. Integrating data from sales, marketing, customer service, and other departments enables organizations to gain insights into customer behavior, preferences, and feedback, allowing for more targeted and effective customer engagement [15].

Data integration helps in optimizing resource allocation and reducing operational costs. By eliminating redundant systems and processes, organizations can achieve cost savings through increased efficiency and improved resource utilization [19].

Many industries face regulatory requirements regarding data management and reporting. Data integration ensures that organizations can compile and report accurate information, helping them stay compliant with industry regulations and standards [19].

In a rapidly changing business environment, organizations need to adapt quickly to market trends and opportunities. Integrated data systems are more adaptable and scalable, allowing businesses to respond promptly to changes and scale their operations without significant disruptions [20].

Integration supports real-time data analytics, providing organizations with insights into current market conditions, customer behavior, and operational performance. Real-time analytics empower businesses to make agile decisions and respond promptly to emerging trends [20].

5 Challenges and Limitations of SAP in Education

Using SAP in education presents several challenges and benefits. One of the main challenges is Acquiring SAP licenses involves an upfront cost, and educational institutions need to consider the number of users and modules required. Implementing SAP may necessitate additional or upgraded hardware to support the software, leading to additional costs [21]. SAP typically charges annual maintenance fees for software updates, patches, and technical support. Periodic upgrades and updates may be necessary to keep the system current, and these can involve costs for testing, implementation, and potential retraining. The cost of implementing and maintaining the system can be high [22].

Another challenge is the complexity of the system, which can require extensive training for users. Develop training programs that are specific to the roles and responsibilities of different staff members. This ensures that employees receive training relevant to their job functions. SAP systems are often updated, and new features are introduced. Provide ongoing training sessions to keep staff up to date with the latest functionalities. Tailor the SAP interface to meet the specific needs of your organization and users [23]. A user-friendly interface reduces the learning curve and encourages adoption. Conduct regular surveys and feedback sessions to understand the challenges faced by users. Use this information to make continuous improvements and address specific pain points. Another Problem is Customization Issue Educational institutions can vary significantly in terms of their structure, processes, and workflows [24]. Universities, schools, vocational institutions, and training centers may have different needs and expectations. Customization efforts need to account for diverse educational models, including variations in academic calendars, grading systems, and course structures. Educational institutions are subject to various regulations and compliance requirements, depending on their location and type. Customizing SAP to meet these regulatory standards can be time-consuming and complex. SAP regularly releases updates and new versions. Customized solutions may require adjustments and retesting to remain compatible with the latest SAP releases [25]. Educational institutions need to have a strategy for keeping their customized SAP systems up to date without causing disruptions, which is very difficult.

6 Future Directions and Recommendations

AI and ML can be integrated into SAP systems to analyze historical data on student performance, engagement, and behavior. This allows educational institutions to predict students at risk of falling behind or dropping out. Early identification enables timely intervention, personalized support, and the implementation of strategies to improve student success rates. By leveraging AI and ML algorithms, SAP systems can analyze individual student learning patterns and preferences. This data can be used to create personalized learning paths, recommend relevant courses or resources, and adapt instructional content to meet the specific needs of each student. This personalization can enhance the overall learning experience and improve academic outcomes. AI-powered chatbots integrated into SAP systems can provide instant support to students for routine queries related to admissions, registration, financial aid, and other administrative processes. This enhances the student experience by providing quick and accurate information 24/7, freeing up administrative staff for more complex tasks. AI can enhance the assessment process by providing adaptive testing that adjusts difficulty levels based on the student's performance. SAP systems can integrate ML algorithms to analyze student responses and dynamically adapt assessments, ensuring that they align with individual learning levels and provide a more accurate representation of student knowledge.

Continued research and development (R&D) in optimizing SAP (Systems, Applications, and Products) for educational contexts is crucial for several reasons. SAP, a leading enterprise resource planning (ERP) software, has the potential to significantly enhance efficiency, effectiveness, and data management in educational institutions. Educational institutions handle sensitive student and staff information. Ongoing research is necessary to strengthen data security and privacy measures within SAP to protect against cyber threats and ensure compliance with data protection regulations. Ongoing advancements in technology, such as artificial intelligence, machine learning, and the Internet of Things, present opportunities for further optimizing SAP in educational contexts. Research is needed to explore and implement innovative features that leverage emerging technologies to enhance educational processes. Educational institutions often operate within budget constraints. Research and development efforts should focus on optimizing SAP to be cost-effective, ensuring that institutions with varying financial resources can benefit from the technology.

7 Conclusion

The use of SAP in education has become increasingly important in managing organizational operations. SAP has been used for various purposes, including student information systems, financial management, and human resource management. Using SAP in education presents challenges and benefits, including cost and complexity, but can provide greater efficiency and data accuracy. It is a leading enterprise software solution that is widely used in various industries for managing business operations and customer relations. While SAP itself may not be directly associated with education, its potential in revolutionizing education lies in its ability to streamline and enhance administrative processes, provide data-driven insights, and support the development of skills needed in the modern workforce. SAP's blockchain capabilities can be used for secure and transparent credential verification, reducing fraud and ensuring the integrity of academic qualifications. Future research and practice should focus on addressing these challenges and maximizing the benefits of using SAP in education, including its potential in supporting teaching and learning processes.

Acknowledgements We would like to thank the International Association of Online Engineering (IAOE) for its financial support, which made the publication of this article possible.

References

1. SAP Team (2024) Higher Education and Research. https://www.sap.com/industries/higher-education-research.html. Accessed 1st Feb 2024
2. SAP Team (2024) SAP For Higher Education and Research. https://www.eunis.org/eunis2008/programme/files/sap. Accessed 1st Feb 2024
3. Allied Market Research, "ERP Software Market: Global Opportunity Analysis and Industry Forecast 2018–2026," 2019. https://www.alliedmarketresearch.com/erp-market Accessed: 3rd February 2024
4. Statista (2024) ERP-Software-Marktanteile der Anbieter weltweit 2017|Statista. https://de.statista.com/statistik/daten/studie/262342/umfrage/marktanteile-der-anbieter-von-erp-software-weltweit/. Accessed 3rd Feb 2024
5. Soellner S (2021) Digital elements for SAP ERP education and training: results from a systematic literature review. Int J Eng Pedagog (iJEP) 11(4):115–129. https://doi.org/10.3991/ijep.v11i4.21843
6. Ambrajei AN, Golovin NM, Valyukhova AV, Rybakova NA, Zorin VY (2019) Use of hybrid learning model for SAP-related technology education. In: Proceedings of the ICID-2019 conference, pp 1–9
7. Jacques S, Ouahabi A, Lequeu T (2020) Remote knowledge acquisition and assessment during the COVID-19 pandemic. Int J Eng Ped 10(6):120. https://doi.org/10.3991/ijep.v10i6.16205
8. McCann DK, Grey D (2009) SAP/ERP Technology in a Higher Education Curriculum and the University Alliance Program, IIS, X, no 1, pp 176–182. https://doi.org/10.48009/1_iis_2009_176-182
9. Govan M (2016) Enterprise resource planning for higher education platform. Int J Bus Trends 2
10. Akkaya M (2017) Educational attitude toward the using of cloud ERP. Int J Commun 540

11. Silver Touch Team (2024) Role of SAP during digital transformation journey, 18th Jan 2024. https://sap.silvertouch.com/blog/role-of-sap-during-digital-transformation-journey. Accessed 5th Feb 2024
12. Australian Deans Council (2010). Learning and teaching academic standards project, business, management and economics: learning and teaching academic standards statement for ACCOUNTING Dec 2010. http://www.abdc.edu.au/data/Accounting_LS/Accounting_Learning_Standards_February_2011.pdf. Accessed 7 Sept 2015
13. Becerra-Fernandez I, Murphy KE, Simon SJ (2000) Enterprise resource planning: Integrating ERP in the business school curriculum. Commun ACM 43(4):39–41. https://doi.org/10.1145/332051.332066
14. Boulianne E (2014) Impact of accounting software utilization on students' knowledge acquisition: an important change in accounting education. J Account Organ Chang 10(1):22–48. https://doi.org/10.1108/JAOC-12-2011-0064
15. Chen CS, Liang WY, Hsu HY (2015) A cloud computing platform for ERP applications. Appl Soft Comput 27:127–136. https://doi.org/10.1016/j.asoc.2014.11.009
16. Corbitt G, Mensching J (2000) Integrating SAP R/3 into a College of Business curriculum: Lessons learned. Inf Technol Manage 1(4):247–258. https://doi.org/10.1023/A:1019181210298
17. G2 Company Team Member (2023) Top 10 SAP for higher education and Research Alternatives and Competitors. https://www.g2.com/products/sap-for-higher-education-andresearch/competitors/alternatives. Accessed 7th Feb 2023.
18. Uskov A, Sekar B (2014) Serious games, gamification and game engines to support framework activities in engineering: case studies, analysis, classifications and outcomes. In: IEEE international conference on electro/information technology, Milwaukee, WI, USA, pp 618–623. https://doi.org/10.1109/eit.2014.6871836
19. Johnson T, Lorents AC, Morgan J, Ozmun J (2004) A customized ERP/SAP model for business curriculum integration. J Inf Syst Educ (JISE) 15(3):245–254
20. Yamani HA (2021) A conceptual framework for integrating gamification in elearning sys-tems based on instructional design model. Int J Emerg Technol Learn 16(04):14. https://doi.org/10.3991/ijet.v16i04.15693
21. Tsiatsos T (2020) Virtual university and gamification to support engineering education. Int J Eng Ped 10(2):4. https://doi.org/10.3991/ijep.v10i2.13771
22. Prifti L, Knigge M, Löffler A, Hecht S, Krcmar H (2017) Emerging business models in education provisioning: a case study on providing learning support as education-as-a-service. Int J Eng Ped 7(3):92. https://doi.org/10.3991/ijep.v7i3.7337
23. Foster S, Hopkins J (2011) ERP simulation game: establishng engagement, collaboration and learning. PACIS 2011 Proceedings, Paper 62
24. Alavi M, Leidner DE (2001) Research commentary: technology-mediated learning—a call for greater depth and breadth of research. Inf Syst Res 12(1):1–10. https://doi.org/10.1287/isre.12.1.1.9720
25. Leger P-M, Robert J, Babin G, Lyle D, Cronan P, Paul Charland (2010) ERP Simula-tion game: a distribution game to teach the value of integrated systems, development in business simulation and experiential learning, vol 37, pp 329–334

Detection of the Creativity Potential of Engineering Students

Luis Vargas-Mendoza, Dulce V. Melo-Máximo, Francisco J. Sandoval-Palafox, and Brenda García-Farrera

Abstract Creativity is a training requirement and a professional skill for every engineer, yet its promotion depends on many factors, including environment, approach, model, teacher, and institution. In engineering, creativity is commonly associated with problem-solving and generating a product. This work evaluated the creativity potential of university students who use engineering-specific tools and methodologies to generate innovative ideas. Students were trained in the use of TRIZ, USIT, and SCAMPER techniques to foster their creativity in problem-solving and projects involving creating new products in mechanics, mechatronics, and nanotechnology. The initial and final creative potential were evaluated using Carter's creative profile and Creatrix assessment, respectively, and the creativity enhancement hypothesis was verified with parametric Z and t-tests. The findings showed that most students had intermediate creative potential at the beginning of the course and improvement in achieving high creative potential at the final evaluation. The latter was evident in some samples when segmented by program, semester, gender, and academic average. The findings indicated that the deliberate use of these methodologies in class improves the creative competency of engineering students, especially the youngest ones in a scientific career.

Keywords Competency-based education · Creativity · Engineering · Higher education · Motivation

1 Introduction

Creativity is the ability to develop new ideas and solve problems. Many researchers consider it a mental process that produces new and useful concepts at the individual, work, or social levels [1]. It has been increasingly accepted in all professional fields,

---

L. Vargas-Mendoza (✉) · D. V. Melo-Máximo · F. J. Sandoval-Palafox · B. García-Farrera
Tecnológico de Monterrey, Monterrey, Mexico
e-mail: lvargas@tec.mx

resulting in various theories with different operating models' fit, making it a complex and challenging concept to define [2].

Methodologically, creativity can involve three perspectives: an original production, a divergent thought, or a personality trait [3]. The first involves generating something novel and appropriate for the intended task. Divergent thinking is the ability to organize mental processes indirectly and use unorthodox strategies. The personality trait perspective postulates that creativity is an element everyone possesses, but some have more developed than others [3, 4].

Creativity evaluation analyzes four main approaches: processes, products, people, and environments [5]. The assessment of creative processes is based on psychometric tests of divergent thinking, mentioning those of Torrance [6] and Artola [7], for example. Assessing the physical manifestation of creativity through products may involve using creative inventory questionnaires such as Taylor's [6]. Judges and experts also appraise and decide whether the product meets the necessary original characteristics [3, 5].

The evaluation of the creative person is perhaps the dimension most extensively measured, employing the widest variety of instruments [3, 6], including personality scales, inventories of experiences, creative styles, or reasoning tests. The Creatrix assessment is an example of the latter [6, 8]; it integrates creative cognitive and motivational dimensions. This type of assessment assumes as a fundamental principle that not everyone is equally creative, but this perspective is not considered socially acceptable in recent times. These two perspectives are reconciled because everyone is somewhat creative and can "potentially" become highly creative with the proper support [3].

The environment can favor or harm creativity through situational variables [9]. This is especially important for engineering because studies indicate that different creativity rankings can occur when applying different environments and metrics to the same design problem [10].

Consequently, the tests used in creativity research must be understood as a measurement of the *probability* of being creative, with the consideration that creative achievement depends on additional factors not measurable by the tests, such as the environment, technical skills, knowledge of an area, independence, attitudes, health, or opportunity, among others [11].

Creativity is a crucial competency for engineers, yet it is widely recognized that engineering students have difficulty developing creativity because their focus is on issues with specific answers. Researchers have attempted to identify the barriers in the learning process by examining various techniques to stimulate students' creativity [12]. Another study evaluated how teachers and students perceived creativity, concluding that engineers strongly lean towards well-established methods and the most effective solutions because correctness and attention to detail are essential in the field [13].

As stated in [14, 15], chosen strategies are necessary to promote creativity, distinguishing two approaches to creativity enhancement: the first entails offering courses on the subject, while the other aims to alter teaching methods to encourage creative thinking in the classroom, employing a creative learning environment and problem-solving as a means of learning. One of the most important needs in engineering pedagogy is the creation of assessments that inspire students to develop their creative abilities and become more conscious of their creative process [16, 17].

Additionally, educators must address barriers to creativity like fear of the unknown, difficult mentoring and measures for developing creativity skills in students, including online activities [18].

The importance of creativity in engineering cannot be overstated because engineers are agents of change and creation; their creativity is crucial for developing the required knowledge and skills [19, 20]. Creativity is one of the outcomes that engineers should embrace, according to the Accreditation Board for Engineering and Technology (ABET) [21]. Several strategies encourage engineering creativity, including problem-solving and using thinking tools. Various thinking tools can provide an inventive methodological structure and develop the fluidity, flexibility, and originality that solving engineering problems requires, such as brainstorming, mind mapping, analogies, morphological analysis, and TRIZ [15, 22].

The use of problem-based learning (PBL) or design projects (POL) is the "natural" way in which students train for their profession, developing intellectual and organizational skills that allow them to simultaneously conceive original and useful solutions that promote functional and pragmatic creativity [23]. The PBL technique allows students to combine production and creativity, converting their knowledge into abilities to cope with the challenges of life and work [24]. POL focuses on solving real-world problems through open-ended solutions that allow the construction of new knowledge via student-created projects that produce a tangible product [25].

TRIZ is a methodology for the inventive analysis and resolution of engineering problems based on generic solutions, created by Genrich Altshuller in 1946. He discovered patterns that drive the creative process of solving engineering problems because problems and their technological solutions reoccurred frequently in industries and sciences [26]. Basically, TRIZ solves contradictions rather than using a conventional approach of adjustments, leveraging inventive principles that inspire an ideal solution. The power of TRIZ is its ability to generate inventive, integrated solutions to a design problem from diverse and seemingly unrelated fields [27].

Despite its great power and usefulness, TRIZ is a complex methodology that requires a mind trained to abstract a real problem into a typical generic one and invent a new solution from its generic and sometimes very cryptic approaches. That is why some propose simplifying its application through the ASIT (Advance Systematic Inventive Thinking) and USIT (Unified Systematic Inventive Thinking) techniques [28]. The first focuses on generating creative ideas by introducing qualitative changes in the object's relationship with its environment. The second one solves engineering problems by optimizing explorative, sequenced operators on the object and its function, producing many response possibilities without suggesting solutions; the user must conceive them [29].

SCAMPER is an inventive method created by Eberle [30], inspired by Alex Osborn's work on the Brainstorming methodology. The technique encourages lateral thinking by generating ideas that optimize a solution to a problem, whether a product, service, or process. The SCAMPER technique aims to create several ideas, as is expected from the divergent thinking process [31], through the concepts that give rise to its name: (S) substitute, (C) combine, (A) adapt, (M) modify, (P) put to another use, (E) eliminate, and (R) reverse. Applying these concepts can produce various questions, leading to exploring modifications to the proposed solution and creating many possibilities for innovating form or function.

This research seeks to assess how much creativity in university students can result from using in-class methodologies specially designed for engineering. This objective led to the following working hypothesis: The deliberate use in-class of tools and methodologies specially designed for the generation of creative ideas in engineering, such as TRIZ, USIT, and SCAMPER, improves the students' creative skills.

2 Method

2.1 Approach

This research employed a quantitative, descriptive, non-probabilistic sampling methodology, making statistical inferences about the results; this allows predictions with a certain level of confidence about how the studied population behaves based on data from population samples [32].

2.2 Population and Sampling

The study focused on engineering students enrolled in a private Mexican university during the February-June 2023 semester. The interventions for creativity occurred in four groups: Group 1: 4th semester of mechatronics (MT), Group 2: 4th semester of nanotechnology (NA), Group 3: 8th semester of mechanics (ME), and Group 4: 4th semester of ME. A group from the 6th semester of ME was the control group, where initial and final measurements were done without creativity interventions.

The total population comprised 80% men and 20% women between 18 and 23 years. By semester, 60% were in the 4th, 23 in the 6th, and 17% in the 8th; by major, ME represented 55%, MT 29%, and NT 16%. Regarding the final score in their corresponding courses, 44% of students attained 70 to 89.9 (out of 100), 43% between 90 and 94.9, and 13% with 95 or above.

At the beginning of the semester, the sample population had 180 students. During the first week of the semester, 157 students voluntarily answered a creative personality test, surpassing the 121 responses needed for a statistically valid sample with 90% confidence and 5% error. A standardized innovative profile test was applied during the last week of each course. By then, the population was 173 students, with

123 completing the test voluntarily; statistically valid because 120 responses were required for 90% confidence and 5% error.

2.3 Measures and Instruments

Two instruments assessed creativity, the variable of interest: Carter's creative profile test [33] as the initial survey and the Creatrix innovative personality assessment [34] as the final, applied at the beginning and end of each class, respectively.

Carter's creative profile questionnaire is a qualitative test that evaluates the creative personality as a person's potential with aspects related to imagination and lateral thinking. It is part of a test bank of imagination and creativity published in *The Complete Book of Intelligence Tests*.

The test comprises 25 items on a Likert scale from 1 to 5, where 1 indicates the person identifies the least agreement, and 5 indicates the person identifies the most. The sum results in a creative profile index classified into three creative level ranges: (1) 95–125 indicates that the person has a high creative potential and tries new things; (2) 70–94 is considered an average potential with uncultivated talents; (3) 25–69 signifies a low creative level; the individual has the potential to be creative but has not explored their talents. The questionnaire's reliability calculation using Cronbach's alpha produced 0.691, indicating a moderate level of reliability for the variable [35].

The Creatrix assessment [34] evaluates innovative potential based on the idea that sometimes a creative person must be daring and determined and often takes risks to materialize an original idea. Thus, a genuinely innovative person must be creative and highly motivated to bring an idea to reality: Innovation = Creativity + Risk Tolerance.

The Creatrix Inventory assesses the ability to produce original ideas because of the interaction between creative thinking and the motivational dimension (risk-taking) [8, 11]. It consists of two groups of 28 questions for each dimension. The questions employ a Likert scale from 1 to 9, ranging from "completely disagree" to "completely agree." The test has no time limit for answering. Depending on the section, the reliability of the test (Cronbach's α) varies from 0.55 to 0.81, with an average value of 0.7, which falls into the acceptable category.

The innovative profile is obtained from the intersection of the results from each group of questions; it is the sum of their corresponding answers and represented in a two-dimensional graph: creativity for the horizontal axis, with values between 99 and 199, and risk-taking for the vertical axis, with values from 113 to 213. The graph displays eight innovative personality profiles: four with extreme scores (*Sustainer, Dreamer, Challenger, Innovator*) and four with intermediate scores (*Planner, Improver, Practicalizer,* and *Synthesizer*).

The first version of this test, currently used worldwide, was developed by Richard Byrd in 1986 [8, 36] to diagnose innovative personalities. It can also be applied to university students because the questions are psychologically oriented on a personal level and do not distinguish between professional and educational fields. In the last decade, the managers of this instrument have worked to strengthen the weaknesses pointed out by some educational researchers [36–38], deepening the factor analysis to reveal the structure of latent variables, or "drivers" for creativity and risk-taking [34].

This work only reports the creativity dimension of the Creatrix assessment.

2.4 Interventions of Creativity

The study employed the same three interventions to promote creativity in all experimental groups, excluding the control. The first, as an introductory trigger, was applied in the first week of the course, along with the Carter test. It used the Nearpod platform so students could write and draw and consisted of ten exercises inspired by De Bono [39] and Artola [7].

The second intervention occurred towards the middle of the course period so that the students would have some notion of the product that would solve their design problem. The TRIZ or USIT methodologies [40, 41] were used to generate solutions to course problems, especially if it was the technical solution to the final design problem.

The third and last intervention occurred towards the end of the period, aiming to explore the concepts to improve the form or functions of their proposal for the final design problem by applying the SCAMPER technique [42]. The exploration continued with either physical or digital mental maps, and the ideas generated were analyzed, combined, and prioritized, resulting in an optimized product concept.

2.5 Data Collection and Analysis

Carter's test data was collected with an online survey after providing the institution's privacy notice and informed consent and obtaining the student's legal authorization to start the questionnaire. The participation was voluntary and individual, and its application occurred in approximately 20-min sessions in students' respective classes during the first week of the semester, except for Group 4, which had a course in the last third of the semester.

The Creatrix assessment took place online on the company website and was accessed through an online capture form with a privacy notice and informed consent protocol similar to the first form. Participation was voluntary, and its application took 15 class minutes. The test was in Spanish because the company offers it in several languages besides English. Test access codes for each group were generated because each group had different application dates. The company processed the consolidated information by group and, together with a database, delivered it to the authors.

The quantitative data analysis employed different statistical functions in Excel and Minitab 21.

3 Results and Discussion

3.1 Carter's Test

Carter's test assessed the incoming creative potential profile. According to the scale above, the general analysis of the scores indicated that most students started the course with a medium creative profile (69% of the total), followed by 29% of students with high creative potential and under 2% with low creative potential. Table 1 shows the distribution of these potentials by semester and group.

Most students fell into a medium creative profile; 29% showed a high potential. The younger they are, the higher the creative potential; in other words, potential decreases the further the students advance in their engineering education. Generally, the students with the most creative potential were in the NA program (Group 2). Between ME and MT, the creative levels showed little difference overall, yet there was a distinguishable variation when examined by course. The ME students with the most creative potential were in the 4th semester, whereas the ones in the 8th semester had the lowest. The students with the most robust creative personality were women in the 4th semester of NA, followed by women in ME. The creativity levels of the Control group matched those of the entire sample and were similar to Group 1 but significantly differed from the 8th-semester mechanics population (Group 3).

These results suggest that the cultural environment of engineering curricula better favors creativity in scientific programs such as nanotechnology. If there is less creative potential in ME and MT, perhaps it is due to the traditional approach of looking for the "correct answer," which privileges the proposal of conventional solutions.

Table 1 Incoming creative potential from Carter's test by semester and group

Creative potential	Semester			Group				
	4th (%)	6th (%)	8th (%)	Control	1 (%)	2 (%)	3 (%)	4 (%)
High	33.7	30.6	11.5	30.6	28.3	37.5	11.5	40.0
Medium	64.2	69.4	88.5	69.4	71.7	58.3	88.5	56.0
Low	2.1	0.0	0.0	0.0	0.0	4.2	0.0	4.0

3.2 The Creatrix Assessment

The Creatrix test assessed the levels of creativity at the end of the semester after the interventions and compared them against the control group to see whether improvements occurred. For the creativity dimension, the scores fell into three categories: low, from 99 to 124 points; average, from 125 to 173 points; and high, from 174 to 199 points.

General results show that the high creativity level rose on average to 31.6%, and the low creativity level reduced to 1%. The overall sample mean was 165 points (SD = 20.3), with most profiles (29%) corresponding to *Synthesizer,* indicating medium to highly creative with moderate risk tolerance. There is also a 27% *Practicalizer* (medium creativity with high-risk tolerance) and a 20% *Innovator* (with the highest creativity and risk tolerance).

The results above break down by group to identify students' levels, scores, and profiles, as shown in Table 2.

Overall, the Creatrix assessment results describe a similar pattern to Carter's test for the proportions at the three levels of creativity. Nevertheless, the number of students with high creativity in the final evaluation was 7.8% greater, while the medium and low levels were reduced by 2.8% and 23%, respectively. Those who demonstrated more creativity after the interventions increased would allow us to infer that the interventions positively affected creativity.

All groups, except Group 3, showed a *Synthesizer* profile, meaning they have a medium–high level of creativity and a medium level of risk-taking. They create by

Table 2 Creativity results by Group from creatrix assessment

Group	Creativity levels			
Control	High	29.6%	Mean	165
	Medium	70.4%	Standard deviation	20.05
	Low	–	Profile	Synthesizer
1	High	32.0%	Mean	165
	Medium	68.0%	Standard deviation	19.78
	Low	–	Profile	Synthesizer
2	High	38.5%	Mean	168
	Medium	61.5%	Standard deviation	20.45
	Low	–	Profile	Synthesizer
3	High	10.0%	Mean	156
	Medium	85.0%	Standard deviation	20.50
	Low	5.0%	Profile	Practicalizer
4	High	34.8%	Mean	164
	Medium	65.2%	Standard deviation	20.24
	Low	–	Profile	Synthesizer

putting things together and promoting ideas, but not at any cost; they are cautious. On the other hand, Group 3 came out with a *Practicalizer* profile characterized by medium creativity and high risk-taking. They carry out the ideas they believe in, trying something new when it is practical. They may have creative ideas but do not consider it their strong point.

Between 7 and 20% of students presented the *Improver* profile (low creativity with moderate risk tolerance), where individuals add value to pre-existing ideas, test the waters before deciding, and take risks only when the benefits outweigh the current state. There are between 1 and 4% of students with a *Challenger* profile, with low creativity and the highest risk-taking; they are not those of ideas but of action and enthusiastically implement the creative ideas of others.

The opposite profile of *Dreamer*, high creativity and the lowest risk-taking, was held by 4 to 7% of the students. They are highly creative but do not know how to materialize their ideas, or they do not dare. Within each group, 12 to 26% of students were identified with high creativity and a high tolerance for risk, corresponding to the *Innovator* profile. They always believe that they can achieve things and are the ones who successfully develop innovation competency.

Compared to the Control group, there were no appreciable differences in creativity levels except that Group 2 (NA students) had a slight incremental difference of 1.8% and a -5.4% difference in Group 3 (the last semester ME students).

3.3 Parametric Tests

A Z-test tested the hypotheses of this research beyond the descriptive data. Before the parametric tests, we verified that creativity had a normal distribution, with the normality parameters presented in Figs. 1 and 2 shows the probability plot on the left (a) and the adjustment to the scatter plot on the right (b) for the probability of normal distribution of creativity.

Figures 1 and 2 illustrate that the creativity distribution is slightly left-skewed (median < mean, and skewness > 0) with a flattened kurtosis, which reduces the normal maximum. However, the scatter diagram presents a clearly linear trend, with a p-value greater than the significance value α of 0.05, so it cannot be concluded that the data does not follow a normal distribution. A dispersion analysis for the normal distribution Fig. 2b allows the data to be fitted to a straight line, so the assumption of normality can be accepted in the distribution of creativity to apply the hypothesis tests.

The level of creativity was expected to improve after the interventions. Regarding the Creatrix assessment, this improvement meant reaching at least the lower limit of the high creativity zone of 174 points because the Carter test showed that 69% of the students already had a medium potential for creativity. The average value of the test for the entire sample of 121 students was 165 points, and its standard deviation was 20.327, using a significance level of 5% as a decision criterion.

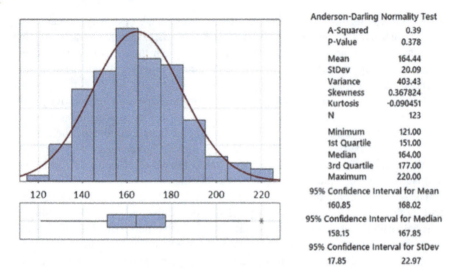

Fig. 1 Frequency distribution and distribution curve, with normality parameters

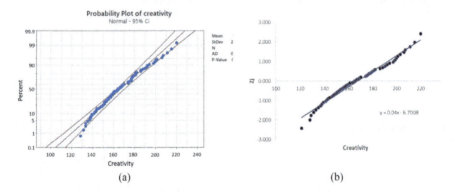

Fig. 2 Probability plot (**a**) and adjustment to scatter plot **b** for the probability of normal distribution of creativity

Z-test

A Z-test to verify the working hypothesis had the following conditions: H0: $\mu = 174$, there is improvement in creativity; and H1: $\mu < 174$, there is NOT improvement in creativity. The decision criterion was: if Zv > Zc, then H0 is accepted. The critical value Zc for the decision rule came from a normal distribution P(x) Table [43]. For a significance level $\alpha = 0.05$, the critical value is Zc = 1.96, and because it is a left-tailed test, Zc = -1.96. The calculated Z-value (Zv) was -4.02 for the sample of 103 students that comprised the experimental groups. As Zv < Zc, the null hypothesis was rejected because there is no statistical evidence to conclude that the level of creativity improved in general.

Student's t-test

Student's t-test was used to analyze creativity in the experimental groups because the sample size was less than 30 individuals in all cases. The decision criteria for a significance level $\alpha = 0.05$ were: if tv > tc and p-value > α, then H0 is accepted. T critical (tc) was obtained from a left-tailed t-test table [44]. Table 3 shows the results.

The Z test for the different groups has better sensitivity and allows two critical results to be identified: (a) It confirms that there is no improvement in creativity because interventions were not carried out in the Control Group, and (b) Group 2 (NA students from the 4th semester) generally improved their creative potential after the interventions.

Table 3 T-test results of creativity segmented by group and gender

Variable		Size	Mean	SD	tv	tc	pv	Criteria	Decision
Control		29	165.48	19.33	−2.37	−1.70	0.012	tv < tc / pv < α	H0 rejected
Group 1		25	165.64	18.38	−2.27	−1.71	0.016	tv < tc / pv < α	H0 rejected
Group 2		26	168.31	22.10	−1.31	−1.71	0.100	tv > tc / pv > α	H0 accepted
Group 3		20	156.05	18.65	−4.30	−1.73	0.000	tv < tc / pv < α	H0 rejected
Group 4		23	164.74	21.28	−2.09	−1.72	0.024	tv < tc / pv < α	H0 rejected
Group 1	M	20	165.75	19.36	−1.44	−1.73	0.083	tv > tc / pv > α	H0 accepted
	W	5	157.20	11.69	−3.21	−2.13	0.016	tv < tc / pv < α	H0 rejected
Group 2	M	17	170.76	24.33	−0.55	−1.75	0.296	tv > tc / pv > α	H0 accepted
	W	9	168.67	17.46	−1.78	−1.86	0.057	tv > tc / pv > α	H0 accepted
Group 3	M	15	157.33	15.33	−4.21	−1.75	0.000	tv < tc / pv < α	H0 rejected
	W	5	152.20	28.37	−1.72	−2.13	0.080	tv > tc / pv > α	H0 accepted
Group 4	M	19	164.32	21.85	−1.93	−1.73	0.035	tv < tc / pv < α	H0 rejected
	W	4	166.75	21.20	−0.68	−2.35	0.272	tv > tc / pv > α	H0 accepted

Notes M = men, W = women, GA = grade average on a 100 pt scale, SD = standard deviation, tc = t critical, tv = t-value calculated, pv = p-value

When groups were segmented by term, the 4th-semester students (Groups 1, 2, and 4) showed an appreciable improvement in their level of creativity, and the 8th-semester students (Group 3) achieved the least. By gender, women developed better creative potential in two of the three 4th-semester groups and the 8th-semester group. The men with the best creative potential were in the 4th-semester groups.

When the entire sample was segmented by their general grade average Table 4, students with scores between 70 and 90 improved their creativity more than those with better grades, as the students with the highest academic performance were the least creative.

Several findings stand out from the previous results. First, the highest creativity values were among NA students, those with more scientific and experimental training, correlating to creativity being essential in scientific work. The nanotechnology program belongs to the area of chemical sciences, and, according to [45], the scientific experience allows them to develop better intuition, which is essential to inspire creativity. The results of the present investigation agree with that.

Second, the creative potential decreases as the student advances in their major, which has been confirmed by several authors [46, 47], showing that creativity and educational level have a curvilinear, inverted-U relationship with maximum creativity at the beginning of the study program and a reduction before the final year. This is because the curricular design of engineering programs usually emphasizes learning specialized skills over reinforcing professional training [48]. The progressive increase in problem complexity and the attempt to find answers within traditional methodological frameworks does not lead to creative thinking [49].

Third, levels of creativity tend to decline for students with high academic averages, which coincides with [46] that creativity declines as the engineer's training becomes more comprehensive and demanding; also, the inflexibility of engineering methodology naturally opposes the flow of creativity. Because creativity requires flexibility in attention [50], which is contrary to the execution of engineering work, being creative does not necessarily mean a better academic performance. Creativity

Table 4 Z-test and t-test for general average analysis

Variable	$70 \leq GA < 90$	$90 \leq GA < 95$	$GA \geq 95$
Mean	172.44	158.41	157.75
SD	19.42	21.49	12.14
Size	39	39	16
	$Zc = -1.96$	$Zc = -1.96$	$tc = -1.75$
	$Zp = -0.50$	$Zp = -4.53$	$tp = -4.53$
Pv	0.308	0.000	0.000
Criteria	$Zp > Zc$ pv $> \alpha$	$Zp < Zc$ pv $< \alpha$	$Zp < Zc$ pv $< \alpha$
Decision	H0 accepted	H0 rejected	H0 rejected

Notes GA = grade average on a 100 pt. scale, SD = standard deviation, Zc = Z critical, Zv = Z-value calculated, tc = t critical, tv = t-value calculated, pv = p-value, α = significance level of 0.05

correlates more with work potential and problem-solving ability than a numerical average [48, 51].

Regarding the differences in creative potential between men and women, women scored higher than men in three of the four groups. However, it is also evident that women in the 4th semester achieved a higher score than those in the 8th semester, consistent with the general result discussed in the previous paragraph. The review in [52] posits that no conclusive evidence exists that one gender is more creative than the other. They are different in their approaches and strengths but not necessarily superior to each other. Experimental evidence suggests that women have more creative flexibility, which agrees with the work of [53].

The fact that engineers are less creative than other disciplines and that it is negatively impacted as they progress in their studies or improve their academic average may be due to the cultural context specific to engineering. Problem-solving methodologies are usually rigid, and there is an inherent demand for developing higher-order intellectual skills necessary for the labor market. Nevertheless, students need more than that to develop their creativity [54]. For creativity to be adopted as part of a university culture, changes must be introduced to the curricular and cultural structures to generate learning environments more conducive to creativity and innovation.

Finally, from a learning perspective, the in-class use of tools and methodologies for generating creative ideas –such as TRIZ, USIT and SCAMPER– helps increase the creative potential of engineering students, which agrees with other works [55, 56].

4 Conclusions

This article presents an approach to assess the creative potential that engineering students develop when using methodologies to solve inventive problems. The results of this study allow us to conclude the following:

1. Three interventions during a course do not seem to be enough to improve the level of creativity of all students significantly.
2. The creative potential of the students decreases as they advance in their engineering education.
3. The cultural environment of curricula better promotes creativity in scientific programs than in hard engineering.
4. Students with average or low academic grades have greater creative potential than those with higher grades.
5. It appears that the deliberate interventions of TRIZ, USIT, and SCAMPER methodologies in class positively promotes the creative potential of engineering students, with the added benefit of sparking their interest in the inventive work of design.

This study acknowledges one limitation: a non-probabilistic convenience sampling had to be carried out due to the availability of the experimental groups in the academic period in which the creativity interventions could be carried out.

For future research, it is proposed to extend the exploratory phase to other engineering programs, including public universities too. It is also suggested to analyze the student's metacognitive creative skills after using these techniques to make them aware of their process and promote their self-regulation.

Acknowledgements This work is for the Psycho-Pedagogical Studies Team of the Educational Innova-tion Research Group of Tecnológico de Monterrey (https://sites.google.com/itesm.mx/gieeiie/). The authors appreciate the financial support of the NOVUS initiative, N22-267, and the technical support of the Writing Lab, Institute for the Future of Education (https://tec.mx/en/ife), Tecnológico de Monterrey, in the production of this work.

References

1. Al-Ababneh M (2020) The concept of creativity: definitions and theories. Int J Tour Hotel Bus Manag 2(1):245–249. https://ssrn.com/abstract=3633647
2. Still A, d'Inverno M (2016) A history of creativity for future AI research. In: Proceedings of the 7th computational creativity conference (ICCC 2016). Université Pierre et Marie Curie. Paris, France. https://research.gold.ac.uk/18616/1/COM-d'Inverno-2016.pdf
3. Morales C (2017) The creativity, a scientific review. Arquitectura y Urbanismo 38(2):53–62. https://www.redalyc.org/pdf/3768/376852683005.pdf
4. Romo M, Sanchez-Ruiz MJ, Alfonso-Benlliure V (2017) Creatividad y personalidad a través de dominios: una revisión crítica. Anuario de Psicología 47(2):57–69. https://doi.org/10.1016/j.anpsic.2017.04.003
5. Plucker JA, Makel MC (2010) Assessment of creativity. In: The Cambridge handbook of creativity. Cambridge University Press, pp 48–73
6. Fields Z, Bisschoff CA (2013) A model to measure creativity in young adults. J Soc Sci 37(1):55–67. https://doi.org/10.1080/09718923.2013.11893204
7. Artola T, Mosteiro P, Poveda B, Barraca J, Ancillo I, Sánchez N (2012) Prueba de imaginación creativa para adultos. TEA Ediciones, Madrid
8. Byrd J, Brown PL (2007) The innovation equation: building creativity and risk-taking in your organization. Pfiffer ed
9. Hunter ST, Bedell KE, Mumford MD (2007) Climate for creativity: a quantitative review. Creat Res J 19(1):69–90. https://doi.org/10.1080/10400410709336883
10. Miller SR, Hunter ST, Starkey E, Ramachandran S, Ahmed F, Fuge M (2021) How should we measure creativity in engineering design? A comparison between social science and engineering approaches. J Mech Des 143(3):031404. https://doi.org/10.1115/1.4049061
11. Cropley, (2000) Defining and measuring creativity: are creativity tests worth using? Roeper Rev 23(2):72–79. https://doi.org/10.1080/02783190009554069
12. Felder RM (1988) Creativity in engineering education, chemical engineering education, vol 22(3), pp 120–125. https://journals.flvc.org/cee/article/download/124277/123290
13. Beaulieu DF (2022) Creativity in science, engineering, and the arts: a study of undergraduate students' perceptions. J Creat 32(3). https://doi.org/10.1016/j.yjoc.2022.100035
14. Erro M, Espinosa E, Dominguez S (2016) Creativity and engineering education: a survey of approaches and current state. In: 9th annual international conference of education, research and innovation, Sevilla, Spain. https://doi.org/10.21125/iceri.2016.0725
15. Tekmen-Araci Y, Mann L (2019) Instructor approaches to creativity in engineering design education. Proc Inst Mech Eng C J Mech Eng Sci 233(2):395–402. https://doi.org/10.1177/0954406218758795

16. Avsec S, Savec VF (2019) Creativity and critical thinking in engineering design: the role of interdisciplinary augmentation. Glob J Eng Educ 21(1):30–36. http://www.wiete.com.au/journals/GJEE/Publish/TOCVol21No1.html
17. Daly SR, Mosyjowski EA, Seifert CM (2014) Teaching creativity in engineering courses. J Eng Educ 103(3):417–449. https://doi.org/10.1002/jee.20048
18. Wolf P, Cormican K, Frederiksen MH, Wilhøft A, Ulus HE, Kunz C, Andiç-Çakır Ö, Sarsar F, Leeuwen M (2024) I think they just logged on and fell asleep: challenges to facilitating creativity online in higher engineering education, creativity and innovation management, 1–23. Wiley
19. Nasrudin D, Setiawan A, Rusdiana D, Liliasari L (2023) Research trends and future works on student creativity in the context of sustainability: a bibliometric analysis. J Pendidik Sains Indones (Indones J Sci Educ), 11(4): 926–936. https://doi.org/10.24815/jpsi.v11i4.33393
20. Nolan EM, (2020) Transcending lockdown: fostering student imagination through computer-supported collaborative learning and creativity in engineering design courses. Creative humanities special issues, 5. https://source.sheridancollege.ca/fhass_creative_humanities/5
21. ABET (2022) Criteria for accrediting engineering programs. Engineering accreditation commission. Accreditation board for engineering and technology. https://www.abet.org/accreditation/accreditation-criteria/criteria-for-accrediting-engineering-programs-2022-2023/
22. Bourgeois-Bougrine S, Buisine S, Vandendriessche C, Glaveanu V, Lubart T (2017) Engineering students' use of creativity and development tools in conceptual product design: what, when, and how? Think Ski Creat 24:104–117. https://doi.org/10.1016/j.tsc.2017.02.016
23. Chang T, Wang H, Haynes AM, Song M, Lai S, Hsieh S (2022) Enhancing student creativity through an interdisciplinary, project-oriented, problem-based learning undergraduate curriculum. Think Ski Creat 46:101173. https://doi.org/10.1016/j.tsc.2022.101173
24. Zakaria MI, Maat SM, Khalid F (2019) A systematic review of problem based learning in education. Creat Educ 10:2671–2688. https://doi.org/10.4236/ce.2019.1012194
25. Bippert K, Espinosa T (2019) Project-oriented learning and teaching: expectations versus reality. Texas Association for Literacy Education Yearbook, vol 6, pp 19–29. http://www.texasreaders.org/uploads/4/4/9/0/44902393/2019_tale_yearbook.pdf
26. Savransky SD (2000) Engineering of creativity: introduction to TRIZ methodology of inventive problem solving. CRC Press
27. Donnici G, Frizziero L, Francia D, Liverani A, Caligiana G (2018) TRIZ method for innovation applied to an hoverboard. Cogent Eng 5(1):1524537. https://doi.org/10.1080/23311916.2018.1524537
28. Horowitz R (1999. Creative problem solving in engineering design [doctoral thesis]. Tel-Aviv University. https://www.asit.info/Creative%20Problem%20Solving%20in%20Engineering%20Design,%20thesis%20by%20Roni%20Horowitz.pdf
29. Nakagawa T (2011) Education and training of creative problem-solving thinking with TRIZ/USIT. Procedia Eng 9:582–595. https://doi.org/10.1016/j.proeng.2011.03.144
30. Eberle RF (1972) Developing imagination through scamper. J Creat Behav 6(3):199–203. https://doi.org/10.1002/j.2162-6057.1972.tb00929.x
31. Özyaprak M (2016) The effectiveness of SCAMPER technique on creative thinking skills. J Educ Gift Young Sci, 4(1):31. https://doi.org/10.17478/jegys.2016116348
32. De la Puente C (2018) Estadística descriptiva e inferencial. Ediciones IDT
33. Carter P (2005) The complete book of intelligence tests, 1st edn. Wiley, England
34. Byrd J (2018) The creatrix impact accelerating the innovative capacity of individuals and teams, organizations, and leaders. The innovation equation and voice of the innovator. http://www.creatrix.com/
35. Taber KS (2018) The use of Cronbach's alpha when developing and reporting research instruments in science education. Res Sci Educ 48:1273–1296. https://doi.org/10.1007/s11165-016-9602-2
36. Galindo O (2006) Estudios encaminados a medir la creatividad orientada al desarrollo económico en organizaciones y ciudades de México [tesis de maestría]. Repositorio institucional del Tecnológico de Monterrey. https://repositorio.tec.mx/bitstream/handle/11285/567632/GalindoHernandez_TesisdeMaestriaPDFA.pdf?sequence=11

37. Hill BA (1997) An analysis of decision styles as a predictor variable for levels of creativity and/or orientations toward risk-taking of knowledge workers [master's thesis]. West Virginia University. https://www.proquest.com/docview/304281186?pq-origsite=gscholar&fromopenview=true
38. Fahed RY (2016) Byrd's creatrix inventory. Int J Pedagog Innov, 4(01). https://journal.uob.edu.bh/bitstream/handle/123456789/927/IJPI040105.pdf?sequence=1
39. De Bono E (2018) Creatividad, 62 ejercicios para desarrollar la mente. Paidós, México
40. Ilevbare IM, Probert D, Phaal R (2013) A review of TRIZ and its benefits and challenges in practice. Technovation 33(2–3):30–37. https://doi.org/10.1016/j.technovation.2012.11.003
41. Nakagawa T (2004) USIT operators for solution generation in TRIZ: clearer guide to solution paths'. In: ETRIA conference proceedings: TRIZ future 2004 conference, Florence, Italy, Nov, pp 3–5. https://www.metodolog.ru/triz-journal/archives/2005/03/05.pdf
42. Delgado-Sanchez C, Tenorio-Alfonso A, Cortés-Triviño E, Martín-Alfonso MJ (2021) Engineering problems resolution by applying the SCAMPER technique to improve students' creativity. CIVINEDU, 459. https://dialnet.unirioja.es/descarga/libro/858261.pdf#page=474
43. Álvarez F (2023) La distribución normal. Studocu. https://www.studocu.com/es-mx/document/grupo-colegio-mexiquense/lengua-espanola/2023-04-20-distribucion-normal-2023/66311816
44. Heagerty P (nd) Critical values (percentiles) for the t distribution. University of Washington. https://faculty.washington.edu/heagerty/Books/Biostatistics/TABLES/t-Tables/
45. Shtulman A, Harrington K (2016) Tensions between science and intuition across the lifespan. Topics in cognitive science, 8(1): 118–137. https://onlinelibrary.wiley.com/doi/pdfdirect/; https://doi.org/10.1111/tops.12174
46. Simonton DK (1983) Formal education, eminence, and dogmatism: the curvilinear relationship. J Creat Behav 17(3):149–162. https://doi.org/10.1002/j.2162-6057.1983.tb00348.x
47. Gajda A, Karwowski M, Beghetto RA (2017) Creativity and academic achievement: a meta-analysis. J Educ Psychol 109(2):269. https://doi.org/10.1037/edu0000133
48. González W (2013) Creativity development in informatics teaching using the project focus. Int J Eng Pedagog (iJEP) 3(1):63–70. https://doi.org/10.3991/ijep.v3i1.2342
49. Almetov N, Zhorabekova A, Sagdullayev I, Abilhairova Z, Tulenova K (2020) Engineering education: problems of modernization in the context of a competency approach. Int J Eng Pedagog (iJEP), 10(6). https://doi.org/10.3991/ijep.v10i6.14043
50. Zabelina DL, Saporeta A, Beeman M (2016) Flexible or leaky attention in creative people? Distinct patterns of attention for different types of creative thinking. Mem Cognit 44(3):488–498. https://doi.org/10.3758/s13421-015-0569-4
51. González W (2016) Detection of potentially creative students for informatics activity. Int J Eng Pedagog (iJEP) 6(1):80–84. https://doi.org/10.3991/ijep.v6i1.5156
52. Abraham A (2016) Gender and creativity: an overview of psychological and neuroscientific literature. Brain Imaging Behav 10(2):609–618. https://doi.org/10.1007/s11682-015-9410-8
53. He WJ, Wong WC (2011) Gender differences in creative thinking revisited: findings from analysis of variability. Personality Individ Differ 51(7):807–811. https://doi.org/10.1016/j.paid.2011.06.027
54. Sopakitiboon T, Tuampoemsab S, Howimanporn S, Chookaew S (2023) Implementation of new-product creativity through an engineering design process to foster engineering students' higher-order thinking skills. Int J Eng Pedagog (iJEP) 13(5). https://doi.org/10.3991/ijep.v13i5.38863
55. Chang YS, Chien YH, Yu KC, Chu YH, Chen MYC (2016) Effect of TRIZ on the creativity of engineering students. Think Ski Creat 19:112–122. https://doi.org/10.1016/j.tsc.2015.10.003
56. Cano-Moreno JD, Arenas Reina JM, Sánchez Martínez FV, Cabanellas Becerra JM (2021) Using TRIZ10 for enhancing creativity in engineering design education. Int J Technol Des Educ 1–26. https://doi.org/10.1007/s10798-021-09704-3

The Next Step in Challenge-Based Learning: Multiple Challenges in a Single Course in Engineering: A New Model in Experiential Education

Miguel Ramírez-Cadena, Juana Méndez-Garduño, Israel Cayetano-Jiménez, Araceli Ortíz-Martínez, and Jorge Membrillo-Hernández

Abstract One of the most used teaching techniques in experiential education in Engineering is Challenge-Based Learning [CBL], which has the characteristic of exposing students to a current, real-life problem, developing both disciplinary and transversal skills. CBL experiences are usually designed from a specific topic that brings together many extents in the course syllabus, a challenge for the entire class group. In many cases, a training partner can generate the challenge to be solved. There is an unwritten rule in the CBL model: a challenge designed by teachers, or a training partner will be solved by an entire class, generating several potential solutions, usually in teams. In this report, we analyze a CBL experience where different training partners were involved for the same class of a subject or course, and several challenges were proposed simultaneously. Here we demonstrate that competencies can be developed using different challenges simultaneously in the same group and with several training partners. Ultimately, the evaluation was carried out with a strict competency development rubric. Thus, here we report a scheme that establishes an expanded model of CBL, where a course is not only governed by a particular challenge but also through the resolution of more than one challenge within the same group class. Our results showed that the discussion was enriched, and the designated competencies were broadly strengthened.

Keywords Challenge-based learning · Experiential learning · Higher education · Educational innovation · Tec21 · Engineering education

---

M. Ramírez-Cadena · J. Méndez-Garduño · I. Cayetano-Jiménez · J. Membrillo-Hernández (✉)
School of Engineering and Sciences, Tecnologico de Monterrey, Monterrey, Mexico
e-mail: jmembrillo@tec.mx

A. Ortíz-Martínez
Klesse College of Engineering and Integral Design, The University of Texas at San Antonio, San Antonio, TX, USA

J. Membrillo-Hernández
Institute for the Future of Education, Tecnologico de Monterrey, Monterrey, Mexico

© The Author(s), under exclusive license to Springer Nature Switzerland AG 2025
E. Vendrell Vidal et al. (eds.), *Advanced Technologies and the University of the Future*, Lecture Notes in Networks and Systems 1140,
https://doi.org/10.1007/978-3-031-71530-3_32

1 Introduction

1.1 Challenge-Based Learning

In higher education, various reports are indicating that experiential learning has dominated the competency-based teaching strategy in recent years [1]. Today, education has gone from being just an information instrument to involving the student in learning through active problem solving, a project, a practice, and, recently, a challenge. A challenge has been described as a more complex structure and requires more interaction between the teacher and students than projects, practices, or problems [2, 3]. The big difference is that challenge-based learning [CBL] involves a much greater amount of uncertainty than other teaching models [1, 4–6].

1.2 One Class, One Challenge

Initially, CBL was designed where engineering students are exposed to a general challenge, and solutions are designed around it [7, 8]. For the design of the challenge, in principle, teachers determine the competencies to be developed in a course and choose an appropriate challenge for the entire class. Normally, the resolution of challenges can last a segment of the course or the entire semester [8]. If the challenge is the product of a collegiate discussion exercise by the teachers. In that case, this guarantees that the process of developing and assigning tasks to students is established simply and by the objectives of the course in question. If this challenge is carried out through the participation of a training partner [company, governmental or non-governmental organization, outside the educational institution, [9], there are always prior talks of coordination and pedagogical design. In any case, the planning of the evaluation must be attached to the competencies development criteria, and the teachers, as well as the participating engineering experts of the training partner, must be prepared themselves in the establishment of evaluation rubrics [10]. It should be emphasized that the participation of a training partner is highly desirable as higher education institutions are challenged to keep up with the demands of the labor market and prepare students with the necessary skills to face an increasingly changing world [10]. For this reason, educational techniques, technological resources, and global knowledge certification approaches change simultaneously and respond to the evolution of employer demands. Hence, teaching practices in higher education in engineering must bring students together and expose them to challenges, often created by employers, that allow them to develop in a quasi-work environment. Until today, teaching practice establishes that a challenge is chosen for the entire group, and specific strategies or tasks are divided to obtain a comprehensive solution to evaluate student competencies' development.

Here, we report evidence of a new and innovative challenge-based learning scheme that uses not one but several challenges with several training partners and encompasses all experiences into comprehensive competency development for an entire group. The scheme was initially very positive for the students and will be very useful in the future for this type of active, experiential teaching.

2 Methodology

2.1 Engineering Multi-challenge Teaching Pathway Framework

To carry out an active learning education experience based on multiple engineering challenges, we designed a roadmap that has three fundamental moments, BEFORE, DURING and AFTER the course. In Fig. 1, we develop the following roadmap:

Before the Course

Be clear about the concepts that are intended to be taught during the course, determine its length and the teachers responsible for it. Set the number of students. Establish a relationship with one or more training partners [TP]. To carry out a teaching experience based on multiple challenges, it is necessary to establish an academic relationship with at least one TP [8]. A TP can be a company, a non-governmental organization, an academic institution, a team of researchers, or an entrepreneur who

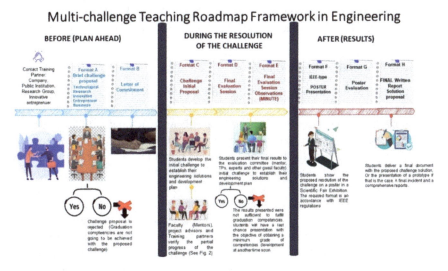

Fig. 1 Multi-Challenge Teaching Roadmap. The chronology of the formats of the experience is depicted in the figure. See text for details

wants to share some challenges that are current and relevant for academic development according to the level of the students. These challenges are collected in **FORMAT A** "BRIEF CHALLENGE PROPOSAL".

For the multi-challenge experience, we look for four fundamental challenges that make up a complete story so that the preparation is comprehensive: [A] Technological Engineering, [B] Research, [C] Innovative Entrepreneurship, and [D] Business. This series of prior visits to the plants or the steering committees of the training partners before the start of the course is essential to highlight the relevance of the challenges and align them with the course objectives. Here, it is important to mention that challenges can arise from the interaction of one or more training partners; we must always remember that what is important for applying CBL is not the resolution per se but the development of competencies in the process of designing and solving the challenge.

The course teaching committee [the teachers responsible for the experience] meets to analyze the challenges proposed by the training partners. This meeting discusses the objectives, competencies, methods, and evaluation instruments [10], teaching strategies, session scheduling, and team formation. Some of the challenges may not be appropriate for the course objectives, or the complexity may not be what is sought within the scope of the course; these challenges are archived for later use and analysis. Only those challenges that have the committee's approval are used for implementation.

Once the challenges have been agreed upon with the training partner and the relevance of the academic and ethical objectives and standards has been analyzed, a commitment letter will be drafted that establishes the guidelines on the standards of intellectual property, protection of personal data, and confidentiality that must be maintained when sensitive data are used. To give more formality and solemnity to the event, a commitment letter is signed to resolve the challenge [**FORMAT B**]. Students have mentioned that this event forces them to increase their level of engagement in the course.

During the Resolution of the Challenge

Once the course has started, students are divided into teams, taking into account their interest in solving one of the 4 chosen challenges [technological, research, entrepreneurship, and business]. Activities are then carried out with each team to determine the envisioned needs. The document **FORMAT C** [Challenge initial proposal] is then generated where the scope and limitations that the team will face are described, as well as the ideas, activities, and sessions that will be carried out to solve the challenge [in Fig. 2 shows each of the phases during the course in detail].

The team responsible for each challenge then embarks on its activities. Data collection, consultations with internal or external experts, or the training partner. The duration of the CBL experience can be from 5 to 15 weeks; in the case of this report, it was only carried out for 5 weeks. Once the team's mentors have endorsed the probable solution proposal to the challenge, it is presented in an academic environment [**FORMAT D**], and a final report is drawn up [FORMAT E], establishing the

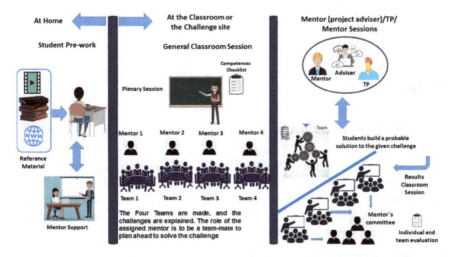

Fig. 2 Activities developed by the academic committee and the students during a multi-challenge experience [see text for details]

observations collected by the evaluators to improve the final document. The evaluators are members of the academic committee of the course, expert engineers from the training partners and some guest teachers, experts who belong to the academic institution. This forum enriches the discussion. If the solution to the challenge involves the production of a prototype, its progress is presented at this stage.

If the challenge resolution proposal does not meet the quality standards of the information and the use of technology and the data are insufficient to demonstrate the development of the competencies. In that case, the students will not be approved. In this case, students will have one last opportunity to present in a later space, making the pertinent modifications suggested by the evaluation committee. In this case, the students will not have access to the maximum grade and the exam will be considered sufficiency.

Once the observations have been made in the evaluation period, oral and written expression and scientific communication skills are developed. To this end, each team is requested to prepare a poster with the strict specifications of the IEEE [**FORMAT F**] and present it at an Academic Fair event before external experts, engineering professors from the academic institution, students, and the public in general. This second evaluation contemplates having made the changes suggested in the first evaluation. The evaluation is reflected in **FORMAT G**.

After the poster evaluations, feedback is provided to the team, and these final observations are added to the final challenge resolution report [**FORMAT H**]. The final delivery of the comprehensive report of the entire multi-challenge experience is done. These observations reflect the comments not only of the challenge in question but also those made to the challenges of the teams that simultaneously solved challenges other than their own [see Fig. 2]. This deliverable includes all observations, corrections, and evaluations of each challenge. If it turns out to be a prototype, it is presented in its final version at this stage.

3 Results

3.1 Before the Challenge

Following the plan described in Fig. 1, the multi-challenge experience [**mCBL**] of the subject *"Design and implementation of mechatronic systems"* was implemented. This class belongs to the Academic Engineering Study Program of several Engineering careers and is taken by students in the final third of their bachelor´s degree. The objective of the subject is that at the end of the course, the student could:

- Propose viable and cutting-edge technological solutions to solve industrial, social, and environmental problems.
- Apply methodologies and technological tools for the design of mechatronic systems.
- Validate automation proposals guaranteeing quality, safety and productivity.
- Implement automation proposals using cutting-edge technologies.
- Develop research on the state of the art based on reliable sources to generate a proposal for a mechatronic system.
- Generate innovative proposals for a mechatronic system according to standards.
- Evaluate the technical and economic feasibility of technological development based on constraints.

The competencies to be developed for the subject are declared in Table 1.

The **mCBL** experience was divided into three moments: *Before* the experience [at home], *During* the experience [At the classroom or the challenge site], and ***after*** the experience [in a classroom]. For clarity, Fig. 2 describes the activities of each phase.

Before the experience. It is the most important moment for the teaching staff. During these meetings, the teachers responsible for the subject carefully plan the strategies for compliance with the development of competencies, verify potential training partners for the development of challenges appropriate to the subject, and establish standardization of the visit to the training partners to learn about the challenges proposed by them Fig. 1.

Training partners. The committee of teachers responsible for the subject visited various training partners who had challenges to be developed Fig. 1. Format A [Brief Challenge Proposal] was compiled and there was a subsequent committee meeting. Several challenges were not suitable and were discarded. After an analysis, validating the competencies to be developed, choosing the necessary mentors and expert advisors for the resolution of the challenge, and having reached an agreement with the

Table 1 Competencies to develop in the subject "Design and implementation of mechatronic systems"

Graduation competence	Description
SMR0103C	Methodologically selects components according to technical specifications
SMR0201C	Proposes feasible and cutting-edge technological solutions to solve industrial, social, and environmental problems
SMR0202C	Applies technological methodologies and tools for the design of mechatronic systems
SMR0303C	Applies technological methodologies and tools for the design of mechatronic systems
SMR0402C	Generates innovative proposals for a mechatronic system in accordance with standards
SMR0403C	Evaluates the technical and economic feasibility of technological development based on constraints
SMR0401C	Prepares research on the state of the art based on reliable sources to generate a proposal for a mechatronic system generate a proposal for a mechatronic system
SEG0202A	Evaluates the impact of entrepreneurial initiatives on a personal level, on the environment, and on different interest groups, from an ethical and sustainability framework
SIIT0302C	Generates proposals for solutions to problems under conditions of uncertainty and different levels of complexity based on engineering and science methodologies

training partners, the challenges and training partners chosen for our multi-challenge experience were those described in Table 2.

Once analyzed, the teachers responsible for the subject decide the challenges that are going to be developed, but very importantly, they carry out a preparation strategy for the students, with readings, content, videos, and explanations of what constitutes challenge-based learning [CBL], the nature of the challenges and above all the role that each of them will play in the resolution of the challenges that will be discussed in the classroom. These previous sessions are to establish homework assignments and establish the Reference Material or means of information, which can be electronic documentation, physical documentation [in libraries], or consultations with the mentor in person or online [Fig. 2, left panel].

During the development of the solution of the challenge. In the first plenary session in the classroom [Fig. 2 central panel], a diagnostic evaluation was carried out to know the students' interests and to guide them toward one of the challenges chosen by the training partners.

Once this evaluation was carried out, the thirty-four students were divided into 11 teams so that each one had a specific challenge; the appropriate mentors were established for each challenge, made up of a member of the training partner, at least one teacher from the subject and experts were sought as external advisors.

Table 2 Training partners whose challenges were approved by the academic committee to be developed in this multi-challenge experience

Training Partner	Challenge	Type of Challenge
Electromobility Lab	Electrostatic Drive	Research
Universal Robot	Training platform interactive	Entrepreneur
Intelligent Automatization Lab	Cybersecurity Device	Entrepreneur
IAMSM	Autonomous vehicle	Technological
ICE	Emergency Aereal Recognition	Business
Mexico Ministry of Defense	Dron kits	Technological
Tech Borregos	Electric Mobile platforms	Research
Petrol Industries SIIP	Automatization of water plant	Technological
Xico	Automatization kit tools	Business
GENERAC	Alternative energy sources	Research
Electromobility Lab	Pothole tracking	Research

Students were then invited to sign the commitment letter [Form B; Fig. 1]. Before the start of the activities, the members of the academic committee met to ensure that each mentor had the necessary reference material to start the experience. This material was made available to students through easily accessible digital platforms Fig. 2.

The 11 teams were classified into **4 *different types of teams*** based on the challenges nature [*research, technology, entrepreneur, and business*] and the Initial Challenge Proposal [ChIP] was announced [Format C; Fig. 1]. The teams met their mentors and support team of professors and experts with whom they discussed challenges and began strategizing their potential solutions.

The teams had sessions at the academic institution with mentors, additional professors, and external advisors and at the training partner's site with engineers and collaborators from the same training partner [Fig. 2, middle panel]. This interaction occurred every week for the duration of the course [Fig. 2].

At the end of the course. Several plenary sessions were held where each team presented their solution proposal and explained the characteristics of the challenge, and the required competencies were evaluated by the mentors of all types of challenges and external professors who served as evaluators [Fig. 2, right panel]. The discussion of each of the solution proposals made the students exchange ideas, improve the solutions, implement new concepts and strategies, and, if they had any structural errors in the planning, correct them. These plenary sessions involved extensive discussion and students could engage in the type of challenges they did not directly develop [Fig. 2]. The students were ready for the final evaluation before the members of the academic committee who would evaluate their graduation competencies in a final presentation [Format D; Fig. 1]. In this evaluation, observations of improvement were made that were recorded in a minute [Format E; Fig. 1]. During the meeting, specifications were given for preparing a poster with the characteristics requested by the IEEE. On this occasion, no team was considered not to have reached the course competencies.

The students then made a Poster for the Engineering academic day [Format F]. They were evaluated [Format G] by a committee of professors from the engineering faculty participating and external to the multi-challenge experience. The evaluations were under a presentation rubric where all team members participated.

With the observations collected, the teams finalized the final details to prepare the final report, which consisted of a structured report with a predetermined format. This document was the final evidence of the experience and the teachers had elements to judge the development of the evidence objectively.

In some cases, a final version of the prototype is presented, and the framework proposed in Fig. 1 is complied with.

4 Discussion

Experiential education, part of active education, is fundamental in the educational framework of higher education, where the skills demanded by employers [now called skills of the future, or industry 4.0] are increasingly complex.

One of the most advanced teaching strategies used in Higher Education in recent years has been Challenge Based Learning [CBL] [5 and references therein]. Within the CBL, it has been established that the plan of having a challenge for a group of students is adequate to monitor the development of both transversal and disciplinary competencies of each student, including the division into teams to have alternative solutions to the challenge is very useful for comparing progress in skill use among the same students.

In this manuscript, we report for the first time [to our knowledge] a multiple-challenge strategy in the same group that is used to develop the required competencies of students who do not have a single objective but several challenges and acquire the competencies not only through of the resolution of their challenge but also of a deep discussion of the solution to the challenges of the students of the same group, but different challenge. The involvement of the students in different challenges, even if they were not responsible, was a good bank of ideas, strategies, and discussion. Constructive criticism towards the work of others invites the analysis of the proposals, the study of the challenge of the classmates, and an integral solution.

The strategy still has some areas of opportunity that can be improved, both in planning and in the evaluation rubrics, perhaps adding a final session of a reflection exam where problems are exposed and an auto-evaluation step is added.

In any case, the innovation presented here is a substantial advance to our conception of CBL; perhaps this new strategy should be added to the one already described by Van dem Bemt et al. [4, 9].

Acknowledgements The authors acknowledge the financial support provided by the Department of Mechatronics of the School of Engineering and Sciencies, Tecnologico de Monterrey, Mexico City Campus, Mexico.

References

1. Burke D (2020) Experiential learning theory. In: Burke D (ed) How Doctors think and learn. Springer International Publishing, Cham, pp 29–37. https://doi.org/10.1007/978-3-030-46279-6_4
2. van den Beemt A, MacLeod MAJ (2021) Tomorrow's challenges for today's students: challenge-based learning and interdisciplinarity. In: Proceedings of the 49th SEFI annual conference: blended learning in engineering education: challenging, enlightening—and lasting? SEFI ISEL, pp 578–587. Accessed 15 Jan 2023. https://research.utwente.nl/en/publications/tomorrows-challenges-for-todays-students-challenge-based-learning
3. Membrillo Hernandéz J (2023) Challenge-based learning an emergent educational model for Engineering Education in the post-COVID era. In: IEEE Teaching Excellence HUB. Accessed 01 Aug 2023. https://teaching.ieee.org/challenge-based-learning-an-emergent-educational-model-for-engineering-education-in-the-post-covid-era/
4. Van Den Beemt A, Van De Watering G, Bots M (2023) Conceptualising variety in challenge-based learning in higher education: the CBL-compass. Eur J Eng Educ 48(1):24–41. https://doi.org/10.1080/03043797.2022.2078181
5. Gallagher SE, Savage T (2020) Challenge-based learning in higher education: an exploratory literature review. Teach High Educ. https://doi.org/10.1080/13562517.2020.1863354
6. Membrillo-Hernández J, Ramírez-Cadena MJ, Caballero-Valdés C, Ganem-Corvera R, Bustamante-Bello R, Benjamín-Ordoñez JA, Elizalde-Siller H (2018) Challenge based learning: the case of sustainable development engineering at the tecnologico de monterrey, Mexico City Campus, vol 715. In: Advances in intelligent systems and computing, vol 715, pp 914. https://doi.org/10.1007/978-3-319-73210-7_103
7. Caratozzolo P, Membrillo-Hernández J (2021) Evaluation of challenge based learning experiences in engineering programs: the case of the Tecnologico de Monterrey, Mexico. In: Auer ME, Centea D (eds) Visions and concepts for education 4.0, vol 1314, in Advances in Intelligent Systems and Computing, vol 1314. Springer International Publishing, Cham, pp 419–428. https://doi.org/10.1007/978-3-030-67209-6_45
8. Membrillo-Hernández J, Ramírez-Cadena MJ, Martínez-Acosta M, Cruz-Gómez E, Muñoz-Díaz E, Elizalde H (2019) Challenge-based learning: the importance of world-leading companies as training partners. Int J Interact Des Manuf 13(3):1103–1113. https://doi.org/10.1007/s12008-019-00569-4
9. Van den Beemt A, Vázquez-Villegas P, Gómez Puente S, O'Riordan F, Gormley C, Chiang FK, Leng C, Caratozzolo P, Zavala G, Membrillo-Hernández J (2023) Taking the challenge: an exploratory study of the challenge-based learning context in higher education institutions across three different continents. Educ Sci 13(3). https://doi.org/10.3390/educsci13030234
10. Membrillo-Hernandez J, Garcia-Garcia R (2020) Challenge-based learning (CBL) in engineering: which evaluation instruments are best suited to evaluate CBL experiences? In: IEEE global engineering education conference, EDUCON, vol 2020, pp 885–893. https://doi.org/10.1109/EDUCON45650.2020.9125364

www.ingramcontent.com/pod-product-compliance
Lightning Source LLC
Chambersburg PA
CBHW072121050125
19958CB00003B/59